Marsupial Biology

Recent Research, New Perspectives

MARSUPIAL BIOLOGY

Recent Research, New Perspectives

Edited by
Norman Saunders
and Lyn Hinds

UNSW PRESS

A UNSW PRESS BOOK
Published by
UNIVERSITY OF NEW SOUTH WALES PRESS LTD
University of New South Wales
Sydney NSW Australia 2052

First published 1997

National Library of Australia
Cataloguing-in-Publication entry:

Marsupial Biology: Recent Research, New Perspectives

Bibliography.
Includes index.
ISBN 0 86840 311 3

1. Marsupialia. I. Saunders, N.R. (Norman Ruthven), 1938– .
II. Hinds, L. A. (Lynette Ann), 1953– .

599.2

Designer Dana Lundmark
Editor Robyn Appleton
Managing Editor Nada Madjar
Production Manager Di Quick
Printer Ligare Printing
Publisher John Elliot

CONTENTS

PREFACE

This book follows from a conference on Recent Advances in Marsupial Biology, held in Hobart in December 1994 to honour Hugh Tyndale-Biscoe on the event of his 65th birthday. Both that conference and this book have their antecedents in the Boden Conference on Marsupial Biology organised by Hugh Tyndale-Biscoe and Peter Jansssens in 1986. Many of the contributors to that conference also attended the conference on Marsupials as Models for Developmental Studies held at the University of Southampton in 1989. The aim of that conference was to bring together a wide spectrum of biologists whose central concern was the biology of marsupials which could be used for developmental studies. The possibility of setting up a regular series of meetings to consider a wide range of aspects of marsupial biology was discussed. Subsequently, the occasion of Hugh Tyndale-Biscoe's 65th birthday in 1994 became the focus for a conference on Recent Advances in Marsupial Biology.

Marsupial Biology: Recent Research, New Perspectives is much more than the proceedings of the 1994 conference. Firstly, a number of major contributions have come from authors who were not able to attend the conference. Furthermore, the guidelines to authors were to write reviews of their field, which were to summarise marsupial biology as it was in 1994 and look at the direction each field was likely to take during the following five to ten years.

The scope of the book is wide, but there are two underlying themes in these diverse works: that marsupials are animals of considerable biological interest in their own right, and—in line with August Krogh's (1929)* dictum that for every physiological problem there will be one species above all others most appropriate for its investigation—that marsupials are being used to study biological questions they are uniquely suited to answer. Many of the articles in this book concern the use of marsupial species to tackle fundamental biological problems in ways that would be difficult or impossible in more conventional laboratory species. There are five main areas in which a substantial amount of progress in marsupial biology has been made during the last few years, namely reproduction and development, genetics, ecology, pathology and homeostasis, and developmental neurobiology.

Like the conference, this book is a tribute to Hugh Tyndale-Biscoe, who is a leading marsupial biologist of his generation. Although much of his research has been in the field of reproduction and lactation, he has been involved in and has encouraged a diverse range of marsupial biological research, and has contributed to many of the fields represented in this volume. Several of his former research students, who now occupy senior positions in research, have contributed to this book. They are: Marilyn Renfree (Head of the Department of Zoology, University of Melbourne), John Hearn (Director of the Primate Research Centre, University of Wisconsin) and Lyn Hinds (Program Leader of Vertebrate Fauna and Pests, CSIRO Division of Wildlife and Ecology, Canberra).

Hugh Tyndale-Biscoe was born in India and educated in New Zealand and England. After obtaining his MSc in Zoology from Canterbury University in New Zealand, he moved to Perth to study for his PhD in the Department of Zoology at the University of Western Australia. He lectured for a time at the

Australian National University, and in 1976 moved to CSIRO to head a Marsupial Biology Group in the Division of Wildlife Research. Over the succeeding 15 years he built a strong research group, concentrating on the biology of reproduction and lactation in the tammar. The reputation of the group became such that numerous overseas scientists came to collaborate with Hugh in Canberra. In the early 1990s, Hugh took up a position in an entirely new research field, becoming the founding director of the Co-operative Research Centre for Biological Control of Vertebrate Pest Populations, based at CSIRO. The aim of the centre was to develop methods of fertility control for vertebrate pests, particularly rabbits, foxes and house mice. Hugh retired in 1995, having seen the whole project come together with a very promising start in developing control methods. He has now returned to bench research on marsupials in the Department of Developmental Neurobiology at the Australian National University.

The editors and a number of the contributors to this book have reasons to be particularly grateful to Hugh Tyndale-Biscoe. For me (Norman Saunders), Hugh was my introduction to marsupial biology. In the early 1980s, Kate Dziegielewska and I had reached the stage in our studies of mechanisms controlling the internal environment of the brain using principally a sheep model where we felt that we were becoming seriously limited by not being able to work with sufficiently early stages of brain development. We wondered whether a marsupial species might be the answer and, after contacting a number of Australian centres, it was Hugh who offered immediate help. Subsequently, we carried out a series of collaborative experiments, with Hugh acting as our guide and mentor in matters marsupial, which resulted in a number of publications and set us well on the way to becoming 'card carrying' marsupial biologists. The research collaboration grew into a personal friendship with Hugh and his family, which Kate and I value greatly.

Norman Saunders

I count myself very fortunate to have been a student of Hugh Tyndale-Biscoe, as an undergraduate, honours and PhD student. Hugh always terrified us undergraduates because he knew so much, could express his ideas clearly and always appeared so unattainably above us. One of the pleasures of doing a PhD with Hugh was learning of the warm and sensitive side of his nature. His love of outdoors is infectious, as is his love of excellent science. Tea time in the Zoology Department was an exciting event, listening to the arguments of Hugh and his senior colleagues. I was Hugh's first PhD student in his area of interest, namely marsupial reproductive physiology, joined some eight months later by John Hearn The small group we had encouraged the excellent interactions and a deep interest in all the work that was going on, as well as a friendly rivalry between us and the 'Great Archimandrite'. I will never forget the excitement with which new ideas were always greeted, although if there were any flaws in the arguments they were soon dissected out. Hugh taught us to always strive for excellence and never to be daunted by what seemed to be impossible. He was always there if we needed him but he encouraged us to seek advice from whichever experts we could find, either locally or at a distance, and so gain the independence which is essential for any graduate student. Hugh also remains a close friend and colleague ever since that time, and is always there to offer wise counsel if asked. My most exciting endeavour with Hugh was the writing of our book some 15 years after I completed my PhD. It was a difficult task to do at a distance, and yet I can honestly say that we never disagreed on a single word, paragraph or chapter. Although Hugh, has moved on to other areas of expertise, to me he will always remain the world's expert on marsupial reproductive physiology.

Marilyn Renfree

Hugh's role in marsupial reproductive physiology, as a leader in research, teaching and graduate training and mentorship, should be stressed. When I worked with him he helped me to develop and define a difficult and timely problem (the role of the pituitary in the regulation of embryonic diapause in the tammar) and then gave me free rein to crack the problem (hypophysectomy and radioimmunoassay for luteinising hormone) in a series of experiments that almost always delivered the opposite to the predicted results. He was away for the year in which most of the experimental work was done, although we kept in close touch by correspondence (he was on sabbatical leave in Colombia). I appreciated the freedom to get on with the job — and to make the necessary mistakes on the way to cracking the problem successfully. This essential part of learning and research was well understood by Hugh. He was able to take a back seat when his students were in the process of working out their projects during the first year when one was to set up the project and attack the problem — often with little data to show for enormous effort.

Hugh's group was a pleasure to belong to in that we had a team approach to shared duties and to assisting each other in research, as well as in discussing results. The department was a marvellous place to be in — the range of real characters and the earthy humour. I feel privileged to have been able to work in that environment, especially when I see the pressures and the less independence afforded to research students and postdocs today. Throughout my career, I have tried to use the personal and the managerial lessons I learned during that time, and I have a lot to thank Hugh for from both of those dimensions. He has remained a firm friend over the years and a prime example of the asset we have in science of friendships around the world that are lasting.

John Hearn

My first introduction to Hugh was during lectures in first year zoology class at the Australian National University. Those lectures were extremely interesting but we suffered from information overload! For the undergraduates, Hugh had a reputation as a serious lecturer and dedicated scientist, and we stood in awe of him. Ever since, he has continued to be my greatest source of useful but stimulating facts and a great one for encouraging lateral thinking. During my honours year in zoology in 1975, Hugh's research in marsupial reproduction and the vagaries of hypophysectomies became a reality. We would creep by the lab when an operation was underway and at the end of the day we knew without asking whether it had been successful or not. In

1976 I was privileged to gain a research position with Hugh when he moved to CSIRO. It was months before I could bring myself to call him Hugh — indeed for a while, because I still held him in such awe, the mouthful of 'Dr Tyndale-Biscoe' was so awkward that I didn't' call him anything at all. But we have progressed since then. Hugh has not only been my colleague but also my PhD supervisor (with Peter Janssens), and has always provided me with encouragement and advice. Throughout my association with him, he has maintained the goal of rigorous scientific endeavour with the eagerness and enthusiasm of a seven-year-old learning to cook at a scout camp. Working with him has always been a challenge and a privilege, with the excitement of breakthroughs always remembered. As with many people who have been associated with Hugh, he has been a wonderful mentor, colleague and friend to me, and will always remain so.

Lyn Hinds

We should like to thank Sharon Monk for the considerable help that she gave us in the preparation of the manuscript and in checking the proofs.

CONTRIBUTORS

PROLOGUE: *MARSUPIALS AND MODERN BIOLOGY: BREAKING THE PARADIGMS*
Hugh Tyndale-Biscoe, CSIRO Division of Wildlife and Ecology, Canberra, Australia.

SECTION 1 • *REPRODUCTION AND DEVELOPMENT*

1 ENDOCRINE SIGNALS IN EMBRYO IMPLANTATION
John P Hearn, Wisconsin Regional Primate Research Center and Department of Physiology, University of Wisconsin, Madison, Wisconsin, USA.

2 CONTROL OF PARTURITION IN THE MARSUPIAL *MACROPUS EUGENII*
Marilyn B Renfree, Laura J Parry, Terry P Fletcher and Geoff Shaw, Department of Zoology, University of Melbourne, Australia.

3 MECHANISMS OF SEASONALITY IN MARSUPIALS – A COMPARATIVE VIEW
Lyn A Hinds, CSIRO Division of Wildlife and Ecology, Canberra, Australia; and Andrew S I Loudon, School of Biological Sciences, University of Manchester, Manchester, UK.

4 DEVELOPMENT OF THE POUCH YOUNG FROM BIRTH TO POUCH VACATION
Peter A Janssens, Division of Biochemistry and Molecular Biology, School of Life Sciences, Australian National University, Canberra, Australia; Anthony J Hulbert, Department of Biological Sciences, University of Wollongong, New South Wales, Australia; and Russell V Baudinette, Department of Zoology, University of Adelaide, Australia.

5 THE DEVELOPMENT OF SEX DIFFERENCES IN BEHAVIOUR: A MARSUPIAL MODEL
Barbara H Fadem, Department of Psychiatry, University of Medicine and Dentistry of New Jersey, New Jersey Medical School, USA.

6 THE USE OF *MONODELPHIS DOMESTICA* AS A MODEL FOR MUSCULO-SKELETAL DEVELOPMENT

Charles W Archer, Heather Morrison, School of Molecular and Medical Biosciences, University of Wales, Cardiff, UK; Michael T Bayliss, Kennedy Institute of Rheumatology, London W6, UK; and Mark W J Ferguson, School of Biological Sciences, University of Manchester, Manchester, UK.

SECTION 2 • *GENETICS*

7 GENETICS OF TAMMAR WALLABIES

Desmond W Cooper and Louise M McKenzie, School of Biological Sciences, Macquarie University, Sydney, Australia.

8 SEXUAL DETERMINATION AND DIFFERENTIATION IN THE MARSUPIAL

Geoffrey Shaw, Jenny L Harry, Deanne J Whitworth and Marilyn B Renfree, Department of Zoology, University of Melbourne, Australia.

SECTION 3 • *ECOLOGY*

9 FIELD ENERGETICS AND WATER FLUXES IN MARSUPIALS

Brian Green, CSIRO Division of Widlife & Ecology, Canberra, Australia.

10 LIVING SLOW AND DYING YOUNG: SENESCENCE IN MARSUPIALS

Andrew Cockburn, Evolutionary Ecology Group, Division of Botany and Zoology, Australian National University, Canberra, Australia.

11 REPRODUCTIVE ECOLOGY OF NEW WORLD MARSUPIALS

John D Harder and David W Fleck, Department of Zoology, The Ohio State University, Columbus, Ohio, USA.

12 DIGESTIVE AND METABOLIC ADAPTATIONS OF ARBOREAL MARSUPIALS FOR DEALING WITH PLANT ANTINUTRIENTS AND TOXINS

Steven J Cork, CSIRO, Division of Wildlife & Ecology, Canberra, Australia; and William J Foley, Division of Botany and Zoology, Australian National University, Canberra, Australia.

13 ABORIGINAL TRADITIONAL KNOWLEDGE AND ECOLOGY EXPLAIN REPRODUCTIVE ANOMALIES IN THE RED KANGAROO IN CENTRAL AUSTRALIA

Alan Newsome, CSIRO Division of Wildlife and Ecology, Canberra, Australia.

SECTION 4 • *PATHOLOGY AND HOMEOSTASIS*

14 THE LABORATORY OPOSSUM (*MONODELPHIS DOMESTICA*) IN BIOMEDICAL RESEARCH

John L VandeBerg and Edward S Robinson, Department of Genetics, Southwest Foundation for Biomedical Research, San Antonio, Texas, USA.

15 ONTOLOGICAL WOUND HEALING STUDIES IN *MONODELPHIS DOMESTICA* FROM BIRTH TO ADULTHOOD

John R Armstrong, West Norwich Hospital, Norwich, UK; and Mark W J Ferguson, School of Biological Sciences, University of Manchester, Manchester, UK.

16 ANTIBODY TRANSPORT ACROSS THE GUT OF THE SUCKLING OPOSSUM

Arthur E Wild, Department of Biology, University of Southampton, Southampton, UK.

17 THE NORTH AMERICAN OPOSSUM *DIDELPHIS VIRGINIANA*: AN IMPORTANT MODEL FOR MECHANISTIC STUDIES OF SECRETORY DIARRHOEA AND FLUID-VOLUME HOMEOSTASIS

William J Krause, Leonard R Forte, F Kent Hamra, and Ronald H Freeman, Departments of Pathology and Anatomical Sciences, Pharmacology, and Physiology, School of Medicine, University of Missouri and the Harry S Truman Memorial VA Medical Center, Columbia, Missouri, USA.

SECTION 5 • *DEVELOPMENTAL NEUROBIOLOGY*

18 HUGH TYNDALE-BISCOE AND THE DEVELOPMENTAL NEUROBIOLOGY OF VISION AT THE ANU

Richard Mark, Developmental Neurobiology Group, Research School of Biological Science, Australian National University, Canberra, Australia.

19 DEVELOPMENT OF SOMATOSENSORY PATHWAYS FROM THE WHISKERS

Phil M E Waite, School of Anatomy, University of New South Wales, Sydney, Australia; and W Lee Weller, Department of Anatomy, University of Tasmania, Hobart, Australia.

20 PROTEINS IN PLASMA, CEREBROSPINAL FLUID AND BRAIN DURING POSTNATAL DEVELOPMENT IN MARSUPIALS

Katarzyna M Dziegielewska, Graham W Knott and Norman R Saunders; Department of Physiology, University of Tasmania, Hobart, Australia.

21 DEVELOPMENT AND DEVELOPMENTAL PLASTICITY OF DESCENDING SPINAL PATHWAYS IN THE NORTH AMERICAN OPOSSUM, *DIDELPHIS VIRGINIANA*

George F Martin and Xian Ming Wang , Department of Cell Biology, Neurobiology and Anatomy, The Ohio State University, Columbus, Ohio, USA.

22 MARSUPIALS AS MODELS FOR STUDIES OF DEVELOPMENT AND REGENERATION OF THE CENTRAL NERVOUS SYSTEM

Norman R Saunders, Department of Physiology, University of Tasmania, Hobart, Australia.

PROLOGUE

MARSUPIALS AND MODERN BIOLOGY—BREAKING THE PARADIGMS

Hugh Tyndale-Biscoe

INTRODUCTION

The modern period in marsupial biology began forty years ago with the publication of two letters to *Nature* on the discoveries being made in Perth on the quokka (*Setonix brachyurus*): Geoff Sharman's (1954) report of the phenomenon of embryonic diapause in the quokka, and the report of ruminant digestion in the quokka by Moir *et al.* (1954). Its beginning thus coincided with that of molecular biology and its flowering has occurred during the greatest revolution in thinking about life on earth. The players who were around at the start are leaving the stage, so it is a convenient time to ponder the achievements of this forty years and to offer some thoughts on where it may go in the next few years.

Four themes can be discerned in this period of marsupial research.

Firstly, from the start, the idea was to study Australian marsupials for their intrinsic worth, as a unique part of the world's faunal heritage; this was a sufficient reason for Harry Waring when he took the Chair of Zoology at the University of Western Australia and attracted young biologists to Perth (Waring 1956).

Simultaneously, in the early 1950s research on the large kangaroos was also started by CSIRO and the Animal Industry Branch of the Northern Territory Administration, primarily for economic reasons, because the kangaroo were seen to be pests, competing with sheep and cattle.

Driven by the discovery that marsupials do things differently rather than less well, these studies challenged the Anglo-centric paradigm that marsupials are poor imitations of real mammals and the only value in studying them is to shed light on evolutionary pathways leading to eutherian mammals. To the contrary, understanding how marsupials work has increased our understanding of processes in other species, including our own.

With the rise of economic rationalism another insidious notion has come to predominate: that if research cannot promise an immediate return on investment it should not be funded. This malign view of research funding is having a particularly serious effect on the health of research into marsupials and monotremes. The pity is that this is happening at just the time when research on several aspects of marsupial biology is coming to fruition.

BREAKING THE PARADIGMS

I suppose everyone contributing to this volume has already answered in the affirmative the question, why study marsupials? From various disciplines we have all come to realise that marsupials of one sort or another offer particularly good models to address questions of mammalian biology that interest us.

So why has it taken so long for marsupials to become respectable research material, in other words why not marsupials? As Peter Janssens and I (Tyndale-Biscoe & Janssens 1988) mentioned in the introduction to *The Developing Marsupial*, two obstacles to the study of marsupials for a long time were the lack of any species that could be bred readily in captivity, and a perception that their separate evolution meant that they were not relevant to the main questions in mammalian development. I think both these obstacles have now been removed.

So far as the first perception is concerned, three species of marsupial are now well established as self-sustaining laboratory animals. Tammar wallabies, *Macropus eugenii*, have now been bred for 20 years in captivity at Canberra and until the last 3 years, the colony was maintained at 500 to 600 animals. A second breeding colony has been established at Macquarie University (Hinds *et al.* 1990a) and, with the formation of the Cooperative Research Centre (CRC) for Conservation and Management of Marsupials, it will grow. In addition there are several smaller colonies in Australia; together, these provide animals for a great variety of studies on the species' physiology, genetics, reproduction and development, so that the tammar is now the best studied species of marsupial. The main disadvantage of the tammar is its size and low reproductive rate and a more suitable small polytocous species was sought for many years. *Sminthopsis macroura* (Stripe-faced dunnart) is now being successfully bred at La Trobe University for studies on early development (Selwood & Woolley 1991), but the big success species is *Monodelphis domestica* (South American grey short-tailed opossum) (VandeBerg 1983), which is well established in many colonies in the USA, Britain and now Tasmania, and seems destined to transform thinking about marsupials as research animals.

The second perception has been around a long time; even our revered mentor and friend, Amoroso (1963), considered that the main interest in marsupial studies was for the light they might shed on evolutionary problems. For those of us who have been studying marsupials for a number of years it is frustrating to find that this perception still persists, particularly among medical scientists. Let us hope that the new wave of research on marsupial development that is evident in this book will help to alter that perception, because there are so many questions in this field that marsupials can uniquely help to elucidate.

To illustrate what I mean, I should like to begin by telling the story I know best;

the quest for the function and control of the macropodid corpus luteum is a sort of morality tale but also most apposite for this book, since many of the people who have participated in the quest are also contributors to other chapters.

THE QUEST FOR THE FUNCTIONS OF THE MACROPODID CORPUS LUTEUM

The quest began with Sandes' (1903) study in which he relates how the idea of the central role of the corpus luteum was bequeathed by Fraenkel, and his own studies on material from the Eastern Quoll *Dasyurus viverrinus*, collected by J.P. Hill, confirmed its probable function as a gland of internal secretion that controls reproductive processes. Later Carl Hartman, who pioneered the experimental study of corpus luteum function with the North American opossum, *Didelphis virginiana* (Hartman 1925), recognised the potential of the monovular marsupials of Australia for this purpose, but 30 years were to elapse before this suggestion was followed up by Geoff Sharman. Sharman's (1955) study on the quokka established that pregnancy occupied almost the whole interval of the oestrous cycle, unlike what was previously known from Hartman's (1923) work, but more significantly it established the extraordinary phenomenon of embryonic diapause in marsupials. Forty years later both phenomena continue to stimulate questions and research into them. I first learnt of Geoff's work when I read the article in *Nature* just after I had spent a year studying reproduction in the brushtail possum, *Trichosurus vulpecula*, in New Zealand (Tyndale-Biscoe 1955). I couldn't reconcile my observations with Geoff's and wrote to him. His response in a letter of 1955 was:

> Briefly what happens is this:
>
> On the day of parturition, or the next day, the female ovulates and it is from this ovulation that the quiescent blastocyst stage results. The ovulating follicle forms the corpus luteum of lactation.
>
> While the young is suckled *in the pouch* and during the period of quiescence this corpus luteum of lactation does not complete development. It remains comparable to the 3 to 4 day corpus luteum of the normal cycle.
>
> If the young be removed from the pouch the corpus luteum of lactation begins to grow and is accompanied by luteal changes in the uterus and with implantation and development of the blastocyst stage.
>
> In all macropods I have looked at I find a corpus luteum of lactation comparable to that of *Setonix* in the ovary of mothers with pouch young.
>
> I have detailed these observations so that no misunderstanding in terminology may crop up.

Four years later I followed up on the quokka work in Perth and Geoff began his studies with Phyllis Pilton on brushtail possums in Adelaide; they resolved the possum's reproductive pattern as being essentially similar to that of *Didelphis*, with pregnancy terminating at the end of the luteal phase (Pilton & Sharman 1962). This suggested that, in the quokka, the influence of the corpus luteum must be extended during pregnancy, as it is in carnivores among the Eutheria. However, when I removed either the corpus luteum or both ovaries from an early stage of pregnancy in quokkas it did not prevent the fetus going to term (Tyndale-Biscoe 1963b). I can't find the correspondence now but my recollection is that Geoff was highly sceptical of my results, but then got similar results in the possum himself (Sharman 1965a).

So here we had the first paradox of the quest; the corpus luteum is a prominent feature histologically in marsupials, which would suggest that it has an important role, but it is dispensable even before it is fully formed.

PROGESTERONE SECRETION BY THE CORPUS LUTEUM

The full significance of this paradox was not evident for 15 years, and then it was my turn to be the sceptic. Don Bradshaw told me in 1978 that Max Cake, Felicity Owen and he, had found that there was a transient pulse of progesterone in the circulation of the quokka on day 4 of pregnancy (Cake *et al.* 1980). I knew that this had to be impossible because the quokka corpus luteum at day 4 is actively dividing and growing (Tyndale-Biscoe 1963a), so it could not be altering its secretion at the same time. But they were right and, as has now been shown in many species, a transient early pulse of progesterone is a constant and essential feature of the oestrous cycle and pregnancy of macropodine and potoroine marsupials (Hinds 1990). Indeed, in the tammar the interval from the pulse to birth is more constant than the interval from oestrus or the start of the delayed cycle to the pulse (Hinds & Tyndale-Biscoe 1982a). Despite its predictive value and clear importance in the control of pregnancy, we do not yet know how the pulse is achieved nor what controls it; certainly there is a brief change in secretion rate by the luteal cells coincident with the pulse (Hinds *et al.*1983) but its control is independent of the pituitary, as Lyn Hinds (1983) showed, and of the embryo, since it occurs in the non-pregnant cycle.

This is an aspect of corpus luteum function that is unknown in other mammals, but it raises another more general question, namely, how cells that are actively dividing can simultaneously alter their synthesis and secretion of progesterone. In the 12 years since those papers were published, nothing further has been done to elucidate this fascinating conundrum.

CENTRAL CONTROL OF THE CORPUS LUTEUM

Another aspect of marsupial endocrinology that defied the paradigm is the extraordinary way in which the corpus luteum is suppressed during reproductive quiescence. In the years after Geoff Sharman described embryonic diapause in the quokka, tammar and other macropodids (Sharman *et al.* 1966), the constant associated feature was the presence of the small, quiescent corpus luteum during lactation. Geoff showed that this was maintained by the sucking stimulus of the pouch young (Sharman 1965b) and Marilyn Renfree, some years later (Renfree 1979), demonstrated that it was transmitted by a neural signal from the teat. Drawing on the eutherian paradigms of pituitary control of the ovary during lactation, our working hypothesis in the 1960s was that, during lactation, prolactin secretion would be elevated and gonadotrophin secretion reduced, the resultant lack of stimulation preventing development of the corpus luteum. Then, on removal of the sucking stimulus, prolactin would decline and gonadotrophin increase.

This hypothesis was shown to be wrong in every respect. John Hearn demolished most of the hypothesis when he demonstrated that removing the pituitary of lactating tammars resulted in reactivation and growth of the corpus luteum, with concomitant growth and development of the fetus to full term (Hearn 1973). He further showed, with one of the first radioimmunoassays for gonadotrophin (Hearn 1974) that the only elevation of luteinising hormone occurred before ovulation and that there was no elevation at the time of reactivation and growth of the corpus luteum.

This led to the quite extraordinary conclusion that the pituitary is engaged in active, tonic suppression of the corpus luteum throughout the long period of quiescence. This is still the only known example of an inhibitory function of the pituitary of any species of mammal.

PROLACTIN AS THE INHIBITOR OF THE CORPUS LUTEUM

The subsequent search for the inhibitor led to further unexpected conclusions. From the begining the two prime candidates were oxytocin and prolactin. After John Hearn left to pursue his career in primate reproductive physiology, Jenny Hawkins and I had to learn his technique for hypophysectomy (Hearn 1975) and, having achieved this eventually, were able to show that prolactin was the prime suspect (Tyndale-Biscoe & Hawkins 1977). However, nailing that down took several years because a large proportion of the sham-operated animals reacted in the same way as hypophysectomised animals and we could not be certain that the effects were due to loss of the pituitary or some effect of the operation. Strong supporting evidence for the importance of prolactin was contributed by Conrad Sernia (Sernia & Tyndale-Biscoe 1979), who showed that the corpus luteum had very high concentrations of specific receptors for prolactin and Francesca Stewart (Stewart & Tyndale-Biscoe 1982), who in addition, showed that it had none for luteinising hormone. By this stage Lyn Hinds was involved and the solution, when it came, was simple; animals taken into the animal house for surgery were being exposed to a constant photoperiod that was sufficiently different from that outside to stimulate reactivation of the corpus luteum. When this matter was attended to the sham-operated females did not reactivate, while the hyophysectomised ones did, unless treated with prolactin (Hinds 1983).

So, once again the results from a marsupial defied prediction, based on eutherian paradigms; prolactin in the tammar is neither luteotrophic, as in the laboratory rat, nor luteolytic as in the domestic sheep, but demonstrates an entirely new function of tonic long-term inhibition, that could be termed luteostatic.

THE PROLACTIN PARADOX

Lyn Hinds also developed the radioimmunoassay for prolactin and this disclosed another paradox; prolactin concentrations in circulation are consistently low through pregnancy and early lactation and do not change when the sucking stimulus of the small pouch young is removed (Hinds & Tyndale-Biscoe 1982b, 1985). If prolactin is the important inhibitor of the corpus luteum how come its secretion does not change? Resolution of this had to await the results of Steve McConnell's study on the role of the pineal gland in the control of seasonal quiescence of the tammar, and was a beautiful piece of serendipity. Steve McConnel had collected blood samples throughout a 24 h period for melatonin and then had refined the assay, so that he required less plasma than he had collected and was about to discard the surplus. Not wanting to see the samples wasted, Lyn Hinds assayed them for prolactin and discovered that there was a transient but large pulse of prolactin in the samples collected around the time of dawn in quiescent tammars that had disappeared in samples taken from the same animals 5 days after a change in the photoperiod they were experiencing (McConnell *et al.* 1986). In a subsequent study Lyn Hinds showed that the prolactin pulse ceased to occur on the fourth morning after a change in photoperiod from 12 h dark to 9 h dark and that, if the pulse were mimicked by an intravenous injection of prolactin each morning, quiescence could be maintained (Hinds 1989). Thus the paradox was solved for the discrepancy between the evident importance of prolactin and the lack of any apparent change in its concentration in plasma; the samples were being taken too late in the day and were missing the early morning pulse. Likewise, lactational quiescence may be maintained in tammars by a similar transient pulse of prolactin once every 24 h but not necessarily at dawn; this would be well-nigh impossible to detect by single daily samples of blood and could only be tested indirectly. The study of Keith Gordon, who took pouch young away from lactating females for increasing intervals of time and found that reactivation only occurred if the interval was greater than 72 h (Gordon *et al.* 1988), provides the best evidence so far, for this hypothesis.

THE CORPUS LUTEUM AND SEASONAL BREEDING

The control of seasonal breeding in the tammar and Bennett's wallaby, *Macropus rufogriseus rufogriseus*, is another fascinating example of a marsupial adaptation that is superficially similar but actually very different to that of other seasonal breeders, like the sheep and the red deer. These two species of wallaby are short-day breeders, which means that they breed after the summer solstice, and Pat Berger showed this very early when she took tammars to the USA and they reversed their time of birth within a few weeks of arrival (Berger 1970). The first experimental investigation of this was done by Richard Sadleir in 1975, when he was on study leave from Canada, and the result of most significance was the most unexpected. At the time it was thought that the important signal must be the longest day itself and so he exposed six quiescent females to long daylength in August. When nothing had happened after six weeks the experiment was abandoned and the tammars returned to outside pens; thirty days later, to everyone's astonishment, all six tammars had new babies in their pouches! This showed that long day per se is not the signal but the change to a shorter day is a potent signal (Sadleir & Tyndale-Biscoe 1977). Many later experiments have confirmed this first observation (Hinds & den Ottolander 1983), and Steve McConnell used it very effectively to examine the role of the pineal gland and its hormone, melatonin, in the response (McConnell & Tyndale-Biscoe 1985; McConnell & Hinds 1985). He showed that, as in most other mammals, the elevated profile of melatonin corresponded closely to the period of darkness (McConnell 1986) and that the change in the profile occurred on the first night after a change in photoperiod. Despite this the tammars would not respond to the change, nor to an artificial elevation of melatonin during the light phase, unless the change in duration was maintained for 5 successive nights; then they would respond irreversibly, leading to birth about 27 days later. In other studies, mainly by Lyn Hinds (1989), the way in which this response is effected has been disclosed, and it is very different from other mammals.

In such species as the sheep and deer that respond to short photoperiod, the hypothalamic pulse generator, which determines the pulse rate of luteinising hormone secretion, is modulated by photoperiod via pineal gland secretions of melatonin, and this determines the breeding season (Karsch *et al.* 1984). In the tammar on the other hand, the hypothalamic pulse generator is unaffected by season and the female is potentially capable of ovulating at any time of the year (Hinds *et al.* 1992). Seasonality is imposed by the tonic inhibition of prolactin, as mentioned above, and it is this morning pulse that appears to be controlled by photoperiod and melatonin secretion. This, again, is a nice example of the marsupial doing it differently, but no less effectively. Indeed, the female tammar is so exquisitely sensitive and responsive to a change in photoperiod that it can detect an increase in the duration of the dark phase (or of elevated melatonin) of one hour from an initial period of nine hours and will respond irreversibly to such a change if it is maintained for only three or four consecutive nights (Tyndale-Biscoe & Hinds 1992).

No other mammal so far investigated, responds so rapidly to experimental photoperiod change as the tammar does. The tammar is thus potentially a very suitable species with which to investigate the site in the mammalian brain where time is measured.

While we are confident of the importance of prolactin in the tammar, the evidence is not so clear for Bennett's wallaby and Andrew Loudon and Jon Curlewis have shown that the response after the summer solstice may be programmed many months before (Curlewis & Loudon 1989; Loudon & Curlewis 1989). Furthermore, they have not been able to detect a pulse of prolactin at dawn, such as occurs in the tammar (Curlewis *et al.* 1986; Curlewis & Loudon 1988; Brinklow & Loudon, 1989; Pearce-Kelly *et al.* 1992). If, as seems likely, seasonal breeding evolved independently in these two species of macropods from the non-seasonal pattern seen in other species of kangaroo and in the mainland subspecies, *Macropus rufogriseus banksianus*, it is perhaps not too surprising that the endocrine control is not the same.

There is no non-seasonal subspecies of the tammar, but, by comparing the seasonal and non-seasonal subspecies of *M. rufogriseus* (Merchant & Calaby 1981) it might be possible to determine how seasonality evolved in the Tasmanian subspecies, and whether or not it is a single mutation that alters the sensitivity of the corpus luteum to transient pulses of prolactin. Here again, marsupials offer an unique opportunity to investigate, in a natural experiment, the evolution of seasonal breeding in a mammal.

THE CORPUS LUTEUM AND PARTURITION

The story of the fetus and the placenta is another case of old paradigms being challenged. In this case the notion is that marsupials don't have a functional placenta and the fetus is too immature to control anything. Marilyn Renfree's first research showed that the placenta is functional (Renfree 1973) and has an influence on the endometrium, which may be endocrine (Renfree & Tyndale-Biscoe 1973). Strangely, this particular area of research has languished for many years and is only now drawing new attention as the process of parturition is understood. Bill Poole and John Kirsch's studies on hybrid grey kangaroos 20 years ago indicated that the fetus must play a role in the timing of parturition and the length of gestation (Kirsch & Poole 1972) and Jim Merchant's careful observations, first on agile wallabies, *Macropus agilis* (Merchant 1976), and then in more detail on the tammar, hinted at an important role for the fetus in determining not only the duration of pregnancy but also the timing of the subsequent post-partum oestrus (Merchant 1979). This stimulated the need to measure oestradiol to understand how the 'Merchant effect' is achieved. Geoff Shaw (Shaw & Renfree 1984) and John Harder (Harder *et al.* 1984) independently achieved the necessary assay sensitivity and demonstrated the interaction between the decline in progesterone from the corpus luteum, oestradiol secretion from the Graafian follicle, which induced oestrous behaviour and, 24 h later, provoked the pre-ovulatory luteinising hormone pulse (Harder *et al.* 1985). While these studies showed that these events were advanced in pregnant, compared to non-pregnant, wallabies, they did not explain how the fetus could initiate them.

Lyn Hinds' discovery of the prepartum pulse of prolactin (quite different from the dawn pulse referred to above) and its close association with the full-term fetus (Hinds & Tyndale-Biscoe 1982b) led us to conclude that the fetus may send a signal to the maternal pituitary, but how was not clear. The prolactin pulse consistently occurred in pregnant females, rarely in non-pregnant females (Tyndale-Biscoe *et al.* 1983), but we showed that its timing was more closely associated with the maturity of the fetus than with the stage of the maternal reproductive cycle (Tyndale-Biscoe *et al.* 1988). It seemed unlikely that the fetus could produce sufficient prolactin to be detected in the maternal circulation and we knew from John Hearn's (1974) results that parturition does not occur after hypophysectomy. From this we concluded that the fetus must be sending a signal to the maternal pituitary, but what that might be was unclear. Prostaglandin was one candidate and we joined forces with Geoff Shaw, Terry Fletcher and Marilyn Renfree to resolve it. While we did not resolve it, we did discover the remarkable way in which prostaglandin induces parturient behaviour in the female tammar (Hinds *et al.* 1990b). Subsequently, as Marilyn describes in chapter 2 of this volume, and in more detail elswhere (Renfree *et al.* 1994a), they have established the essential role of placental prostaglandin in the initiation of parturition and how that is achieved via fetal cortisol and luteal mesotocin.

Thus the original paradigm has been completely overturned and the onset and control of parturition in the tammar is now shown to be remarkably similar to that quintessential eutherian mammal, the sheep.

I think these results should finally lay to rest the notion that Amoroso (1955) and others favoured, that the marsupials did not exploit gestation as a reproductive strategy because of inadequacies in their endocrine system; to the contrary the emphasis on lactation was an evolutionary pathway with equal potential. This is a theme that Marilyn and I espoused in our book *Reproductive Physiology of Marsupials* (1987), and for which she has been a strong advocate in her recent publications, particularly her chapter in *Marshall's Physiology of Reproduction* (Renfree 1994a). Let's hope the notion is now finally put to rest.

The marsupial corpus luteum, that insignificant, evanescent structure, has yielded some of its secrets in the past 40 years but, as I have tried to indicate, many questions still remain. Some of these, when answered, may help in the understanding of corpus luteum function and control in other mammals as well. I particularly think of the way in which the macropodid corpus luteum is tonically suppressed for many months and of how the luteal cells regulate their own rate of secretion of progesterone. But the new evidence for the secretion by the tammar corpus luteum of mesotocin and relaxin shows that there are many other surprises still to come.

LACTATION — ANOTHER PARADIGM SHIFT

From the fetus to the pouch young and lactation. Here again there have been big and exciting developments in our understanding of mammary physiology and biochemistry, of lactation itself and of early mammalian development that would not have occurred, but for the particular features disclosed in marsupials. In the 1950s and early 1960s Adolph Bolliger and his colleagues made the first analyses of marsupial milk, using the common brushtail possum that lives around the University of Sydney. They found high levels of oligosaccharides in early milk but not in late stage milk and changes in lipids between early and late milk (Bolliger & Gross 1959). Malcolm Peaker in England also reported on differences in the ionic composition of wallaby milk between early and late stages of lactation, even in the same animal suckling young of different ages. Then for about 15 years Brian Green and his colleagues at CSIRO and the University of Sydney showed that, in all marsupials, substantial changes occurred through lactation in all components of the milk, which reflected the changing demands of the pouch young. However, between species there was not much difference in the composition of late stage milk (Green & Merchant 1988). These results contrasted markedly with eutherian mammals in which the composition of milk does not change much at all during lactation, but varies very greatly between species (Oftedal 1984). Steve Cork and Hugh Dove took Brian Green's data on the energetic cost of early lactation and added their observations on the energetics of late lactation and weaning in the tammar to examine the total energetic costs involved in lactation and compared it to the total cost of lactation in eutherian species. They came up with the interesting conclusion (Cork & Dove 1989) that the total investment of the two strategies was the same, but that the marsupial mother spread her investment over a longer period of time than the eutherian; in the context of the Australian environment the tammar strategy is one that has been adopted by many other animals and plants.

These conclusions about the changing composition of marsupial milk raised the interesting question of how the mammary epithelial cells, which are highly differentiated secretory cells, are capable of these profound changes in what they secrete. Can individual cells change during the course of lactation, or do new cells arise from undifferentiated precursor cells? Ros Joseph and Chris Collet (1994) have shown by *in situ* hybridisation that the genes for different milk proteins are being expressed in the same cells, which suggests that the regulation of secretion resides in the mammary epithelial cell. For the past ten years Kevin Nicholas has investigated the endocrine control of milk protein synthesis in the tammar mammary gland. He has shown that at the start of lactation it can synthesise several important milk proteins in response to prolactin alone (Nicholas 1988), whereas in eutherian mammals the same proteins require stimulation by a suite of adrenal and pituitary hormones.

Sex determination and differentiation, as discussed by Geoff Shaw *et al*. (Chapter 8 in this volume) is another area where research on marsupials has fully challenged the eutherian paradigm; in this case the convincing evidence from marsupials of a genetic, as well as an hormonal, component in sexual differentiation has led to a re-examination of the paradigm in regard to eutherian species as well.

In this brief review I have only touched on a very few examples from the research on marsupials of the last 40 years and it is necessarily biased to those I know best. I think, however, that it is sufficient to show that it has been very rewarding to work on marsupials, to learn how they have solved common problems in different ways, or in the same way by different routes, and how many of these findings have challenged existing paradigms, built on eutherian species. I always experience a special pleasure when one or another of these paradoxes has been resolved. While tackling a research topic I have this feeling that we are in a personal discourse with Nature; each time we think we have found the answer the results show it to be wrong; then another try and a far more beautiful, often simpler, solution is disclosed. Although Han Suin had a different idea, I think Francis Thompson must have had the experience of scientific discovery in mind when he wrote:

The angels keep their ancient places:
Turn but a stone and start a wing
'Tis ye, 'tis your estranged faces,
That miss the many splendoured thing.

ECONOMIC VALUE OF MARSUPIAL RESEARCH

So, the economic rationalists will say, you have indulged your curiosity at the expense of the Australian taxpayer, but what useful outcome can you show for all this? Can it be justified? This question can be asked of any creative endeavour but, for some reason, is more frequently directed at scientists than at any other creative group. Perhaps this is because society has come to expect scientists to produce cargo for our enjoyment and science and technology have been singularly successful at this. The odd thing is that some scientific research that is driven purely by the wish to know, curiosity driven research, is accepted by our society without question. Nevertheless, one might ask why it is more justifiable for some Australian scientists to indulge their curiosity about the stars of the southern hemisphere, but not for others to understand the ways the indigenous plants and animals have adapted to this continent over aeons of time; both enlarge our understanding of the place we live in and, because no one else has the opportunity to do it, satisfy, in a unique way our quest for knowledge.

As we are all aware as scientists, deeper understanding does frequently lead to knowledge that is beneficial to society, but the order can seldom be reversed. The study of marsupials is no exception; the greater understanding that has come from the research on these animals during the past 40 years now offers developments of great potential benefit that have followed from many unexpected discoveries and that could not have been foreseen in advance or planned for. Let us consider a few of these.

Embryonic diapause

A large industry has grown up in the past 30 years to exploit the fundamental knowledge of gamete biology and fertilisation of mammals, and one of the major obstacles has been the storage of gametes and fertilised eggs. Various methods of cryopreservation have been developed to achieve this and yet we have two species of wallaby that can store 100 cell embryos for 11 months at body temperature, during which no growth, cell division or differentiation occurs. How is this done? Our colleague, Toshi Kojima, who works on bovine *in vitro* fertilization and egg transfer in Tokyo, spent a year in Canberra and Sydney hoping to answer this question so that he could exploit it for use with cattle (Kojima *et al.* 1993). It eluded him but remains as pertinent today as it was then. With the advance of molecular techniques the question is becoming more tractable.

Immunocontraception

The burgeoning population of possums in New Zealand and their susceptibility to bovine tuberculosis is posing a problem of great magnitude. It has led to the search for new methods of control, including immunocontraception. This program is building on the earlier and present work of John Rodger, Lynne Selwood and Peter Temple-Smith. Indeed, the fact that it can even be contemplated and a major research program be started on gamete antigens for the interruption of fertilisation is only possible because of their work. Now the CRC for the Conservation and Management of Marsupials has been established with this work as one of its primary objectives.

Biorhythms

There is today much interest in biorhythms and the effect they can have on health of people who do shiftwork or work on international airlines. For this we need to understand the way our bodies measure time and also how we record and perceive intervals in the passage of time. As I indicated before, the work with tammars and Bennett's wallabies has shown us that these species are uniquely sensitive to the light dark cycle and they are probably of all mammals, the best models to further the understanding of the mammalian timekeeper.

Marsupial milk and fostering

The knowledge gained in recent years about the composition of the milk of marsupials and the way in which it changes through lactation, in relation to the changing needs of the developing young, has been exploited successfully in the burgeoning interest in rearing orphan marsupials and in the care of rare species; indeed a minor industry has grown up to service this need. Jim Merchant and Meredith Clark pioneered the fostering of pouch young between macropodid species (Merchant & Sharman 1966, Clark 1968), but the use of common species of marsupial as foster mothers for the young of rare or endangered species has not been developed to any extent, although this would be a much simpler and more fruitful way to rescue rare species than by gamete preservation and *in vitro* fertilisation technologies.

The marsupial mammary gland as a protein biofactory

Of all the recent discoveries in marsupials that might have a commercial future, the work of Kevin Nicholas and Chris Collet has the greatest promise. Marsupials secrete lactose-free milk and there is a huge potential market for such a milk if the genes encoding for these enzymes in the marsupial mammary gland could be introduced to transgenic cows; late stage wallaby milk is much more concentrated than cow's milk with a higher lipid content and if this feature could be introduced to cows it could revolutionise the dairying industry; and the genes encoding several of the proteins respond to a much simpler hormonal signal than do other species, so that if this expression system could be introduced to a transgenic sheep or cows, along with the gene for a commercial protein, it could be secreted selectively by the mammary gland of the transgenic animal. But will we exploit this unique opportunity or let it lapse?

Very young marsupials as models for human development

The idea of using the very immature young of marsupials as models for intrauterine stages of human development was first entertained in the 1930s but did not take off until recently. The 1986 Boden Conference was the first time it was seriously discussed and *The Developing Marsupial* was the first attempt to bring the potential to the attention of a wider audience. Now, in this book two recent developments in the use of the immature marsupial to investigate the central nervous system are described. Norman Saunders' (Chapter 22 this volume) gives us an exciting account of the work on the repair of the spinal cord in *Monodelphis* and the potential that it may have for the treatment of paraplegia. In years to come we will all remember the day we read of this work when the first paraplegic walks again. Richard Mark (Chapter 18 this volume) also describes something about the exciting work that his group is doing in understanding the way the mammalian retina is wired up to the cerebral cortex, work that can be done in no other kind of mammal because the eye and brain are inaccessible to investigation. The work of Robert Holland and his colleagues (Calvert *et al.* 1993; 1994; Tibben *et al.* 1991) on the changing forms of haemoglobin between fetal, neonatal and older young has disclosed important differences in the molecular structure and oxygen-carrying capacity of embryonic, fetal and adult haemoglobin in the tammar; which have relevance to the phenomena in other mammals. And Marilyn Renfree, Geoff Shaw and Roger Short are challenging conventional ideas about mammalian sexual differentiation and development through their work on fetal and neonatal tammars (Renfree 1994b; Shaw et al. Chapter 8 in this volume).

These are just a few of the ways in which the research on marsupials has proved, or has the potential to prove, its usefulness in the terms that economic rationalists can understand; some might even earn export dollars! But there is another aspect of the research on marsupials that should not be ignored; that is its cultural value.

OUR UNIQUE MARSUPIAL HERITAGE

As new discoveries have been made some have been put into films and these have had an enormous impact on the thinking and perceptions of young people. The first of these was *Birth of the Kangaroo* and I heard recently someone remark that the only sex education that he got as a boy in the 1960s was to be shown that film at school. It was followed by the two films that Mervyn Griffiths made on *Echidnas* and *Lactation,* the latter winning a prize in Japan. More recently there has been the evocative and highly informative *Kangaroos: faces in the mob*, based on the studies by Robyn Stuart-Dick (Stuart-Dick & Higginbottom 1989), and the superb film on the birth of the tammar by David Parer, based on the work of Marilyn Renfree and colleagues (Renfree *et al.* 1989). All of these have helped to make the modern research on marsupials known around Australia and elsewhere.

Other aspects of modern research have helped in the understanding of earlier knowledge. Some years ago I was approached by an anthropologist who was intrigued by what he had read about embryonic diapause in red kangaroos and wondered whether this was known in some way to Aborigines and might account for their ideas about conception. Alan Newsome has given us an appreciation of the way ecological understanding of the importance of refugia for red kangaroos is reflected in Aboriginal sacred sites of the kangaroo (Newsome 1980); this has now been extended by Tunbridge (1991) in understanding the relationship between species now extinct in the Flinders Ranges and Aboriginal Dreaming; it has even been suggested that it may be possible to infer from sacred sites for Yamuti something about the palaeoecology of the extinct *Diprotodon.* In another way the study of the dentition, internal anatomy and chromosomes of *Lagostrophus faciatus* (banded hare-wallaby) in the 1960s led David Ride, Geoff Sharman (1961) and me (1965) to consider that it was a very unusual wallaby; now Tim Flannery (1989) has concluded that it is the only living member of the Sthenurinae, all the rest of which became extinct more than 10 000 years ago.

But when all else has been said we must acknowledge the sheer beauty of such creatures as *Tarsipes spencerae* (honey possum) and the extraordinary way that they are adapted to the heath lands of Western Australia; or the wonderful social structure of the yellow-bellied glider, *Petaurus australis* (Russell 1984) and the complex olfactory communication that keeps the group together; or the alert vivacity of the quoll, *Dasyurus viverrinus*, and the numbat, *Myrmecobius fasciatus*. These and many other species of marsupial adorn and enrich this land, and the knowledge that has been learnt about them in the last 40 years should be seen to have contributed to the culture of our society as much as the other cultural achievements of this time.

ADEQUACY OF RESOURCES TO ACHIEVE THESE

It is a pity that one has to end a celebration of the achievements of 40 years of research on marsupials on a sombre note, but it cannot be denied that the support for research on marsupials is declining and in the last few years several productive and original research workers have ceased to work on marsupials. As Peter Janssens pointed out, of the 28 authors of the 15 chapters in *The Developing Marsupial*, published in 1988, half have ceased to work on marsupials, seven of them because they are no longer funded for this field; eight others have reduced their efforts in marsupial research, leaving only six of those authors still engaged in active research on marsupials. This is particularly disturbing because the people invited to write for the book represented many of the major practitioners; if they no longer work in this area it also means that students are not being attracted to it either.

One reason for the decline in funding support by the major agencies — Australian Research Council and National Health and Medical Research Council — could be because they still perceive marsupials to be off the mainstream of intellectual endeavour and, in the case of National Health and Medical Research Council, not relevant to medical science. In the case of the Australian Research Council the reason may also be due to the perception that the outcomes of the research are uncertain and the economic benefits slight or non-existent. The one bright spot is the new Cooperative Research Centre for the Conservation and Management of Marsupials and, ironically, the funds provided by the New Zealand Government to several Australian scientists to work on gamete biology of marsupials. The arguments that I have tried to put here are not, it seems, understood or appreciated; nevertheless, when marsupial studies are conducted rigorously they can and do enlarge general knowledge in unpredictable and important ways, and the outcomes can have considerable potential benefit to the country. If these perceptions cannot be changed, and we must continue to press the arguments, then the cultural value of marsupial studies must be demonstrated more vigorously. If a prime minister could provide large infusions of federal funds to encourage cultural activities in Australia, we should press the claims of marsupial research for this support. The record of the last 40 years has demonstrated the beauty of marsupial adaptations and offered the promise of much more to come. All that is needed to make it reality is the will to do it. As Goethe wrote:

If you can do something,
Or believe you can,
Begin it,
Boldness has genius, power
And magic in it.
Advance Marsupials fair!

References

Amoroso E C (1955). Endocrinology of pregnancy. *British Medical Bulletin*, 11, 117–125.

Amoroso E C (1963). Summary of the Conference. In *Delayed Implantation*. Ed A C Enders. University of Chicago Press, Chicago. pp 305–309.

Berger P J (1970). *Reproductive biology of the tammar wallaby*, *Macropus eugenii*. PhD. thesis, Tulane University, New Orleans. Dissertation Abstracts International 31B, 3760–3761.

Bolliger A and Goss R (1959). Composition of milk of the marsupial, *Trichosurus vulpecula*. *AMA Journal of Diseases of Children*, 98, 768–775.

Brinklow B R and Loudon A S I (1989). Effect of exogenous prolactin and bromocriptine on seasonal reproductive quiescence in the Bennett's wallaby (*Macropus rufogriseus rufogriseus*). *Journal of Endocrinology*, 120, 189–193.

Cake M H, Owen F J and Bradshaw S D (1980). Difference in concentration of progesterone in the plasma between pregnant and non-pregnant quokkas (*Setonix brachyurus*). *Journal of Endocrinology*, 84, 153–158.

Calvert S J, Holland R A B and Hinds L A (1993). Blood O_2 transport and Hb types in the embryonic tammar wallaby (Marsupialia, *Macropus eugenii*). *Respiratory Physiology*, 91, 99–109.

Calvert S J, Holland R A B and Gemmell R T (1994). Respiratory properties of the neonatal blood of the common brush-tail possum (*Trichosurus vulpecula*). *Physiological Zoology*, 67, 407–417.

Clark M J (1968). Growth of pouch-young of the red kangaroo *Megaleia rufa* in the pouches of foster mothers of the same species. *International Zoo Yearbook*, 8, 102–106.

Cork S J and Dove H (1989). Lactation in the tammar wallaby (*Macropus eugenii*). II. Intake of milk components and maternal allocation of energy. *Journal of Zoology*, 219, 399–409.

Curlewis J D and Loudon A S I (1988). Experimental manipulations of prolactin following removal of pouch young or bromocriptine treatment during lactational quiescence in the Bennett's wallaby. *Journal of Endocrinology*, 119, 405–411.

Curlewis J D and Loudon A S I (1989). The role of refractoriness to long daylength in the annual reproductive cycle of the female Bennett's wallaby (*Macropus rufogriseus rufogriseus*). *Journal of Experimental Zoology*, 252, 200–206.

Curlewis J D, White A S, Loudon A S I and McNeilly A S (1986). Effects of lactation and season on plasma prolactin concentrations and response to bromocriptine during lactation in the Bennett's wallaby (*Macropus rufogriseus rufogriseus*). *Journal of Endocrinology*, 110, 59–66.

Flannery T F (1989). Phylogeny of the Macropodoidea; a study in convergence. In *Kangaroos, Wallabies and Rat-Kangaroos*. Eds G Grigg, P Jarman and I Hume. Surrey Beatty & Sons Pty Ltd, Sydney. pp. 1–46.

Gordon K, Fletcher T P and Renfree M B (1988). Reactivation of the quiescent corpus luteum and diapausing embryo after temporary removal of the sucking stimulus in the tammar wallaby (*Macropus eugenii*). *Journal of Reproduction and Fertility*, 83, 401–406.

Green B and Merchant J C (1988). The composition of marsupial milk. In *The Developing Marsupial: Models for Biomedical Research*. Eds C H Tyndale-Biscoe and P A Janssens. Springer-Verlag, Berlin. pp 41–54.

Harder J D, Hinds L A, Horn C A and Tyndale-Biscoe C H (1984). Oestradiol in follicular flluid and in utero-ovarian venous and peripheral plasma during parturition and post-partum oestrus in the tammar, *Macropus eugenii*. *Journal of Reproduction and Fertility*, 72, 551–558.

Harder J D, Hinds L A, Horn C A and Tyndale-Biscoe C H (1985). Effects of removal in late pregnancy of the corpus luteum , Graafian follicle or ovaries on plasma progesterone, oestradiol, LH, parturition and post-partum oestrus in the tammar, *Macropus eugenii*. *Journal of Reproduction and Fertility*, 75, 449–459.

Hartman C G (1923). The oestrous cycle in the opossum. *American Journal of Anatomy*, 32, 353–421.

Hartman C G (1925). The interruption of pregnancy by ovariectomy in the aplacental opossum: a study in the physiology of implantation. *American Journal of Physiology*, 71, 436–454.

Hearn J P (1973). Pituitary inhibition of pregnancy. *Nature* 241, 207–208.

Hearn J P (1974). The pituitary gland and implantation in the tammar wallaby, *Macropus eugenii*. *Journal of Reproduction and Fertility*, 39, 235–241.

Hearn J P (1975). Hypophysectomy of the tammar wallaby, *Macropus eugenii*: surgical approach and general effects. *Journal of Endocrinology*, 64, 403–416.

Hinds L A (1983). *Progesterone and Prolactin in Marsupial Reproduction*. PhD thesis, Australian National University, Canberra.

Hinds L A (1989). Morning prolactin pulse maintains seasonal quiescence in the tammar wallaby, *Macropus eugenii*. *Journal of Reproduction and Fertility*, 87, 735–744.

Hinds L A (1990). The control of pregnancy, parturition and luteolysis in marsupials. *Reproduction, Fertility and Development*, 2, 535–552.

Hinds L A and den Ottolander R C (1983). Effects of changing photoperiod on peripheral plasma prolactin and progesterone concentrations in the tammar wallaby, *Macropus eugenii*. *Journal of Reproduction and Fertility*, 69, 631–639.

Hinds L A and Tyndale-Biscoe C H (1982a). Plasma progesterone levels in the pregnant and non-pregnant tammar, *Macropus eugenii*. *Journal of Endocrinology*, 93, 99–107.

Hinds L A and Tyndale-Biscoe C H (1982b). Prolactin in the marsupial, *Macropus eugenii*, during the estrous cycle, pregnancy and lactation. *Biology of Reproduction*, 26, 391–398.

Hinds L A and Tyndale-Biscoe C H (1985). Seasonal and circadian patterns of circulating prolactin during lactation and seasonal quiescence in the tammar, *Macropus eugenii*. *Journal of Reproduction and Fertility*, 74, 173–183.

Hinds L A, Diggle P J and Tyndale-Biscoe C H (1992). Effects of the ovary, sucking stimulus and season on the pattern of LH and FSH release in the female tammar, *Macropus eugenii*. *Reproduction, Fertility and Development*, 4, 25–34.

Hinds L A, Evans S M and Tyndale-Biscoe C H (1983). In-vitro secretion of progesterone by the corpus luteum of the tammar wallaby, *Macropus eugenii*. *Journal of Reproduction and Fertility*, 67, 57–63.

Hinds L A, Poole W E, Tyndale-Biscoe C H, van Oorschot R A H and Cooper D W (1990a). Reproductive biology and the potential for genetic studies in the tammar wallaby, *Macropus eugenii*. *Australian Journal of Zoology*, 37, 223–234.

Hinds L A, Tyndale-Biscoe C H, Shaw G, Fletcher T P and Renfree M B (1990b). Effects of prostaglandin and prolactin on luteolysis and parturient behaviour in the non-pregnant tammar, *Macropus eugenii*. *Journal of Reproduction and Fertility*, 88, 323–333.

Joseph R and Collet C (1994). Double staining in situ study of mRNAs encoding milk proteins in the mammary gland of the tammar wallaby (*Macropus eugenii*). *Journal of Reproduction and Fertility*, 101, 241–246.

Karsch F J, Bittman E L, Foster D L, Goodman R L, Legan S J and Robinson J E (1984). Neuroendocrine basis of seasonal reproduction. *Recent Progress In Hormone Research*, 40, 185–232.

Kirsch J A W and Poole W E (1972). Taxonomy and distribution of the grey kangaroos, *Macropus giganteus* Shaw and *Macropus fuliginosus* (Desmarest), and their subspecies (Marsupialia: Macropodidae). *Australian Journal of Zoology*, 20, 315–339.

Kojima T, Hinds L A, Muller W J, O'Neill C and Tyndale-Biscoe C H (1993). Production and secretion in vitro and presence of platelet activating factor (PAF) in early pregnancy of the marsupial, *Macropus eugenii*. *Reproduction, Fertility and Development*, 5, 15–25.

Loudon A S I and Curlewis J D (1989). Evidence that the seasonally breeding Bennett's wallaby (*Macropus rufogriseus rufogriseus*) does not exhibit short-day photorefractoriness. *Journal of Reproduction and Fertility*, 87, 641–648.

McConnell S J (1986). Seasonal changes in the circadian plasma profile of the tammar, *Macropus eugenii. Journal of Pineal Research*, 3, 119–125.

McConnell S J and Hinds L A (1985). Effect of pinealectomy on plasma melatonin, prolactin and progesterone concentrations during seasonal reproductive quiescence in the tammar, *Macropus eugenii. Journal of Reproduction and Fertility*, 7, 433–440.

McConnell S J and Tyndale-Biscoe C H (1985). Response in peripheral plasma melatonin to photoperiod change and the effects of exogenous melatonin on seasonal quiescence in the tammar, *Macropus eugenii. Journal of Reproduction and Fertility*, 73, 529–538.

McConnell S J, Tyndale-Biscoe C H and Hinds L A (1986). Change in duration of elevated concentrations of melatonin is the major factor in photoperiod response of the tammar, *Macropus eugenii. Journal of Reproduction and Fertility*, 77, 623–632.

Merchant J C (1976). Breeding biology of the Agile wallaby, *Macropus agilis* (Gould) (Marsupialia: Macropodidae), in captivity. *Australian Wildlife Research*, 3, 93–103.

Merchant J C (1979). The effect of pregnancy on the interval between one oestrus and the next in the tammar wallaby, *Macropus eugenii. Journal of Reproduction and Fertility*, 56, 459–463.

Merchant J C and Calaby J H (1981). Reproductive biology of the red-necked wallaby (*Macropus rufogriseus banksianus*) and Bennett's wallaby (*M.r. rufogriseus*) in captivity. *Journal of Zoology*, 194, 203–217.

Merchant J C and Sharman G B (1966). Observations on the attachment of marsupial pouch young to the teats and on the rearing of pouch young by foster-mothers of the same or different species. *Australian Journal of Zoology*, 14, 593–609.

Moir R J, Somers M, Sharman G B and Waring H (1954). Ruminant-like digestion in a marsupial. *Nature*, 173, 269–270.

Nicholas, K R (1988). Control of milk protein synthesis in the marsupial Macropus eugenii: a model system to study prolactin-dependent development. In *The Developing Marsupial. Models for Biomedical Research*. Eds C H Tyndale-Biscoe and P A Janssens. Springer-Verlag, Berlin. pp 68–85.

Newsome A E (1980). The ecomythology of the red kangaroo in Central Australia. *Mankind*, 12, 327–333.

Oftedal O T (1984). Milk composition, milk yield and energy output at peak lactation: a comparative review. *Symposium of the Zoological Society of London*, 51, 33–85.

Pearce-Kelly A S, Loudon A S I and Curlewis J D (1992). Seasonal and lactational effects on the prolactin response to a dopamine antagonist and TRH in the Bennett's wallaby (*Maacropus rufogrisus rufogriseus*). *General and Comparative Endocrinology*, 86, 323–331.

Pilton P E and Sharman G B (1962). Reproduction in the marsupial *Trichosurus vulpecula. Journal of Endocrinology*, 25, 119–136.

Renfree, M.B. (1973) The composition of fetal fluids of the marsupial *Macropus eugenii. Developmental Biology*, 33, 62–79.

Renfree M B (1979). Initiation of development of diapausing embryo by mammary denervation during lactation in a marsupial. *Nature*, 278, 549–551.

Renfree M B (1994a). Endocrinology of pregnancy, parturition and lactation of marsupials. In *Marshall's Physiology of Reproduction*. 4th edn. Ed G E Lamming. Chapman & Hall, London. Vol.3. pp 677–766.

Renfree M B (1994b). Sexual dimorphisms in the gonads and reproductive tract of marsupial mammals. In *The Differences between the Sexes*. Eds R V Short and E Balban. Cambridge University Press, Cambridge. pp 213–230.

Renfree M B, Shaw G and Fletcher T P (1994). Evidence for the essential role of prostaglandins for parturition in a marsupial, *Macropus eugenii. Journal of Reproduction and Fertility*, 102, 433–446.

Renfree M B and Tyndale-Biscoe C H (1973). Intra-uterine development after diapause in the marsupial, Macropus eugenii. *Developmental Biology*, 32, 28–40.

Renfree M B, Fletcher T P, Blanden D R, Lewi P R, Shaw G, Gordon K, Short R V, Parer-Cook E and Parer D (1989). Physiological and behavioural events around the time of birth in macropodid marsupials. In *Kangaroos, Wallabies and Rat-Kangaroos*. Eds G Grigg, P Jarman and I Hume. Surrey Beatty & Sons Pty Ltd, Sydney. pp 323–337.

Russell R (1984). Social behaviour of the yellow-bellied glider Petaurus australis in north Queensland. In *Possums and Gliders*. Eds A P Smith and I D Hume, Australian Mammal Society and Surrey Beatty & Sons Pty Ltd, Sydney. pp 343–353.

Sadleir R M F S and Tyndale-Biscoe C H (1977). Photoperiod and the termination of embryonic diapause in the marsupial *Macropus eugenii*. *Biology of Reproduction*, 16, 605–608.

Sandes F P (1903). The corpus luteum of *Dasyurus viverrinus*, with observations on the growth and atrophy of the Graafian follicle. *Proceedings of the Linnean Society of New South Wales*, 28, 364–405.

Selwood L and Woolley P A (1991). A timetable of embryonic development, and ovarian and uterine changes during pregnancy, in the stripe-faced dunnart, Sminthopsis macroura (Marsupialia: Dasyuridae). *Journal of Reproduction and Fertility*, 91, 213–227.

Sernia C and Tyndale-Biscoe C H (1979). Prolactin receptors in the mammary gland, corpus luteum and other tissues of the tammar wallaby, *Macropus eugenii*. *Journal of Endocrinology*, 83, 79–89.

Sharman G B (1954). Reproduction in marsupials. *Nature*, 173, 302–303.

Sharman G B (1955). Studies onmarsupial reproduction. III. Normal and delayed pregnancy in *Setonix brachyurus*. *Australian Journal of Zoology*, 3, 56–70.

Sharman G B (1961). The mitotic chromosomes of marsupials and their bearing on taxonomy and phylogeny. *Australian Journal of Zoology*, 9, 38–60.

Sharman G B (1965a). The effects of the suckling stimulus and oxytocin injection on the corpus luteum of delayed implantation in the red kangaroo. *Excerpta Medica*, 83, 669–674.

Sharman G B (1965b). The effects of suckling on normal and delayed cycles of reproduction in the red kangaroo. *Sonderdruck aus Zeitschrift fur Saugertierkunde*, 30, 10–20.

Sharman G B and Berger P J (1969). Embryonic diapause in marsupials. *Advances in Reproductive Physiology*, 4, 211–240.

Sharman G B, Calaby J H and Poole W E (1966). Patterns of reproduction in female diprotodont marsupials. *Symposium of the Zoological Society of London*, 15, 205–32.

Shaw G and Renfree M B (1984). Concentrations of oestradiol-17B in plasma and corpora lutea throughout pregnancy in the tammar, *Macropus eugenii*. *Journal of Reproduction and Fertility*, 72, 29–37.

Stewart F and Tyndale-Biscoe C H (1982). Prolactin and luteinizing hormone receptors in marsupial corpora lutea: relationship to control of luteal function. *Journal of Endocrinology*, 92, 63–72.

Stuart-Dick R I and Higginbottom K B (1989). Strategies of parental investment in Macropodoids. In *Kangaroos, Wallabies and Rat-Kangaroos*. Eds G Grigg, P Jarman and I Hume. Surrey Beatty & Sons Pty Ltd, Sydney. pp 571–592.

Tibben E A, Holland R A B and Tyndale-Biscoe C H (1991). Blood oxygen carriage in the marsupial, tammar wallaby (*Macropus eugenii*), at prenatal and neonatal stages. *Respiratory Physiology*, 84, 93–104.

Tunbridge D.(1991). *The Story of the Flinders Ranges Mammals*. Kangaroo Press, Sydney. 96 pp.

Tyndale-Biscoe C H (1955). Observations on the reproduction and ecology of the brush tailed possum, *Trichosurus vulpecula* (Marsupialia). *Australian Journal of Zoology*, 3, 162–184.

Tyndale-Biscoe C H (1963a). The role of the corpus luteum in the delayed implantation in marsupials. In *Delayed Implantation*. Ed A C Enders. University of Chicago Press, Chicago. pp 15–32.

Tyndale-Biscoe C H (1963b). Effects of ovariectomy in the marsupial, *Setonix brachyurus*. *Journal of Reproduction and Fertility*, 6, 25–40.

Tyndale-Biscoe C H (1965). The female urogenital system and reproduction of the marsupial, *Lagostrophus fasciatus*. *Australian Journal of Zoology*, 13, 255–267.

Tyndale-Biscoe, C H and Hawkins J (1977). The corpora lutea of marsupials; aspects of function and control. In *Reproduction and Evolution*. Eds J H Calaby and C H Tyndale-Biscoe. Australian Academy of Science, Canberra. pp 245–252.

Tyndale-Biscoe C H and Hinds L A (1992). Components of the melatonin message in the response to photoperiod of the tammar wallaby (*Macropus eugenii*). *Journal of Pineal Research* 12, 155–166.

Tyndale-Biscoe C H, Hinds L A, Horn C A and Jenkin G (1983). Hormonal changes at oestrus, parturition and post-partum oestrus in the tammar wallaby (*Macropus eugenii*). *Journal of Endocrinology*, 96, 155–161.

Tyndale-Biscoe C H and Janssens P A (1988). In *The Developing Marsupial. Models for Biomedical Research*, Springer-Verlag, Berlin.

Tyndale-Biscoe C H and Renfree M B (1987). *Reproductive Physiology of Marsupials*. Cambridge University Press, Cambridge.

Tyndale-Biscoe C H, Hinds L A and Horn C A (1988). Fetal role in the control of parturition in the tammar, *Macropus eugenii*. *Journal of Reproduction and Fertility*, 82, 419–428.

VandeBerg J L (1983). The grey short-tailed opossum: a new laboratory animal. *Institute of Laboratory Animal Resources* News, 26, 9–12.

Waring H (1956). Marsupial studies in Western Australia. *Australian Journal of Science*, 18, 66–76.

SECTION

1

REPRODUCTION AND DEVELOPMENT

CHAPTER 1

ENDOCRINE SIGNALS IN EMBRYO IMPLANTATION

John P Hearn

INTRODUCTION

It is a pleasure to contribute to this volume and to honour the substantial contribution that Hugh Tyndale-Biscoe has made to marsupial biology and to broader aspects of reproductive biology in marsupials and other species. I have enjoyed a most stimulating, challenging and productive association with Hugh over many years, commencing with his mentorship during my PhD research in the endocrine regulation of embryonic diapause in the tammar wallaby. Among the many lessons I learned were never to take the dogma for granted and always to design experiments for the unexpected. These were important lessons because in virtually every experiment we did, probing the hypothalamic and pituitary control of embryonic diapause, the results were almost entirely the opposite of those predicted (Hearn 1973; Tyndale-Biscoe *et al.* 1974). This experience is perhaps just one example of the fascination of marsupial biology, with the alternative regulatory mechanisms that have evolved in the unique development of marsupial reproduction.

Implantation, the process whereby the embryo first attaches and then invades the maternal-endometrial epithelium, varies widely in its timing, requirements for synchrony and in the emphasis of the hormonal signals that are employed across species (Table 1.1). Preimplantation development can take 4 to 6 days in mice, up to 20 days in marsupials and 25 to 30 days in sheep. Steroid or protein signals that are engaged in the embryo-maternal dialogue, and on which the survival of the embryo and pregnancy depend, are diverse, as are the morphological and biochemical interactions that regulate implantation and early placentation. There are fundamental differences in the ways in which maternal recognition of pregnancy and the support of the corpus luteum of early pregnancy is achieved (Flint *et al.* 1990). Embryonic diapause in marsupials and the varied systems for delayed implantation in eutherians present added complexity and are related to a separation of the most advantageous times for mating and that for birth, designed to synchronise the time for birth with the environmental constraints that the young will face at weaning. Diversity is also apparent in the structure

Table 1.1 The timing of entry into the uterus, implantation and gestation in a selection of mammalian species. Time (days) after ovulation is estimated.

SPECIES	ENTER UTERUS	IMPLANTATION	GESTATION
human	3–4	8–10	252–274
rhesus	3–4	8–10	159–174
marmoset	3–4	11–12	144–150
rat	3	5	20–22
mouse	3	4	19–20
rabbit	3–4	7–8	30–32
horse	4–5	28	335–345
cow	3–4	30–35	275–290
sheep	2–4	15–16	145–155
kangaroos and wallabies	2–4	20	27–30

and components of the preimplantation embryo (for example, the marsupials lack an inner cell mass), the location of the implantation site, the degree of invasiveness of trophoblast, and the extent of the endometrial decidual reaction. Table 1.2 lists some of the endocrine signals identified in the embryo-maternal dialogue at the start of pregnancy in mammals (Seshagiri & Hearn 1995).

Although one might expect, or look for, a conservative influence on the evolution of embryo implantation and the establishment of pregnancy, this is hard to find, even among many closely related groups. It is clear that a wide range of strategies have been evolved and that the plasticity of endocrine regulatory mechanisms has been remarkably fluid through evolution.

In this discussion, I present some of our recent results in studying the regulation of embryo implantation in primates. The primate embryo is endocrinologically active early on and is distinct from other groups in that chorionic gonadotropin is a principal endocrine signal from the trophoblast that sustains the failing corpus luteum of the cycle until the luteo-placental shift is complete (Hearn 1986). Our data suggest that this hormone is essential for early pregnancy to proceed, and may be required both at the implantation site, and perhaps within the embryo itself, for normal differentiation. In recent studies, we have shown that gonadotrophin-releasing hormone (GnRH) is secreted by the preattachment blastocyst, suggesting a causal relationship between GnRH and chorionic gonadotrophin that may be analagous to that between GnRH and luteinising hormone at the level of the hypothalamus and pituitary (Seshagiri *et al.* 1994). This analogy is emphasised by the fact that chorionic gonadotrophin is a very similar molecule to luteinising hormone, consisting of alpha and beta subunits, with the main difference

Table 1.2 Secretion of embryonic signals during the peri-implantation period

TYPE SIGNAL	SPECIES	EARLIEST STAGE DETECTED
chorionic gonadotrophin	marmoset	prehatching blastocyst
	rhesus	prehatching blastocyst
	baboon	attached blastocyst
	human	prehatching blastocyst
gonadotrophin-releasing hormone	rhesus	prehatching blastocyst
oestrogen	mouse	prehatching blastocyst
	rat	prehatching blastocyst
	human	preattached blastocyst
progesterone	mouse	prehatching blastocyst
	rat	prehatching blastocyst
	human	preattached blastocyst
inhibin	mouse	cleavage stage embryo
	human	cleavage stage embryo
activin	mouse	cleavage stage embryo
platelet activating factor	mouse	prehatching blastocyst
	marmoset	preattaching blastocyst
	human	prehatching blastocyst
histamine	mouse	prehatching blastocyst
prostaglandin	mouse	prehatching blastocyst
	human	prehatching blastocyst
trophoblastic protein	sheep	prehatching blastocyst
	cattle	prehatching blastocyst
growth factors	mouse	cleavage stage embryo
	cattle	cleavage stage embryo

For full references see Seshagiri & Hearn (1995).

being an additional 32 amino acid polypeptide chain at the C-terminal end of the chorionic gonadotrophin beta subunit. Thus, it appears that in primates, the only group that has so far been found to express the genes for chorionic gonadotrophin, the survival of early pregnancy may depend on the establishment of a GnRH-chorionic gonadotrophin mechanism before attachment takes place, that may influence the release of chorionic gonadotrophin and the maternal recognition of the presence of a viable embryo. Since we know that approximately 50% of embryos in women may be lost up to this stage, the mechanism may be an important one in determining the health of the embryo and its functional ability for future differentiation and survival.

With the above in mind, the hypotheses that we addressed in this work were that chorionic gonadotrophin

a) is essential for early pregnancy to proceed,
b) is primarily responsible for the early support of the corpus luteum,
c) is required at the site of implantation for attachment and invasion to proceed, and
d) depends on a regulation by GnRH that may also require its presence in the preimplantation embryo for normal differentiation and synchronisation to be ready for implantation.

The first two hypotheses are generally accepted (Hearn 1986), although there may still be outstanding questions as to the absolute requirements and level of hormone needed. The third and fourth hypotheses still require substantial work before they can be proven.

The questions that we posed in challenging the hypotheses were

a) when are the genes for chorionic gonadotrophin activated?
b) what is the timing of release of biologically active chorionic gonadotrophin from the preimplantation embryo?
c) can antibodies to chorionic gonadotrophin prevent implantation or early pregnancy?
d) how quickly can chorionic gonadotrophin be recognised at the corpus luteum?
e) how is chorionic gonadotrophin secretion regulated within the embryo?

Our progress to date indicates only partial answers to these questions, as summarised below.

METHODOLOGICAL CONSIDERATIONS

In order to approach these experiments, we had to overcome a number of substantial methodological problems. The first of these was in developing a non-invasive recovery system for morula and blastocyst stage embryos from the uteri of rhesus and marmoset monkeys. This was achieved (Seshagiri *et al.* 1993; Thomson *et al.* 1994) giving us the ability to obtain high quality embryos without surgery. Therefore, a limited number of monkeys can be retained and monitored for their ovarian cycles, with embryos being recovered on the appropriate day each month by trans-cervical flushing. Second, due to the scarcity of primate embryos, even with this system, it was necessary to develop microassays for chorionic gonadotrophin and GnRH to measure the production of these hormones from individual embryos (Seshagiri *et al.* 1994). Third, we developed a culture system for growing morula and blastocyst stage embryos through the stages of hatching, attachment and outgrowth *in vitro* (Hearn *et al.* 1988). These were compared at light and electron microscope levels with embryos recovered *in vivo* to show that development was 'normal' up to 10 days post-attachment, when the yolk sac is well differentiated. This became the endpoint of our experiments (Moore *et al.* 1985; Smith *et al.* 1987).

WHEN ARE THE GENES FOR CHORIONIC GONADOTROPHIN ACTIVATED?

In the mammalian species studied to date, the embryonic genome begins to be expressed by the 16 cell stage. There is little information in primates, but reports indicate that mRNA for chorionic gonadotrophin can be detected in human embryos between 8 and 16 cell stages (Braude 1988; Bonduelle 1988). The analyses were performed on trisomic embryos, so that the extrapolation to normal embryos may not be entirely valid. However, the indication is that by this time the message for chorionic gonadotrophin is available in the embryo. Chorionic gonadotrophin, like the other gonadotropins and thyroid-stimulating hormone, is composed of a structural alpha subunit and a beta subunit in which the biological activity is retained. As yet, we have very little information as to when the subunits are expressed or secreted. In non-human primates this is because we have lacked the specific reagents to measure these subunits.

WHAT IS THE TIMING OF RELEASE OF BIOLOGICALLY ACTIVE CHORIONIC GONADOTROPHIN FROM THE PREIMPLANTATION EMBRYO?

There are no reports to date of biologically active hormone being secreted before the blastocyst stage of development. Our own studies in the marmoset monkey (Hearn *et al.* 1988, 1991) and in the rhesus monkey (Seshagiri *et al.* 1994) show that chorionic gonadotrophin can be measured at very low levels from the mid-blastocyst stage. Once the embryo hatches, attaches and begins outgrowth *in vitro*, the levels of chorionic gonadotrophin secreted into the culture medium increase exponentially. Lopata & Oliva (1993) have shown a similar profile in human embryos, and Dokras *et al.* (1991) also reported similar results. In the latter two cases, as the embryos studied were those cultured from *in vitro* fertilisation, only a small proportion survived through hatching and attachment. The embryos reported by Lopata & Oliva (1993) responded well to incubation with fetal cord plasma or insulin-like growth factor, showing an exponential rise in chorionic gonadotrophin release after attachment. The embryos studied by Dokras *et al.* (1991) may have been in a rather more debilitated state, since the release of chorionic gonadotrophin was low and transient. The results from these three sources do suggest that in these 3 species there is a similar pattern of chorionic gonadotrophin production, with very low levels measurable before embryo attachment and an exponential rise once attachment and outgrowth commence.

These studies were all based on embryos cultured *in vitro*, but the measurement of hormone varied between laboratories in being either immunologically based (radioimmunoassay or receptor assay) or biologically based (bioassay). There are variables in detail that prevent direct comparison of the results, although the overall pattern seems similar in the 3 species (Seshagiri & Hearn 1995).

As expected, chorionic gonadotrophin can be measured earlier *in vitro* than in maternal plasma *in vivo*. Studies *in vivo* (Table 1.3), indicate that chorionic gonadotrophin appears in the maternal plasma 2 to 3 days after embryo attachment. This may be consistent with the time required for the embryo to attach, and the trophoblast to invade the maternal epithelium and make contact with the maternal blood supply.

Table 1.3 *In vivo* and *in vitro* secretion of chorionic gonadotrophin in various primate species*

	IN VIVO EMBRYO		IN VITRO EMBRYO	
Species	Day of attachment	Day of CG secretion	Day of attachment	Day of CG secretion
human	7-9	9-12 (IA)	8-10	6 (1A)
chimpanzee	7-9?	10-11 (IA)		
rhesus	8-10	11-12 (IA)	6-8	4 (BA)
baboon	8-10	11-12 (IA)	7-12	8 (IA)
marmoset	11-12	14-17 (IA)	7-10	6 (BA)

*The information provided here is obtained from Hearn *et al*. (1988, 1991), Seshagiri & Hearn (1995). Abbreviations used are CG: chorionic gonadotrophin; IA: radioimmunoassay or immunoradiometric assay; BA: mouse Leydig cell bioassay.

CAN ANTIBODIES TO CHORIONIC GONADOTROPHIN PREVENT IMPLANTATION OR EARLY PREGNANCY?

There is a substantial body of work from the marmoset (Hearn 1976), rhesus monkey (Talwar *et al*. 1976) and baboon (Stevens 1976) that in total presents compelling evidence for the inhibitory effects of antibodies to the beta subunit of chorionic gonadotropin in either blocking implantation or in terminating early pregnancy during the luteo-placental shift. These data show that once antibody levels are at a sufficiently high level, implantation and early pregnancy can be prevented as long as these levels are maintained. The same holds for both active or passive immunisation against chorionic gonadotrophin. These results have given rise to clinical trials carried out by the World Health Organisation and the Population Council to examine the immunogenicity of natural or synthetic beta subunit in preventing pregnancy in the human. The objective is to develop a long-term, easily administered vaccine against pregnancy. These studies are still developing and are inherently slow because of the needs for precisely controlled vaccine formulation, compliance with the regulatory requirements before they can be applied to potentially pregnant women, and the variability that is found in the immune response of individual women to such antigens.

In total, these data suggest that in all of the primate species studied with a range of antigens based on chorionic gonadotrophin conjugated to different carriers, immunological neutralisation of this hormone will interrupt implantation and early pregnancy. The implication is that chorionic gonadotrophin is necessary for implantation to proceed and that neutralisation or disruption of its secretion can inhibit the establishment of pregnancy.

The mechanism of action whereby chorionic gonadotrophin is required for the establishment of pregnancy is still not understood and is the basis for our further studies outlined below. The accepted mechanism is that this hormone is the essential signal (perhaps among others) involved in informing the mother of the presence of a viable embryo. The receipt of this signal by the corpus luteum replaces the need for lutenising hormone as a luteotrophin and rescues the corpus luteum by transforming it into the longer-lived corpus luteum of pregnancy. This mechanism, while generally accepted, tells us little about the levels of chorionic gonadotrophin required to enable pregnancy to proceed, or the extent of the critical window during which time the corpus luteum must be rescued in order for implantation to occur.

Data which tend to suggest a direct influence of chorionic gonadotrophin at the attachment site are sparse. It is known that receptors for chorionic gonadotrophin can be found on the endometrial epithelium, but as yet no clear data are available to show the presence of chorionic gonadotrophin receptors within the embryo. These are experiments to which we will proceed. We hypothesise that chorionic gonadotrophin may indeed be necessary at the implantation site and also within the embryo during this period.

How quickly can chorionic gonadotrophin be recognised by the corpus luteum?

In two earlier papers we showed that when chorionic gonadotrophin was perfused at supra-physiological levels through cannulae placed in the corpora lutea of marmoset monkeys during the mid-luteal phase, the levels of progesterone secreted into the perfusate rose very rapidly, commencing within 10 min of the start of chorionic gonadotrophin infusion. These results, albeit in a rather coarse experimental system, suggest that the corpus luteum is exquisitely sensitive to chorionic gonadotrophin and is ready to react with the secretion of progesterone almost immediately the chorionic gonadotrophin signal is received (Webley & Hearn 1987; Hearn & Webley 1987).

How is chorionic gonadotrophin secretion regulated within the embryo?

In recent studies we have tested the hypothesis that chorionic gonadotrophin, as a substitute and analogue of luteinising hormone, might be released from the embryo by intra-embryonic secretion of GnRH (Figure 1.1). We therefore measured GnRH in the culture medium from individual embryos incubated for 48 hour periods through blastocyst, hatched blastocyst and attached blastocyst stages. We showed (Seshagiri *et al.* 1994) that GnRH could indeed be measured in these preparations, with its secretion beginning before that of chorionic gonadotrophin and increasing slowly throughout the period of study, while chorionic gonadotrophin rose exponentially once attachment was achieved.

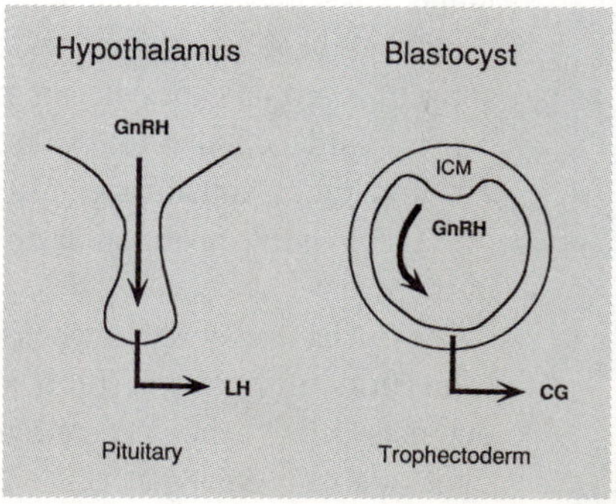

Figure 1.1 Our hypothesis is that intra-embryonic secretion of gonadotrophin-releasing hormone (GnRH) may be responsible for the initial regulation of chorionic gonadotrophin (CG) secretion from the primate embryo, analogous to the relationship between hypothalamic GnRH and the control of luteinising hormone (LH) secretion from the pituitary gland. ICM: inner cell mass.

These studies show that GnRH and chorionic gonadotrophin are secreted concurrently through this period, but the results in themselves do not prove a causal relationship between the two hormones as the secretions could be coincidental.

In order to explore a possible causal relationship, we have recently carried out studies where individual embryos (15 per treatment) were incubated either with Nal-Glu antagonist or with the agonist [DTrp^6,Pro^9-NHEt]-GnRH. The results showed that when embryos were incubated with antagonist, chorionic gonadotrophin levels were reduced, the time to embryo attachment was delayed, and there was also some disruption in embryo differentiation. The numbers of embryos proceeding to attachment and outgrowth were reduced. When individual embryos were incubated with agonist, the release of chorionic gonadotrophin was accelerated and enhanced, there was a transient rise of chorionic gonadotrophin and then a rapid decline. The embryos were disorganised and the vast majority failed to attach (Hearn & Becker, unpublished observations).

These results, still being analysed in detail for each embryo and each stage of differentiation, suggest that there may indeed be a causal relationship between chorionic gonadotrophin and GnRH in the preimplantation embryo. Manipulating chorionic gonadotrophin release has detrimental effects on embryo attachment and outgrowth. We are proceeding to examine the direct effects of GnRH on embryos, and any potential feedback mechanisms involved. It is also possible that GnRH may have other effects on embryonic growth and differentiation, as well as direct effects on the endometrium.

While our approach to these experiments has been simple, limited to the relationship between GnRH and chorionic gonadotrophin, we might expect that this relationship is not a direct one and could involve other factors, including hormones such as inhibin, and proteins such as the attachment proteins, growth factors and cytokines that are known to be expressed during this very dynamic period of embryo differentiation.

Acknowledgments

I thank Hugh Tyndale-Biscoe as mentor, colleague and friend throughout my career. I acknowledge the many colleagues listed in the references from our laboratory in the course of these studies. I thank Polani Seshagiri, Robert Becker and Ei Terasawa, who are partners in our recent work; Steve Eisele and Scott Kudia for their expertise in embryo flushing; and Fritz Wegner for expert hormone assay of gonadotropins. I am grateful to Susan Carlson and to Edi Chan for their help in the preparation of the manuscript. Nal-Glu antagonist and [DTrp^6,Pro^9-NHEt]-GnRH agonist were synthesised at the Salk Institute (under Contract NO1-HD-0-2906 with National Institute of Health) and made available by the Contraceptive Development Branch, Center for Population Research, National Institute of Child Health and Diseases. These studies were supported by National Institute of Health, grant RR00167, with a supplemental grant from the Office of Research on Women's Health Initiatives (USA); program grants from the Medical Research Council and the Agriculture and Food Research Council (UK); and project support from the Special Program for Research, Training and Research Development in Human Reproduction of the World Health Organisation. This paper is publication number 34-030 of the Wisconsin Regional Primate Research Center.

References

Bonduelle M L, Dodd R, Liebaers I, Van Steirteghem A, Williamson R and Akhurst R (1988). Chorionic gonadotrophin-β mRNA, a trophoblastic marker, is expressed in human 8-cell embryos derived from tripronucleate zygotes. *Human Reproduction*, 3, 909–914.

Braude P R, Bolton V and Moore S (1988). Human gene expression first occurs between the four- and eight-cell stages of preimplantation development. *Nature*, 332, 459–461.

Dokras A, Sargent I L, Ross C, Gardner R L and Barlow D H (1991). The human blastocyst: morphology and human chorionic gonadotrophin secretion *in vitro*. *Human Reproduction*, 6, 1143–1151.

Flint A P F, Hearn J P, Michael A E (1990). The maternal recognition of pregnancy in mammals. *Journal of Zoology*, 221, 327–341.

Hearn J P (1973). Pituitary inhibition of pregnancy. *Nature*, 241, 292–293.

Hearn J P (1986). The embryo-maternal dialogue during early pregnancy in primates. *Journal of Reproduction and Fertility*, 76, 809–819.

Hearn J P (1976). Immunization against pregnancy. *Proceedings of the Royal Society of London* B, 195, 149–160.

Hearn J P and Webley G E (1987). Regulation of the corpus luteum of early pregnancy in the marmoset monkey: Local interactions of the luteotrophic and luteolytic hormones *in vivo* and their effects on the secretion of progesterone. *Journal of Endocrinology*, 114, 231–239.

Hearn J P, Hodges J K and Gems S (1988). Early secretion of chorionic gonadotrophin by marmoset embryos *in vivo* and *in vitro*. *Journal of Endocrinology*, 119, 249–255.

Hearn J P, Webley G E and Gidley-Baird A A (1991). Chorionic gonadotrophin and embryo-maternal recognition during the peri-implantation period in primates. *Journal of Reproduction and Fertility*, 92, 497–509.

Lopata A and Oliva K (1993). Regulation of chorionic gonadotrophin secretion by cultured human blastocysts. In *Preimplantation Embryo Development*. Ed B D Bavister. Springer-Verlag, New York. pp 276–295.

Moore H D M, Gems S and Hearn J P (1985). Early implantation stages in the marmoset monkey (Callithrix jacchus). *American Journal of Anatomy*, 172, 265–278.

Seshagiri P B, Bridson W E, Dierschke D J, Eisele S G and Hearn J P (1993). Non- surgical uterine flushing for the recovery of preimplantation embryos in rhesus monkeys: lack of seasonal infertility. *American Journal of Primatology*, 29, 81–91.

Seshagiri P B, Terasawa E and Hearn J P (1994). The secretion of gonadotrophin releasing hormone by peri-implantation embryos of the rhesus monkey: Comparison with the secretion of chorionic gonadotrophin. *Human Reproduction*, 9, 1300–1307.

Seshagiri P B and Hearn J P (1995). Secretion of endocrine signals by the primate embryo during the peri-implantation period. *Current Science*, 68, 373–379.

Smith C A, Moore H D M and Hearn J P (1987). The ultrastructure of early implantation in the marmoset monkey (*Callithrix jacchus*). *Anatomy and Embryology*, 175, 399–410.

Stevens V C (1976). Perspectives of development of a fertility control vaccine from hormonal antigens of the trophoblast. In *Development of Vaccines for Fertility Regulation* (*World Health Organisation meeting*). Scriptor, Copenhagen. pp 93–110.

Talwar G P, Dubey S K, Salahuddin M and Shastri N (1976). Kinetics of antibody response in animals injected with processed beta hCG conjugated to tetanus toxoid. *Contraception*, 13, 153–162.

Thomson J A, Kalishman J and Hearn J P (1994). Non-surgical uterine stage preimplantation embryo collection from the common marmoset. *Journal of Medical Primatology*, 23, 333–336.

Tyndale-Biscoe C H, Hearn J P and Renfree M B (1974). Control of reproduction in macropodid marsupials. *Journal of Endocrinology*, 63, 589–614.

Webley G E and Hearn J P (1987). The local production of progesterone by the corpus luteum of the marmoset monkey in response to perfusion with chorionic gonadotrophin and melatonin *in vivo*. *Journal of Endocrinology*, 112, 449–457.

CHAPTER 2

CONTROL OF PARTURITION IN THE MARSUPIAL *MACROPUS EUGENII*

Marilyn B Renfree, Laura J Parry, Terry P Fletcher and Geoff Shaw

INTRODUCTION

In marsupials successful delivery of live young depends on the corpus luteum and the pituitary, but, as in many eutherian mammals, the fetus may determine the time of birth (reviewed in Tyndale-Biscoe & Renfree 1987; Tyndale-Biscoe *et al.* 1988). The fetus is delivered through the cervix, then via the median vagina to the urogenital sinus. In most marsupials the median vagina only becomes patent immediately before birth. Clearly, a loosening of the connective tissue of the cervix, urogenital sinus and median vagina is necessary before birth. Although several events in the process of parturition are known for some non-macropodid marsupials (reviewed in Renfree 1994), most experimental effort has been directed to the tammar, *Macropus eugenii*, a small macropodid marsupial that is abundant on Kangaroo Island, South Australia. The endocrine control of parturition and the initiation of lactation are now relatively well understood in the tammar.

In many eutherian species parturition depends critically on declining progesterone and increasing oestrogen (Liggins & Thorburn 1994). In tammars progesterone does fall precipitately at birth (Hinds & Tyndale-Biscoe 1982a; Lewis *et al.* 1986; Fletcher *et al.* 1990), and oestradiol levels rise at postpartum oestrus (Flint & Renfree 1982; Shaw & Renfree 1984; Harder *et al.* 1984; 1985). However, these changes are not needed for parturition in tammars. Maintaining high progesterone levels using silastic implants does not prolong gestation or inhibit oestrus (Ward & Renfree 1984), nor does removal of the corpus luteum precipitate parturition (Young & Renfree 1979). Females passively immunised against gonadotrophin-releasing hormone maintain pregnancies and give birth normally, despite the absence of follicular development in their ovaries (Short *et al.* 1985). There was also no postpartum mating in these immunised animals. In another macropodid, the swamp wallaby *Wallabia bicolor*, oestrus occurs pre-partum, also suggesting that a decrease in the progesterone:oestrogen ratio is not essential for birth.

Hinds *et al*. (1990) have proposed that luteolysis at birth is induced by prolactin. There is a pulse of prolactin at term (Hinds & Tyndale-Biscoe 1982b; Tyndale-Biscoe *et al*. 1983), which is apparently induced by prostaglandins (Hinds *et al*. 1990). However, treatment with bromocriptine before birth, which completely abolishes the prolactin pulse, does not affect the timing of parturition or the peripartum surge of prostaglandins (Fletcher *et al*. 1990). The release of prolactin may be essential for the initiation of lactation since all the young born in the bromocriptine-treated females were dead in the pouch within 24 h post-partum (Fletcher *et al*. 1990). Thus the peripartum release of prolactin appears to be incidental to, or a consequence of, birth.

The control of pregnancy and parturition in marsupials has recently been reviewed in considerable detail (Hinds 1990; Renfree 1994) and so this chapter concentrates on recent studies on the role of prostaglandins, the fetal trigger for birth, and finally the peptide hormones mesotocin and relaxin in the tammar.

PROSTAGLANDINS

Prostaglandins play a central role in parturition in tammars. As in other mammals, prostaglandins induce smooth muscle contractions in the tammar. Prostaglandin F2a (PGF2α) and prostaglandin E2 (PGE2) each induce strong uterine contractions both in vitro and in vivo in the tammar. Myometrial activity is increased after injection of the PGF2α agonist Cloprostenol during late pregnancy, but the myometrium is refractory to prostaglandins after parturition (Shaw 1983a, b). Little spontaneous uterine activity is seen during mid- to late gestation (Young 1978; Shaw 1983a, b) so there must be a rapid increase in uterine contractility at parturition in the tammar. It seems unlikely that uterine activity is controlled by placental steroids because placental steroidogenesis occurs at a very low level and oestrogens are virtually undetectable (Heap *et al*. 1980), so prostaglandins are the likely stimulators.

Prostaglandins also regulate parturient behaviour in tammars (Hinds *et al*. 1990; Shaw 1990). This characteristic behaviour is only expressed at parturition but exogenous PGF2α or PGE2 in 'physiological' doses rapidly induces this behaviour in both females and males. The latency and duration are dose-dependent.

Uterine PGF2α, PGE2 and 6-keto-prostaglandin F1α content rises over the last 4 days of pregnancy (Shaw 1983a). Explants of endometrium produce PGF2α and PGE2 *in vitro* and the production rate rises three-fold in the last 2 days of gestation (Figure 2.1)(Shaw *et al*. 1994a). The yolk sac placenta also synthesises appreciable quantities of PGF2α and PGE2 during the last 3 days of pregnancy even though it is a thin membrane of only 3 cell layers weighing about 100 mg (Figure 2.1)(Shaw *et al*. 1994b).

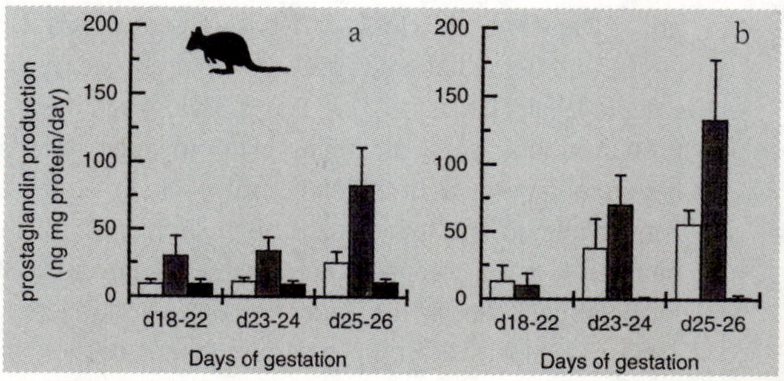

Figure 2.1 Prostaglandins synthesised by the endometrium (a) and the yolk sac placenta (b) of the tammar during the last few days of gestation.
Key: [Open bars] PGE2 [shaded bars] PGF2α [solid bars] PGFM
Source: Redrawn from Shaw *et al*. (1994a,b).

This gradual rise in synthesis in the days leading to birth cannot account for the dramatic pulse of 13,14-dihydro-15-keto-prostaglandin F2α (PGFM) at birth. Maternal plasma PGFM rises minutes before birth and has fallen to near basal levels by 15 min after birth (Lewis *et al.* 1986). Thus, it seems that an acute signal causes the dramatic increase in prostaglandin production at parturition.

Direct evidence for the essential role of prostaglandins comes from studies involving prostaglandin-synthase inhibitors such as indomethacin. Administration of this inhibitor through late gestation prevents parturition and post-mature fetuses were found *in utero* 3 days after control-treated animals gave birth (Renfree *et al.* 1995). Although there is no plasma PGFM pulse in these animals, low transient elevations were seen in some treated females.

A ROLE FOR CORTICOSTEROID IN THE TIMING OF BIRTH

The tammar fetus has a major influence on the timing of parturition (Tyndale-Biscoe *et al.* 1988). In many eutherian species, production of steroids by the fetal adrenal is a key step in the early stages of initiation of parturition. For instance, in the sheep, cortisol from the fetal adrenal induces an increase in synthesis of oestrogens by the placenta, which, in turn, induce uterine prostaglandin release and uterine contractions (Liggins 1982). Despite the vast difference in both the size and the relative maturity of sheep and tammar fetuses, there is growing evidence that cortisol also acts as the fetal trigger for parturition in tammars. The adrenal cortex of several marsupial species is structurally well developed by the time of birth and is able to produce cortisol (reviewed in Renfree 1994). However, this is not direct evidence that cortisol is the trigger for birth, since this hormone is also essential for the prepartum maturation of the lung, for surfactant release, lung liquid resorption, glucose storage, lactogenesis and closure of the ductus arteriosus (Liggins & Thorburn 1994).

Stronger evidence that cortisol is a trigger for parturition is the observation that the synthetic glucocorticoid dexamethasone induces premature delivery. Female tammars treated prepartum with dexamethasone at midnight on day 24 of gestation gave birth early. In dexamethasone-treated animals births occurred 20 to 26 h after injection, significantly ($p<0.05$) earlier than in saline-treated control females which gave birth 45 to 48 h after injection (Shaw *et al.* 1996). The premature young of the dexamethasone-treated animals were born live, although they were lighter than those of control-treated females.

The hormonal profiles of control and treated animals were similar when synchronised around the time of birth. In both groups of animals progesterone concentrations fell steeply at parturition, coincident with pulses of prolactin and PGFM (Figure 2.2). The only difference between control and treated animals was in the timing of the events, which were significantly advanced by the dexamethasone.

ROLE OF MESOTOCIN

Oxytocin, like prostaglandin, plays an important role in birth in many mammalian species, and can promote delivery of the young via its direct action on uterine smooth muscle. Marsupials may produce either (or both) oxytocin or mesotocin, closely related nonapeptides which differ from each other by a substitution of leucine for isoleucine at position 8. The tammar produces only mesotocin (Chauvet *et al.* 1981).

Spontaneous uterine contractile activity is minimal throughout most of pregnancy in tammars (Renfree *et al.* 1996). Mesotocin and oxytocin both stimulate uterine

Figure 2.2 Peripheral plasma hormone changes at parturition in tammars treated with the synthetic corticosteroid dexamethasone. Plasma concentrations of the prostaglandin F2α metabolite (PGFM), prolactin and progesterone were synchronised around the time of parturition (time 0). Values are mean ± sem, n=6. Arrows show the exact time of dexamethasone injection relative to the time of parturition.

Note: Profiles in control animals (not shown) are identical but occur 24 h later (day 26).

Source: Redrawn from Shaw *et al*. (1996).

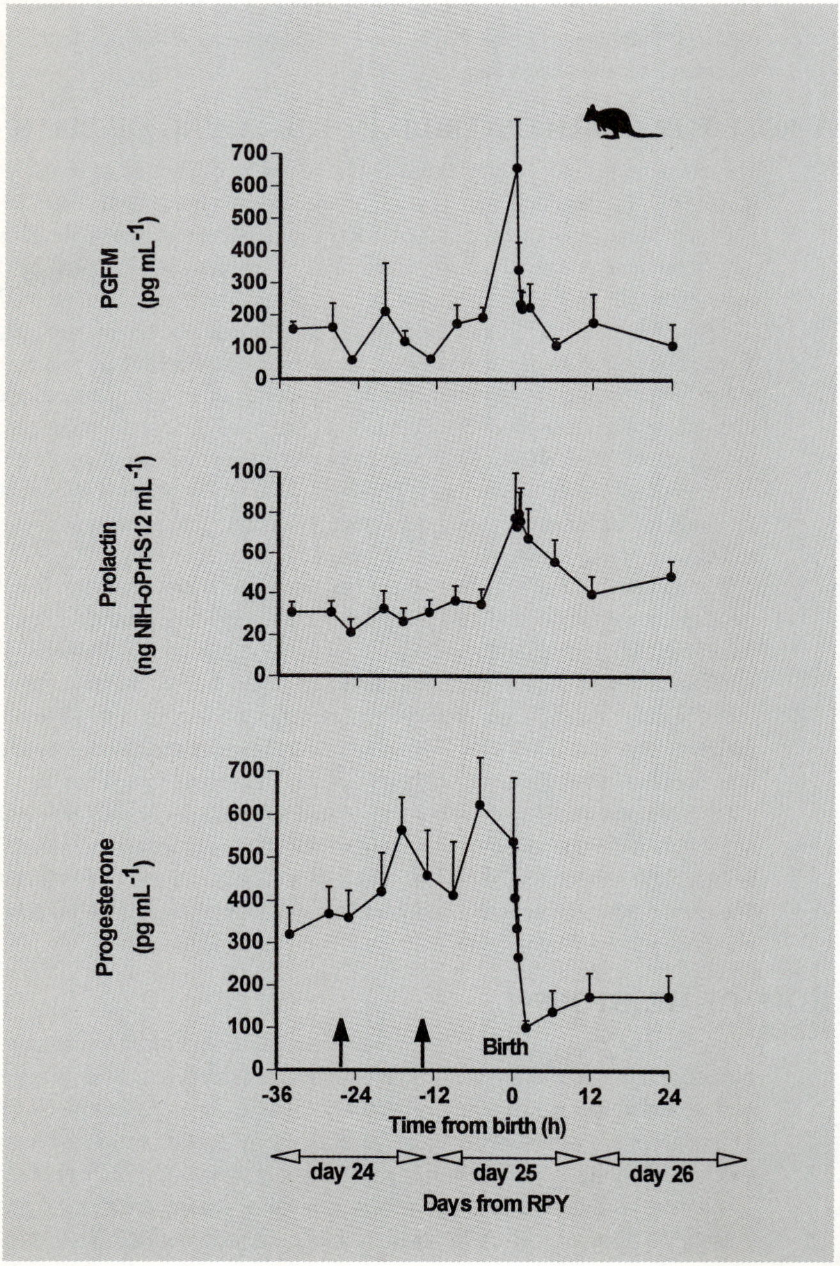

contractions *in vivo* and *in vitro* (Renfree & Young 1979; Shaw, 1983a & b). The sensitivity of both the gravid and non-gravid uteri to mesotocin/oxytocin increases during gestation, reaching a maximum shortly before birth. One day after birth, the uteri are refractory to oxytocin. Pretreatment with an oxytocin receptor antagonist, atösiban (Ferring Pharmaceuticals) prevents both mesotocin and oxytocin from having an effect on *in vitro* uterine contractions. Infusion of atösiban for either 3 or 7 days in late pregnant tammars resulted in delayed birth. All animals delivered normal live young 2 to 3 days later than control animals (Fig. 2.3) and all survived (Renfree *et al.* 1996). There is a surge in plasma PGFM at birth in both saline and atösiban-treated animals similar to that shown in Figure 2.2. These studies demonstrate that mesotocin is important, but not essential for parturition. Mesotocin probably stimulates uterine contractions directly, but this hormone may also affect the release of other hormones at birth such as prostaglandins and prolactin.

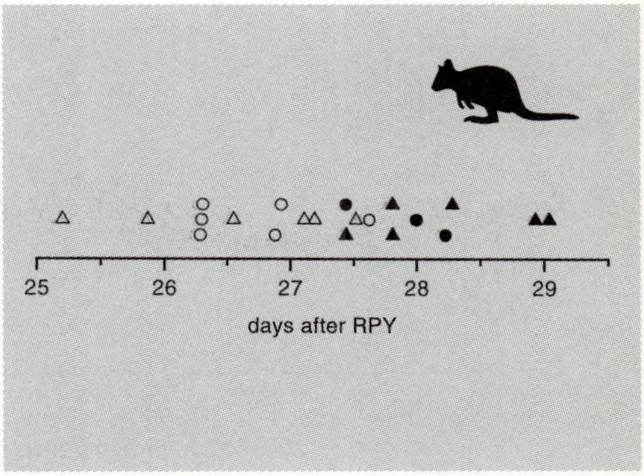

Figure 2.3 Time of birth after treatment with saline (O, Δ) or the oxytocin receptor antagonist, atösiban (●, ▲) for 3 or 7 days respectively. Source: Redrawn from Renfree *et al.* (1996).

The source of the oxytocic peptides which facilitate parturition is likely to be the maternal pituitary, which is essential for delivery of live young in tammars (Hearn 1973). After hypophysectomy fetuses are carried to full term, but die in the uterus (Hearn 1973, 1974). In 2 females in which the anterior pituitary was removed leaving the posterior pituitaty intact, the uteri were evacuated, but no fetuses or pouch young were found at autopsy (Hearn 1972).

Peripheral plasma mesotocin remains low throughout gestation (15 pg/ml) and were low in a single sample taken serendipitously 5 minutes pre-partum. There is a sharp increase just after birth (Figure 2.4) reaching peak levels of about 500 pg/ml, falling to half this level within 20 min postpartum (Parry *et al.* 1996). The pattern of release of mesotocin at delivery is similar to oxytocin release in rats, rabbits and humans. It is not yet clear whether mesotocin release is a cause, or a consequence, of birth.

RELAXIN AND THE BIRTH CANAL

Relaxin was first investigated in marsupials by Tyndale-Biscoe (1966, 1969, 1981), who showed that relaxin activity, measured by bioassay, increased during late gestation in the corpora lutea of tammars and brushtail possums. Krause & Sherman (1992) reported the immunohistochemical localisation of relaxin in corpora lutea of the North American opossum, *Didelphis virginiana*, during late pregnancy and early lactation. Immunoreactive relaxin is present in corpora lutea of tammars in the second-half of pregnancy (Parry & Renfree 1993). In several eutherian species relaxin is stored in electron-dense, membrane-bound cytoplasmic granules (Sherwood 1994) and similar granules are a prominent feature of the luteal cell cytoplasm throughout pregnancy in marsupials (Tyndale-Biscoe 1981). Using colloidal gold immunostaining we have shown that relaxin is associated with the electron dense granules (Parry & Renfree 1994) (Figure 2.5). Although the number of granules in each cell remains relatively constant throughout pregnancy, the number of gold particles bound to the granules increases significantly from day 22 of gestation, which coincides with the peak concentrations of bioassayable relaxin in the corpus luteum. Thus relaxin concentrations seem to be highest at the end of pregnancy, suggesting a role at parturition.

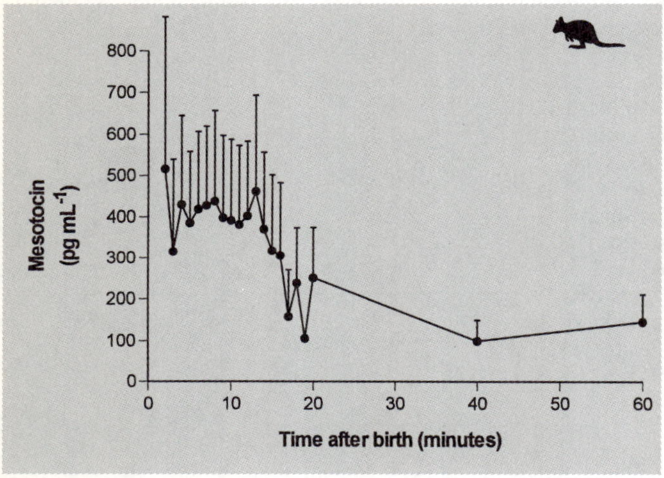

Figure 2.4
Concentrations of mesotocin in peripheral plasma at parturition in tammars.
Note: Plasma changes were synchronised around the time of parturition (time 0). Values are mean ± sem, n=5.
Source: Redrawn from Parry *et al.* (1996).

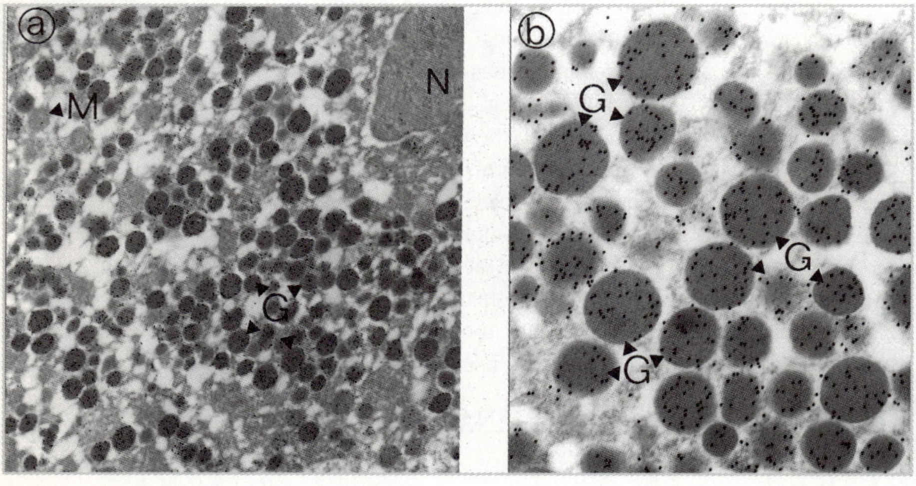

Figure 2.5 Electron micrographs of a luteal cell of a day 26 pregnant tammar.
Notes:(a) Relaxin immunogold particles were present on membrane bound, electron dense granules (G) but not in mitochondria (M) or the nucleus (N).
(b) Analysis at high magnification showed that all granules were immunostained and the density of gold particles per granule was uniform.
Source: Parry and Renfree (1994).

Although relaxin softens the cervix and pubic ligaments in some eutherian species (Sherwood 1994), no convincing effect of porcine relaxin on softening the cervix and vaginal canal has been demonstrated in the tammar (Young, 1978; Renfree & Young 1979). Progesterone treatment can cause substantial loosening of the connective tissue of the urogenital strand in the brushtail possum and the tammar (reviewed in Tyndale-Biscoe & Renfree 1987). It is likely that luteal progesterone, together with relaxin, is necessary for preparation of the birth canal.

CONCLUSIONS AND FUTURE DIRECTIONS

It is now clear that birth in the tammar clearly requires the interaction of several hormones. Prostaglandins are the key regulators of parturition in the tammar wallaby. Mesotocin also has a role, but it appears to facilitate, rather than initiate, parturition. PGF2α and PGE2 stimulate contractions of the myometrium and median vagina *in vitro*. Concentrations of the PGF2α metabolite PGFM in the plasma increase just before birth, and PGF2α and PGE2 synthesis by both the endometrium and the yolk sac placenta is highest on days 25 and 26 of gestation just before birth. Treatment with prostaglandin synthase inhibitors throughout late pregnancy prevents birth. Premature birth can be induced with the glucocorticoid agonist dexamethasone. In these animals the prostaglandin peak occurred early, coincident with birth. This result supports the idea that cortisol might be the fetal signal that initiates birth.

The active oxytocic peptide in tammars is mesotocin. Both oxytocin and mesotocin cause uterine contractions and the oxytocin receptor antagonist, atösiban, inhibits mesotocin-induced contractions of late pregnant tammar myometria *in vitro*. Infusion of atösiban at the end of gestation delays birth for 2 to 3 days. In atösiban-treated animals, the prostaglandin surge was also delayed until birth. Plasma concentrations of mesotocin remain low throughout pregnancy, but mesotocin increases rapidly within 2 min of birth suggesting that mesotocin is only released once parturition has been initiated.

On days 25 and 26 of gestation the cervices of the uteri and the median vagina become increasingly dilated in preparation for birth. This may be caused by relaxin, which is present in electron-dense granules in the luteal cell cytoplasm. Relaxin immunostaining is maximal one day before birth.

Thus, the control of marsupial parturition has many features in common with the mechanisms of control in eutherians, but in the one species that has been most studied, the tammar, there is still a need for much more information (Table 2.1). One enigma which remains is the role of the pre-partum prolactin pulse. The current data are consistent with the model shown in Figure 2.6. As the fetus approaches term prostaglandin production in both the endometrium and yolk-sac placenta increases. Relaxin, together with progesterone and possibly locally elevated prostaglandins may cause cervical softening and preparation of the birth canal. The timing of birth appears to be controlled by the feto-placental unit. The fetal pituitary may stimulate the fetal adrenal to release cortisol which in turn may increase prostaglandin production by the uterus or placenta eventually leading to the parturient surge of this hormone in the plasma. Rising prostaglandin levels, either alone or in concert with mesotocin, presumably lead to increased uterine contractions, and set up a positive feedback loop to ensure completion of the birth process once it is started. It is likely that there is a direct interaction between prostaglandins and mesotocin and it is also possible that relaxin may influence mesotocin release. Prostaglandins also induce parturient behaviour starting a few minutes before emergence of the fetus, so that it can safely climb from the urogenital sinus to the pouch. It is clear that there must be a rapid onset of high levels of prostaglandin production by either endometrium or yolk sac or both. The mechanisms controlling this increased prostaglandin production and the interactions between cortisol, prostaglandin, mesotocin and relaxin are currently being investigated.

Table 2.1. Experimental evidence for endocrine influences at parturition in tammars

PARTURITION IN THE TAMMAR IS:	INFERRED FACTOR
induced by dexamethasone	? fetal cortisol
inhibited by indomethacin	? maternal prostaglandins
delayed by oxytocin receptor antagonist	? maternal oxytocin
unaffected by bromocriptine	? role of prolactin
unaffected by progesterone	

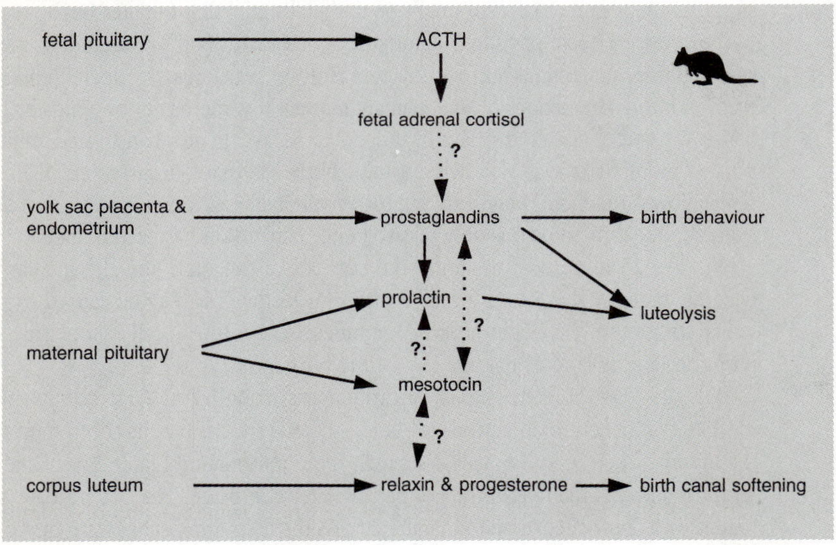

Figure 2.6 A summary of known and proposed mechanisms involved in the control of parturition in the tammar.
Key: [Pathways shown in solid lines] Pathways for which evidence exists. [dotted lines] Pathways which are speculative.
Note: ACTH—adrenocorticotrophic hormone.

References

Chauvet M T, Hurpet D, Chauvet J and Archer R (1981). A reptilian neurohypophysial hormone, mesotocin (Ile^8-oxytocin), in Australian marsupials. *Federation of European Biological Societies Letters*, 129, 120–122.

Fletcher T. P, Shaw G and Renfree M B (1990). Effects of bromocriptine at parturition in the tammar wallaby, *Macropus eugenii*. *Journal of Reproduction, Fertility and Development*, 2, 79–88.

Flint A P F and Renfree M B (1982). Oestradiol-17β in the circulation during seasonal reactivation of the diapausing blastocyst in a wild population of tammar wallabies. *Journal of Endocrinology*, 95, 293–300.

Harder J D, Hinds L A, Horn C A and Tyndale-Biscoe C H (1984). Oestradiol in follicular fluid and in utero-ovarian venous and peripheral plasma during parturition and post-partum oestrus in the tammar, *Macropus eugenii*. *Journal of Reproduction and Fertility*, 72, 551–558.

Harder J D, Hinds L A, Horn C A and Tyndale-Biscoe C H (1985). Effects of removal in late pregnancy of the corpus luteum, Graafian follicle or ovaries on plasma progesterone, oestradiol, LH, parturition and post-partum oestrus in the tammar, *Macropus eugenii*. *Journal of Reproduction and Fertility*, 75, 449–459.

Heap R B, Renfree M B and Burton R D (1980). Steroid metabolism in the yolk sac placenta and endometrium of the tammar wallaby, *Macropus eugenii*. *Journal of Endocrinology*, 87, 339–349.

Hearn J P (1972). Effect of advanced photoperiod on termination of embryonic diapause in the marsupial, *Macropus eugenii* (Macropodidae). *Australian Mammalogy*, 1, 40–42.

Hearn J P (1973). Pituitary inhibition of pregnancy. *Nature*, 241, 207–208.

Hearn J P (1974). The pituitary gland and implantation in the tammar wallaby, *Macropus eugenii*. *Journal of Reproduction and Fertility*, 39, 235–241.

Hinds L A (1990). Control of pregnancy, parturition and luteolysis in marsupials. *Journal of Reproduction, Fertility and Development*, 2, 535–552.

Hinds L A and Tyndale-Biscoe C H (1982a). Plasma progesterone levels in the pregnant and non-pregnant tammar, *Macropus eugenii*. *Journal of Endocrinology*, 9, 99–107.

Hinds L A and Tyndale-Biscoe C H (1982b). Prolactin in the marsupial, *Macropus eugenii* during the estrous cycle, pregnancy and lactation. *Biology of Reproduction*, 26, 391–398.

Hinds L A, Tyndale-Biscoe C H, Shaw G, Fletcher T P and Renfree M B (1990). Effects of prostaglandin and prolactin on luteolysis and parturient behaviour in the non-pregnant tammar, *Macropus eugenii*. *Journal of Reproduction and Fertility*, 88, 325–333.

Jones S A and Summerlee A J S (1986). Relaxin acts centrally to inhibit oxytocin release during parturition: an effect that is reversed by naloxone. *Journal of Endocrinology*, 111, 99–102.

Krause W J and Sherman D M (1992). Immunohistochemical localization of relaxin in the reproductive system of the female opossum (*Didelphis virginiana*). *Annals of Anatomy*, 174, 341–344.

Lewis P R, Fletcher T P and Renfree M B (1986). Prostaglandin in the peripheral plasma of tammar wallabies during parturition. *Journal of Endocrinology*, 111, 103–109.

Liggins G C (1982). *The Fetus and Birth*. Cambridge University Press, Cambridge.

Liggins G C and Thorburn G D (1994). Initiation of parturition. In *Marshall's Physiology of Reproduction*. Ed G E Lamming, 4th edn, Chapman & Hall, London. Vol. 3, pp 863–1002

Parry L J and Renfree M B (1993). Immunohistochemical localisation of relaxin in a marsupial, the tammar wallaby. *Journal of Reproduction* and *Fertility Abstract Series*, 11, 10.

Parry L J and Renfree M B (1994). Ultrastructural localization of relaxin in the corpus luteum of the tammar wallaby. *Journal of Reproduction and Fertility Abstract Series*, 14, 27.

Parry L J, Guymer F, Fletcher T P and Renfree M B (1996). Release of an oxytocic peptide at parturition in a marsupial, *Macropus eugenii*. *Journal of Reproduction and Fertility*, 107, 191–198.

Renfree M B (1994). Endocrinology of pregnancy, parturition and lactation of marsupials. *Marshall's Physiology of Reproduction*, Ed G E Lamming, 4th edn, Chapman & Hall, London. Vol 3, pp 677–766.

Renfree M B and Young I R (1979). Steroids in pregnancy and parturition in the marsupial, *Macropus eugenii*. *Journal of Steroid Biochemistry*, 11, 515–522.

Renfree M B, Shaw G and Fletcher T P (1995). Evidence for the essential role of prostaglandins for parturition in a marsupial, *Macropus eugenii*. *Journal of Reproduction and Fertility*, 102, 433–446.

Renfree M B, Parry L J and Shaw G (1996). Infusion with an oxytocin receptor antagonist delays parturition in a marsupial, the tammar wallaby. *Journal of Reproduction and Fertility*, 108, 131–137.

Shaw G (1983a). *Pregnancy after Diapause in the Tammar Wallaby, Macropus eugenii*. PhD thesis. Murdoch University, Perth.

Shaw G (1983b). Effect of PGF2a on uterine activity, and concentrations of 13,14-dihydro-15-keto-PGF2a in peripheral plasma during parturition in the tammar wallaby (*Macropus eugenii*). *Journal of Reproduction and Fertility*, 69, 429–436.

Shaw G (1990). Control of parturient behaviour by prostaglandin F2a in the tammar wallaby (*Macropus eugenii*). *Journal of Reproduction and Fertility*, 88, 335–342.

Shaw G and Renfree M B (1984). Concentrations of oestradiol-17β in plasma and corpora lutea throughout pregnancy in the tammar, *Macropus eugenii*. *Journal of Reproduction and Fertility*, 72, 29–37.

Shaw G, Gehring H and Bell E C (1994a). Prostaglandin production by placental tissues in late pregnant tammar wallabies. *Proceedings of 26th Annual Meeting of the Australian Society for Reproductive Biology*, 45.

Shaw G, Gehring H and Bell E C (1994b). Prostaglandin production by endometrium in late pregnant tammar wallabies. *Proceedings of 26th Annual Meeting of the Australian Society for Reproductive Biology*, 46.

Shaw G, Fletcher T P and Renfree M B (1996). A role for corticosteroids in parturition in a marsupial, *Macropus eugenii*. *Biology of Reproduction*, 54, 728–733.

Sherwood O D (1994). Relaxin. In: *The Physiology of Reproduction*. Eds E Knobil and J D Neill 2nd edn, Raven Press Ltd., New York. pp 861–1010.

Short R V, Flint A P F and Renfree M B (1985). Influence of passive immunization against GnRH on pregnancy and parturition in the tammar wallaby, *Macropus eugenii*. *Journal of Reproduction and Fertility*, 75, 567–575.

Tyndale-Biscoe C H (1966). The marsupial birth canal. *Symposium of the Zoological Society of London*, 15, 233–250.

Tyndale-Biscoe C H (1969). Relaxin activity during the oestrous cycle of the marsupial, *Trichosurus vulpecula* (Kerr). *Journal of Reproduction and Fertility*, 19, 191–193.

Tyndale-Biscoe C H (1981). Evidence for relaxin in marsupials. In *Relaxin*, Eds G Bryant-Greenwood, H Niall and F Greenwood. North Holland/Elsevier Biomedical Press, Amsterdam. pp 225–232

Tyndale-Biscoe C H and Renfree M B (1987). *Reproductive Physiology of Marsupials*. Cambridge University Press, Cambridge.

Tyndale-Biscoe C H, Hinds L A, Horn C A and Jenkin G (1983). Hormonal changes at oestrus, parturition and post-partum oestrus in the tammar wallaby (*Macropus eugenii*). *Journal of Endocrinology*, 96, 155–161.

Tyndale-Biscoe C H, Hinds L A and Horn C A (1988). Fetal role in the control of parturition in the tammar, *Macropus eugenii*. *Journal of Reproduction and Fertility*, 82, 419–428.

Ward K L and Renfree M B (1984). Effects of progesterone on parturition in the tammar, *Macropus eugenii*. *Journal of Reproduction and Fertility*, 72, 21–28.

Young I R (1978). *The Physiology of Parturition in the Macropodid Macropus eugenii (Desmarest)*. PhD thesis, Murdoch University, Perth.

Young I R and Renfree M B (1979). The effects of corpus luteum removal during gestation on parturition in the tammar wallaby (*Macropus eugenii*). *Journal of Reproduction and Fertility*, 56, 249–254.

CHAPTER 3

MECHANISMS OF SEASONALITY IN MARSUPIALS: A COMPARATIVE VIEW

Lyn A Hinds and Andrew S I Loudon

INTRODUCTION

Seasonality of breeding is a widespread strategy of reproduction in many species of mammals and birds. In temperate zone species, the annual pattern of changing daylength provides the key environmental cue for the appropriate timing of reproductive processes such as gamete production, territorial behaviour, conception and pregnancy. As a result, birth or weaning of young coincides with optimal environmental conditions for survival (Lincoln & Short 1980; Karsch *et al.*1984; Lee & Cockburn 1985; Tyndale-Biscoe & Renfree 1987; Tyndale-Biscoe & Janssens 1988; Nicholls *et al.* 1988; Hinds 1994; Loudon 1994).

Many marsupials also exhibit seasonal patterns of reproduction, but in only a few species has a role for photoperiodic regulation been demonstrated. In marsupials the reproductive strategy emphasises lactation rather than pregnancy and, as a consequence, in some species, particularly the kangaroos, wallabies and rat-kangaroos, the hormone of lactation, prolactin, has evolved a critical role. Here, prolactin has an inhibitory or luteostatic effect on the function of the corpus luteum formed at post-partum oestrus.

In contrast to eutherian mammals, investigations of the neuroendocrine mechanisms involved in the seasonal regulation of fertility have been confined to relatively few marsupial families. Two of the macropodid species, the tammar wallaby, *Macropus eugenii* and the Bennett's wallaby, *Macropus rufogriseus rufogriseus*, are obligate seasonal breeders and much is now known of the endocrine mechanisms involved in timing reproduction. Here, seasonality appears to be confined to one sex, the female, and is expressed as a control over the timing of reactivation of the corpus luteum and associated diapausing embryo. Unlike eutherians, photoperiodic time measurement appears to be primarily restricted to the action of melatonin on the secretion of prolactin, rather than other hormones of the neurohypophysis.

In this review, we contrast the physiological strategies that have evolved in eutherian species for life in seasonal environments with what is known of seasonal marsupials. The effects of melatonin on the function of the hypothalamic-pituitary-ovarian axis are described, including recent studies of the distribution of melatonin receptors in the developing marsupial brain. Finally, the role of an underlying circannual clock controlling seasonality in marsupials and eutherians is reviewed. The operation of this clock on only part of the neuroendocrine axis in macropodids is discussed in the light of current theories developed from the eutherian literature for melatonin-tuned regulation of seasonality.

PATTERNS OF SEASONAL REPRODUCTIVE CYCLES IN MARSUPIALS

Marsupials are characterised by short pregnancies and long lactations (reviewed by Russell 1982). Several features of reproduction are common to all marsupials; most are polyoestrous and most ovulate spontaneously. When conception occurs, pregnancy generally occupies a period less than, or similar to, the length of the oestrous cycle, and after parturition subsequent ovarian activity is suppressed by the sucking stimulus during lactation (reviewed in Tyndale-Biscoe & Renfree 1987).

In some species, the oestrous cycle is significantly longer than the length of pregnancy and here, the sucking stimulus serves to inhibit folliculogenesis and ovulation (for example, brushtail possum, *Trichosurus vulpecula*, Pilton & Sharman 1962; North American opossum, *Didelphis virginiana*; Hartman 1923). In others, gestation is of similar length to the oestrous cycle and as a result the onset of lactation occurs at the same time as ovulation and conception. In these species (most macropodid marsupials) there is a post-partum oestrus and the sucking stimulus serves to inhibit subsequent embryonic development (Sharman *et al.* 1966). Subtle differences in the relative lengths of the oestrous cycle and gestation may therefore have profound effects on the type of neuroendocrine mechanism employed to suppress fertility in the breeding female. In species where the oestrous cycle is longer than pregnancy, inhibition of ovulation during lactation and in the non-breeding season is likely to involve the suppression of hypothalamic gonadotrophin-releasing hormone (GnRH) and pituitary gonadotrophins, as has been described for seasonally breeding eutherians such as sheep (Karsch *et al.* 1984). For example, in brushtail possums the interval from photoperiod change or melatonin treatment to resumption of ovarian cyclicity is of the order of 30 to 60 days (Gemmell 1987a,b, 1990). This is similar to the interval of 50 to 60 days seen in sheep and deer (Arendt *et al.* 1983; Adam & Atkinson 1984; Milne *et al.* 1990; Loudon & Brinklow 1992).

Where a postpartum oestrus and conception occurs, inhibition of subsequent embryonic development is known to involve the action of prolactin on the corpus luteum (see below). In the superfamily Macropodoidea (kangaroos, wallabies and rat-kangaroos), with the exception of the two species of grey kangaroos, parturition is followed by a postpartum oestrus and ovulation, and here the development of the newly formed corpus luteum is suppressed by the sucking stimulus of the new young. If conception occurs, the embryo develops to a unilaminar blastocyst of around 100 cells and then enters diapause. This period, when the sucking stimulus of the young maintains the corpus luteum in a quiescent state and the blastocyst in diapause, is known as lactational quiescence.

During lactational quiescence, loss or removal of the pouch young induces reactivation of the corpus luteum and the quiescent blastocyst, and birth and postpartum oestrus occur within 26 to 27 days (tammar; Merchant 1979; Hinds & Tyndale-Biscoe 1982a; Tyndale-Biscoe *et al.* 1983) or 27 to 28 days (Bennett's wallaby; Merchant &

Calaby 1981; Curlewis *et al.* 1987). In both species, reactivation of the corpus luteum is characterised by a bimodal progesterone profile; an early increase occurs on days 5 to 6 in the tammar (Hinds & Tyndale-Biscoe 1982a) and days 4 to 5 in the Bennett's wallaby (Walker & Gemmell 1983; Curlewis et al. 1987; Curlewis & Loudon 1988a), and is followed by a more prolonged period of high concentrations occurring in the second-half of the cycle (Figure 3.1). This early progesterone peak, which occurs in both the oestrous cycle and pregnancy, represents irreversible reactivation of the quiescent corpus luteum, and it is essential in the pregnant animal for the maintenance of embryonic development (Berger & Sharman 1969; Moore 1978; Tyndale-Biscoe 1986; Tyndale-Biscoe & Hinds 1989).

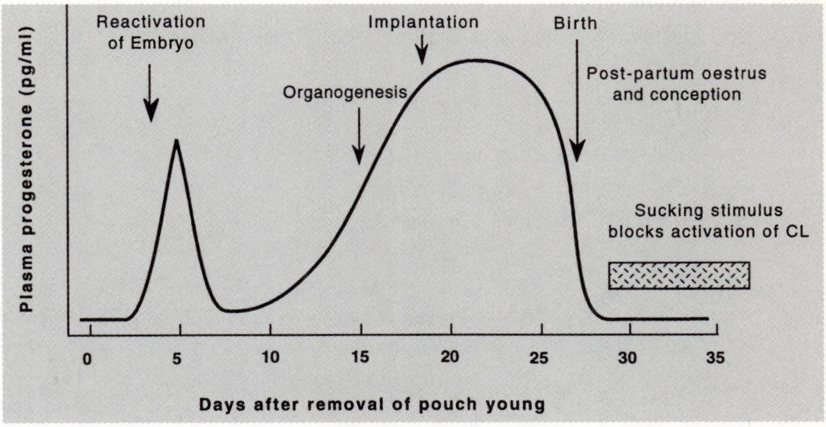

Figure 3.1 Diagrammatic representation of mean changes in plasma progesterone concentrations following reactivation of the quiescent corpus luteum (CL) after removal of the pouch young in the tammar or Bennett's wallaby.
Note: During the quiescent phase, concentrations are typically 50 to 150 pg/ml, rising to about 500 pg/ml at the time of the early peak. Note the relatively short postimplantation period.
Source: Adapted from Hinds & Tyndale-Biscoe (1982a); Walker & Gemmell (1983), with permission of the Journal of Endocrinology Ltd.

The early progesterone pulse commits the embryo to reactivation in both tammars and Bennett's wallabies. Indeed, a single injection of progesterone is sufficient to induce reactivation of the blastocyst within 2 days (Moore 1978). The commitment to reactivation following an increase in progesterone means that there is a fixed time interval between the progesterone pulse and subsequent birth (Hinds & Tyndale-Biscoe 1982a; Curlewis *et al.* 1987; Hinds 1989). Variation in the time to birth following removal of the sucking stimulus is therefore attributable to the time taken for the corpus luteum to recover from inhibition. In the tammar, the corpus luteum remains suppressed for about 3 days after the removal of pouch young and the associated sucking stimulus. This has been confirmed by Gordon *et al.* (1988) who showed that returning the tammar young to the pouch within 72 hours of removal held both the corpus luteum in quiescence and the embryo in diapuase. If the removal interval was longer than 3 days, reactivation occurred and was irreversible.

SEASONAL MACROPODIDS

Although many of the 47 species of Macropodoidea exhibit some form of seasonal breeding (Tyndale-Biscoe 1989), only the tammar and Bennett's wallaby show obligate seasonal breeding. These species occur in the most southerly parts of Australia (Figure 3.2), on Kangaroo Island and Tasmania, respectively, where cool winters and seasonal rainfall are a predictable feature of the climate and selection for seasonal reproduction strong. The mainland subspecies of *Macropus rufogriseus*, *M. r. banksianus*, does not exhibit a seasonal pattern of breeding in Queensland (J C Merchant, personal communication), although there have been no studies of its breeding pattern in the more southern areas of Australia. This suggests that, at least for the Bennett's wallaby, obligate seasonal quiescence may have evolved as recently as 11 000 years ago after Tasmania became separated from mainland Australia (Tyndale-Biscoe & Hinds 1990; Hinds 1994).

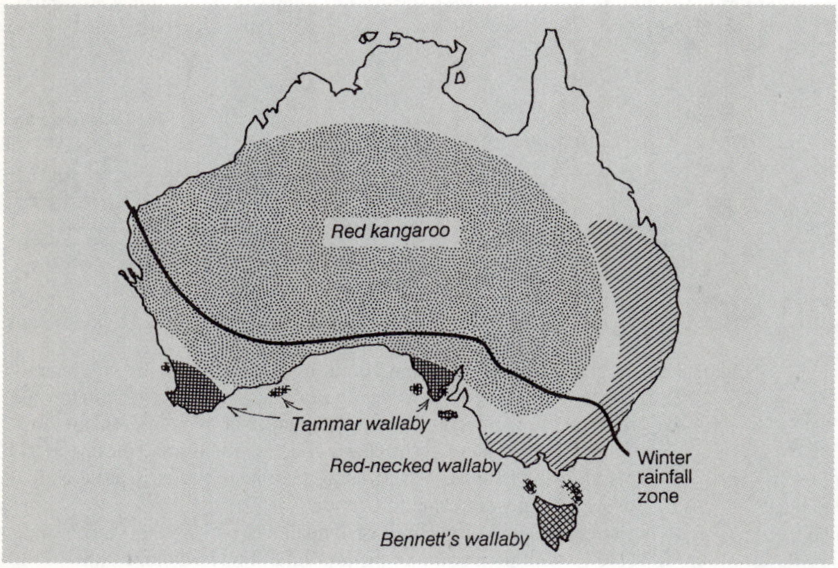

Figure 3.2 The distribution of the tammar, *Macropus eugenii*, (on mainland Australia and Kangaroo Island), and the two subspecies of *Macropus rufogriseus*—the red-necked wallaby, *Macropus rufogriseus* banksianus (on the mainland) and the Bennett's wallaby, *Macropus rufogriseus rufogriseus* (in Tasmania).

In seasonal wallabies, embryonic diapause lasts approximately 11 months (Berger 1966; Merchant & Calaby 1981). Breeding is highly synchronised, with births occurring within 6 weeks of the summer solstice in tammars (late January to early February in the southern hemisphere; Andrewartha & Barker 1969; Tyndale-Biscoe *et al.* 1986) and 2 to 4 months after the summer solstice in Bennett's wallabies (Catt 1977; Fleming *et al.* 1983; Curlewis & Loudon 1989; Curlewis 1989; J C Merchant, personal communication). In both species, there is a period of several months extending approximately from the winter to the summer solstice during which the quiescent corpus luteum and diapausing blastocyst remain in a viable state, but the removal or loss of the pouch young does not result in reactivation of the corpus luteum and birth. This is the period of obligate embryonic diapause, termed seasonal quiescence, and known to be regulated by photoperiod. As with other seasonally breeding species (sheep and red deer) transferred across the equator (Lincoln & Short 1980), tammars and Bennett's wallabies transferred to the northern hemisphere retain their seasonality and adjust their annual cycle so that they give birth in July or August after the boreal summer solstice (Berger 1970; Fleming *et al.* 1983; Dressen *et al.* 1990).

HOW IS SEASONALITY MAINTAINED IN MACROPODIDS?

The seasonal regulation of fertility in macropodids appears to be confined to one sex, the female. In both tammars and Bennett's wallabies, males do not undergo significant seasonal changes in testis size although some changes are observed in the size of accessory glands (Inns 1982; Curlewis 1991). It is also clear from experiments in which out-of-season breeding has been induced in females that males are capable of fertile matings at any time of the year. Studies of female tammars indicate that the hypothalamo-hypophysial-follicular axis functions as in other mammals during the breeding season but, in contrast to species like the sheep, it remains potentially fully functional in the non-breeding season. Injections of oestradiol-17β elicit the same negative and positive feedback responses in follicle-stimulating hormone and luteinising hormone secretion in ovariectomised tammars during the breeding and non-breeding season (Horn *et al.* 1985), as does a standard dose of GnRH (C A Horn & C H Tyndale-Biscoe, personal communication). In addition, the short-term pattern of release of both luteinising hormone and follicle-stimulating hormone is not different between the periods of lactational and seasonal quiescence in ovariectomised females (Hinds *et al.* 1992). This is in marked contrast to seasonally breeding eutherian mammals. Here, changes in gonadotrophin secretion provide the primary drive on the reproductive axis. For instance, in ovariectomised sheep treated with a subcutaneous implant secreting a low physiological dose of oestradiol, there are profound seasonal changes in the feedback effects of oestrogen on luteinising hormone secretion (Legan *et al.* 1977). In eutherians, these changes in response to steroid feedback are sensitive to photoperiod and represent a key pathway whereby reproductive seasonality is maintained.

It appears that the quiescent corpus luteum plays a key role in suppressing follicular development in the contralateral ovary. Surgical removal of the corpus luteum induces ovulation within 10 to 14 days and this can occur at any time of the year (Tyndale-Biscoe & Hawkins 1977). However, for 11 months of the year, the hypothalamo-hypophyseal-follicular axis is prevented from being functionally active by the presence of the quiescent corpus luteum on the ovary. A key to understanding seasonality in wallabies therefore depends on unravelling the mechanisms responsible for maintaining the corpus luteum in this suppressed state.

The role of photoperiod

Detailed responses to photoperiod change have not been determined for many seasonally breeding marsupial species. In the northern brown bandicoot, *Isoodon macrourus*, which normally breeds in spring as daylength is increasing, the onset of breeding can be delayed by melatonin treatment before the winter solstice (Gemmell 1987a). The brown antechinus, *Antechinus stuartii*, also normally breeds during periods of increasing daylength. When females are transferred to short days (10L:14D) before the winter solstice the onset of the breeding season is significantly advanced (McAllen *et al.* 1991), whereas transfer to long days (14L:10D) inhibits ovulation and in males the normal increase in plasma testosterone concentrations is suppressed (Scott 1986). In the brushtail possum exposure to artificial long daylength inhibits reproduction and exposure to short daylength stimulates reproduction (Gemmell 1990; Gemmell & Sernia 1992; Gemmell *et al.* 1993). The physiological mechanisms involved in these responses to photoperiod change have yet to be determined.

There is clear evidence that photoperiod change provides the primary environmental signal regulating seasonality in macropodids. The 5 to 6 month period of declining daylength from mid-summer to mid-winter is analagous to the breeding season of many seasonal eutherian mammals. Over this time, non-pregnant and unmated females undergo repeated 28 to 30 day oestrous cycles (Merchant 1979; Tyndale-Biscoe & Hinds 1984; Curlewis *et al.* 1986; Brinklow & Loudon 1989). The onset of seasonal quiescence occurs in both species at, or shortly after, the winter solstice at a time when daylength is increasing (Sadleir & Tyndale-Biscoe 1977; Merchant & Calaby 1981; Curlewis *et al.* 1987; Hinds 1994). At this time oestrous cycles cease and the most recently formed corpus luteum in the ovary becomes quiescent.

The role of photoperiod in timing reproduction has been studied in both tammars and Bennett's wallabies. The physiological response to various manipulations of photoperiod depends strongly upon the direction of change of the photoregimen (Figures 3.3 and 3.4). Animals exposed to reductions in photoperiod during seasonal quiescence reactivate and undergo oestrous cycles (Sadleir & Tyndale-Biscoe 1977; Hinds & den Ottolander 1983; McConnell & Tyndale-Biscoe 1985; McConnell *et al.* 1986; Loudon & Curlewis 1987; Brinklow & Loudon 1989; Hinds 1989; Tyndale-Biscoe & Hinds 1992). Conversely, exposure to artificial long photoperiods during early lactational quiescence can cause both species to re-enter seasonal quiescence (Curlewis & Loudon 1989; Hinds 1989; Tyndale-Biscoe & Hinds 1992). Thus, the photoperiodic mechanisms of seasonal quiescence are additional to the primary macropodid feature of lactational quiescence.

Both species of wallaby respond remarkably rapidly to photoperiod change: the first endocrine changes can be measured within 3 days (Hinds & den Ottolander 1983; McConnell & Tyndale-Biscoe 1985; Loudon & Curlewis 1987; Brinklow & Loudon 1989; Hinds 1989). However the interval from photoperiodic change to birth is 3 to 10 days longer than the interval from pouch young removal to birth in the breeding season. This longer interval is due to a delay in the appearance of the early plasma progesterone peak which occurs on days 8 to 14 after the reduction in daylength (Hinds & den Ottolander 1983; Brinklow & Loudon 1989).

Tammars and Bennett's wallabies show differences in the timing of entry into seasonal quiescence, and this may reflect differences in the way in which photoperiod change engages the neuroendocrine axis (see below). In the tammar, most females naturally enter seasonal quiescence before the winter solstice (that is, before daylength has started to increase; Hinds 1994). In contrast, for Bennett's wallabies an increase in daylength appears to be essential. Animals maintained on winter solstice daylength of 9L:15D from 3 weeks before the solstice continued to exhibit oestrous cycles for up to 8 months after the time at which control animals had entered quiescence (Loudon & Curlewis 1989). In further studies, animals maintained on a simulation of natural photoperiod increase after the winter solstice entered quiescence within 5 weeks, indicating that the reproductive axis of the Bennett's wallaby is remarkably sensitive to a photoperiod change of

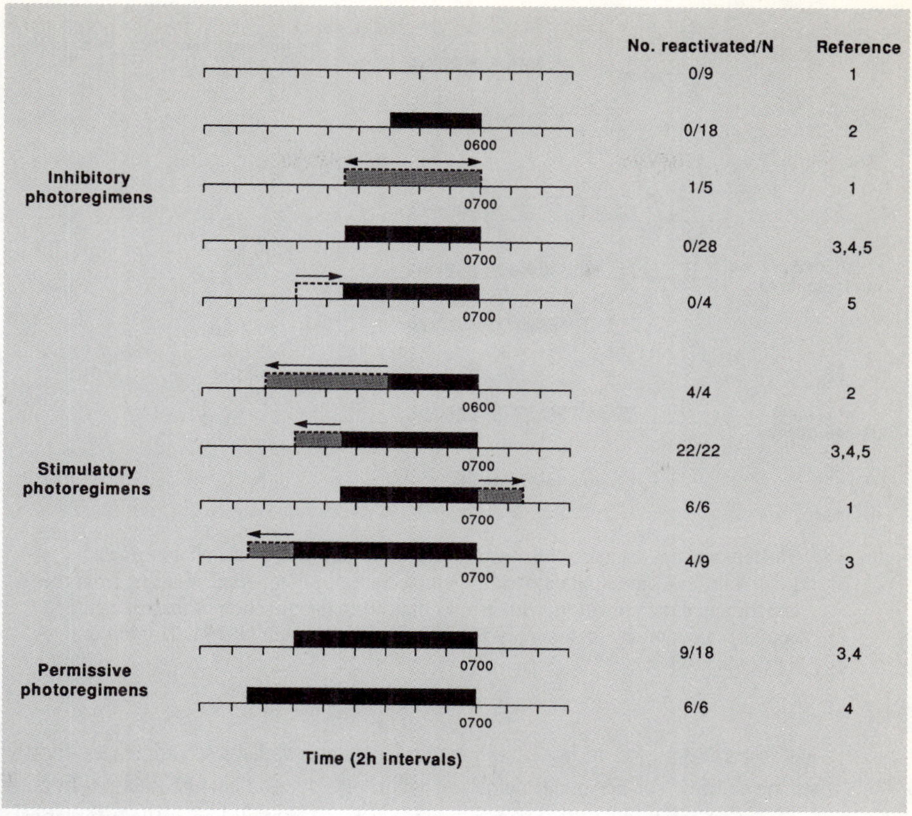

Figure 3.3. The reproductive response of tammars to various artificial photoregimen. Key: [Solid bars] initial photoperiod; [stippled areas] photoperiod resulting from the experimental manipulation, with arrows indicating the direction of that change. Sources: [1]McConnell *et al.* (1986); [2]Tyndale-Biscoe & Hinds (1992); [3]Sadleir & Tyndale-Biscoe (1977); [4]Hinds & den Ottolander (1983); [5]McConnell & Tyndale-Biscoe (1985).

under 1 hour over a period of 1 month after the solstice (Loudon & Curlewis 1989). In this species, fixed winter solstice daylengths and melatonin treatment all induce the same response, namely, a continuation of reproductive cyclicity (Loudon & Curlewis 1989; Loudon & Brinklow 1990). In both species, the onset of seasonal quiescence occurs at a time when reactivation of the quiescent corpus luteum cannot be induced by further reductions in photoperiod, or melatonin treatment (Loudon & Curlewis 1987; L A Hinds & C H Tyndale-Biscoe unpublished observations).

At present, it is not possible to disentangle the various influences of photoperiodic history in the above experiments, since combinations of natural and abrupt artificial changes in daylength have been employed. The simplest explanation is that in natural photoperiod environments, wallabies need to experience a period of several weeks of declining photoperiod before they once again become sensitive to the inhibitory effect of long daylengths and during early seasonal quiescence, they require a similar length of time before they are able to respond to the stimulatory effects of short photoperiods (Loudon & Curlewis 1987; Hinds 1989; Tyndale-Biscoe & Hinds 1992). (See Figure 3.5).

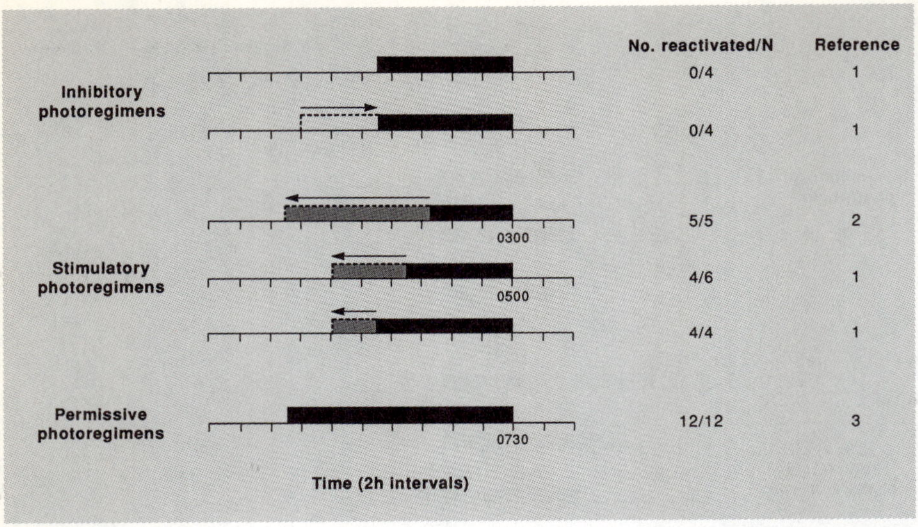

Figure 3.4 The reproductive response of Bennett's wallabies to various artificial photoregimen. Key: [Solid bars] initial photoperiod; [stippled areas] photoperiod resulting from the experimental manipulation, with arrows indicating the direction of that change. Sources: [1]Loudon & Curlewis (1987); [2]Brinklow & Loudon (1989); [3]Loudon & Curlewis (1989).

The breeding season in both tammars and Bennett's wallabies commences shortly after the summer solstice at a time of year when photoperiod change is slight. In both species, onset of breeding can occur in some individual animals before the summer solstice (Hearn 1972a; Sadleir & Tyndale-Biscoe1977; Brinklow & Loudon 1992). In contrast to the timing of the onset of seasonal quiescence (see above), the onset of the breeding season does not require a change in photoperiod. Animals held on fixed long summer solstitial photoperiods can commence breeding at approximately the same time as animals maintained on natural photoperiods (Sadleir & Tyndale-Biscoe 1977; Curlewis & Loudon 1989; Brinklow & Loudon 1992). Indeed, in Bennett's wallabies the timing of breeding can be advanced by approximately 1 month by exposing animals to long photoperiods from 2 months after the winter solstice (Curlewis & Loudon 1989). These experiments suggest that in wallabies the onset of the breeding season commences as a result of long-day photorefractoriness, a phenomenon that is widespread in seasonally breeding animals (Robinson *et al.* 1985; Nicholls *et al.* 1988).

PROLACTIN AS THE LUTEOSTATIC HORMONE

The early studies of John Hearn have provided remarkable evidence that the pituitary of the tammar exerts an inhibitory, rather than stimulatory, effect on the ovary. Surgical removal of the pituitary gland of tammars induces reactivation of the quiescent corpus luteum and the dormant blastocyst (Hearn 1972b; 1973; 1974; Tyndale-Biscoe & Hawkins 1977; Hinds 1983). However, normal follicular maturation and ovulation does not occur, as might be expected, because gonadotrophin secretion would be ablated by such a procedure. In hypophysectomised animals, reactivation of the corpus luteum can be delayed by injections of prolactin (Tyndale-Biscoe & Hawkins 1977), providing strong support for the hypothesis that prolactin exerts an inhibitory effect on the ovary (Hearn 1973). It is also clear that in contrast to eutherians, the corpus luteum does not require pituitary support during the luteal phase, since normal luteal function and steroidogenesis can proceed following hypophysectomy (Hinds 1983). Since embryonic reactivation can be induced by progesterone injection (see above), it is a reasonable assumption that prolactin serves to maintain the embryo in a quiescent state by suppressing progesterone secretion from the corpus luteum. Indeed, prolactin receptors have been described in both quiescent and active corpora lutea of the tammar (Sernia & Tyndale-Biscoe 1979; Stewart & Tyndale-Biscoe 1982), where they exhibit higher specific binding than in the lactating mammary gland (Stewart & Tyndale-Biscoe 1982).

Further support for the prolactin hypothesis comes from studies in which reactivation of the corpus luteum following removal of the pouch young in the breeding season is blocked if injections of prolactin or domperidone (a dopamine antagonist that elicits the release of prolactin) are administered at the time of pouch young removal (Tyndale-Biscoe & Hawkins 1977; Curlewis & Loudon 1988a). Recently we have shown in the tammar that the frequency and duration of the inhibitory prolactin signal is precise. After removal of pouch young, a short-duration intravenous pulse of <2 h given at intervals of 72 h or less, delays reactivation (L A Hinds & C H Tyndale-Biscoe, unpublished observations). These results, together with the observation that the corpus luteum is not committed to reactivation until 72 h after removal of the sucking stimulus (Gordon *et al.* 1988), suggest that there is a remarkable latency of response to removal of the inhibitory action of prolactin.

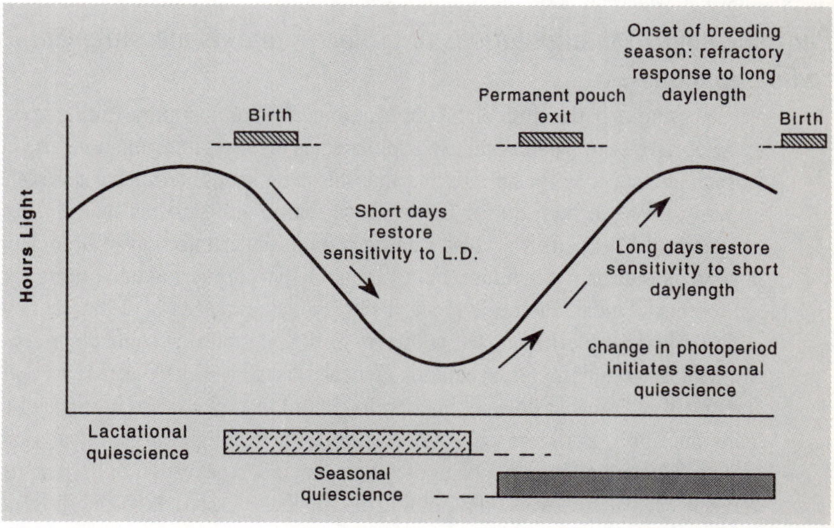

Figure 3.5 The annual cycle of reproduction of the tammar and Bennett's wallaby in relation to the changing daylength.
Note: The period of lactational quiescence occupies the period of decreasing daylength and the period of seasonal quiescence the period of increasing daylength. The entry into seasonal quiescence may be triggered by different mechanisms in each wallaby (see text) but the onset of the each breeding season appears to be due to refractoriness to increasing or long daylengths.

Dual role of prolactin in wallabies

It is clear from studies of wallabies that prolactin serves two distinct roles, support of lactation and inhibition of the activity of the corpus luteum. It is an essential hormone for the establishment and maintenance of lactation. Prolactin receptors have been described in the mammary gland of the tammar (Stewart 1984). These receptors are present in both the pregnant and non-pregnant cycle, with a marked increase in the last 10 days before birth or oestrus (Stewart 1984). A number of studies have demonstrated that, in marsupials, successful lactation can be established when new-born young are transferred to the teats of non-pregnant animals which are at a comparable stage of development post-oestrus (Sharman 1962; Sharman & Calaby 1964; Merchant & Sharman 1966). Thus, the initiation of lactation does not depend upon pregnancy.

Successful lactogenesis is dependent on prolactin. Treatment of tammars and Bennett's wallabies with the dopamine agonist bromocriptine (5 mg/kg) which, in eutherians, induces suppression of prolactin, causes cessation of lactation, dehydration and death of the pouch young (Tyndale-Biscoe & Hinds 1984; Tyndale-Biscoe *et al.* 1986; Curlewis *et al.* 1986). Simultaneous administration of ovine prolactin (once-daily injection) is sufficient to protect the mammary gland from the effects of bromocriptine, and pouch young survival and growth is normal (Curlewis *et al.* 1986). In this latter study, daily prolactin injections were effective in maintaining suppression of the corpus luteum and normal lactation for the duration of the treatment, but cessation of prolactin treatment resulted in reactivation of the corpus luteum 4 to 6 days later, even though lactation persisted unimpaired. Thus, it is possible to separate the effect of the suckling stimulus from the biological action of prolactin on the corpus luteum and lactogenesis.

Pharmacological manipulations of prolactin and its measurement by radioimmunoassay

To date, studies of the effect of the dopamine agonist bromocriptine during seasonal quiescence have been paradoxical. Since prolactin is involved in maintaining the quiescent corpus luteum, it might be expected that this drug should terminate quiescence at all times of the year, both during the breeding season and seasonal quiescence. In both tammars and Bennett's wallabies a single intramuscular injection of bromocriptine (5 mg/kg) is sufficient to induce reactivation of the corpus luteum during lactational quiescence (Tyndale-Biscoe & Hinds 1984; Tyndale-Biscoe *et al.* 1986; Curlewis *et al.* 1986), but both species become refractory to bromocriptine at some point between the winter solstice and the vernal equinox (Tyndale-Biscoe *et al.* 1986; Curlewis *et al.* 1986). Treatment with higher doses of bromocriptine (50 mg/kg) are also insufficient to terminate diapause (Brinklow & Loudon 1989). Interestingly, in seasonal quiescence, although lactation failed completely in some females, reactivation of the corpus luteum did not occur (Curlewis *et al.* 1986). This suggests that although the drug may have inhibited suckling induced hyperprolactinaemia, the component of the prolactin signal inhibiting the corpus luteum was not affected. An alternative hypothesis is that prolactin is not the common agent inhibiting the corpus luteum in lactational and seasonal quiescence. Although it is not possible to resolve these two possibilities at present, it is necessary to consider that there may be two distinct mechanisms regulating the secretion of prolactin in seasonal macropodids, one involving lactation, the sucking stimulus and dopamine inhibition, and the other involving photoperiod and the melatonin signal.

This is discussed in more detail below.

Until recently studies of prolactin secretion in wallabies have employed heterologous radioimmunoassays, using an antibody raised against human prolactin (McNeilly & Freisen 1978). In all seasonal eutherian species so far examined, prolactin concentrations exhibit marked seasonal changes (for review, Curlewis 1992), with high concentrations on long photoperiods and low on short photoperiods (Lincoln 1984; Karsch *et al.* 1989; Loudon *et al.* 1989). In seasonal wallabies there are also significant seasonal changes in plasma prolactin concentrations, but these are associated primarily with lactation (see below) and not photoperiod. In late lactation, when the energy demands of the pouch young are high (Cork & Dove 1989), plasma prolactin concentrations are elevated (Hinds & Tyndale-Biscoe 1982b; 1985; Curlewis *et al.* 1986; Loudon 1987) (Figure 3.6). Hyperprolactinaemia during late lactation is associated with the presence of the young in the pouch since prolactin concentrations fall abruptly at the time of natural pouch exit and can be induced by temporary removal of the pouch young (Hinds & Tyndale-Biscoe 1985; Curlewis *et al.* 1986).

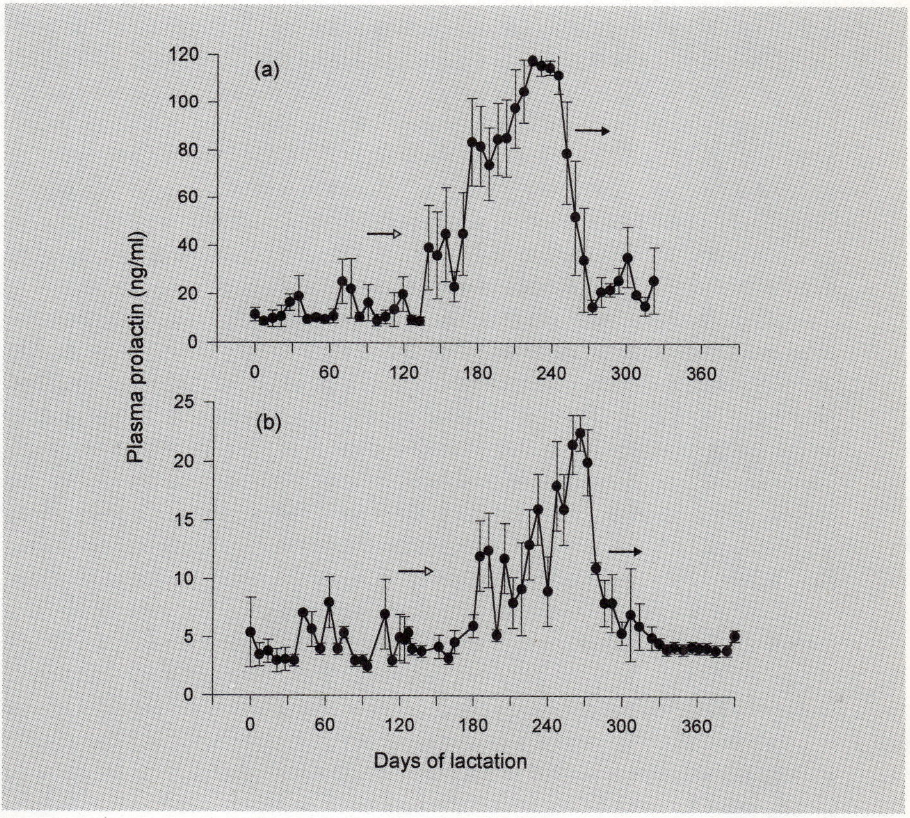

Figure 3.6 Changes in plasma prolactin concentrations (mean ± sem) throughout lactation in (a) tammars and (b) Bennett's wallabies.
Key: [open arrow] Approximate time of detatchment from the teat and establishment of an intermittent suckling pattern; [solid arrow] The time of permanent pouch exit.
Source: Redrawn from Hinds & Tyndale-Biscoe (1985); Curlewis *et al.* (1986), with permission of the Journal of Endocrinology Ltd.

In contrast to the effects of lactation, most studies have failed to show significant seasonal variation in plasma concentrations in non-lactating animals over the course of a year (Curlewis *et al.* 1986; Hinds & Tyndale-Biscoe 1985; McConnell & Hinds 1985), although some studies have reported small seasonal differences in prolactin concentrations (Hinds & den Ottolander,1983; Tyndale-Biscoe & Hinds 1984; Curlewis 1994). Using the heterologous assay of McNeilly & Freisen (1978), natural and artificial changes in photoperiod do not appear to alter plasma concentrations, nor are concentrations suppressed by bromocriptine in non-lactating animals (Tyndale-Biscoe & Hinds 1984; Curlewis et al. 1986). However, using a more recently developed homologous assay (Curlewis & McNeilly 1993), Curlewis (1994) has shown elevations in prolactin in non-lactating animals at the onset of seasonal quiescence — a pattern that is clearly different from that observed in eutherian species (Curlewis 1992).

Daily patterns of prolactin secretion

Absence of marked seasonal differences in prolactin secretion represents something of a paradox, since prolactin is implicated as the hormone responsible for the maintenance of luteal quiescence. These apparent inconsistencies were partially resolved when an early morning plasma prolactin pulse was detected in tammars held on photoinhibitory daylengths (15L:9D) in seasonal quiescence (McConnell *et al.* 1986). This pulse of 1 to 2 hour duration represented a 2- to 6-fold elevation in plasma concentrations above basal levels and was coincident with the end of the dark phase in animals held on photoinhibitory daylengths (Figure 3.7). This 'dawn pulse' was not detected 5 days after the photoperiod was reduced to a stimulatory photoperiod of 12L:12D (McConnell *et al.* 1986). The earlier discrepancies between studies in plasma prolactin concentrations could now be explained in terms of inappropriate sampling times in relation to the occurrence of this diurnal prolactin pulse.

Could this prolactin pulse maintain the inhibition of the corpus luteum during seasonal quiescence while prolactin was at basal concentrations for the rest of the day? In one experiment photoperiod was reduced from 15L:9D to 12L:12D and frequent blood samples taken each day. The pulse was present until day 3, but absent in most animals by the fourth morning (Hinds 1989), about 5 days in advance of the plasma progesterone peak. Since the normal interval from removal of the sucking stimulus to the progesterone peak in the breeding season is 5 days in the tammar, these observations are consistent with a release of the corpus luteum from the inhibitory prolactin signal after 3 days. Replacement of the daily prolactin signal following exposure to stimulatory short photoperiods is effective in maintaining quiescence. For instance, in tammars treated daily with either thyrotrophin-releasing hormone to induce a transient release of prolactin, or ovine prolactin, quiescence is maintained for the duration of treatment. The early progesterone peak in animals treated with thyrotrophin-releasing hormone or prolactin occurred 5 days after the last treatment (Hinds 1989). Similarly, in Bennett's wallabies subjected to a stimulatory photoperiod change, treatment with the dopamine antagonist domperidone, or with ovine prolactin, causes a delay in reactivation of the corpus luteum for the duration of treatment (Brinklow & Loudon 1989). In Bennett's wallabies, the exact nature of the photoperiodically mediated prolactin signal is yet to be determined. In two separate studies employing controlled photoperiod environments, a dawn pulse of prolactin has not been detected (Pearce-Kelly *et al.* 1992; A M Paterson & A S I Loudon, unpublished results).

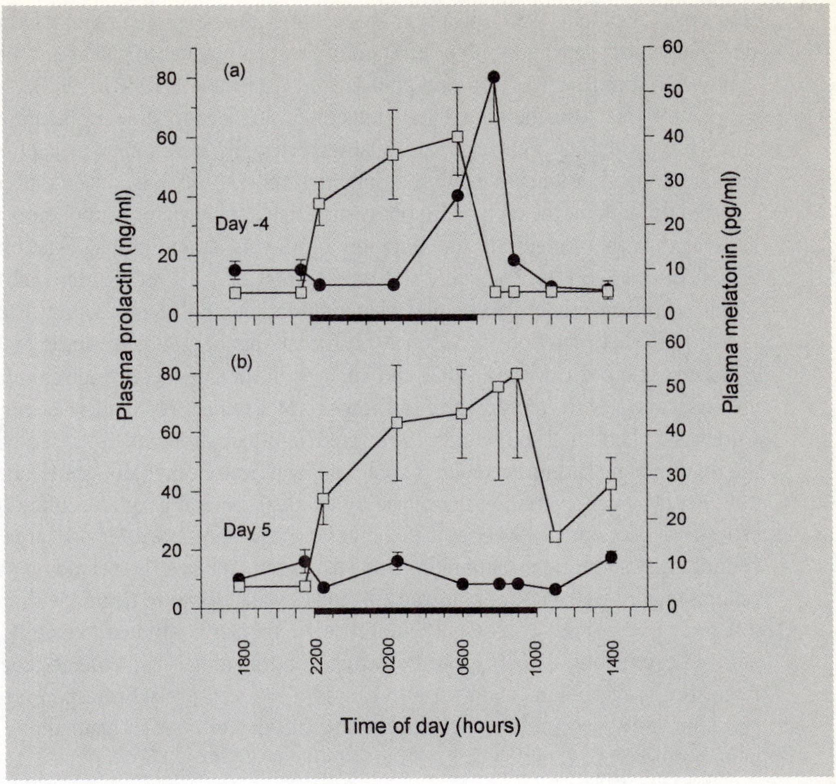

Figure 3.7 Profiles of plasma melatonin (pg/ml, □—□) and prolactin (ng/ml, ●—●) in tammars maintained on (a) an inhibitory photoperiod of 15L:9D and (b) 5 days after a change to 12L:12D.
Note: Solid bars indicate the length of the dark phase.
Source: Redrawn from McConnell *et al.* (1986).

Finally, the dawn prolactin pulse is not just a feature of reproduction in the female tammar. Although male tammars are not seasonal breeders (Tyndale-Biscoe 1980; Catling & Sutherland 1980; Inns 1982), a short-duration prolactin pulse is seen at the end of the dark phase in males maintained on long daylengths (15L:9D). This disappears shortly after a photoperiod change to 12L:12D (L A Hinds & C H Tyndale-Biscoe, unpublished observations). Thus the prolactin pulse has not evolved independently in the sexes, but a specific use of the hormone has evolved in the females: the luteal cells of the corpus luteum have become sensitive to its influence, but unlike eutherians, it is luteostatic.

THE ROLE OF THE PINEAL GLAND: THE MELATONIN MESSAGE

The pineal gland and the nocturnal secretion of its hormone melatonin is recognised as the key neuroendocrine transducer of photoperiodic information in seasonally breeding mammals (Reiter 1980; Arendt 1986; Cassone 1990). Early experiments in pinealectomised animals have examined the role of the melatonin signal by providing an artificial profile by infusion. For instance, in Djungarian hamsters, short-duration signals (mimicking long daylengths) are effective in stimulating gonadal development

(Carter & Goldman 1983), while in sheep, long-duration signals (mimicking short daylengths) drive gonadotrophin secretion in a similar manner to pineal-intact controls exposed to stimulatory (short-day) photoperiods (Bittman & Karsch 1984).

It is now clear that the same neuroendocrine transduction system operates in marsupials. In studies of both seasonal wallaby species, the secretion of melatonin in the circulation of adult females exhibits a daily rhythm with plasma concentrations being elevated throughout the entire dark phase such that annual changes in photoperiod are associated with changes in the duration of the melatonin signal (McConnell & TyndaleBiscoe 1985; McConnell & Hinds 1985; McConnell 1986; Curlewis & Loudon 1988b). Artificial manipulations of daylength induce a rapid change within 1 to 2 days in the profile of secretion (McConnell & Tyndale-Biscoe 1985; Curlewis & Loudon 1988b; Loudon & Curlewis 1987) and changes in the endogenous melatonin profile are associated with reproductive responses (McConnell & Tyndale-Biscoe 1985; Curlewis & Loudon 1988; Hinds 1989; Tyndale-Biscoe & Hinds 1992). In tammars, the rhythm of melatonin secretion is abolished by pinealectomy (McConnell and Hinds 1985) and by denervation of the pineal by bilateral cervical ganglionectomy (Renfree et al. 1981; McConnell 1984), as it is in other mammals (Arendt 1986). Treatment with continuously secreting melatonin implants may serve to block the pervading photoperiodic signal in wallabies. In one study, Bennett's wallabies were treated with implants just prior to the onset of seasonal quiescence — animals continued to exhibit regular oestrous cycles long after the period at which control animals had entered quiescence. Interestingly, melatonin treatment also caused a reduced prolactin response to domperidone, a response in keeping with the role of this hormone in maintaining corpus luteum quiescence (Loudon & Brinklow 1990).

In both wallaby species subcutaneous implants of melatonin or daily injections mimic the effects of short photoperiods in animals maintained on photoinhibitory daylengths and reactivation occurs within 1 to 6 days of treatment (McConnell & Tyndale-Biscoe 1985; Loudon *et al.* 1985; Loudon & Curlewis 1987;Tyndale-Biscoe & Hinds 1992). In the tammar an increase in duration of the endogenous melatonin signal is a key component in determining the reproductive response (McConnell et al. 1986). Thus, in animals maintained on photoinhibitory daylengths (15L:9D), melatonin injections administered between 0.5 to 2.5 hours in advance of a 9 h dark phase induce reactivation in most animals (Tyndale-Biscoe & Hinds 1992). However, a change in duration alone is not sufficient to induce reactivation — if animals are kept on longer photoperiods of 18L:6D, an injection 3 hours before dark is ineffective in breaking seasonal quiescence (Tyndale-Biscoe & Hinds 1992). Where 2 injections were administered, one at 3 hours and another at 6 h before dark, all treated animals reactivated (Tyndale-Biscoe & Hinds 1992). Comparable studies to those undertaken in pinealectomised melatonin-infused eutherian models present major technical difficulties and have not been performed in wallabies. However, it is clear from the above study that a change in the critical duration of the melatonin signal to more than 9 hours may be required to suppress the inhibitory prolactin pulse.

The events which occur in response to a stimulatory photoperiod change can now be summarised (Figure 3.8). A change in photoperiod from 15L:9D to 12L:12D causes an immediate change in the duration of melatonin secretion and (in the tammar) this leads to an abolition of the prolactin pulse by the fourth day. Prolactin needs to be suppressed for a further 3 to 4 days in order to commit the corpus luteum to reactivation.

Once the corpus luteum has reactivated the following events occur: within one day the corpus luteum stimulates the endometrium to produce a signal (possibly Platelet Activating Factor; Kojima *et al.* 1993), which releases the blastocyst from diapause. One to two days later there is an early peak of progesterone that is essential for the stimulation of the endometrium and subsequent maintenance of pregnancy. Birth occurs 21 to 22 days (tammar) or 23 to 24 days (Bennett's wallaby) after this progesterone peak.

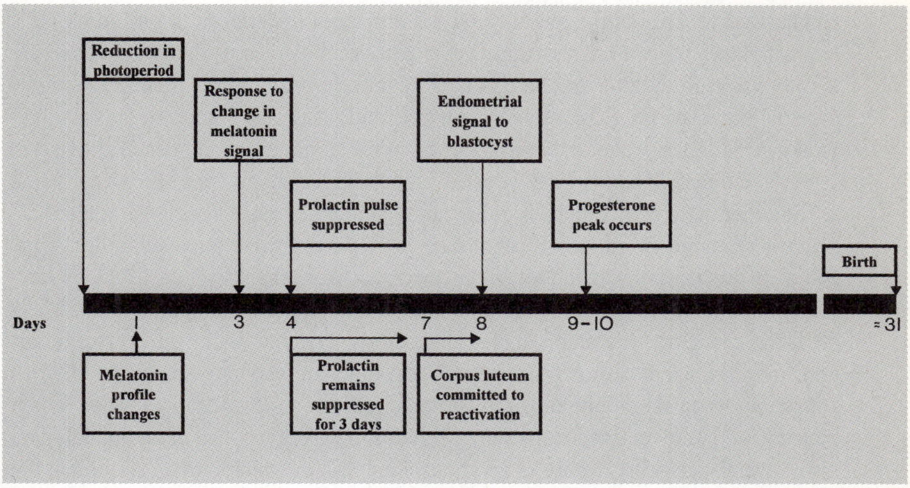

Figure 3.8 Summary of the proposed sequence of events following exposure to short (stimulatory) photoperiods.
Note: A reduction in photoperiod (for example, from 15L:9D to 12L:12D) results in a rapid change in pineal melatonin secretion by 1 day, with (in tammars) suppression of the early morning prolactin pulse within 3 days. This commits the corpus luteum to reactivation and initiates subsequent embryonic development and birth by day 30 to 32. In Bennett's wallabies, the morning pulse of prolactin secretion has not been detected.

Pinealectomy, seasonality and circannual rhythms.

Early studies of pineal function in eutherian species demonstrated that an intact pineal is an essential pre-requisite for a photoperiodic response. The reproductive response to pinealectomy varies considerably between species and an early view was formed that the primary role of the pineal was to inhibit reproductive function. This arose because pinealectomised male Syrian hamsters remained in reproductive condition even when exposed to daylengths that were normally inhibitory (Hoffman & Reiter 1965). It is now clear that there are differences between short-lived species such as seasonal rodents, in which pinealectomy generally results in persistent reproductive activity, and longer-lived species such as sheep. In those species programmed to live for two or more breeding seasons, pinealectomy prevents the animal from generating a reproductive response to photoperiod change, but persistent changes in reproductive function do nonetheless occur. The spontaneous changes in reproductive function often 'free-run' out of synchrony with the external photoperiodic environment with a periodicity close to a year. In the sheep the response to pinealectomy varies with the time of year that surgery is performed, but with the general result that there is a disorganisation of the timing of the reproductive cycle. Such animals exhibit free-running persistent alternating periods of gonadotrophin secretion that free-run out of synchrony with pineal-intact sheep (Karsch & Woodfill 1990; Woodfill *et al.* 1994).

Directly comparable long-term studies of the effects of pinealctomy or pineal denervation (by superior cervical ganglionectomy) have not been performed in marsupials and surgical removal of the pineal gland has only been undertaken in one species, the tammar. As in eutherian species the effect of pinealectomy on seasonality in the tammar is complex, with the time of year that surgery is performed and the time since surgery affecting the response. Removal of the pineal gland or its denervation by superior cervical ganglionectomy abolished the secretion of melatonin in the tammar (Renfree *et al* 1981; McConnell & Hinds 1985). Renfree *et al.* (1981) examined the effects of removal of the superior cervical ganglia from lactating tammars in early lactational quiescence. Animals remained in quiescence until their pouch young were removed at the vernal equinox; reactivation was induced immediately, with births occurring 27 to 28 days later. These observations are in keeping with the hypothesis that a photoperiodically mediated changing melatonin signal is an essential pre-requisite for the normal transition from lactational to seasonal quiescence. A different response was observed when animals were pinealectomised in the late autumn/early winter (April to June). When pouch young were removed at the vernal equinox in September, half of the animals gave birth several weeks in advance of the control animals in which births occurred at the normal time after the summer solstice. The remaining pinealectomised animals also gave birth at the normal time (McConnell 1984; McConnell & Hinds 1985). These results suggest that tammars may require an intact pineal to interpret the increasing daylengths after the winter solstice in order to enter or to maintain seasonal quiescence.

The immediate and longer-term response to pinealectomy performed in seasonal quiescence (July to December) is highly variable and can only be explained if we consider the neuroendocrine axis is capable of exhibiting changes associated with the onset and subsequent termination of the breeding season in the absence of external photoperiodic input. Immediately after the operation one-third of the pinealectomised females and half of the ganglionectomised females reactivated. None of the sham-treated animals responded. In the second year after operation, a similar array of responses was seen when the animals were challenged with bromocriptine after the vernal equinox. In the third year when the groups were exposed to an artificial photoperiod change (15L:9D to 12L:12D) some operated females responded, as did all sham-operated controls, while other treated females remained quiescent. That is, under each of the manipulations some pinealectomised animals were aseasonal, while others retained seasonality, although none showed any evidence of melatonin secretion (McConnell 1984; C H Tyndale-Biscoe & L A Hinds, unpublished observations). The absence of a melatonin message has clearly disrupted downstream responses in the hypothalamus and pituitary in some, but not all, pinealectomised animals. Thus the neuroendocrine axis is capable of exhibiting changes associated with the onset and subsequent termination of the breeding season in the absence of external photoperiodic input. One interpretation of this is that the animals have an intrinsic endogenous rhythm of reproductive activity.

Although the responses of seasonal wallabies to pinealectomy are not necessarily directly comparable to similar studies of eutherians with quite different modes of reproduction, it is none the less striking that the tammar exhibits many of the characteristics of the seasonal sheep. Extensive studies by Karsch and colleagues have demonstrated that pinealectomised sheep continue to exhibit free-running rhythms of gonadotrophin secretion. The intervals between periods of 'reproductive activity' in castrated oestradiol-implanted animals vary widely, but on average are close to a periodicity of a year (Karsch *et al.* 1989). Recently, melatonin-replacement studies have been performed in which daily patterns of the hormone have been administered for short periods of 90 days each year, but at different times of year in different groups of animals (Woodfill et al. 1994). In this study, melatonin synchronised the endogenous circannual rhythm to periodicities close to 365 days, although not all melatonin patterns were equally effective. The most effective patterns mimicked those of summer (Woodfill *et al.* 1994), leading to the conclusion that there is a seasonal specificity with regard to the cues that synchronise the circannual rhythm of neuroendocrine function. These experiments provide the most convincing evidence to date that the neuroendocrine axis of seasonal mammals possesses an endogenous rhythmical capacity which can be entrained by a zeitgeber signal of melatonin to a periodicity of a year.

Further evidence for the existence of a free-running 'circannual' reproductive rhythm in seasonal macropodids was obtained in an experiment in pineal-intact Bennett's wallabies (Brinklow & Loudon 1992). Here, animals were maintained under ambient temperature conditions in photoperiod controlled rooms for 16 months and exposed to either fixed summer solstitial photoperiod (18L:6D) or a mimic of the natural changes in photoperiod. In both groups the onset of the breeding season occurred shortly after the summer solstice. The duration of the breeding season was similar in both groups, despite an unchanging long-day signal in the treatment group. Similarly, the prolactin response to domperidone fell sharply at the onset of the breeding season in both groups, rising with the onset of seasonal quiescence. This change occurred spontaneously in long-day held animals and thus provides evidence that the neuroendocrine axis is capable of exhibiting changes associated with the onset and subsequent termination of the breeding season in the absence of external photoperiodic input.

SITE OF ACTION OF MELATONIN IN THE BRAIN

Research on the distribution of melatonin receptors is now permitted by the discovery of Vakkuri and co-workers that the melatonin molecule could be iodinated to produce a label of high specific activity which can then be used to localise receptors in unfixed frozen sections of tissue (Vakkuri *et al.* 1984). In the last 5 years a large number of studies of the melatonin receptor have been undertaken in different vertebrate groups (Morgan & Williams 1989; Morgan *et al.* 1994). In lower vertebrates binding is widely distributed throughout the brain, mainly in the visual, auditory and limbic systems. In contrast, the distribution of binding sites in the mammalian brain is less widespread and commonly is localised to specific neural nuclei. There is species variation in the distribution of melatonin binding in eutherian mammals, but a feature of all studies to date is that in the adult brain one of the regions consistently labelled is found in the *pars tuberalis* of the pituitary gland, adjacent to the mediobasal hypothalamus. Indeed, in one group of seasonally breeding mammals, the mustelids, the *pars tuberalis* is the only area of the brain where melatonin binding is observed (Weaver & Reppert 1990; Duncan & Mead 1992). The *pars tuberalis* has the highest concentration of melatonin receptors in the mammalian nervous system (Bittman &

Weaver 1990). The localisation of melatonin receptors in the adult eutherian brain is in marked contrast to the situation observed in the fetus. Here, receptors have a widespread distribution, including peripheral tissues and within the nervous system, the hypothalamus, cerebral cortex and hippocampus (for review, Morgan *et al.* 1994). The role of this hormone in the developing brain remains to be determined. An intriguing feature of many of these studies is the observation that binding occurs at specific developmental stages in the nasal epithelium. It is this area of the developing mammalian brain that gives birth to the origin of the GnRH neurons, which then migrate to their destination in the hypothalamus (Schwanzel-Fukuda *et al.* 1992).

The pharmacological characterisitics of the melatonin receptor have been described in brain tissues of the Bennett's wallaby. Competititon studies (Paterson *et al.* 1992) reveal that the brain contains a single class of high affinity receptor with binding in the picomolar range. In addition, the receptor appears to be G-protein coupled with a change in affinity following guanosine 5'-o-(3-thiotriphosphate) treatment (A M Paterson & A S I Loudon, unpublished results). We have established that the Bennett's wallaby melatonin receptor exhibits pharmacological characteristics similar to those reported in a number of other studies, including eutherians, reptiles and birds (for review, Morgan *et al.* 1994). Our studies also reveal a widespread distribution of melatonin receptors in the central nervous system and peripheral tissues in early pouch life (Figure 3.9). The developing pouch young exhibits a receptor distribution similar to that observed in many fetal eutherian species (Table 3.1), with strong melatonin binding in the nasal epithelium of 2 month old pouch young (A M Paterson and A S I Loudon, unpublished observations). In later pouch life (5 months of age), binding is much more localised, being confined to the pars tuberalis of the pituitary gland (Figure 3.9) and from this age on we have been unable to find evidence of binding in any sites in the brain. In this respect wallabies exhibit a similar restriction in distribution in the adult to that observed in mustelids (Duncan & Mead 1992).

The role of the pars tuberalis in the control of prolactin secretion.

Our studies of the distribution of melatonin receptors in wallabies are of particular significance in relation to a recent experiment on the seasonally breeding Soay ram by Lincoln and Clarke (1994). In this study, the hypothalamic portal vasculature to the pituitary gland was surgically disconnected. This procedure permits the survival of the majority of the pituitary gland, but deprives it of endocrine support from the hypothalamus. The procedure resulted in the collapse of gonadotrophin secretion, as would be expected following the destruction of the terminals of the hypothalamic GnRH neurones. Testicular atrophy occurred and the testes were reduced to a smaller size than that normally observed in the non-breeding season (Lincoln & Clarke 1994, G.A. Lincoln, personal communication). Remarkably, one pituitary hormone, prolactin, exhibited a significant response to photoperiod change, with high concentrations on long days and low concentrations on short photoperiods (Lincoln & Clarke 1994). Although prolactin concentrations were higher on both short and long photoperiods in disconnected animals, photoperiod change induced a change in secretion of similar magnitude to that observed in unoperated control animals. Pituitary disconnected animals were unable to respond to a range of standard provocation tests (for example, exposure to a barking dog or injection of serotonin and N-methyl-D-aspartate), all of which elicit a prolactin response in intact animals (Lincoln & Clarke 1994). Dopamine is implicated in many studies as the

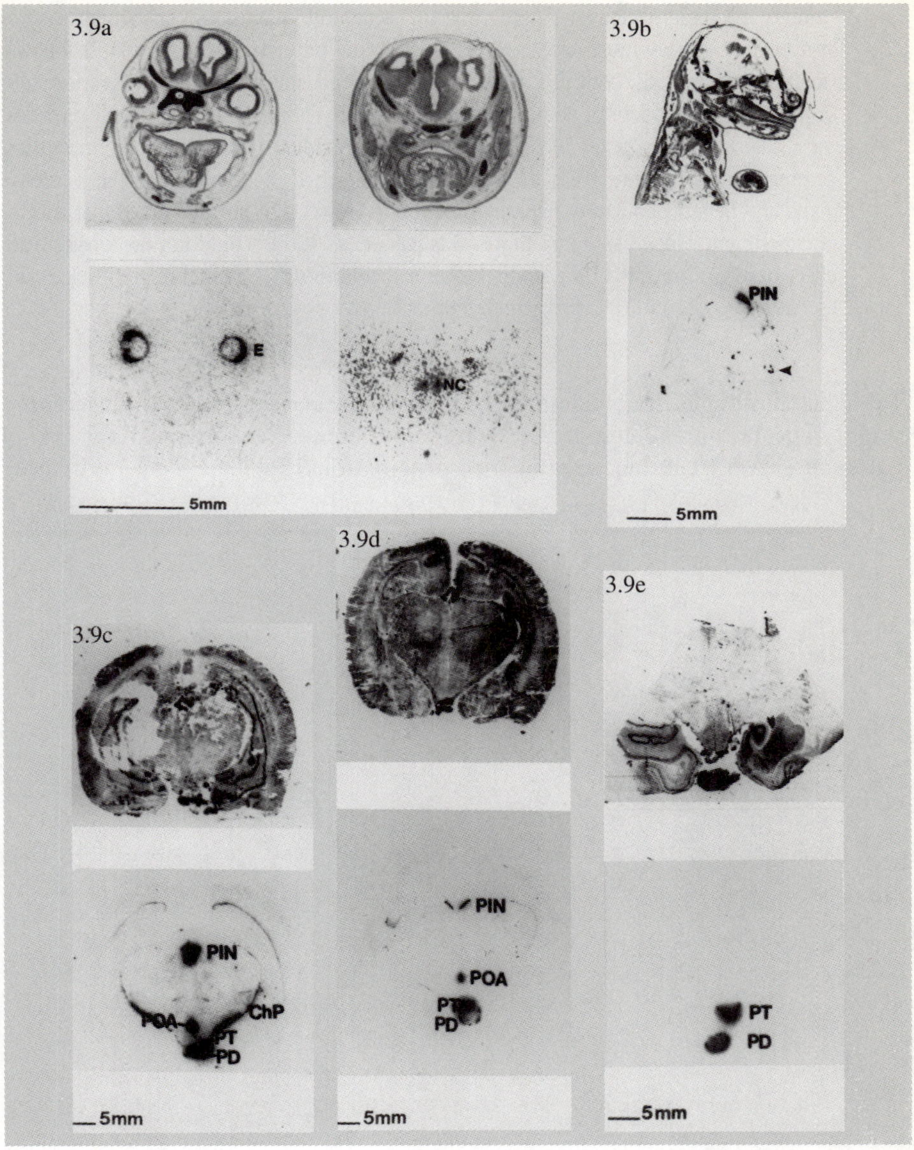

Figure 3.9

Autoradiographical images of 2[125]Iodomelatonin binding in the brains of the Bennett's wallaby at 1week of age (*plate a*), 1 month (*plate b*), 3 months (*plate c*), 6 months (*plate d*) and when adult (*plate e*).

Notes: Sections were counterstained with toluidine blue for histological comparison (see *upper sections* of each plate). Alternate sections were also incubated with excess cold melatonin as a control (data not shown). All sections are coronal, except *plate b* which is sagittal. At 1 week of age (*plate a*), binding is apparent in the nasal cavity (NC; *bottom right*) and eyes (E, *bottom left*). At 1 month, binding is observed in the nasal cavity (arrow) and pineal gland (PIN). Binding at 3 months of age is widespread and includes the pineal, pre-optic area (POA), choroid plexus (ChP) and *pars tuberalis* (PT) and *pars distalis* (PD). At 6-months, binding is confined to the pineal, pre-optic area and pituitary stalk and in the adult (*plate e*) to only the pituitary.
Source: A Paterson & A S I Loudon (unpublished observations).

primary inhibitory factor regulating prolactin secretion in mammals. Since hypothalamic dopamine supply must have been blocked by the pituitary disconnection, it follows that the photoperiodic control of prolactin secretion may not depend on dopamine secretion and is independent of neural input to the hypothalamus. This experiment provides the clearest demonstration to date that melatonin-mediated photoperiodic time measurement may, as far as prolactin secretion is concerned, be confined to the pituitary gland and be independent of hypothalamic input. However, it must be noted that another experiment, albeit using intact ewes (Malpaux *et al.* 1994), could not demonstrate an effect of microimplants of melatonin placed subcutaneously, in the third ventricle or in the *pars tuberalis* on LH or prolactin concentrations.

Table 3.1 Binding of iodomelatonin in brain of eutherian species at different stages of gestation and comparison with a marsupial, Bennett's wallaby (A Paterson & A S I Loudon, unpublished observations)

RAT[1]	HAMSTER[2,3,4]	SHEEP[5]	BENNETT'S WALLABY
E 15 pituitary	E 10-14 oral/nasal cavities developing	E 30-31 thyroid, pituitary gland, spinal nerves, nasal cavity, bronchi	1 week nasal cavities, eyes
E 17 paraventricular thalamus	E 16 paraventricular thalamic nuclei, mediobasal hypothalamus, pineal	E 40 cochlear epithelium, pars distalis and pars tuberalis	1 month pineal, nasal cavities
E 18 SCN			3 months pars tuberalis, pars distalis, pineal, choroid plexus and pre-optic area 6 month pars tuberalis, pars distalis, pineal, preoptic area 9 month pars tuberalis

E: embryonic/gestation day,
SCN: suprachiasmatic nucleus.
[1]Williams *et al.* (1991); [2]Weaver *et al.* (1983); [3]Carlson *et al.* (1991); [4]Rivkees & Reppert (1992); [5]Morgan *et al.* (1994)

The localisation of melatonin receptors to the pars tuberalis in the Bennett's wallaby suggests a similar mechanism as proposed for Soay rams (Lincoln & Clarke 1994) may operate in this group of marsupials. Here, prolactin appears to be the only hormone of the neurohypophysis to undergo photoperiodic change and seasonal marsupials lack many of the adaptive physiological rhythms observed in seasonal eutherian species which are known to depend on changes in secretion of a number of pituitary hormones. The results of Lincoln & Clarke (1994) may offer a solution to the paradox discussed above relating to the lack of response of wallabies in seasonal quiescence to the dopamine agonist, bromocriptine. Photoperiodic drive on prolactin secretion may be dopamine-independent, relying instead on a direct action of melatonin on the *pars tuberalis* and subsequent modification (by an unknown mechanism) of prolactin secretion by the lactotroph cells. If this is the case (for sheep or wallabies) it follows that discrimination of a time-based signal must reside within the pituitary. Thus, the photoperiodic transduction system may effectively bypass the hypothalamus. It is striking that the only other group of mammals in which melatonin binding appears to be confined to the *pars tuberalis* are the mustelids. In this group, there is also a clear reproductive role for prolactin, and the phenomenon of embryonic diapause is a feature of the reproductive biology of many mustelid species (Martinet & Allain 1985).

The current understanding of how photoperiod controls prolactin secretion in any mammal is poor and it is now clear that in wallabies we need to undertake further studies of this hormone. Our observations of a diurnal 'long-day' pulse (in tammars) together with the localisation of melatonin receptors only to the *pars tuberalis* in the Bennett's wallaby suggest that while prolactin may be photoperiodically regulated at the pituitary level, we do need to obtain more information on the apparent diurnal nature of the inhibitory signal. Further development of more sensitive homologous radioimmunoassays for marsupial pituitary peptides (Curlewis & McNeilly 1993) may prove fruitful. However, at present there is no *a priori* reason to propose that any hormone other than prolactin is involved in the photoperiodic regulation of reproduction.

Research on the reproductive biology of these two seasonal macropodids now reveals that a number of aspects of their seasonal biology are common to those of other seasonal vertebrate groups. Wallabies exhibit the phenomenon of photorefractoriness, and may exhibit endogenous circannual rhythmicity in reproductive function when maintained on long photoperiods which may persist in pinealectomised animals. In view of the recent experiments described above, it would be fascinating to know whether photoperiodic prolactin regulation as well as the endogenous circannual rhythmicity is all maintained within the pituitary gland and is entirely independent of hypothalamic input. This is of particular importance since in eutherian models, research on 'circannual rhythmicity' currently focuses on a possible hypothalamic origin for the annual clock. There is the possibility that one component of this circannual clock might reside within the pituitary itself.

Lactational and photoperiodic control of prolactin secretion

Our current experimental data are consistent with the hypothesis that hyperprolactinaemia associated with lactation and the sucking stimulus is induced as a result of changes in hypothalamic dopamine secretion. Dopamine agonists disrupt lactation at all stages of lactation (Curlewis *et al.* 1986) and in late lactation when tonic levels are significantly elevated, bromocriptine causes an abrupt decline in concentrations and impairment of lactation. Further, there is a reduced sensitivity to the dopamine antagonist domperidone in late lactation when prolactin concentrations are significantly elevated (Pearce-Kelly *et al.* 1990). This observation is consistent with the hypothesis that late lactation and hyperprolactinaemia are associated with an increased endogenous dopamine turnover (Ben-Jonathan *et al.* 1985) such that higher doses of an antagonist are required to elicit a response. This may be regarded as the primary mechanism of control for all macropodid marsupials, based on lactation and the inhibitory role of the sucking stimulus of the pouch young. The evolution of a photoperiodic mechanism in two species is clearly a more recent evolutionary event. It may not depend on dopamine modulation of prolactin secretion, but perhaps a direct action of the melatonin signal on the pituitary (Figure 3.10). If this is the case then we might expect to find that aseasonal species of macropodid marsupials may lack melatonin binding in the *pars tuberalis*. Further, species in which the primary photoperiodic drive is via the action of melatonin on GnRH secretion may be expected to exhibit melatonin receptors elsewhere in the central nervous system, including the hypothalamus.

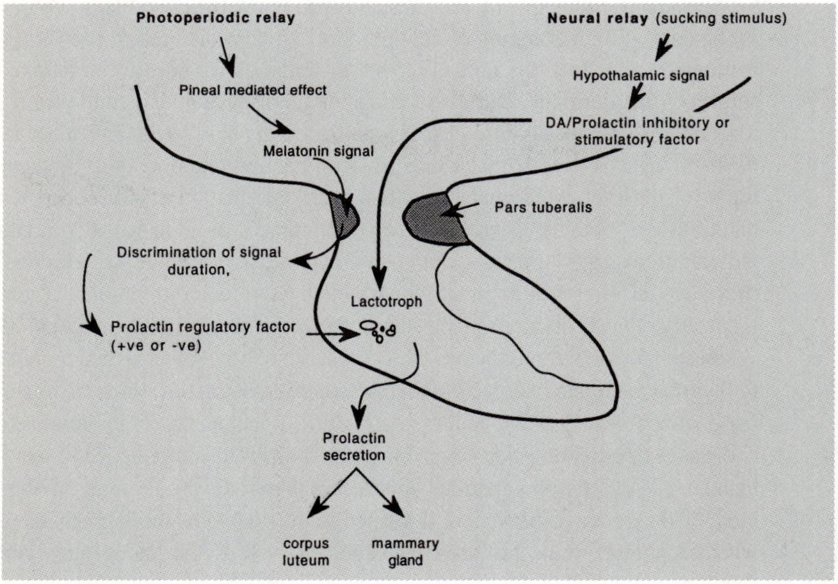

Figure 3.10 A model for the control of prolactin secretion.
Notes: The primary control mechanism is via inhibitory regulation from the hypothalamus (heavy arrow). This neural pathway includes suckling-induced hyperprolactinaemia, via dopamine (DA). A second relay system is based on the photoperiodic-pineal axis. Here, melatonin acts directly on the *pars tuberalis* of the pituitary. These cells are presumed to possess the capacity to discriminate the duration of the melatonin signal independently of hypothalamic input and produce a *pars tuberalis*-specific product that acts on the lactotroph cells. No assumptions are made concerning the inhibitory or stimulatory nature of the *pars tuberalis*-specific signal. In the model presented, the neural (sucking stimulus) relay is the over-riding pathway in most macropodid species — in the tammar and Bennett's wallaby the photoperiodic input via melatonin over-rides the sucking input during periods of long daylength.

FUTURE DIRECTIONS: HOW DIFFERENT ARE SEASONAL MARSUPIALS AND EUTHERIANS?

The basic pattern of opportunistic breeding seen in the red kangaroo has been adapted by the tammar and Bennett's wallaby for seasonal breeding in the more southern areas of Australia. The use of prolactin during seasonal quiescence represents an extension of its use as a mechanism controlling fertility during lactational quiescence. In these species, seasonality is confined to relatively subtle effects on the secretion of prolactin and this has clearly evolved rather recently. In marked contrast, seasonal eutherian species are characterised by a wide range of different physiological adaptations. These processes are clock-based and regulated by photoperiod and pineal melatonin (Loudon 1994). Most eutherian species breed seasonally and almost all aspects of their regulatory physiology are under environmental control (Mrosovsky 1990). Here, reproductive changes are part of a wider range of physiological changes that can include changes in sexual behaviour (Bittman *et al.* 1988), metabolism, hibernation, seasonal hyperphagia (Loudon *et al.* 1989; Heydon *et al.* 1994), body fat content and the periodicity of growth of many tissues, including antlers or horns, hair and pelage (Lincoln 1984; Loudon 1994). Although comparable aspects have not been investigated in detail in marsupials, in one species, the Bennett's wallaby, metabolism and voluntary food intake do not exhibit comparable changes to those observed in seasonal eutherians (Loudon 1987).

There are many species of marsupial that breed seasonally. Although the concentration of research on seasonal macropidids has unravelled many important features of reproductive adaptation, there is still very little known of the environmental regulation of fertility in other marsupial families. One species of marsupial is even known to be capable of hibernating (Geiser 1985). All seasonal eutherians exhibit conspicuous moult cycles, which are known to be prolactin-dependent (Duncan & Goldman 1984, 1985), but we still have no information on the moult cycle (if it exists) in marsupials. Furthermore, sexual behaviour and the associated production of secondary sexual characteristics are one of the most conspicuous seasonal features of many eutherian species, yet, to date there is no comparable knowledge of these aspects of the biology of many marsupials.

It is now clear that marsupials have a valuable contribution to make in many areas of biology, including the study of development (Tyndale-Biscoe & Janssens 1988). The production of a neotenous offspring that completes much of its development postnatally offers a great tool to the biologist. However, perhaps the greatest value that studies of marsupials can offer to biology is to provide much-needed comparative phylogenetic insight. Indeed, the most valuable insights in physiology must be rooted in comparative and evolutionary biology, for it is only at this level that the universal nature of regulatory processes can be placed in context. We therefore complete this brief review, with its emphasis on macropodids, with a plea that the pioneering marsupial research of Hugh Tyndale-Biscoe and his many colleagues be extended to other marsupial species and, perhaps, that biologists working on eutherian species should take more notice of the growing and voluminous scientific literature of the 'other half' of the mammalian kingdom.

Acknowledgments

We would like to thank our many colleagues for their help over the years, but specific mention must go to Drs Hugh Tyndale-Biscoe, Peter Janssens, Steve McConnell, Jon

Curlewis, Barry Brinklow and Tony Flint. We also thank Alison Patterson for permission to cite from her unpublished studies. The patience and willingness of the staff of the CSIRO Division of Wildlife and Ecology, as well as of the Zoological Society of London are gratefully acknowledged. Their provision, maintenance and assistance with experimental animals is highly appreciated. Research on the Bennett's wallaby has been supported by grants from the Science and Engineering Research Council UK and a joint program grant from the Medical Research Council UK and Agricultural and Food Research Council UK.
Copyright permission received for all previously published figures.

References

Adam C L and Atkinson T (1984). Effect of feeding melatonin to red deer (*Cervus elaphus*) on the onset of the breeding season. *Journal of Reproduction and Fertility*, 72, 463–466.

Andrewartha H G and Barker S (1969). Introduction to a study of the ecology of the Kangaroo Island wallaby, *Protemnodon eugenii* (Desmarest) within Flinders Chase, Kangaroo Island, South Australia. *Transactions of the Royal Society of South Australia*, 93, 127–132.

Arendt J (1986). Role of the pineal gland and melatonin in seasonal reproductive function in mammals. *Oxford Reviews of Reproductive Biology*, 8, 266–320.

Arendt J, Symons A M, Laud C A and Pryde S J (1983). Melatonin can induce early onset of the breeding season in sheep. *Journal of Endocrinology*, 97, 395–400.

Ben-Jonathan N (1985). Dopamine: a prolactin inhibiting hormone. *Endocrine Reviews*, 6, 564–589.

Berger P J (1966). Eleven-month 'Embryonic Diapause' in a marsupial. *Nature*, 211, 435–436.

Berger P J (1970). Reproductive biology of the tammar wallaby, *Macropus eugenii*. PhD thesis, Tulane University. *Dissertation Abstracts International*, 31B, 3760–3761.

Berger P J and Sharman G B (1969). Progesterone-induced development of dormant blastocysts in the tammar wallaby, *Macropus eugenii* Desmarest: Marsupialia. *Journal of Reproduction and Fertility*, 20, 201–210.

Bittman E L and Karsch F J (1984). Nightly duration of pineal melatonin secretion determines the reproductive response to inhibitory day length in the ewe. *Biology of Reproduction*, 30, 585–593.

Bittman E L and Weaver D R (1990). The distribution of melatonin binding sites in neuroendocrine tissues in the ewe. *Biology of Reproduction*, 43, 986–993.

Bittman E L, Hegarty C M, Layden M Q and Jonassen JA (1990). Influences of photoperiod on sexual behaviour, neuroendocrine steroid receptors and adenohypophysial hormone secretion and gene expression in the female golden hamster. *Journal of Molecular Endocrinology*, 5, 15–25.

Brinklow B R and Loudon A S I (1989). Effect of exogenous prolactin and bromocriptine on seasonal reproductive quiescence in the Bennett's wallaby (*Macropus rufogriseus rufogriseus*). *Journal of Endocrinology*, 120, 189–193.

Brinklow, B R and Loudon, A S I (1992) Evidence of a circannual rhythm of reproduction and prolactin secretion in a seasonally breeding macropodid marsupial, the Bennetts's wallaby. *Journal of Reproduction and Fertility*, 98, 625–630.

Carlson L L, Weaver D R and Reppert S M (1991). Melatonin receptors and signal transduction during development in Siberian hamsters (*Phodopus sungorus*). *Developmental Brain Research*, 59, 83–88.

Carter D S and Goldman B G (1983). Antigonadal effects of timed melatonin infusion in pinealectomized male Djungarian hamsters (*Phodopus sungorus sungorus*): duration is the critical parameter. *Endocrinology*, 113, 1261–1267.

Cassone V M (1990). Melatonin: Time in a bottle. *Oxford Reviews of Reproductive Biology*, 12, 319–367.

Catling P C and Sutherland R L (1980). Effect of gonadectomy, seasons, and the presence of females on plasma testosterone, luteinizing hormone, and follicle stimulating hormone levels in male tammar wallabies (*Macropus eugenii*). Journal of Endocrinology, 86, 25–33.

Catt D C (1977). The breeding biology of Bennett's wallaby (*Macropus rufogriseus fruticus*) in South Canterbury, New Zealand. *New Zealand Journal of Zoology*, 4, 401–411.

Cork S and Dove H (1989). Lactation in the tammar wallaby (*Macropus eugenii*). II. Intake of milk components and maternal allocation of energy. *Journal of Zoology*, 219, 399–409.

Curlewis J D (1989). The breeding season of Bennett's wallaby (*Macropus rufogriseus rufogriseus*) in Tasmania. *Journal of Zoology*, 218, 337–339.

Curlewis J D (1991). Seasonal changes in the reproductive organs and plasma and pituitary hormone content of the male Bennett's wallaby (*Macropus rufogriseus rufogriseus*). *Journal of Zoology*, 223, 223–231.

Curlewis J D (1992). Seasonal prolactin secretion and its role in seasonal reproduction: a review. *Reproduction, Fertility and Development*, 4, 1–23.

Curlewis J D (1994). Reproductive responses to photoperiod in the female marsupial. In *Perspectives in Comparative Endocrinology.* Eds K G Davey, R E Peter and S S Tobe. National Research Council of Canada. pp 555–560

Curlewis J D and Loudon A S I (1988a). Experimental manipulations of prolactin following removal of pouch young or bromocriptine treatment during lactational quiescence in the Bennett's wallaby. *Journal of Endocrinology*, 119, 405–411.

Curlewis J D and Loudon A S I (1988b). Effects of photoperiod on the 24-hour melatonin profiles of the Bennett's wallaby (*Macropus rufogriseus rufogriseus*). *Journal of Pineal Research*, 5, 373–383.

Curlewis J D and Loudon A S I (1989). The role of refractoriness to long daylength in the annual reproductive cycle of the female Bennett's wallaby (*Macropus rufogriseus rufogriseus*). *Journal of Experimental Zoology*, 252, 200–206.

Curlewis J D and McNeilly A S (1993). Purification and partial characterization of prolactin and growth hormone from the Bennett's wallaby. *General and Comparative Endocrinology*, 88, 341–350.

Curlewis J D, White A S, Loudon A S I and McNeilly A S (1986). Effects of lactation and season on plasma prolactin concentrations and response to bromocriptine during lactation in the Bennett's wallaby (*Macropus rufogriseus rufogriseus*). *Journal of Endocrinology*, 110, 59–66.

Curlewis J D, White A S and Loudon A S I (1987). The onset of seasonal quiescence in the female Bennett's wallaby (*Macropus rufogriseus rufogriseus*). *Journal of Reproduction and Fertility*, 80, 119–124.

Dressen W, Grun H and Hendrichs H (1990). Radio telemetry of heart rate in male tammar wallabies (Marsupialia: Macropodidae): temporal variations and behavioural correlates. *Australian Journal of Zoology*, 38, 89–103.

Duncan M J and Goldman B D (1984). Hormonal regulation of the annual pelage colour cycle in the Djungarian hamster, *Phodopus sungorus*. II Role of Prolactin. *Journal of Experimental Zoology*, 230, 97–103.

Duncan M J and Goldman B D (1985). Physiological doses of prolactin stimulate pelage pigmentation in Djungarian hamster. *American Journal of Physiology*, 248, 664–667

Duncan M J and Mead R A (1992). Autoradiographic localisation of binding sites for I^{125} iodomelatonin in the pars tuberalis of the Western Spotted Skunk (*Spilogale putorius latifrons*). *Brain Research*, 569, 152–155.

Fleming D, Cinderey R N and Hearn J P (1983). The reproductive biology of Bennett's wallaby (*Macropus rufogriseus rufogriseus*) ranging free at Whipsnade Park. *Journal of Zoology*, 201, 283–291.

Geiser F (1985). Hibernation in pygmy possums (Marsupialia: Burramyidae). *Comparative Biochemistry and Physiology*, 81A, 459–463.

Gemmell R T (1987a). Effect of melatonin and removal of pouch young on the seasonality of births in the marsupial possum, *Trichosurus vulpecula*. *Journal of Reproduction and Fertility*, 80, 301–307.

Gemmell R T (1987b). Influence of melatonin on the initiation of the breeding season of the marsupial bandicoot, *Isoodon macrourus*. *Journal of Reproduction and Fertility*, 79, 261–265.

Gemmell R T (1990). Influence of daylength on the initiation of the breeding season of the marsupial possum, *Trichosurus vulpecula*. *Journal of Reproduction and Fertility*, 88, 605–609.

Gemmell R T and Sernia C (1992). The role of photoperiod on the initiation of the breeding season of the possum, *Trichosurus vulpecula*. *Journal of Reproduction and Fertility*, 95, 701–708.

Gemmell R T, Cepon G and Sernia C (1993). Effect of photoperiod on the breeding season of the marsupial possum *Trichosurus vulpecula*. *Journal of Reproduction and Fertility*, 98, 515–520.

Godfrey G K (1969). The influence of increased photoperiod on reproduction in the dasyurid marsupial, *Sminthopsis crassicaudata*. *Journal of Mammalogy*, 50, 132–133.

Gordon K, Fletcher T P and Renfree M B (1988). Reactivation of the quiescent corpus luteum and diapausing embryo after temporary removal of the sucking stimulus in the tammar wallaby (*Macropus eugenii*). *Journal of Reproduction and Fertility*, 83, 401–406.

Hartman C G (1923). The oestrous cycle in the opossum. *American Journal of Anatomy*, 32, 353–421.

Hearn J P (1972a). *The Pituitary Gland and Reproduction in the Marsupial, Macropus eugenii (Desmarest)*. PhD thesis, Australian National Univeristy, Canberra.

Hearn J P (1972b). Effect of advanced photoperiod on termination of embryonic diapause in the marsupial *Macropus eugenii* (*Macropodidae*). *Australian Mammalogy*, 1, 40–42.

Hearn J P (1973). Pituitary inhibition of pregnancy. *Nature*, 241, 207–208.

Hearn J P (1974). The pituitary gland and implantation in the tammar wallaby, *Macropus eugenii*. *Journal of Reproduction and Fertility*, 39, 235–241.

Heydon M J, Sibbald A M, Milne J A, Brinklow B R and Loudon A S I (1993). The interaction of food availability and endogenous physiological cycles on the grazing ecology of red deer (*Cervus elaphus*). *Functional Ecology*, 7, 216–222.

Hinds L A (1983). *Progesterone and Prolactin in Marsupial Reproduction*. PhD thesis, Australian National University, Canberra.

Hinds L A (1989). Morning prolactin pulse maintains seasonal quiescence in the tammar wallaby, *Macropus eugenii*. *Journal of Reproduction and Fertility*, 87, 735–744.

Hinds L A (1994). Prolactin, a hormone for all seasons: endocrine regulation of seasonal breeding in the Macropodidae. *Oxford Reviews of Reproductive Biology*, 16, 247–299.

Hinds L A and den Ottolander R C (1983). Effects of changing photoperiod on peripheral plasma prolactin and progesterone concentrations in the tammar wallaby, *Macropus eugenii*. *Journal of Reproduction and Fertility*, 69, 631–639.

Hinds L A and Tyndale-Biscoe C H (1982a). Plasma progesterone levels in the pregnant and non-pregnant tammar, *Macropus eugenii*. *Journal of Endocrinology*, 93, 99–107.

Hinds L A and Tyndale-Biscoe C H (1982b). Prolactin in the marsupial, *Macropus eugenii*, during the estrous cycle, pregnancy and lactation. *Biology of Reproduction*, 26, 391–398.

Hinds L A and Tyndale-Biscoe C H (1985). Seasonal and circadian patterns of circulating prolactin during lactation and seasonal quiescence in the tammar, *Macropus eugenii*. *Journal of Reproduction and Fertility*, 74, 173–183.

Hinds L A, Diggle P J and Tyndale-Biscoe C H (1992). Effects of the ovary, sucking stimulus and season on the pattern of LH and FSH release in the female tammar, *Macropus eugenii*. *Reproduction, Fertility and Development*, 4, 25–34.

Hoffman RA and Reiter R J (1965). Pineal gland: influence on gonads of male hamsters. *Science*, 148, 1609–1610.

Horn C A, Fletcher T P and Carpenter S (1985). Effects of oestradiol-17β on peripheral plasma concentrations of LH and FSH in ovariectomized tammars (*Macropus eugenii*). *Journal of Reproduction and Fertility*, 73, 585–592.

Inns R W (1982). Seasonal changes in the accessory reproductive system and plasma testosterone levels of the male tammar wallaby, *Macropus eugenii*, in the wild. *Journal of Reproduction and Fertility*, 66, 675–80.

Karsch F J and Woodfill C J I (1990). Neuroendocrinology of seasonal breeding: mode of action of melatonin. In *Neuroendocrine Regulation of Reproduction*. Eds S S C Yen, W W Vale. Serono Symposia, USA, Norwell, Massachusetts. pp 9–17.

Karsch F J, Bittman E L, Foster D L, Goodman R L, Legan S J and Robinson J E (1984). Neuroendocrine basis of seasonal reproduction. *Recent Progress In Hormone Research*, 40, 185–232.

Karsch F J, Robinson J E, Woodfill C J I and Brown M B (1989). Circannual cycles of luteinizing hormone and prolactin secretion in ewes during prolonged exposure to a fixed photoperiod: evidence for an endogenous reproductive rhythm. *Biology of Reproduction*, 41, 1034–1046.

Kojima T, Hinds L A, Muller W J, O'Neill C and Tyndale-Biscoe C H (1993). In vitro production and secretion of progesterone and presence of platelet activating factor (PAF) in early pregnancy of the marsupial, *Macropus eugenii*. *Reproduction, Fertility and Development*, 5, 15–25.

Lee A K and Cockburn A (1985). *The Evolutionary Ecology of Marsupials*. Cambridge University Press, Cambridge.

Legan S J, Karsch F J and Foster D L (1977). The endocrine control of seasonal reproductive function in the ewe: a marked change in response to the negative feedback action of estradiol on luteinizing hormone secretion. *Endocrinology*, 101, 818–824.

Lincoln G A (1984). Seasonal Breeding in Deer. *In Biology of Deer Production*. Eds P F Fennessey and K R Drew. Bulletin New Zealand Royal Society, No. 22.

Lincoln G A and Clarke I J (1994). Photoperiodically-induced cycles in the secretion of prolactin in hypothalamo-pituitary disconnected rams: Evidence for translation of the melatonin signal in the pituitary gland. *Journal of Neuroendocrinology*, 6, 251–260.

Lincoln G A and Short R V (1980). Seasonal breeding: Nature's Contraceptive. *Recent Progress in Hormone Research*, 36, 1–52.

Loudon A S I (1987). Nutritional effects on puberty and lactational infertility in mammals: some interspecies considerations. *Proceedings of the Nutrition Society*, 46, 203–216.

Loudon A S I (1994). Photoperiod and the regulation of annual and circannual cycles of food intake. *Proceedings of the Nutrition Society*, 53, 495–507.

Loudon A S I and Brinklow B R (1990). Melatonin implants prevent the onset of seasonal quiescence and suppress the release of prolactin in response to a dopamine antagonist in the Bennett's wallaby (*Macropus rufogriseus rufogriseus*). *Journal of Reproduction and Fertility*, 90, 611–618.

Loudon A S I and Brinklow B R (1992). Reproduction in deer: adaptations for life in seasonal environments. In *The Biology of Deer*. Ed R D Brown, Springer-Verlag, Berlin. pp 261–278.

Loudon A S I and Curlewis J D (1987). Refractoriness to melatonin and short daylengths in early seasonal quiescence in the Bennett's wallaby (*Macropus rufogriseus rufogriseus*). *Journal of Reproduction and Fertility*, 81, 543–552.

Loudon A S I and Curlewis J D (1989). Evidence that the seasonally breeding Bennett's wallaby (*Macropus rufogriseus rufogriseus*) does not exhibit short-day photorefractoriness. *Journal of Reproduction and Fertility*, 87, 641–648.

Loudon A S I, Curlewis J D and English J (1985). The effect of melatonin on the seasonal embryonic diapause of the Bennett's wallaby (*Macropus rufogriseus rufogriseus*). *Journal of Zoology*, 206, 35–39.

Loudon A S I, Milne J A, Curlewis J D and McNeilly A S (1989). A comparison of the seasonal hormone changes and patterns of growth, voluntary food intake and reproduction in juvenile and adult red deer (*Cervus elaphus*) and Pere David's deer (*Elaphurus davidianus*) hinds. *Journal of Endocrinology*, 122, 733–745.

Malpaux B, Daveau A, Maurice F, Locatelli A and Thiery J-C (1994). Evidence that melatonin binding sites in the pars tuberalis do not mediate the photoperiodic actions of melatonin on LH and prolactin in ewes. *Journal of Reproduction and Fertility*, 101, 625–632.

Martinet L and Allain D (1985). Role of the pineal gland in the photoperiodic control of reproductive and non-reproductive functions in the mink (*Mustela vison*). In: *Photoperiodism, Melatonin and the Pineal Gland*. Eds D Evered and S Clark. Pitman Publishing, London. pp 170–187.

McAllen B M, Joss J M P and Firth B T (1991). Phase delay of the natural photoperiod alters reproductive timing in the marsupial *Antechinus stuartii*. *Journal of Zoology*, 25, 633–646.

McConnell S J (1984). *The Pineal Gland and Reproduction in the Tammar Wallaby, Macropus eugenii*. PhD thesis, Australian National University, Canberra.

McConnell S J (1986). Seasonal changes in the circadian plasma profile of the tammar, *Macropus eugenii*. *Journal of Pineal Research*, 3, 119–125.

McConnell S J and Hinds L A (1985). Effect of pinealectomy on plasma melatonin, prolactin and progesterone concentrations during seasonal reproductive quiescence in the tammar, *Macropus eugenii*. *Journal of Reproduction and Fertility*, 7, 433–440.

McConnell S J and Tyndale-Biscoe C H (1985). Response in peripheral plasma melatonin to photoperiod change and the effects of exogenous melatonin on seasonal quiescence in the tammar, *Macropus eugenii*. *Journal of Reproduction and Fertility*, 73, 529–538.

McConnell S J, Tyndale-Biscoe C H and Hinds L A (1986). Change in duration of elevated concentrations of melatonin is the major factor in photoperiod response of the tammar, *Macropus eugenii*. *Journal of Reproduction and Fertility*, 77, 623–632.

McNeilly A S and Friesen H (1978). Prolactin during pregnancy and lactation in the rabbit. *Endocrinology*, 102, 1539–1547.

Merchant J C (1979). The effect of pregnancy on the interval between one oestrus and the next in the tammar wallaby, *Macropus eugenii*. *Journal of Reproduction and Fertility*, 56, 459–463.

Merchant J C and Calaby J H (1981). Reproductive biology of the red-necked wallaby (*Macropus rufogriseus banksianus*) and Bennett's wallaby (*M.r. rufogriseus*) in captivity. *Journal of Zoology*, 194, 203–217.

Merchant J C and Sharman G B (1966). Observations on the attachment of marsupial pouch young to the teats and on the rearing of pouch young by foster-mothers of the same or different species. *Australian Journal of Zoology*, 14, 593–609.

Milne J A, Loudon A S I, Sibbald A M, Curlewis J D and Mc Neilly A S (1990). Effects of melatonin and a dopamine agonist and antagonist on seasonal changes in voluntary food intake, reproductive activity, and plasma concentrations of prolactin and tri-iodothyronine in red deer hinds. *Journal of Endocrinology*, 125, 241–249.

Moore G P M (1978). Embryonic diapause in the marsupial Macropus eugenii. Stimulation of nuclear RNA polymerase activity in the blastocyst during resumption of development. *Journal of Cellular Physiology*, 94, 31–36.

Morgan P J and Williams L M (1989). Central melatonin receptors: Implications for mode of action. *Experientia*, 45, 955–965.

Morgan P J, Barrett P, Howell H E and Helliwell R (1994). Melatonin receptors: localisation, molecular pharmacology and physiological significance. *Neurochemistry International*, 24, 101–146.

Mrosovsky N (1990). *Rheostasis: the physiology of change*. Oxford University Press, Oxford.

Nicholls T J, Goldsmith A R and Dawson A (1988). Photorefractoriness in birds and comparison with mammals. *Physiological Reviews*, 68, 133–176.

Paterson A, Chong N W, Brinklow B R, Loudon A S I and Sugden D (1992). Characterisation of 2-[125] iodomelatonin binding sites in the brain of a marsupial, Bennetts's wallaby (*Macropus rufogriseus rufogriseus*). *Comparative Physiology and Biochemistry*, 102A, 55–58.

Pearce-Kelly A S, Loudon A S I and Curlewis J D (1992). Seasonal and lactational effects on the prolactin response to a dopamine antagonist and TRH in the Bennett's wallaby (*Macropus rufogriseus rufogriseus*). *General and Comparative Endocrinology*, 86, 323–331.

Pilton P E and Sharman G B (1962). Reproduction in the marsupial *Trichosurus vulpecula*. *Journal of Endocrinology*, 25, 119–136.

Reiter R J (1980). The pineal and its hormones in the control of reproduction in mammals. *Endocrine Reviews*, 1, 109–131.

Renfree M B, Lincoln D W, Almeida O F X and Short R V (1981). Abolition of seasonal embryonic diapause in a wallaby by pineal denervation. *Nature*, 293, 138–139.

Rivkees S A and Reppert S M (1992). Appearance of melatonin receptors during embryonic life in Siberian hamsters (*Phodoporus sungorus*). *Brain Research*, 568, 345–349.

Robinson J E, Wayne N L and Karsch F J (1985). Refractoriness to inhibitory day lengths initiates the breeding season of the Suffolk ewe. *Biology of Reproduction*, 32, 1024–1030.

Russell E M (1982). Patterns of parental care and parental investment in marsupials. *Biological Reviews*, 57, 423–486.

Sadleir R M F S and Tyndale-Biscoe C H (1977). Photoperiod and the termination of embryonic diapause in the marsupial *Macropus eugenii*. *Biology of Reproduction*, 16, 605–608.

Schwandel-Fukuda M, Jorgenson K L, Bergen H T, Weesner G D and Pfaff D W (1992). Biology of normal luteinizing hormone-releasing hormone neurons during and after their migration from the olfactory placode. *Endocrine Reviews*, 13, 623–633.

Scott M P (1986). The timing and synchrony of seasonal breeding in the marsupial, *Antechinus stuartii*: interaction of environmental; and social cues. *Journal of Mammalogy*, 67, 551–560.

Sernia C and Tyndale-Biscoe C H (1979). Prolactin receptors in the mammary gland, corpus luteum and other tissues of the tammar wallaby, *Macropus eugenii*. *Journal of Endocrinology*, 83, 79–89.

Sharman G B (1962). The initiation and maintenance of lactation in the marsupial, *Trichosurus vulpecula*. *Journal of Endocrinology*, 25, 375–385.

Sharman G B and Calaby J H (1964). Reproductive behaviour in the red kangaroo, *Megaleia rufa*, in captivity. *CSIRO Wildlife Research*, 9, 58–85.

Sharman G B, Calaby J H and Poole W E (1966). Patterns of reproduction in female diprotodont marsupials. *Symposium of the Zoological Society of London*, 15, 205–232.

Stewart F (1984). Mammogenesis and changing prolactin receptorconcentrations in the mammary glands of the tammar wallaby (*Macropus eugenii*). *Journal of Reproduction and Fertility*, 71, 141–148.

Stewart F and Tyndale-Biscoe C H (1982). Prolactin and luteinizing hormone receptors in marsupial corpora lutea: relationship to control of luteal function. *Journal of Endocrinology*, 92, 63–72.

Tyndale-Biscoe C H (1980). Photoperiod and the control of seasonal reproduction in marsupials. *In Proceedings of the VI International Congress of Endocrinology*. Eds I A Cumming, J W Funder and F A O Mendelsohn. Australian Academy of Science, Canberra. pp 277–282.

Tyndale-Biscoe C H (1986). Embryonic diapause in a marsupial: roles of the corpus luteum and pituitary in its control. *In Comparative Endocrinology: Developments And Directions*. Alan R Liss Inc., New York. pp 137–155.

Tyndale-Biscoe C H (1989). The adaptiveness of reproductive processes. In *Kangaroos, Wallabies and Rat-kangaroos*. Eds G Grigg, P Jarman and I Hume. Surrey Beatty & Sons Pty Ltd, Sydney. pp 277–285.

Tyndale-Biscoe C H and Hawkins J (1977). The corpora lutea of marsupials, aspects of function and control. In *Reproduction and Evolution*. Eds J H Calaby and C H Tyndale-Biscoe. Australian Academy of Science, Canberra. pp 245–252.

Tyndale-Biscoe C H and Hinds L A (1984). Seasonal patterns of circulating progesterone and prolactin and response to bromocriptine in the female tammar, *Macropus eugenii*. *General and Comparative Endocrinology*, 53, 58–68.

Tyndale-Biscoe C H and Hinds L A (1989). The hormonal milieu during development in marsupials. In *Development of preimplantation embryos and their environment*. Eds K Yoshinaga and T Mori. Alan R Liss Inc., New York. pp 237–246

Tyndale-Biscoe C H and Hinds L A (1990). Control of seasonal reproduction in the tammar and Bennett's wallabies. In *Progress in Comparative Endocrinology*. Eds A Epple, C G Scanes and M H Stetson. Wiley-Liss Inc., New York. pp 659–667.

Tyndale-Biscoe C H and Hinds L A (1992). Components of the melatonin message in the response to photoperiod of the tammar wallaby,

Macropus eugenii. *Journal of Pineal Research*, 12, 155–166.

Tyndale-Biscoe C H and Janssens P A (1988). *The Developing Marsupial. Models for Biomedical Research*. Springer-Verlag, Berlin.

Tyndale-Biscoe C H and Renfree M B (1987). *Reproductive Physiology of Marsupials*. Cambridge University Press, Cambridge.

Tyndale-Biscoe C H, Hinds L A, Horn C A and Jenkin G (1983). Hormonal changes at oestrus, parturition and post-partum oestrus in the tammar wallaby (*Macropus eugenii*). *Journal of Endocrinology*, 96, 155–161.

Tyndale-Biscoe C H, Hinds L A and McConnell S J (1986). Seasonal breeding in a marsupial: opportunities of a new species for an old problem. *Recent Progress in Hormone Research*, 42, 471–512.

Vakkuri O, Lamsa E, Rahkamaa E, Ruotsalinien H and Leppaluoto J (1984). Iodinated melatonin: preparation and characterisation of molecular structure by mass of ^{1}H NMR spectroscopy. *Analytical Biochemistry*, 142, 284–289.

Walker M T and Gemmell R T (1983). Plasma concentrations of progesterone, oestradiol-17β and 13, 14-dihydro-15-oxo-prostaglandin F2α in the pregnant wallaby (*Macropus rufogriseus rufogriseus*). *Journal of Endocrinology*, 97, 369–377.

Weaver D R and Reppert S M (1990). Melatonin receptors are present in the ferret *pars tuberalis* and *pars distalis*, but not in brain. *Endocrinology*, 127, 2607–2609.

Weaver D R, Namboodiri M A A and Reppert S M (1988). Iodinated melatonin mimics melatonin action and reveals discrete binding sites in fetal brain. *FEBS Letters*, 228, 123–127.

Williams L M, Martonoli M G, Tichener L T and Pelletier G (1991). The ontogeny of central melatonin binding sites in the rat. *Endocrinology*, 128, 2083–2090.

Woodfill C J I, Wayne N L, Moenter S M and Karsch F J (1994). Photoperiodic synchronisation of a cirannual reproductive rhythm in sheep: identification of season-specific time cues. *Biology of Reproduction*, 50, 965–976

CHAPTER 4

DEVELOPMENT OF THE POUCH YOUNG FROM BIRTH TO POUCH VACATION

Peter A Janssens, Anthony J Hulbert and Russell V Baudinette

INTRODUCTION

Neonatal marsupial young are extremely small, with body weights 4 to 5 orders of magnitude less than the mother. Thus a young grey kangaroo weighs less than one gram where its mother will weigh tens of kilograms; a young Tasmanian devil weighs about 30 mg compared with the maternal weight of about 10 kg. In contrast the body weights of most eutherian young are only 1 to 2 orders of magnitude less than the mother. Even the smallest eutherian neonate in relative terms, the giant panda, weighs about 100 g at birth compared with a maternal weight of about 100 kg, only 3 orders of magnitude difference (Ramsay & Dunbrack 1986).

Marsupial neonates also appear undeveloped when compared with eutherians and many organ systems are relatively poorly developed at birth; this has led to pouch young being described as 'exteriorised fetuses' or 'embryonic'. Such descriptions may, in many ways, be counterproductive because, although they are apt descriptions of the state of many neonatal marsupial organ systems, the neonate still must breathe air, suck milk, crawl in a directional manner and respond to sensory inputs, all of which are postnatal functions in eutherians. It may be preferable, therefore, to regard pouch life as a different developmental system from that used by eutherians and not to place value judgements on the relative merits of each. At the end of pouch life the young moves from a stable to an unstable environment, especially in terms of temperature, and it must be independent to the extent that it can cope physiologically with this instability.

One other characteristic of the marsupial system is that during most of its development the pouch young receives only a single input of nutrients, this being the milk from the mother. The milk does not have a constant composition throughout pouch life. For example, milk fat content rises several-fold during the course of lactation (Green & Merchant 1988). Knowing both milk composition and the amount of milk passing from mother to young (Cork & Dove 1989), one can determine the intake of various nutrients and messengers during this early part of development. And when changes occur during marsupial development, it is possible to have an idea as to whether the change is being directed by the mother via the milk or is an autonomous activity of the pouch young.

The aim of this chapter is to discuss recent findings that relate to the development of physiological independence in marsupial young. We concentrate on studies on the tammar (*Macropus eugenii*) because of the large amount of information that is available for this species, but we draw comparisons with other species where possible.

GROWTH AND DEVELOPMENT

Body growth and growth of tissues

Although growth curves of marsupial young vary among species they are sigmoidal in shape and may be broadly described by the Gompertz equation (Cockburn & Johnson 1988), although other analyses have been used (Hall 1990; Gemmell & Hendrikz 1993). During early pouch life, the absolute rate of growth is slow; there is an increase in the absolute rate of growth towards the end of pouch life after which growth progressively slows until adult weight is reached. However, growth is indeterminate in several marsupial species and Cockburn and Johnson (1988) have argued that the major deviation from the Gompertz model in macropodids is attributable to their indeterminate growth pattern.

Tammars are born weighing approximately 0.4 g and grow to a final adult size of approximately 5 to 8 kg. When the body weight of the tammar is plotted against age, a sigmoidal relationship is clear (Figure 4.1a). A re-plot of these data showing the rate of growth for each successive 14-day interval against age (Figure 4.1b) gives a clearer picture of the timing of the growth spurt. The rate of increase slows between 74 and about 150 days of pouch life; this is followed by an increase to a maximum at about 220 days, after which the rate of gain again decreases. The growth spurt seen between 150 days and 220 days of age is present in both male and female young. A similar pattern of change is also seen in the limb bones but not in the skull (Figure 4.2a, Figure 4.2b), which indicates that there is a real change in growth rate and not just in body mass.

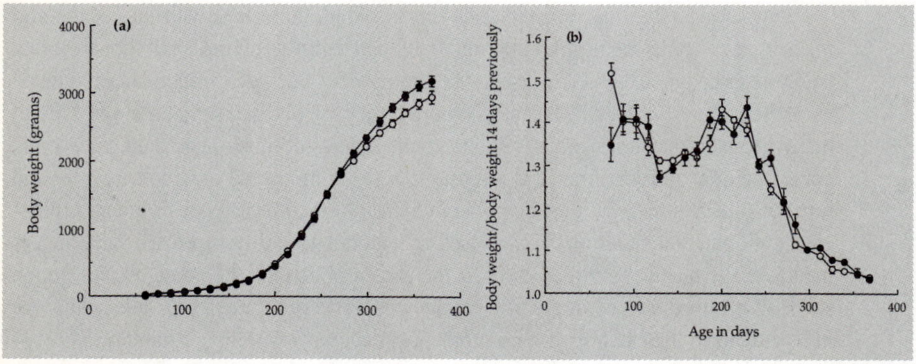

Figure 4.1 Changes in body weight of tammar pouch young with age in days: (a) cumulative and b) incremental.
Notes: In (b) the values are the body weight at the time plotted divided by the body weight 14 days previously. Values are mean ± sem for 14 male (•) and 13 female (o) young. Where no sem is seen, the value is smaller than the symbol depicting the mean.
Source: Data kindly supplied by W E Poole.

We have also measured growth rates of a number of internal organs of tammar pouch young. A useful way of examining tissue growth is to plot the change in tissue weight allometrically against body weight, where the allometric slopes indicate the rate of tissue growth relative to total body size. When such a plot is made of the tissues from the tammar (Figure 4.3), liver and heart remain a constant proportion of total body weight throughout development (allometric slopes of 1.00), whereas brain and kidney grow relatively faster than the total body at first (allometric slopes >1), and then relatively more slowly (allometric slopes <1). In both these latter tissues, the change is abrupt in developmental terms and appears to occur at a similar age, about 120 to 140 days in each. Although the allometric plot is a convenient way of presenting such data, it underemphasises the effects of a change in growth rate. For example, the brain weighs 7 g in a 100 g wallaby (7% of body weight), at about the time of the change in growth rate, but only 20 g in a 5000 g wallaby (0.4% of body weight).

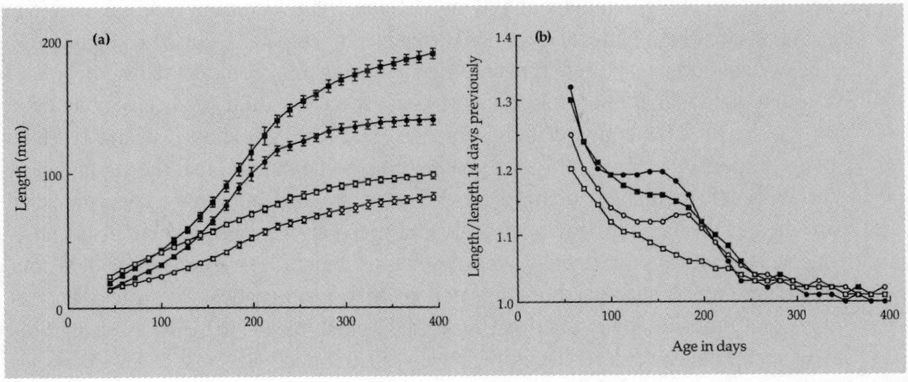

Figure 4.2 Changes in the dimension of several skeletal structures (■ leg; • pes; □ head; o arm) in tammar pouch young plotted against age in days: (a) cumulative and (b) incremental.
Note: In (b) the values are calculated as in Figure 4.1. Values are mean ± sem for 10 male young. sems were consistently small in b) and are not shown.
Source: Data kindly supplied by W E Poole.

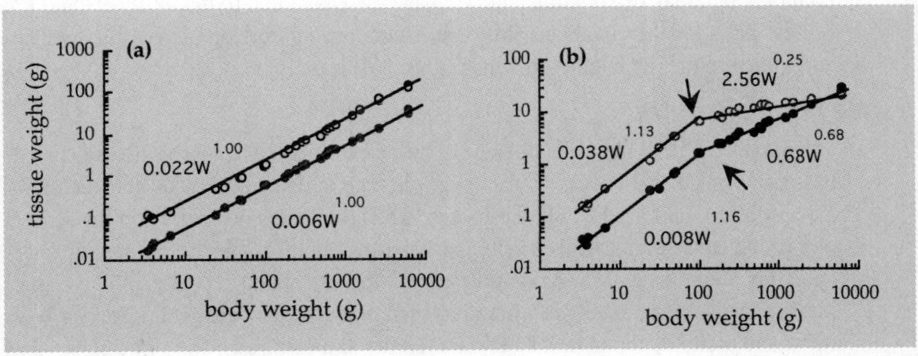

Figure 4.3. The allometric growth of (a) liver o and heart •, and (b) brain o and kidney • during the development of the tammar.
Note: The allometric equations are written next to appropriate line. The arrows indicate the break between the two allometric relationships.
Source: Hulbert *et al.* (1991).

These analyses, and those of others, demonstrate that growth is a complex phenomenon that must be highly regulated. One aim of this study has been to better understand this regulation.

DEVELOPMENT OF ADULT METABOLIC RATE

Whole animal metabolic rate

A clear picture of control of metabolic rate in the mammalian fetus is not currently available, largely because of the great difficulty of separating the metabolism of the fetus from that of its mother. No such difficulties occur in marsupials. In macropodid marsupials, in which pouch life lasts until the young are physiologically independent from their mothers, the development of homeothermy, and of an adult standard metabolic rate (SMR), is delayed until late in pouch life. In colonies in Australia, young tammars are normally born in late January or early February, and make their first exit from the pouch at about 200 days of age in late August/early September when they weigh about 400 to 500 g. Until about 180 days of pouch life, tammar young are unable to maintain a constant body temperature at an ambient temperature of 20°C (Figure 4.4). The weight-specific oxygen consumption of young remains essentially unchanged through pouch life in the tammar and does not reach the value predicted for an adult of that size until the body weight has reached 100 to 200 g at about 130 and 165 days of age respectively (Figure 4.5). It should also be borne in mind that the young whose metabolic rates are plotted in Figure 4.5 were growing and, since they remained attached to the teat during the period of measurement, were also feeding. As a consequence, the measured metabolic rate would be attributable to a combination of factors, including growth and digestion, and not solely to maintenance. The true SMR was, therefore, probably lower than that measured, which means that pouch young SMR may not approach adult levels until relatively late in pouch life, close to the time when isolated young can maintain their body temperature.

We believe that this delayed development of an adult metabolic rate may be critical to survival of tammar young. If it occurs too early, the young may overheat; if it occurs too late, the young will not be able to maintain its body temperature during periods spent outside the pouch. It is therefore of interest to know how the SMR of the young is controlled, but the whole body SMR is a sum of the metabolism of individual tissues, each of which may be very different. We have therefore been trying to understand the cellular basis for the change in SMR that occurs during development in the tammar, and to determine the manner in which it is regulated.

Tissue metabolic rates

The measurement of the metabolic rate of individual tissues is fraught with difficulty. It has been argued that, since cellular metabolic rate is dependent upon the adenosine triphosphate (ATP) production of each cell, which in turn is dependent on the activity of enzymes that are located on mitochondrial membranes, the surface area of mitochondrial membrane is a good indicator of the metabolic capacity of the tissue. Measurement of the total mitochondrial membrane surface area (MMSA) has been extremely useful in comparisons between reptiles and mammals, and of mammals and reptiles of a variety of sizes (Else & Hulbert 1981; 1985; Hulbert & Else 1981), and has permitted an explanation of the differences in metabolic rates between animals and a distinction between tissues within a single animal. The MMSA reflects the metabolic rate of the cells in the tissues of the living animal, undisturbed by the preparation of perfused

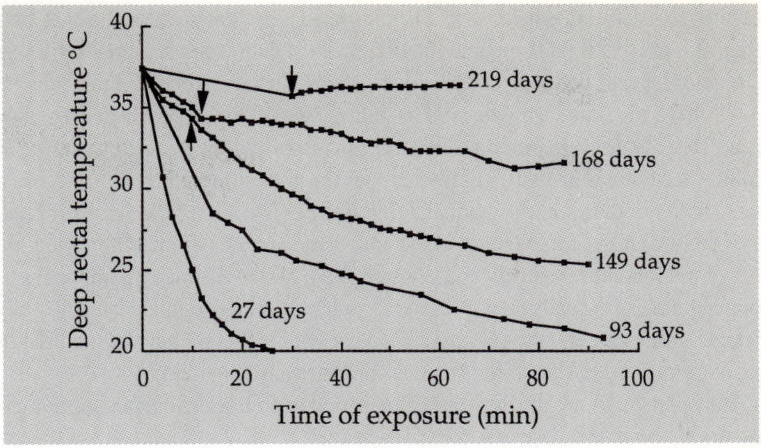

Figure 4.4 Representative cooling curves of variously aged tammar pouch young at an ambient temperature of 20±1.5°C.
Note: Arrows indicate the first appearance of shivering; the age of each young is given on the figure. Rectal temperatures were recorded from copper/constantan thermocouples, calibrated to 0.1°C, inserted in the rectum to a depth of 0.5 cm to 4.5 cm depending upon the size of the young. Limbs of older young were taped to limit struggling and prevent removal of the thermocouples. Ambient temperature was recorded from a thermocouple fixed 25 cm above the animal.
Source: Janssens & Rogers (1989).

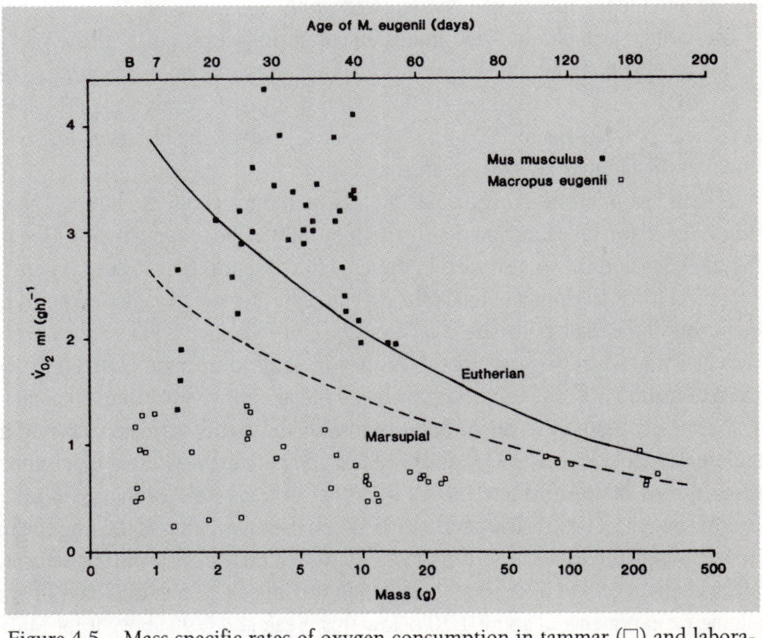

Figure 4.5 Mass specific rates of oxygen consumption in tammar (□) and laboratory mice (■) as a function of age and body mass.
Key:[The solid line] level of VO_2 predicted for eutherians (Kleiber 1961)[the dotted line] level of VO_2 predicted for marsupials (Hayssen & Lacy 1985)
Source: Baudinette *et al.* (1988).

organs, tissue slices, isolated cells or homogenates and, as such, is likely to be a better indicator of what is happening in the tissues of the living animal than is the oxygen consumption of cells or slices, or the activity of enzymes measured in a test tube.

We have measured growth rates of brain, heart, kidney and liver (see above) and used stereological (quantitative electron microscopy) techniques to measure the total MMSA for these four tissues in the tammar throughout development; this has allowed us to better understand the gradual development of an adult SMR (Hulbert *et al.* 1991).

Total MMSA is a combination of the area of membrane per mitochondrion, the number of mitochondria per cell, and the size of the tissue; the major component in developing animals is usually change in tissue weight. Change in tissue weight is, of course, related to increase in body weight and to correct for this we have plotted the development of total MMSA in all four tissues allometrically (Figure 4.6).

The brain grows rapidly during the first period of pouch life until a transition point that occurs at about 140 days at a body weight of about 100 g (Figure 4.3,) (Renfree *et al.* 1982). If the total MMSA plot for brain (Figure 4.6a) is analysed in the same way as the growth of the brain, then it too can be divided into an early period of rapid increase followed by a later period of much slower increase. This early period of rapid development of the brain, in both size and metabolic potential, occurs at a time when brain structure is developing rapidly (Reynolds & Saunders 1988), and is soon followed by the opening of the eyes and the development of the ability to stand. A mature brain and nervous system is clearly an essential pre-requisite for such functions as vision and balance and it is tempting to speculate that the development of maturity of the nervous system must also precede the maturation of physiological functions such as the mounting of a thermogenic response to cold and the production of a concentrated urine.

The kidney also shows some indication of an inflection point in the plot of total MMSA against body weight (Figure 4.6b), although it is by no means as clear as for the brain. In the case of the kidney, the inflection point is at a body weight of 300 to 500 g, equivalent to an age of 180 to 200 days, an age where the kidney is first fully functional in the tammar (Wilkes & Janssens 1988).

In both heart and liver, organs that are essential throughout the lifespan, there is little evidence for inflection points in the total MMSA/body weight plot (Figure 4.6c, Figure 4.6d) and this is also seen in the growth curve for these tissues (Figure 4.3a).

Thus, during development total MMSAs gradually increase to values expected for adult mammals of the same body size, but the shape of the allometric plots differs between tissues in a way which suggests that there may be some correlation between the development of maturity of structure and function, a not altogether surprising outcome.

These four tissues are major contributors to the standard metabolic rate of adult mammals (Aschoff *et al.* 1971) and we have compared the allometric development of the summed MMSA of these tissues with that of the whole animal SMR taken from Baudinette et al. (1988). The similarity between these two plots is striking (Figure 4.7). In both cases, the data can be described by two allometric relationships separated by a break at about 300 to 600 g body weight (approximately 175 to 200 days of age), with allometric exponents of about 1.10 below this break and 0.95 above it for MMSA, and 1.02 below and 0.75 above the break for SMR. This is the age at which young tammars become fully homeothermic, become physiologically independent of their mothers, and make their first exit from the pouch. Brody (1974) found a similar pattern for oxygen consumption for rats, humans and cattle, with the break occurring at body weights

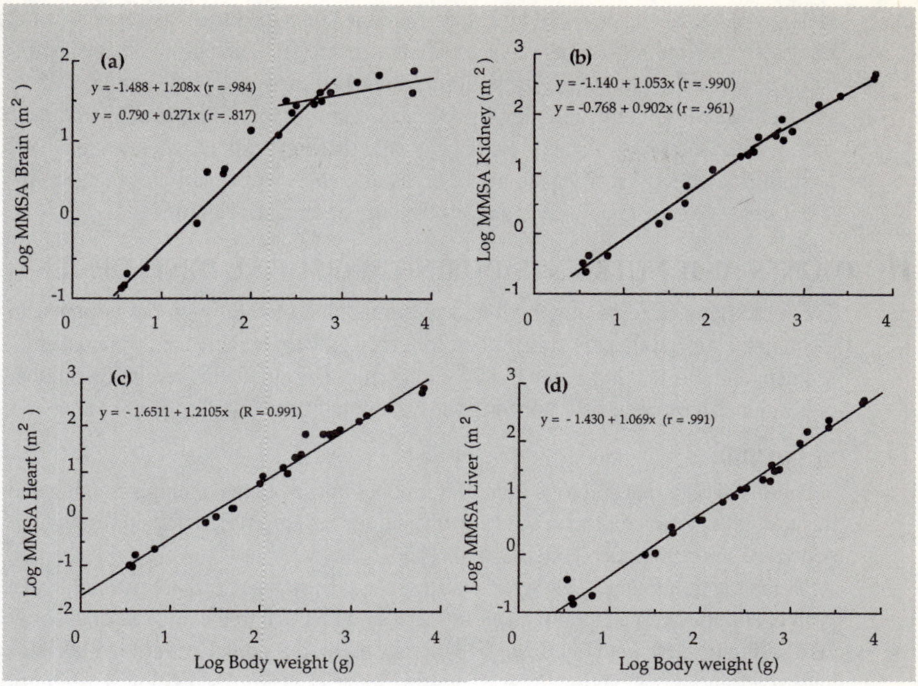

Figure 4.6 The allometric growth of the total mitochondrial membrane surface area (MMSA) of (a) brain (b) kidney (c) heart and (d) liver during development of the tammar. Note: The calculated equations for the two parts of the curve for brain and kidney, and for the single curve in heart and liver, are given on the figures.
Source: Redrawn from Hulbert *et al.* (1991).

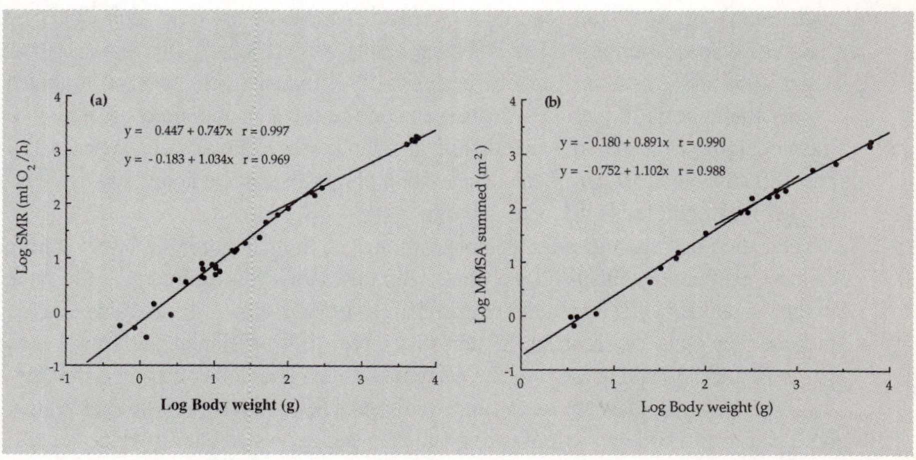

Figure 4.7 The allometric growth of (a) 'standard' metabolic rate and (b) summed mitochondrial membrane surface area (MMSA) of the liver, heart, kidney and brain during development of the tammar. Note: The calculated equations for the two parts of the curve are given on the figures.
Source: Redrawn from Hulbert *et al.* 1991.

of 100 g, 12 to 15 kg and 120 kg respectively. He concluded that the break was not associated with either weaning or lactation in these species. There is no obvious correlation between the time of the break in the allometric curve for whole animal SMR or summed MMSA and any specific physiological event in any of the species that have been studied; however, it should be remembered that the break in whole animal SMR or summed MMSA is a summation of a series of breaks, which vary in timing in different tissues, and it may well be dangerous to try and make too much of it.

HORMONES AND NUTRIENTS DURING MARSUPIAL DEVELOPMENT

These changing patterns of growth and metabolic rate seen through development in the tammar suggest that the systems that influence growth, notably the hormonal environment and nutrient input, must also be changing. The circulating levels of a number of hormones have been determined in the developing tammar.

Thyroid hormones

The levels of both free and total thyroid hormones in the plasma of tammar young and the activity of 5'-deiodinase in the liver and kidney have been determined throughout pouch life in the tammar (Janssens et al. 1990).

Plasma total thyroxine (tT4) levels rose from less than 10 nM early in pouch life to a peak of about 80 nM at about 160 to 180 days and then fell to the adult level of about 20 nM (Figure 4.8). Free thyroxine (fT4) levels followed a similar pattern but at much lower concentrations. Plasma total triiodothyronine (T3) remained below 1 nM until about 140 days, rose to 4 to 5 nM by about 220 days, and then fell to the adult level of about 1 nM (Figure 4.8). Free plasma triiodothyronine (fT3) concentration followed a similar pattern. In all cases the rise in concentration was best described by a logistic function and the fall by an exponential function.

The conversion of T4 to T3 is thought to occur peripherally by 5'-deiodination of T4, primarily in the liver and kidney. In both liver and kidney, 5'-deiodinase activity was undetectable before 80 days and then increased sigmoidally reaching adult values by about 250 days (Figure 4.9). The inflection points of these logistic functions occur at about 205 days. These increases in peripheral 5'-deiodinase activity parallel almost exactly the increase in plasma T3 concentration (Figure 4.8) and are consistent with the model that T4 secretion by the thyroid gland increases to a maximum between 160 and 180 days, and that peripheral deiodination of this T4 starts at about 100 days and reaches maximum levels by about 220 days.

These findings are the most complete profile of thyroid hormone levels during development in a mammal so far obtained and this, in itself, is testimony to the value of the marsupial model for studies of mammalian development. It is intriguing that the increase in plasma T4 occurs at the time of the growth spurt (Figures 4.1 and 4.2 on pages 72 and 73.) and at the time the pelage is forming and homeothermy is developing. Our findings to date are consistent with those of Setchell (1974), who showed that both pelage and body growth of young tammars was depressed after thyroidectomy.

Although Setchell (1974) has shown that the thyroid gland in very young tammars is afollicular and does not specifically sequester iodine, there were still measurable quantities of thyroid hormones in the blood of tammar young less than 50 days of age (Figure 4.8). The concentration of total T4 in maternal plasma remains relatively unchanged through lactation at about 30 nM (Nicholas 1988), somewhat higher than

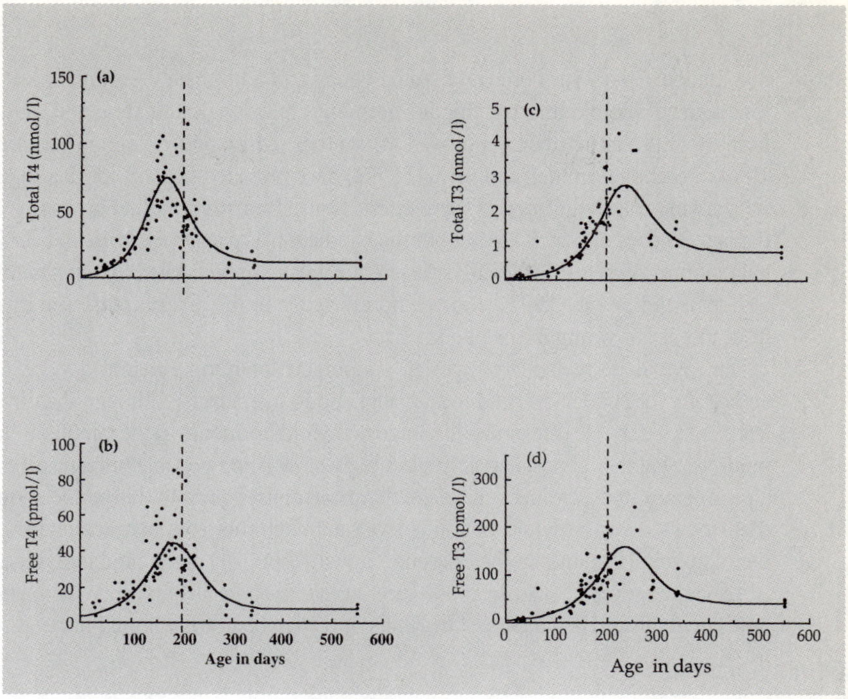

Figure 4.8 Plasma concentrations of (a) total T4 (b) free T4 (c) total T3 and (d) free T3 in the tammar plotted against age in days.
Key: [The solid lines] fitted curves derived as described in the original paper [The dashed line] 200th day of pouch life, demonstrating the separation between the peaks in T3 and T4 concentration
Note:Each point is the mean of duplicate determinations from one animal.
Source: Janssens *et al.* (1990), reproduced with permission of the Journal of Endocrinology Ltd.

Figure 4.9 Changes in 5'-deiodinase activity in (a) liver and (b) kidney in the tammar against age in days.
Key:[The solid lines] fitted curves derived as described in the original paper.
Note: Each point is the mean of duplicate determinations from one animal.
Source: Janssens *et al.* (1990), reproduced with permission of the Journal of Endocrinology Ltd.

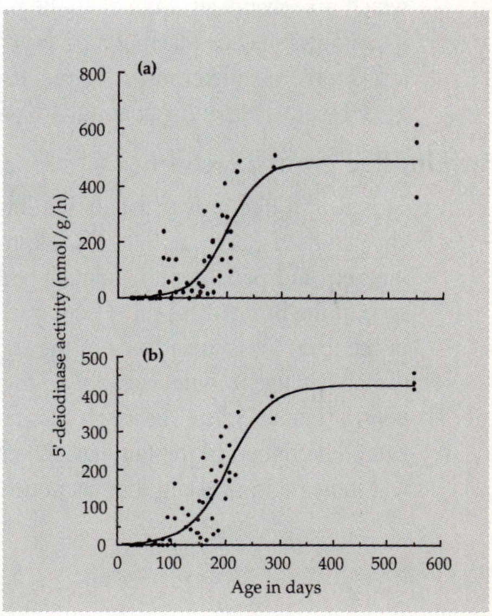

our value of 15 nM for both early pouch young and adult males (Figure 4.8). It therefore seemed feasible that the thyroid hormones in young tammars could be derived from the milk, rather than (or as well as) synthesised by pouch young (Janssens *et al.* 1990). Recently Gemmell and Sernia (1992) have provided support for this suggestion by showing that radiolabelled thyroxine is transferred from mother to young during the first 20 days of pouch life in both the bandicoot, *Isoodon macrourus*, and the brushtail possum, *Trichosurus vulpecula*. Despite the greater growth rate of young bandicoots, Gemmell and Sernia (1992) reported no difference in the rate of transfer of thyroxine from mother to young in the two species.

The thyroid hormones are especially important for normal neural development in eutherians (Legrand 1986) and presumably also in marsupials. Although the physiological effects of the mid-late pouch life surge in thyroid hormones in the tammar is not currently known, they almost certainly affect body growth and possibly brain maturation. In a preliminary study, tammars that were thyroidectomised prior to this period showed no dramatic increase in thyroid hormone levels although they still had measurable quantities of thyroid hormones in their plasma. Growth, especially of the hind foot, was diminished and changes in plasma appearance suggest that there were dramatic changes in lipid metabolism (P A Janssens, L A Hinds & A J Hulbert, unpublished findings).

Other hormones

Growth hormone

We have measured growth hormone (GH) levels in single blood samples taken at 2 to 4 week intervals from 182 days of pouch life through to the adult and in sequential blood samples taken each 15 min for 6 h at intervals through development (E M Bugledich, L A Hinds & P A Janssens, unpublished findings). GH levels averaged 2000 to 5000 pg/ml plasma towards the end of pouch life and gradually fell to about 1000 pg/ml in adults (Figure 4.10a). In general, females had higher levels than males. There was a great deal of variation in plasma GH concentration at all stages of development, which is probably attributable to the pulsatile nature of GH release (Figure 4.10b); single samples may be taken at a peak or trough of the release profile. Unfortunately, GH levels were not determined before 180 days of pouch life and so no link with the growth spurt beginning at about 150 days of age can be made.

Insulin-like growth factors

The insulin-like growth factor binding proteins (IGFBPs) in the tammar have been characterised (Carr *et al.* 1995). Plasma from both male and female adults, lactating mothers and pouch young contain several IGFBPs with a pattern very similar to that seen in the human and the sheep; transcripts for IGFBP-1 and IGFBP-2 were detected in the liver, the same as seen in the sheep and the rat. The level of circulating IGFBPs did not change through lactation in the mother but increased through pouch life in the young (Figure 4.11). Although plasma levels of insulin-like growth factors have not been determined in the tammar, the changes in circulating IGFBPs suggests that they will increase in concentration as pouch life progresses.

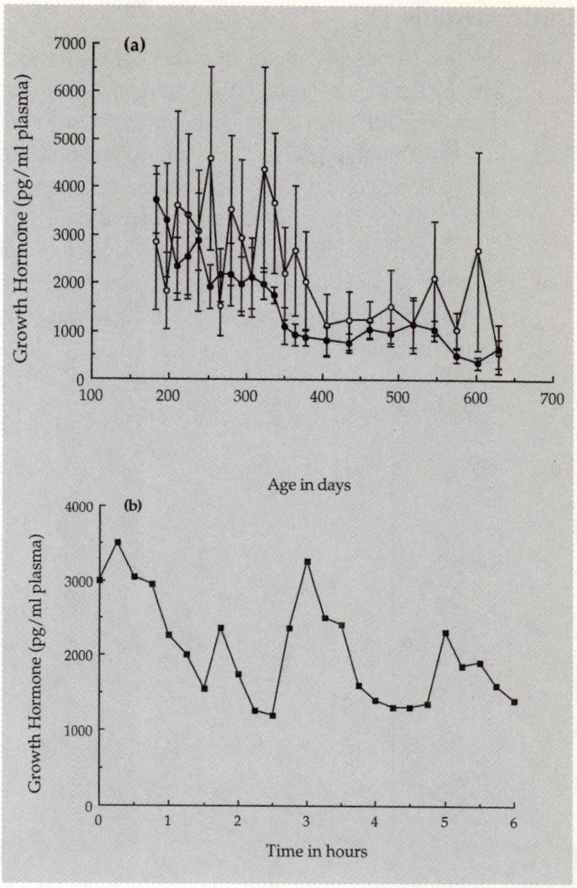

Figure 4.10 Plasma growth hormone levels in developing tammars: (a) in males • (n = 8–9) and females o (n = 5–7) sampled at 2 to 4 weekly intervals from 180 to 630 days of age (Mean ± sem), and (b) in a single young aged 190 days from which sequential samples were taken at 15 min intervals, showing pulsatile release of the hormone.
Note: Radioimmunoassays were performed with a heterologous antiserum (NIADDK-anti-ratGH-S-5) validated for the tammar.
Source: E M Bugledich, L A Hinds & P A Janssens (unpublished findings).

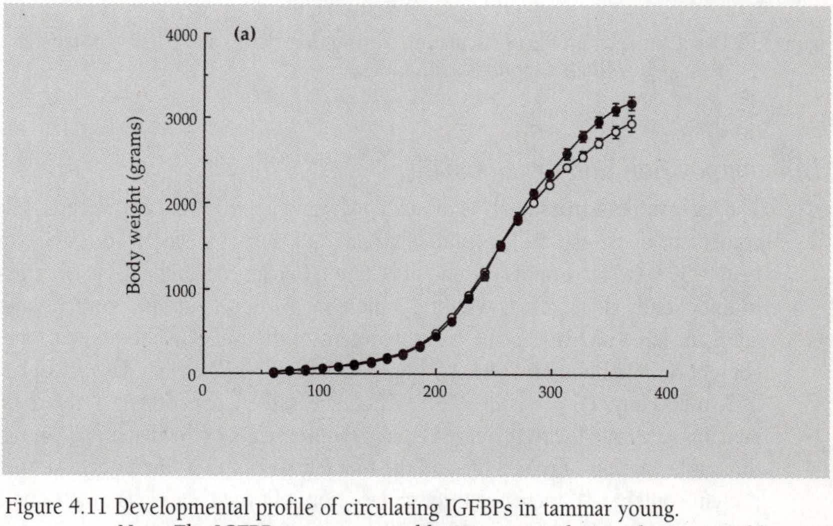

Figure 4.11 Developmental profile of circulating IGFBPs in tammar young.
Note: The IGFBPs were separated by Western analysis and quantitated by laser densitometry.
Source: Carr *et al*. (1995).

Corticosteroids

The first studies of patterns of corticosteroid secretion during pouch life in the tammar were by Catling & Vinson (1976) who reported a plasma corticosteroid concentration of about 3 ng/100 ml in pouch young from 40 to 140 days of age and a concentration of 6 to 10µg/100 ml in adults. These values were obtained with competitive protein binding assays. More recently, L A Hinds, using a radioimmunoassay, reported levels of cortisol of about 100 ng/ml in adult tammars, a value similar to that of Catling and Vinson (1976) for total corticosteroids; in pouch young levels rose from about 20 ng/ml at 50 days reaching adult values by about 160 days (Figure 4.12, L A Hinds, unpublished findings). On the basis of these findings it seems unlikely that cortisol is a major factor in the growth spurt occurring in the final third of pouch life in the tammar.

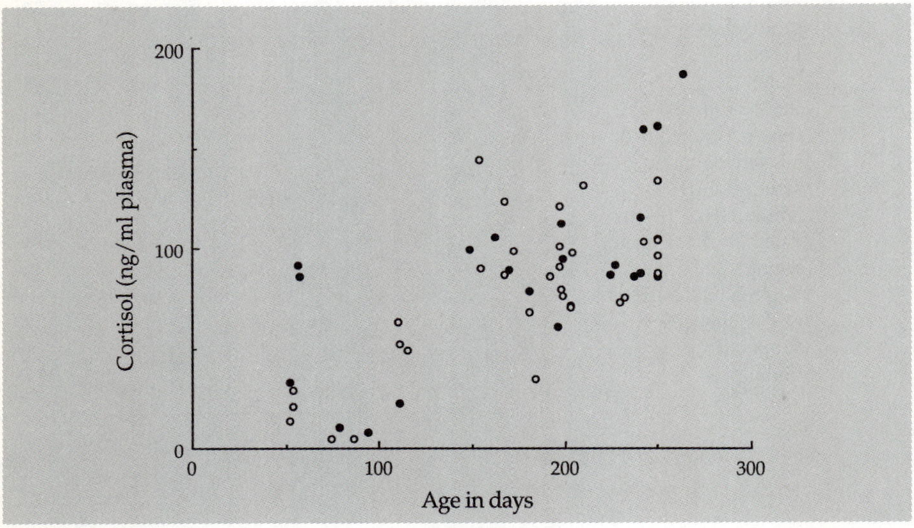

Figure 4.12 Developmental profile of circulating cortisol in male (•) and female (o) tammar young. Source: L A Hinds (unpublished findings).

Milk composition and development

The early milk of marsupials is dilute, containing about 10% (w/w) solids. The solids content of marsupial milk continually increases during lactation to average approximately 30% (w/w) towards the end of lactation (Green & Merchant 1988). This change in milk solids is associated with a dramatic increase in total lipid content and a concomitant large increase in the energy content of the milk. In the tammar, the energy content of milk increases from 2.5 kJ/g at 7 days to 5 kJ/g at 180 days and 11.5 kJ/g at 250 to 280 days (Green *et al.* 1988). The increased energy content of the milk is presumably associated with increased energy requirements of the marsupial young during the last few weeks of pouch life and the first few weeks after first pouch vacation. Milk protein content also increases through lactation, but the increase is much smaller than for lipid (Green & Merchant 1988; Nicholas 1988). Although there are only limited data, it appears that the efficiency of conversion of both milk energy and protein to body mass and protein is similar in marsupials and eutherians and it is unlikely that a low and inefficient conversion efficiency is the reason for the slow growth of marsupials (Green & Merchant 1988). Indeed, Green *et al.* (1988) have argued that the growth rate of marsupial young is limited by the supply of milk from the mother, an argument supported by the finding that if macropodid pouch young of different ages are exchanged, the growth of the smaller young is accelerated, while that of the larger young is reduced (Findlay 1982; Green & Merchant 1988).

The composition of milk lipids also changes during lactation. The vast majority of lipids in the milk are triglycerides, but early marsupial milk also contains significant amounts of phospholipid (Green & Merchant 1988). Apart from the very first milk sample, the fatty acid composition of tammar milk was relatively constant throughout lactation (Green et al. 1983). About 40% to 50% were saturated fatty acids (mainly palmitic acid 16:0 and stearic acid 18:0), while 50% to 60% were unsaturated. Of the unsaturated fatty acids, the majority were mono-unsaturates (predominantly oleic acid 18:1 and palmitoleic acid 16:1), with the remainder being polyunsaturates (primarily linoleic acid 18:2 and linolenic acid 18:3).

Fatty acids in milk are not only a source of energy but are also an important source of membrane components for the growth and development of tissues. This is especially true for the polyunsaturates as they are an important component of phospholipids (and thus membranes), and it is believed no vertebrate can synthesise them. They must thus be obtained in the diet. There are two types of such dietary essential fatty acids; the omega-6 polyunsaturates and the omega-3 polyunsaturates. In tammar milk, the omega-6 polyunsaturated linoleic acid (18:2) is present at 4 to 8% of total milk triglyceride fatty acids throughout pouch life, whilst the omega-3 polyunsaturated linolenic acid (18:3) is present as 1 to 4% of total fatty acids during pouch life (Green et al. 1983), levels normally considered adequate to provide essential dietary requirements in eutherians (Holman 1992).

One of the omega-3 polyunsaturates, docosahexanoic acid (22:6), is particularly interesting. It is the most polyunsaturated membrane component and is found in high concentrations in the brain and especially the visual system. It is known to be necessary for the normal development of the brain and especially for visual acuity in rats and humans (Salem & Niebylski 1992). It is not reported present in tammar milk at any time during pouch life (Green et al. 1983), although its precursor, linolenic acid (18:3) is present throughout pouch life. It is found in measurable amounts in tammar pouch young liver, kidney and heart, with a peak during mid- to late-pouch life (Figure 4.13, A J Hulbert, D A Pan & P A Janssens, unpublished findings). The liver has a higher proportion of 22:6 in its phospholipids but shows essentially the same profile as the other tissues. In the brain, 22:6 has a unique developmental profile starting at a level less than the liver but gradually increasing throughout pouch life until, in the adult, it accounts for about 20% of phospholipid fatty acids compared to 1 to 4% in other tissues. The functional significance of these patterns, especially the peak during mid–late pouch life has yet to be fully ascertained but might have important implications for the development of physiological independence.

RECAPITULATION OF POUCH LIFE IN THE TAMMAR

In the first 100 or so days of pouch life, the tammar young grows slowly, its weight-specific SMR is low for its body weight, although normal for its adult body weight, and growth hormone, cortisol and thyroid hormones are present; however, the latter are at

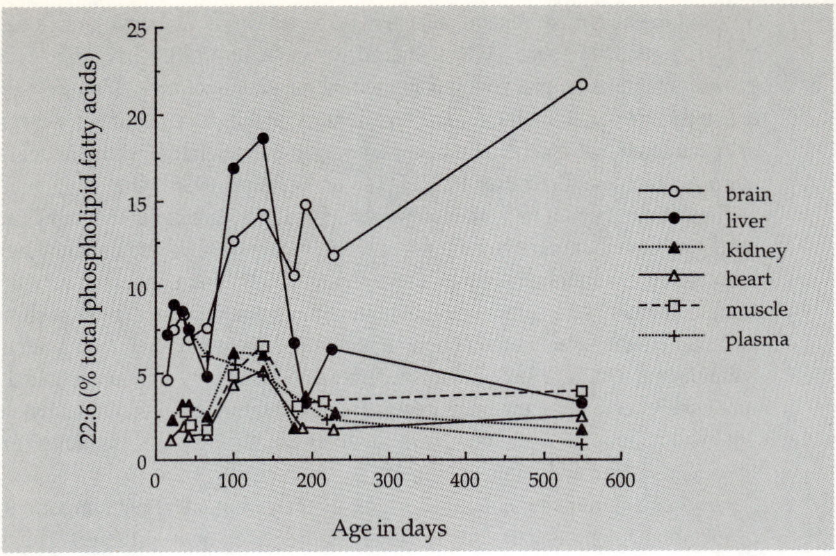

Figure 4.13 Docosahexanoic acid (22:6) content of tissue phospholipids in several tissues through development in the tammar.
Source: A J Hulbert, D A Pan and P A Janssens (unpublished findings).

low concentrations and are probably derived from the milk in the first few weeks of pouch life. The brain and kidney are growing more rapidly than the rest of the body and the brain reaches its maximum size relative to body weight at about 120 days. At about 140 days the eyes open and by 160 days the young can stand.

In the second period of pouch life, from 120 days until the first pouch vacation at about 200 days, the kidney function matures and the young experiences a period of rapid growth that extends from about 150 to 220 days of age. As the young grows, the weight-specific SMR approaches that of an adult marsupial of the same body weight and by 180 days the isolated young is able to maintain its body temperature. During this period of rapid growth, there is a large increase of both T3 and T4 in the blood and deiodinase activity increases in peripheral tissues. The developmental changes during the first 200 days of pouch life lead to the production of an individual that can maintain its body temperature, concentrate its urine, hop, and can thus survive for short periods outside the pouch and get back into the pouch when its mother permits it to. One as yet unresolved question is why the young gets out of the pouch when it does.

POUCH VACATION

There is consensus that the tammar young first leaves the pouch at about 200 days and is permanently excluded from the pouch some weeks later. During the first few weeks after the first pouch exit, the young spends progressively more time out of the pouch until, at permanent pouch exit, it is prevented from re-entering the pouch by the mother. The underlying reason why the young leave the pouch when they do is not at all clear. It has been argued that, with the development of full homeothermy, young macropodids may be producing heat at a rate that makes them physiologically incompatible with their mothers; there is substantial supporting evidence for this in seasonally breeding macropodids (reviewed in Janssens & Rogers 1989).

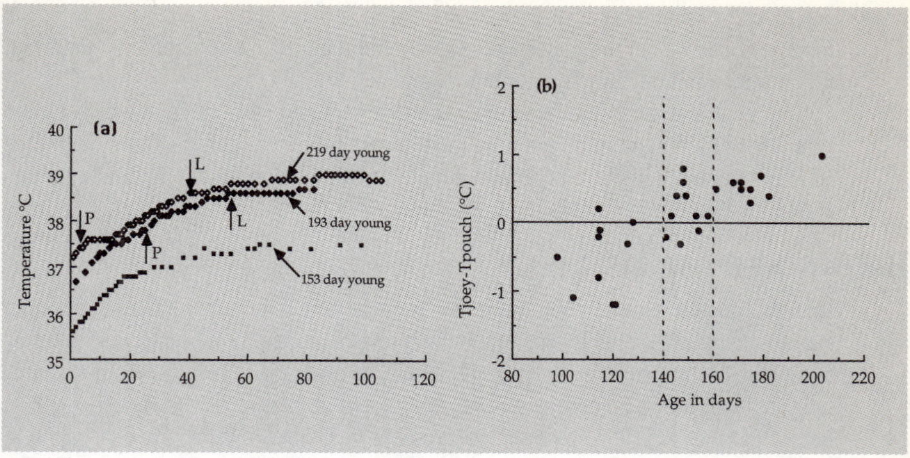

Figure 4.14 (a) Representative heating curves of variously aged tammar pouch young at an ambient temperature of 37.5±1.0°C.
Key: [P] first appearance of panting [L] first appearance of licking the extremities. (b) Pouch young deep body temperature/pouch temperature differential for young between 100 and 200 days of age.
Note: Rectal temperatures were recorded from copper/constantan thermocouples, calibrated to 0.1°C, inserted in the rectum to a depth of 0.5 cm to 4.5 cm depending upon the size of the young. Limbs of older young were taped to limit struggling and prevent removal of the thermocouples. Ambient temperature was recorded from a thermocouple fixed 25 cm above the animal.
Source: (a) Janssens & Rogers (1989). (b) Rogers (1979).

First, pouch young of the parma wallaby (*Macropus parma*, Wallis & Maynes 1973), the quokka (*Setonix brachyurus*, Loh & Shield 1977) and the tammar (Figure 4.14a) (Janssens & Rogers 1989) all developed higher body temperatures than their mothers towards the end of pouch life. Second, late pouch young quokkas (Loh & Shield 1977) and tammars (Figure 4.14b) (Janssens & Rogers 1989) exhibited behavioural signs of over-heating, such as panting and salivation, if kept at pouch temperature in a humid atmosphere; this finding led Loh & Shield (1977) to suggest that 'mutual thermal intolerance between mother and joey may lead to the joey's first ventures from the pouch'. This is consistent with our observations that tammar young are first seen out of the pouch at an earlier stage of development if they are exposed to high ambient temperatures. In tammars where birth has been induced in July, the young are growing large during summer when ambient temperatures are high. At this time of year, young make their first exit from the pouch at about 170 to 180 days when they weigh about 300 g (Dendrinos 1988), very close to the time when homeothermy develops (Janssens & Rogers 1989). And mothers carrying large young, which are not in the pouch at the time, partially evert the pouch so that the soft skin lining the pouch is exposed. This may be a behaviour to reduce body temperature.

Several recent observations are counter to this hypothesis. Gemmell and co-workers have reported that the body temperature of pouch young bandicoots, Isoodon macrourus (Gemmell & Johnston 1985), potoroos, Potorous tridactylus (Gemmell *et al.* 1987) and, most recently, brushtail possums, Trichosurus vulpecula (Gemmell & Cepon 1993) is lower than that of the mother. They have argued that the young leaves the pouch because of a need for solid food and for curiosity. However, most observers agree, in the macropodids at least, that the young is ejected from the pouch by the mother or falls out of the pouch by accident, and the first reaction of the young when it first finds itself outside the pouch is to get back in again.

The position is clearly different among species and the only cases in which a clear reason for pouch exit can be given are the bettongs, where the current young is excluded from the pouch by the mother at the time of birth of the next one (*Bettongia gaimardi*, Rose 1986; Bettongia penicillata, Smith 1989).

BROWN ADIPOSE TISSUE

It might be expected that when the young leave the pouch they have a need to generate heat to maintain their body temperature and several groups have been trying to determine whether brown adipose tissue is present in marsupials. Early work in this area indicated that brown adipose tissue was not present in marsupials (Rowlatt *et al.* 1971) but Loudon et al. (1985) reported its presence in Bennett's wallaby pouch young at about 135 days of age. Their identification was based upon a stimulation of metabolic rate following injection of noradrenaline, the presence of two mitochondrial purine nucleotide binding sites in adipose tissue from young with a body weight of 350 to 500 g, and the finding that electron microscopy of intrascapular fat from such young revealed cells with multi-locular fat droplets and large numbers of mitochondria.

More recently, Hayward & Lisson (1992) subjected 38 species of marsupial at several stages of development, particularly late pouch young and young adults, to light, fluorescence and electronmicroscopical analysis and and were unable to find evidence for the presence of brown adipose tissue. They argued that brown adipose tissue was unique to eutherian mammals and went so far as to suggest a new name, Thermolipia, for the subclass Eutheria.

Brown adipose tissue is characterised by the presence of a specific mitochondrial uncoupling protein that is not present in other tissues (Nicholls & Locke 1984; Himms-Hagen 1989). We have recently looked for this specific mitochondrial uncoupling protein in fat samples collected from pericardial, perirenal and interscapular sites from tammar young ranging in age from 148 to 250 days of age (P Trayhurn & P A Janssens, unpublished findings). All samples had a very low mitochondrial content, much lower than would be expected in brown adipose tissue, but there was some evidence for low levels of the specific uncoupling protein in pericardial fat samples taken from 3 tammar young at 200 to 230 days of age. Uncoupling protein was not, however, found consistently in this fat depot, nor was there evidence for the protein in fat samples from the other regions. Although this evidence is preliminary, it suggests the possibility that uncoupling protein may be present in pericardial fat of tammar young at about the time of pouch vacation. Whether this is indicative of the presence of brown adipose tissue is debatable. In any case, the small amount of the protein and the very low mitochondrial content of the tissue indicates that heat production by brown adipose tissue is unlikely to be a significant contributor to thermogenesis during development in this species.

CONCLUDING REMARKS

It is clear that, although some of the questions raised in The Developing Marsupial have been answered, many more have not. We do not know the significance of the changes in fatty acid composition of tammar pouch young tissues during development, we are not yet certain whether brown adipose tissue is present in marsupials and, except in a limited number of species, we do not know what stimulates young to leave the pouch. There is clearly a great deal to be done.

Acknowledgments

The authors thank Ms E M Bugledich, Dr J M Carr, Dr L A Hinds, Dr J A Owens, Dr D A Pan, Mr W E Poole, Dr P Trayhurn and Dr J C Wallace for permission to use their unpublished findings and Ms J Grigg and Mr W Mantaj for excellent technical assistance. The authors' work has been supported by the Australian Research Council, the National Health and Medical Research Council, the Australian National University Faculties Research Fund, the Board of Research of Flinders University, and the Australian Flora and Fauna Research Program at the University of Wollongong.

References

Aschoff J, Gunther B and Kramer K (1971). *Energiehaushalt und Temperaturregulation.* Urban and Schwartzenberg, Munich.

Baudinette R V, Runciman S I C, Frappell P F and Gannon B J (1988). Development of the marsupial cardiorespiratory system. In *The Developing Marsupial*. Models for *Biomedical Research*. Eds C H Tyndale-Biscoe and P A Janssens. Springer-Verlag, Berlin. pp 132–147.

Brody R (1974). *Bioenergetics and Growth.* Hafner, New York.

Carr J M, Owens J A, Baudinette R V and Wallace J C (1995). Analysis of insulin-like growth factor binding proteins in the tammar wallaby, *Macropus eugenii*. General and *Comparative Endocrinology*, 99, 50–59.

Catling P C and Vinson G P (1976). Adrenocortical hormones in the neonate and pouch young of the tammar wallaby, *Macropus eugenii. Journal of Endocrinology*, 69, 447–448.

Cockburn A and Johnson C N (1988). Patterns of growth. In *The Developing Marsupial. Models for Biomedical Research.* Eds C H Tyndale-Biscoe and P A Janssens. Springer-Verlag, Berlin. pp 28–40.

Cork S and Dove H (1989). Lactation in the tammar wallaby (*Macropus eugenii*). II. Intake of milk components and maternal allocation of energy. *Journal of Zoology*, 219, 399–409.

Dendrinos G (1988). *Pouch Exit and Thermal Relation Between Mother and Young in the Tammar Wallaby, Macropus eugenii.* Honours thesis, Australian National University, Canberra.

Else P L and Hulbert A J (1981). Comparison of the "mammal machine" and the "reptile machine": energy production. *American Journal of Physiology*, 240, R3-R9.

Else P L and Hulbert A J (1985). An allometric comparision of the mitochondria of mammalian and reptilian tissues: the implications for the evolution of endothermy. *American Journal of Physiology*, 156, 3–11.

Findlay L (1982). The mammary glands of the tammar wallaby (*Macropus eugenii*) during pregnancy and lactation. *Journal of Reproduction and Fertility*, 65, 59–66.

Gemmell R T and Cepon G (1993). The development of thermoregulation in the marsupial brushtail possum, *Trichosurus vulpecula. Comparative Biochemistry and Physiology*, 106A, 167–173.

Gemmell R T, Cepon G and Barnes A (1987). The development of thermoregulation in the rat kangaroo, *Potorous tridactylus. Comparative Biochemistry and Physiology*, 88A, 257–261.

Gemmell R T and Hendrikz J K (1993). Growth rates of the bandicoot *Isoodon macrourus* and the brushtail possum *Trichosurus vulpecula. Australian Journal of Zoology*, 41, 141–149.

Gemmell R T and Johnston G M (1985). The development of thermoregulation and the emergence from the pouch of the marsupial bandicoot, *Isoodon macrourus. Physiological Zoology*, 58, 299–302.

Gemmell R T and Sernia C (1992). The transfer of thyroxine from the mother to the young of the marsupials, the bandicoot (*Isoodon macrourus*) and the possum (*Trichosurus vulpecula*). *Comparative Biochemistry and Physiology*, 103A, 541–543.

Green B, Griffiths M and Leckie R M C (1983). Qualitative and quantitative changes in milk fat during lactation in the tammar wallaby (*Macropus eugenii*). *Australian Journal of Biological Science*, 36, 455–461.

Green B and Merchant J (1988). The composition of marsupial milk. In *The Developing Marsupial. Models for Biomedical Research.* Eds C H Tyndale-Biscoe and P A Janssens. Springer-Verlag, Berlin. pp 41–54.

Green B, Merchant J C and Newgrain K (1988). Milk consumption and energetics of growth in pouch young of the tammar wallaby, *Macropus eugenii. Australian Journal of Zoology*, 36, 217–227.

Hall L S (1990). Growth and a description of the development of external features of pouch young of captive *Isoodon macrourus*. In *Bandicoots and Bilbies*. Eds J H Seebeck, P R Brown, R L Wallis and C M Kempner. Surrey Beatty & Son Pty Ltd, Sydney. pp 123–133.

Hayssen V and Lacy R C (1985). Basal metabolic rates in mammals: taxonomic differences in the allometry of BMR and body mass. *Comparative Biochemistry and Physiology*, 81A, 741–754.

Hayward J S and Lisson P A (1992). Evolution of brown fat: its absence in marsupials and monotremes. *Canadian Journal of Zoology*, 70, 171–179.

Himms-Hagen J (1989). Brown adipose tissue thermogenesis and obesity. *Progress in Lipid Research*, 28, 67–115.

Holman R T (1992). A long scaly tale — the study of essential fatty acid deficiency at the University of Minnesota. In *Essential Fatty*

Acids and Eicosanoids. Eds A Sinclair and R Gibson. American Oil Chemists Society, Champaign, Illinois. pp 3–17.

Hulbert A J (1988). Metabolism and the development of endothermy. In *The Developing Marsupial, Models for Biomedical Research*. Eds C H Tyndale-Biscoe and P A Janssens. Springer-Verlag, Berlin. pp 148–161.

Hulbert A J and Else P L (1981). Comparison of the "mammal machine" and the "reptile machine": energy use and thyroid activity. *American Journal of Physiology*, 241, R350-R356.

Hulbert A J, Mantaj W and Janssens P A (1991). Development of mammalian endothermic metabolism: quantitative changes in tissue mitochondria. *American Journal of Physiology*, 261, R561-R568.

Janssens P A, Grigg J A, Dove H and Hulbert A J (1990). Thyroid hormones during development of a marsupial, the tammar wallaby, *Macropus eugenii*. *Journal of Endocrinology*, 127, 427–436.

Janssens P A and Rogers A M T (1989). Metabolic changes during pouch vacation and weaning. In *Kangaroos, Wallabies and Rat-kangaroos*. Eds G C Grigg, P J Jarman and I D Hume. Surrey Beatty & Son Pty Ltd, Sydney. pp 367–376.

Kleiber M (1961). *The Fire of Life: an introduction to animal energetics*. John Wiley & Son, Inc, New York. 454pp.

Legrand J (1986). Thyroid hormone effects on growth and development. In *Thyroid Hormone Metabolism*. Ed G Henneman. Marcel Dekker, New York. pp 503–534.

Loh T T and Shield J (1977). Temperature regulation and oxygen consumption in the developing macropod marsupial, *Setonix brachyurus*. *Journal of Physiology* 269, 677–686.

Loudon A, Rothwell N and Stock M (1985). Brown fat, thermogenesis and physiological birth in a marsupial. *Comparative Biochemistry and Physiology*, 81A, 815–819.

Nicholas K R (1988). Control of milk protein synthesis in the marsupial *Macropus eugenii*: a model system to study prolactin-dependent development. In *The Developing Marsupial. Models for Biomedical Research*. Eds C H Tyndale-Biscoe and P A Janssens. Springer-Verlag, Berlin. pp 68–85.

Nicholls D G and Locke R M (1984). Thermogenic mechanisms in brown fat. *Physiological Reviews*, 64, 1–64.

Ramsay M A and Dunbrack R L (1986). Physiological constraints on life history phenomena: the example of small bear cubs at birth. *The American Naturalist*, 127, 735–743.

Renfree M B, Holt A B, Green S W, Carr J P and Cheek D B (1982). Ontogeny of the brain in a marsupial (*Macropus eugenii*) throughout pouch life. 1. Brain growth. *Brain, Behaviour and Evolution*, 20, 57–71.

Reynolds M L and Saunders N R (1988). Differentiation of the neocortex. In *The Developing Marsupial. Models for Biomedical Research*. Eds C H Tyndale-Biscoe and P A Janssens. Springer-Verlag, Berlin. pp 101–116.

Rogers A M T (1979). *Maternal-Pouch Young Thermal Relationships in the Tammar Wallaby, Macropus eugenii. (Desmarest, 1817)*. Honours thesis, Australian National University, Canberra.

Rose R W (1986). The control of pouch vacation in the Tasmanian bettong, *Bettongia gaimardii*. *Australian Journal of Zoology*, 35, 485–491.

Rowlatt U, Mrosovsky N and English A (1971). A comparative survey of brown fat in the neck and axilla of mammals at birth. *Biology of the Neonate*, 17, 53–84.

Salem N Jr and Niebylski C D (1992) An evaluation of alternative hypotheses involved in the biological function of docosahexanoic acid in nervous system. In *Essential Fatty Acids and Eicosanoids*. Eds A Sinclair and R Gibson. American Oil Chemists Society, Champaign, Illinois. pp 84–86.

Setchell P J (1974). The development of thermoregulation and thyroid function in the marsupial *Macropus eugenii* (*Desmarest*). *Comparative Biochemistry and Physiology*, 47A, 1115–1121.

Smith M J (1989). Release of embryonic diapause in *Bettongia penicillata. In Kangaroos, Wallabies and Rat-Kangaroos*. Eds G C Grigg, P J Jarman and I D Hume. Surrey Beatty & Son Pty Ltd, Sydney. pp 317–322.

Wallis R L and Maynes G M (1973). Ontogeny of thermoregulation in *Macropus parma* (Marsupialia: Macropodidae). *Journal of Mammalogy*, 54, 278–281.

Wilkes G E and Janssens P A (1988). Development of renal function. In *The Developing Marsupial. Models for Biomedical Research*. Eds C H Tyndale-Biscoe and P A Janssens. Springer-Verlag, Berlin. pp 176–189.

CHAPTER 5

THE DEVELOPMENT OF SEX DIFFERENCES IN BEHAVIOUR: A MARSUPIAL MODEL

Barbara H Fadem

INTRODUCTION

Regardless of genetic sex, mammals have the potential to display both masculine and feminine behaviour patterns. Whether or not such patterns are displayed depends to a large extent on the presence or absence of gonadal hormones in adulthood. In eutherian species, the display of certain masculine or feminine behaviour patterns require only activation by androgens, such as testosterone, or oestrogens, such as oestradiol, respectively in adulthood. Expression of other behaviour patterns is dependent not only on the presence of circulating gonadal hormones, but also on the presence of gonadal hormones during critical periods of development (Arnold & Breedlove 1985). For example, male rodents castrated on the day of birth have limited ability to show masculine behaviour patterns in adulthood despite replacement therapy with testosterone (Gerall *et al.* 1967). Similarly, treatment of newborn female rodents with testosterone results in masculinisation of adult behaviour patterns (Goy *et al.* 1962). In some species, androgens may exert their masculinizing effects on the organisation of sex-specific behaviour during development only after intracellular aromatisation to oestrogen in the brain (McEwen *et al.* 1977). In rats, although a well-established 'critical period' for the differentiation of masculine sexual behaviour by gonadal hormones occurs on the first day of postnatal life (Gerall *et al.* 1967), prenatal testicular secretions or their oestrogenic metabolites (McEwen *et al.* 1977; Fadem & Barfield 1981) may sensitise the developing brain to the early postnatal action of these hormones. For example, in ferrets, oestrogen and testosterone act sequentially prenatally and postnatally to effect behavioural masculinisation (Baum & Erskine 1984). In some eutherian species, oestrogen acts over a long period to organise behaviour (Arnold & Breedlove 1985); in pigs this period extends to early prepuberty (Adkins-Regan *et al.* 1989). There is evidence also that oestrogens play an active role in the differentiation of feminine behaviour patterns (Dunlap *et al.* 1978; Toran-Allerand 1984) and exposure to low levels of oestrogen perinatally may feminise behaviour in males (Yalom *et al.* 1973;

Stewart & Cygnan 1980; vom Saal *et al.* 1983). Feminisation of behaviour by oestrogen also may occur over a prolonged period; injection of male rats with oestrogen even as late as 20 to 30 days of age increases feminine sexual behaviour in adulthood (Hendricks & Weltin 1976).

Although oestrogen is involved in the differentiation of a sexually dimorphic brain and behaviour during perinatal life, its role has been difficult to study in traditional placental mammalian models such as rodents and primates (Baum 1984). The major problem with such models is that the experimental subject (the fetus) is not directly available. Consequently, experimental manipulation of the subject must be accomplished indirectly by manipulation of the maternal environment. This is a particular problem for studies about oestrogen since in some eutherian species, fetuses have physiological systems to protect them from circulating maternal oestrogens (Wilson *et al.* 1981; Davidson *et al.* 1983).

In contrast to eutherian species, marsupial neonates are available to the experimenter for an extended period of their development and differentiation since they are born in an immature state (Tyndale-Biscoe & Janssens 1988; Janssens *et al.* see Chapter 4 in this volume). The marsupial neonate is attached to the teats for a period which corresponds roughly to the latter part of intrauterine life in some eutherian mammals (Sharman 1970), a stage when oestrogenic effects on the developing brain probably occur (Baum 1984). At birth, the gonads and accessory structures of the marsupial are at a relatively sexually indifferent stage (McCrady 1938). The sensitivity of the marsupial neonate to oestrogenic stimulation was first demonstrated by treatment of newborn male North American opossums *Didelphis virginiana* with oestradiol dipropionate. Such treatment resulted in ovarian development and feminisation of the internal and external genitalia (Burns 1961). The newborn marsupial thus presents an accessible experimental animal which:

(a) is at an immature stage of development with relatively undifferentiated gonads and genitalia;

(b) is relatively independent of maternal metabolic and placental influences; and

(c) can be observed and manipulated directly.

Since the mid-1980s, we have been studying the development of sex differences in behaviour using South American grey short-tailed opossums, *Monodelphis domestica* (hereinafter referred to as grey opossums), small (60 to 150 gm) members of the Didelphidae or American opossums. Studies from this and collaborators' laboratories have demonstrated in grey opossums:

(a) a role for pheromones provided by males in the activation of oestrus in adult females and the stimulus control of this phenomenon (Fadem 1985, 1987, 1989a);

(b) sex differences in behaviour (Fadem & Cole 1985; Fadem 1989b) and associated morphology (Fadem & Schwartz 1986);

(c) the activational effects of gonadal hormones on these sex differences (Fadem *et al.* 1989; Fadem 1990; Fadem & Corbett 1993);

(d) neural oestrogen receptor and neural aromatase activity profiles over development and in adulthood (Etgen & Fadem 1987, 1989; Fadem *et al.* 1993);

(e) luteinising-hormone releasing hormone neuron distribution in neonatal and adult brain (Schwanzel-Fukuda *et al.* 1988); and

(f) plasma gonadal hormone levels and gonadal development in adults and neonates (Fadem & Harder 1992a, b; Fadem *et al.* 1992). In addition, our studies have essentially replicated and also expanded those of Burns (1961) with the findings that testis development is blocked and genitalia are feminised in male grey opossums treated with oestradiol benzoate within a few days of birth (Fadem & Tesoriero 1986) and that the luteinising hormone-releasing hormone system is essentially normal in these feminized males (Fadem & Schwanzel-Fukuda 1990). Studies currently under way are examining the effects of manipulation of the hormonal environment during early development on the display in adulthood of sexually dimorphic behaviour. The results of the work catalogued above and described in some detail below, suggest a role for oestrogen in the development of sex differences in behaviour in grey opossums. These results demonstrate further that studies utilising this marsupial species can provide unique information concerning the role of gonadal hormones in mammalian sexual differentiation.

Reproduction in grey opossums

In order to provide subjects for these research studies it was necessary to maximise breeding in grey opossums. To this end, observations and experimental studies were conducted. It was found that in grey opossums, sexual maturity is reached at 4 to 5 months and breeding occurs throughout the year (VandeBerg 1983; Kraus & Fadem 1987).

Initially, by examining breeding records, and later, by experimentation, our studies demonstrated that female grey opossums failed to show vaginal oestrus (identified by cornified vaginal smears) when housed in a room containing only other females and only showed vaginal oestrus 4 to 12 days after they were exposed to males (Fadem 1985). Because in rodents, males produce a pheromone that acts on the female vomeronasal system resulting in secretion of gonadotropins and oestrus (Feder 1981), we conducted a series of studies in grey opossums to identify the mechanisms involved in induction of oestrus. We found that induction of oestrus in this species is mediated also through pheromonal cues provided by males and that production of these cues is, at least in part, androgen dependent (see below); intact males were more likely to induce oestrus in females than castrated males or other females (Fadem 1987) (Figure 5.1). Stonerook &

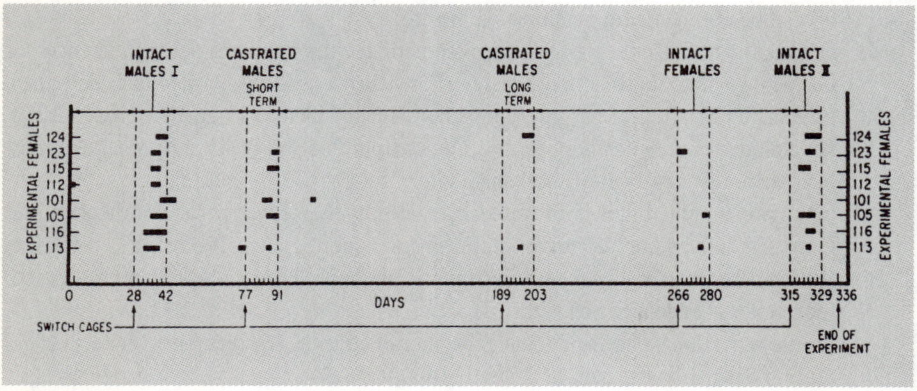

Figure 5.1 Effects of pheromones produced by intact males, castrated males and intact females on oestrus in female grey opossums.
Notes: Each bar represents a period of oestrus. The first vertical dotted line for each test situation is the first day that the experimental females were exposed to cages that had previously housed intact males, castrated males or intact females. The second vertical line for each test situation is the day that the experimental females were placed in clean cages.
Source: Fadem (1987).

Harder (1992) showed recently that female grey opossums only enter puberty after exposure to male pheromones. Our studies demonstrated also that the source of the pheromone inducing oestrus is likely to be male skin glands and that oestrogen levels in females that enter oestrus following exposure to skin gland secretions of male are higher than in females that fail to show oestrus (Fadem 1989a). Hinds *et al.* (1992) demonstrated recently that plasma progesterone levels also were higher in females exposed to males than in unexposed females.

The responsiveness of female grey opossums to pheromonal cues is likely to be mediated as in eutherian mammals, through the hypothalamic-pituitary-gonadal axis (Feder 1981). Because manipulation of the perinatal hormonal environment in eutherian species affects neural structures important also in induction of oestrus (Smale *et al.* 1985), the finding that male grey opossums induce oestrus in females has an additional benefit for studies of neural sexual differentiation. By observing the effects of pheromonal cues provided by males on induction of oestrus in females in which the neonatal hormonal environment has been manipulated, it will be possible to assess sexual differentiation of the hypothalamus.

Adult studies Sexually dimorphic behaviour

In order to examine the organising role of gonadal hormones in the differentiation of sexually dimorphic behaviour in grey opossums it was necessary to first catalogue sex differences in behaviour patterns in this species. This was done by conducting behavioural observations and experiments.

The Didelphidae are solitary animals which show complex patterns of olfactory (scent marking) and vocal communication and high levels of conspecific aggression (Hunsaker & Shupe 1977; Eisenberg & Golani 1977; Fadem 1986). In a study of social behaviour using 16 male and 16 female grey opossums tested in same sex and in mixed-sex pairs (Fadem 1989b), males were highly aggressive, biting ('bite'), locking together with front and hind legs ('lock') and tumbling over ('tumble') when tested with other males but showed little aggression toward females. In tests with females, males followed or chased ('follow'), circled head to tail ('circle'), and grabbed the female with a forepaw ('grab') while sniffing her genital area ('sniff'). Frequencies of follow, circle, grab and sniff are hereinafter summed and collectively referred to as 'precopulatory score'. Females showed high levels of threat behaviour (open mouth threats and screeching vocalisations) towards newly introduced males (Figure 5.2). Males paired with females used strategies including clicking vocalisations which, as in other marsupials (Russell 1970; Hunsaker & Shupe 1977; Wemmer & Collins 1978; Biggins 1984), apparently are aimed at reducing this female aggression.

When grey opossums were placed alone in an unfamiliar testing chamber, we found that they scent marked with the head, flank, chest and hip, and that this scent marking was sexually dimorphic; males marked significantly more frequently than females, older males marked more than younger males, and only males marked with the chest (Fadem & Cole 1985). This is probably related to the fact that only adult males possess a testosterone-dependent suprasternal scent gland (Fadem & Schwartz 1986).

Because sex differences in behaviour were observed in adult animals, it was of interest to determine if the display of these sex differences was related to sexual maturity. For this, male and female grey opossums aged 10, 13, 16 (juvenile) and 19 and 22 weeks (adult), received behaviour tests when alone and when paired with animals of

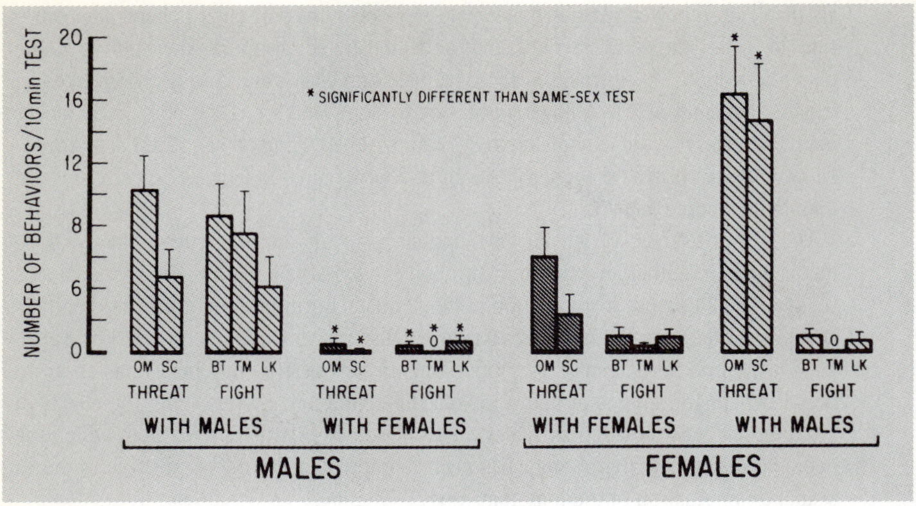

Figure 5.2 Sex differences in threat (open-mouth postures [OM], screeching vocalisations [SC]) and fight (bites [BT], tumbles [TM] and locking [LK]) behaviours in grey opossums in same-sex and mixed-sex tests.
* Significant differences (p<0.0001 to p<0.01) between tests with animals of the opposite sex and tests with animals of the same sex.
Source: Fadem (1989a).

each sex. The results indicated that the frequency of open mouth threats and screeching vocalisations by females in tests with male partners was higher in adulthood (22 weeks of age) than at younger ages and was higher in adult females than in adult males (19 and 22 weeks of age). Because in tests with female partners, females showed more threat behaviour than males at 13 and at 16 weeks of age, this sex difference is apparent before puberty. Scent marking and clicking behaviour occurred only in males and then only when the animals were 16 weeks of age or older (Fadem & Corbett 1997) (Figure 5.3).

These findings identified sexually dimorphic behaviour patterns and associated morphology, that is, scent marking, clicking vocalisations, screeching vocalisations, precopulatory score and presence of a suprasternal scent gland in grey opossums. While screeching vocalisations showed a sex difference before puberty; these other characteristics were likely to be activated by gonadal hormones in adulthood. We then sought to determine the characteristics of and specific hormones involved in such activation.

The hormonal activation of sexually dimorphic behaviour

As noted previously, the display of sexually dimorphic behaviour in adult mammals is dependent largely on the presence of circulating gonadal hormones. While in general, androgens activate masculine behaviour and oestrogens activate feminine behaviour, aromatisation of androgens to oestrogens via neural aromatase activity may be involved also in the activation of certain behaviour patterns (Arnold & Breedlove 1985). In fact, sexually dimorphic behaviour patterns that require aromatisation for their activation frequently are organised perinatally by oestrogen resulting from aromatisation of androgens (Goy & McEwen 1980).

To examine the role of gonadal hormones in the activation of scent marking and

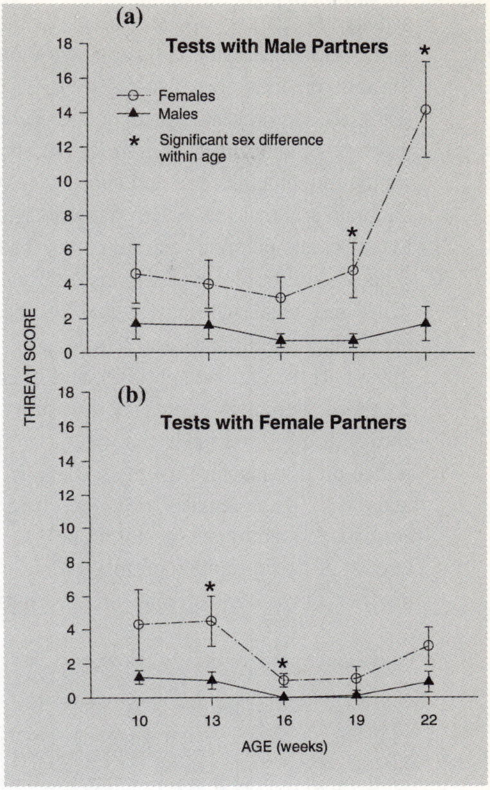

Figure 5.3 Threat score (frequency of open mouth postures with screeching vocalisations) in male and female grey opossums aged 10, 13, 16, 19 and 22 weeks in tests with (a)male partners and (b)female partners.
Source: Fadem (unpublished observations).

related behaviour in grey opossums, behaviour was examined in 26 adult males before and for 3 weeks after castration, and for 4 weeks following receipt of 5 mm silastic capsules of 5% crystalline oestradiol and 95% cholesterol, 100% testosterone or 100% cholesterol (control) under the skin of the neck. While all males scent marked prior to castration, castration resulted in significant declines in chest, hip/flank and head marking. Both testosterone and oestradiol stimulated flank/hip marking in castrated males; only testosterone stimulated chest marking (Fadem *et al.* 1989). In tests with females, precopulatory score declined postcastration and increased following testosterone treatment. Using a similar experimental design and 37 adult female grey opossums, no females had scent glands and few females scent marked before or after ovariectomy. Treatment with testosterone, but not with oestradiol stimulated chest marking while both testosterone and oestrogen stimulated head marking. Also, testosterone but not oestradiol stimulated growth of the suprasternal scent gland and the phallus in ovariectomised females (Fadem 1990). In a recent study, we found also that females ovariectomised and treated with oestradiol show less screeching behaviour toward males than females that had not received oestradiol (controls), and that gonadally intact introduced males had higher precopulatory scores when they were tested with oestradiol-treated females than with control females. Furthermore, some oestradiol-treated females but no control females allowed males to copulate with them (Fadem *et al.* 1996).

Because oestradiol activated aspects of scent marking behaviour in male and female grey opossums it was of interest to characterise oestrogen binding sites and aromatase activity in adult grey opossum brain, particularly in a brain area known to be important in the hormonal organisation and activation of sexually dimorphic behaviour, the hypothalamic-preoptic area (Baum 1984). We found that in adult grey opossums of both sexes, the hypothalamic-preoptic area contained macromolecules with steroid binding properties characteristic of oestrogen receptors (Etgen & Fadem 1987). In another study we found that adult male and female grey opossums demonstrated aromatase activity in the brain and that this activity showed significant sex differences (males higher) in both hypothalamic-preoptic area and cortex (Fadem *et al.* 1993).

Because oestradiol as well as testosterone activated aspects of masculine behaviour in males and females (Fadem *et al.* 1989; Fadem 1990), because the hypothalamic-preoptic area of both males and females contained high affinity saturable oestrogen binding sites (Etgen & Fadem 1987), and because aromatase activity showing a sex difference was present in adult grey opossum hypothalamic-preoptic area and cortex (see above), it appeared likely that neural aromatisation of androgen was involved in the activation of behaviour in this species. To test this hypothesis, gonadectomised adult males (n=29) and females (n=17) received subcutaneous implants containing either an aromatisable androgen (testosterone), a non-aromatisable androgen (dihydrotestosterone) or no hormone (control). The results of this study indicated that testosterone, but not dihydrotestosterone or the control, activated hip and head marking and

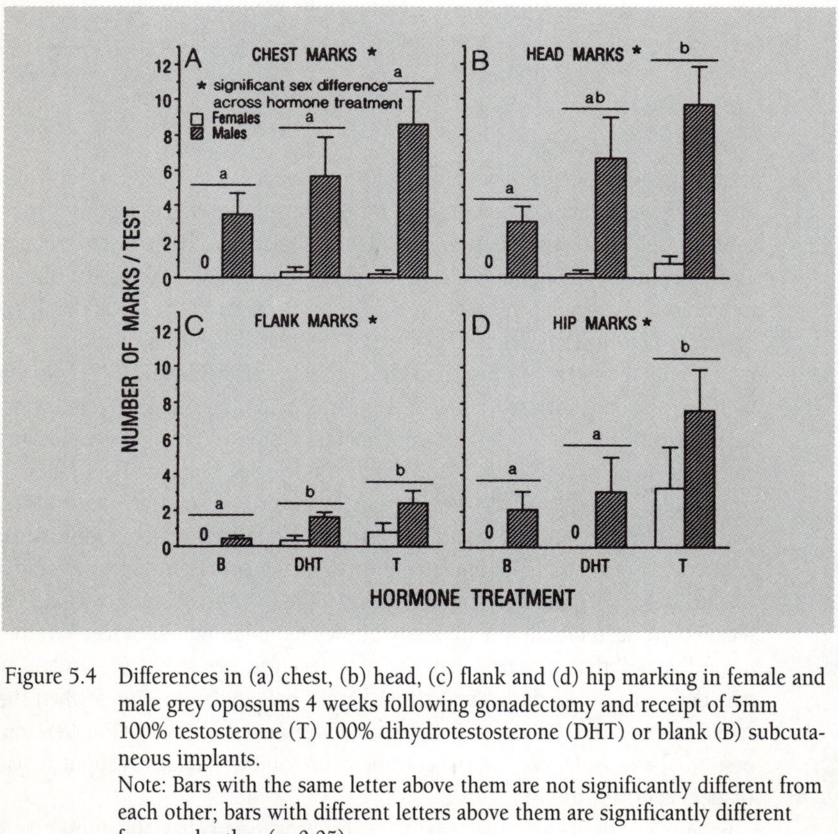

Figure 5.4 Differences in (a) chest, (b) head, (c) flank and (d) hip marking in female and male grey opossums 4 weeks following gonadectomy and receipt of 5mm 100% testosterone (T) 100% dihydrotestosterone (DHT) or blank (B) subcutaneous implants.
Note: Bars with the same letter above them are not significantly different from each other; bars with different letters above them are significantly different from each other ($p<0.05$).
Source: Fadem & Corbett (1993).

clicking vocalisations; testosterone also stimulated significantly more scent gland growth than dihydrotestosterone or the control treatment. In comparing the behaviour of males and females in this study, we found also that sex differences in scent marking (Figure 5.4), screeching vocalisations in tests with males (Figure 5.5) and clicking vocalisations in tests with females persisted despite the fact that males and females had received the same hormone treatment (Fadem & Corbett 1993).

These findings suggested that in grey opossums: a) androgens activate scent marking and clicking vocalisations; b) head and hip marking and clicking vocalisations require aromatisation of testosterone to oestradiol for their activation; c) screeching vocalisation is not affected by gonadectomy or by adult androgen treatment and 4) sex differences in scent marking, clicking and screeching persist following gonadectomy and treatment with the same gonadal hormones in adulthood. The last finding suggests that the perinatal hormonal organisation of the brain is likely to be different between males and females. In the next group of studies, possible mechanisms for this differential organisation were examined.

DEVELOPMENTAL STUDIES NORMAL SEXUAL DIFFERENTIATION

The gestation period of grey opossums is 14 to 15 days, most litters contain between 5 and 12 offspring each weighing 100 mg and post-partum dependence is about 50 days (VandeBerg 1983; Kraus & Fadem 1987). Neurally (Schwanzel-Fukuda *et al.* 1988; Saunders *et al.* 1989) and reproductively (Moore & Thurstan 1990; Baker *et al.* 1990; Fadem *et al.* 1992), grey opossums are in an immature state (when compared to eutherian species) from the day of birth until about 16 days postnatal when they begin to relinquish the teat.

One way to identify critical periods for the action of gonadal hormones on the brain during early development is to detect developmental periods when levels of circulating gonadal hormones and neural aromatase activity in brain areas known to be important in behavioural sexual differentiation of rats are high and/or show a sex

Figure 5.5 Differences in percentage of male and female grey opossums showing screeching vocalisations in tests with introduced intact males following gonadectomy and receipt of 100% testosterone (T) 100% dihydrotestosterone (DHT) or blank (B) implants.
*Significant sex difference across hormone treatment (x2 = 11.6, p<0.001).
Note: Bars with the same letter above them are not significantly different from each other.
Source: Fadem & Corbett (1993).

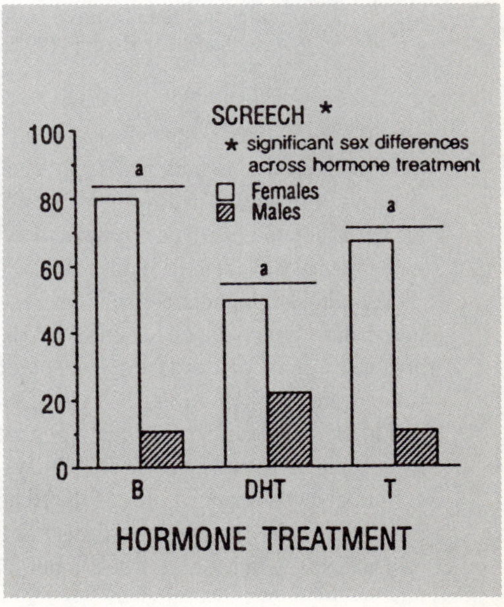

difference (Baum 1984; Perakis & Stylianopoulou 1986). In these studies, we assayed androgen and oestrogen levels in peripheral plasma and neural aromatase activity in the whole brain, the hypothalamic-preoptic area and the cortex, and characterised the oestrogen receptor system in brain in grey opossums from the day of birth through adulthood. Because maternal milk could be a source of gonadal hormones in developing neonates, hormone levels were assessed also in blood samples obtained from mothers nursing litters aged 1, 4, 8, 16 and 30 days.

We found that levels of androgen in mixed-sex plasma pools of animals 4 and 8 days postnatal were higher than in either sex at all other ages examined. Androgen levels at

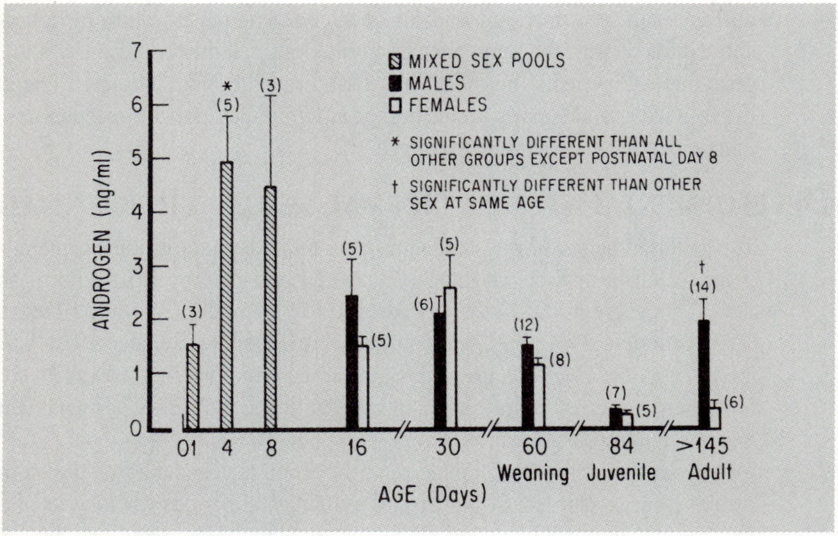

Figure 5.6 Plasma androgen concentrations (mean ± sem) of developing grey opossums.
Note: Each bar represents the mean value for plasma pools for young aged 1 to 30 days and for individual animals 60 days and older. Number of plasma pools or individual samples are given in parentheses.
Source: Fadem & Harder (1992a).

16, 30 and 60 days postnatal were as high as in adult males and were equivalent in males and females. In both males and females, androgen levels were lower in juveniles (84 days postnatal) than at younger ages. Sex differences in androgen levels were only seen in adulthood (>145 days postnatal), when levels were significantly higher in males than in females (Fadem & Harder 1992a) (Figure 5.6).

Oestradiol levels could not be measured accurately in 1-day-old neonates because of the small amount of blood obtained. Levels of oestradiol in one mixed-sex plasma pool obtained from 4 day neonates and one of two mixed-sex plasma pools from 8 day young were very high (over 300 pg/ml). While no significant sex differences in oestradiol levels were seen 30 or 60 days postnatal, mean oestradiol levels were significantly higher at 30 days than at 60 or 145 days of age (Figure 5.7). Maternal androgen and oestradiol levels obtained on 1, 4, 8, 16 and 30 days postnatal were low and did not differ significantly by litter age (Fadem & Harder 1992a; Fadem & Harder 1992b).

In mixed-sex samples of whole brain, aromatase activity was present at birth and increased until 16 days postnatal. In the hypothalamic-preoptic area, significantly higher levels of aromatase activity were seen in 16-day-old young than in 30-day-old young and the latter animals had higher aromatase activity than all older ages including adults. Significant sex differences were seen: 16-day-old male young and adult males had higher levels of aromatase activity than females. While lower overall than in hypothalamic-preoptic area, aromatase activity was present also in the cortex where it was higher on 16 and 30 days than at older ages (Fadem *et al.* 1993) (Figure 5.8).

Results of studies examining binding of 3H-oestradiol to brain cytosols from grey opossums indicated that while oestrogen binding was undetectable in 1-day neonates, by 4 days of age high affinity saturable oestrogen binding sites were present, and binding remained low but detectable through to 63 days postnatal. Oestrogen binding sites in the hypothalamic-preoptic area increased substantially from 16 to 63 days postnatal (Etgen & Fadem 1989).

These findings of high levels of gonadal hormones, of neural aromatase activity and oestrogen receptor binding suggest that 4 to 8 days, 16 days and 30 days postnatal may be critical periods for the action of oestrogen and testosterone on the brain in grey opossums. Because levels of aromatase activity are high and show a sex difference at 16 days postnatal, aromatisation of testosterone to oestradiol may be involved in androgen action on brain sexual differentiation at this time. Furthermore, masculine sexual differentiation of the rat brain by androgens occurs around the day of birth (Goy *et al.* 1962). Because 16-day-old grey opossum young are neurally and reproductively equivalent to 1-day-old rats, there thus may be a similarity between this

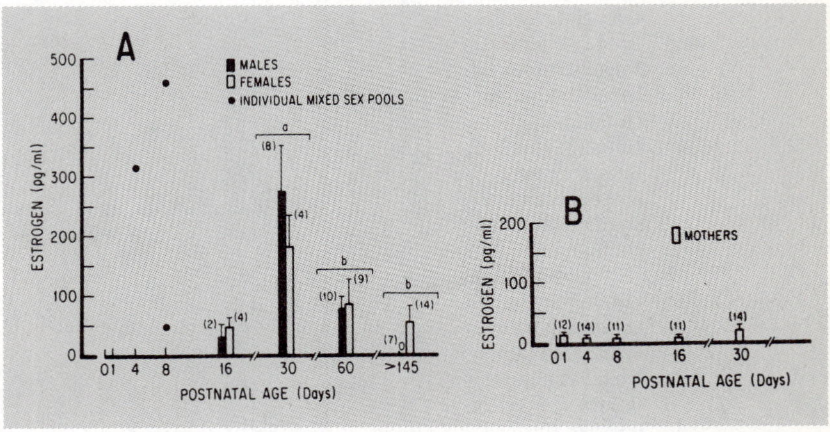

Figure 5.7 (a)Plasma oestrogen (mean ± sem) in (a) developing grey opossums. Values for individual plasma pools are shown for 4 and 8 days postnatal.
(b) Plasma oestrogen (mean ± sem) in nursing mothers of litters of different postnatal ages.
Notes:
(1) In Figure 5.7(a) bars represent mean values for plasma pools for animals 16 and 30 days postnatal and for individual animals at 60 days and >145 days old (adult); and different letters above the bars represent significant differences between age groups (16 days was not included in data analyses) ($p<0.05$).
(2) Number of plasma pools or individual samples are given in parentheses, in both diagrams.
Source: Fadem & Harder (1992b).

marsupial species and rats in the occurrence of at least one critical period for androgen action on the developing brain.

The source of oestrogen or testosterone early in postnatal development in grey opossums is not obvious since the testes are relatively undeveloped during the first week of postnatal life and the ovaries are undeveloped until at least the third week of postnatal life (Fadem *et al.*1992) (in contrast to the tammar wallaby, *Macropus eugenii*; Shaw *et al.* Chapter 8, in this volume and references therein). Also, because maternal testosterone and oestrogen levels are low during lactation (Fadem & Harder 1992a, b), milk is not likely to be the source of high levels of gonadal hormones seen in neonatal grey opossums. Because in some marsupials the adrenal glands can synthesise steroids by the time of birth (Call *et al.* 1980) the fetal adrenals could be the source of these steroids. Whatever the source, these results suggested that gonadal hormones could act over an extended period during the first two months of postnatal life to effect sexual differentiation of the grey opossum brain. In a group of studies currently under way, one of which is described below, we are manipulating perinatal hormonal environment in grey opossums and observing the behavioural results of these manipulations in adulthood.

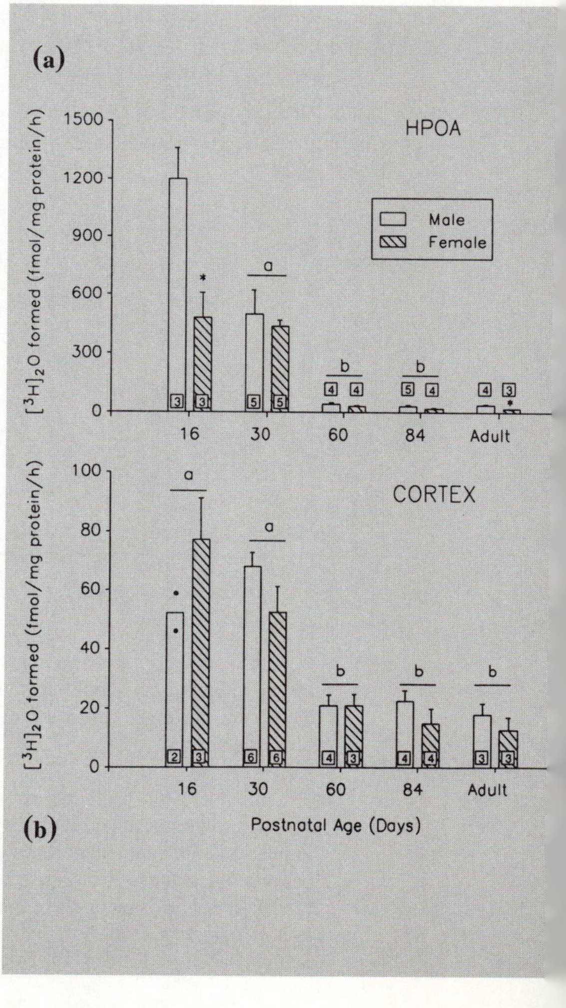

Figure 5.8 Aromatase activity (mean ± sem) in (a) the hypothalamic-preoptic area and (b) cerebral cortex of male and female grey opossums from 16 days postnatal through to adulthood (>145 days). *Significant sex difference within age ($p<0.05$). Notes:(1) Bars with the same letter above them are not significantly different from each other. (2) Number of samples appears on or above each bar. While both the hypothalamic-preoptic area and the cortex were obtained from each animal, differences in the number of samples from each area and each sex at each age is due to loss of tissue during processing. Source: Fadem *et al.* (1993).

The role of gonadal hormones in the organisation of sexually dimorphic behaviour

One useful method for examining the role of gonadal hormones in the development of behaviour is to block the action of such hormones on brain during critical developmental periods and to observe the behaviour of treated animals versus untreated animals in adulthood.

In order to block the action of oestrogen on the brain at a time (16 days postnatal) when neural aromatase activity was found to be high (Fadem *et al.* 1993), 61 neonatal grey opossums received 5 injections of 5 μg of tamoxifen, an oestrogen receptor blocker, or corn oil (control) at 12, 14, 16, 18 and 20 days postnatal. At 10 weeks of age (prepuberty), all animals were gonadectomised. At this time it was noted that males treated neonatally with tamoxifen had undescended testicles and smaller phalluses than control males. This finding was not unexpected since, in some eutherian species, tamoxifen has oestrogenic as well as anti-oestrogenic morphological effects

Table 5.1 Phallus length and body weight of adult grey opossums four weeks following receipt of 5 mm long 5% oestradiol or 50% testosterone implants.

NEONATAL TREATMENT	ADULT TREATMENT	PHALLUS LENGTH (MM ± SEM)	BODY WEIGHT (G ± SEM)
Females: tamoxifen	intact	<1.0	35 ± 1.2
Females: oil	intact	<1.0	29 ± 0.6
Males: tamoxifen	intact	1.4 ± 0.1	37 ± 1.1
Males: oil	intact	2.0 ± 0.1	31 ± 0.8
Females: tamoxifen	oestradiol	<1.0	60 ± 2.3
Females: oil	oestradiol	<1.0	52 ± 2.2
Males: tamoxifen	oestradiol	2.2 ± 0.2	71 ± 3.0
Males: oil	oestradiol	2.0 ± 0.2	65 ± 2.2
Females: tamoxifen	testosterone	1.4 ± 0.3	77 ± 2.8
Females: oil	testosterone	1.1 ± 0.4	69 ± 3.4
Males: tamoxifen	testosterone	3.9 ± 0.3	92 ± 3.6
Males: oil	testosterone	4.1 ± 0.2	87 ± 2.9

Notes: These opossums had been treated at 12, 14, 16, 18 and 20 days postnatal with tamoxifen (n=16 males and 14 females) or corn oil (n=19 males and 12 females) when intact, and were gonadectomised at 70 days as juveniles. Phallus length differed significantly between tamoxifen- and oil-treated males only when they were intact juveniles ($p<0.0003$). Tamoxifen-treated animals weighed more than oil-treated animals as juveniles ($p<0.001$) and following oestrogen treatment ($p<0.001$); the difference in body weight approached significance following testosterone treatment ($p<0.06$).
Source: Fadem (1995).

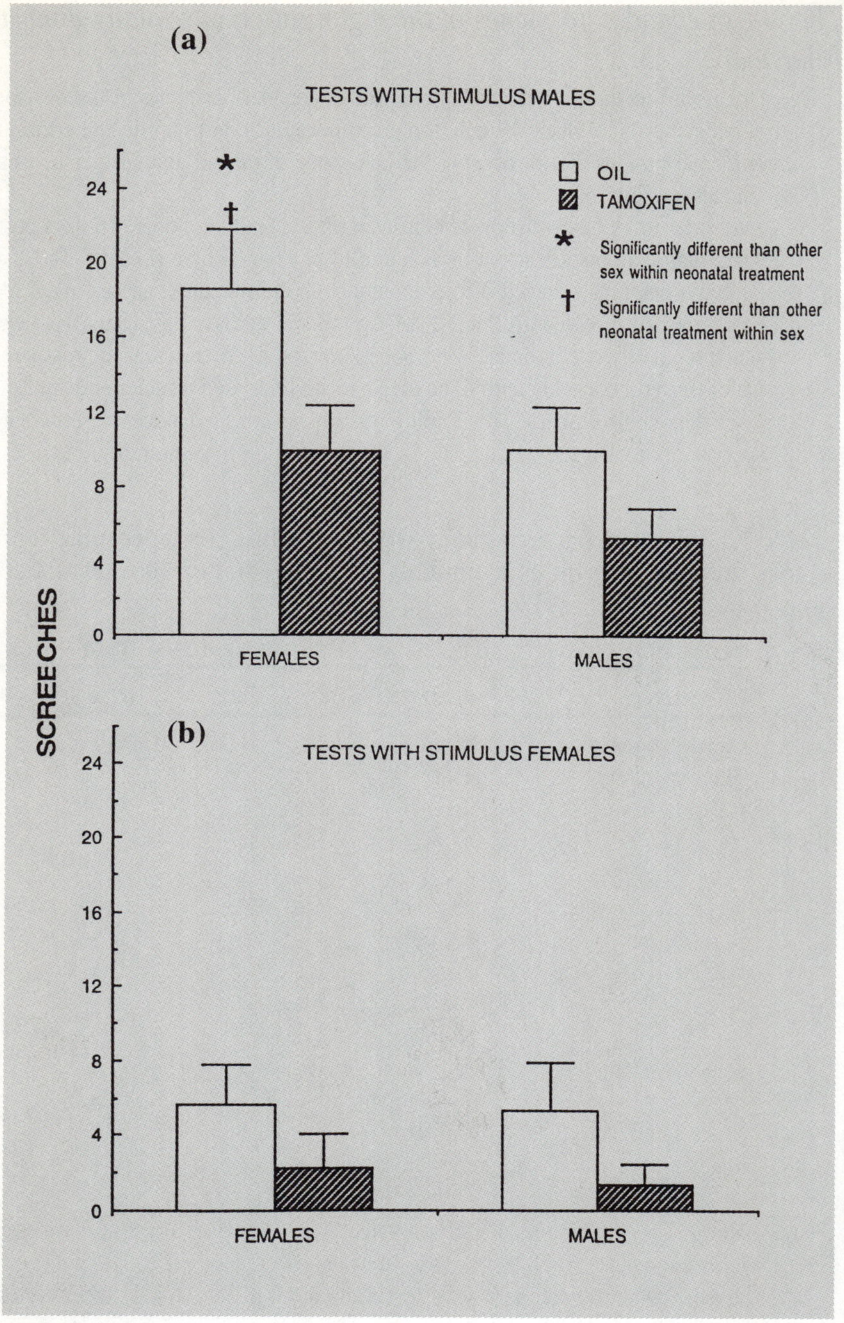

Figure 5.9 Effect of treatment of male and female grey opossums on 12, 14, 16,18 and 20 days postnatal with tamoxifen or corn oil on screeching threat behaviour in adulthood.
Notes: In adulthood, neonatally hormone-treated males and females were gonadectomised, given 5 mm long 5% oestradiol implants and tested 4 weeks later with intact, introduced (stimulus) males (a) or females (b).
Source: Fadem (1996).

(Wade *et al.* 1993). This neonatal treatment-related difference in phallus size did not persist following adult treatment with oestrogen or testosterone. Tamoxifen also had an anti-oestrogenic morphological effect on body weight in this study. Male grey opossums weigh significantly more than females and both males and females treated neonatally with tamoxifen weighed more than same-sex control animals (Table 5.1).

At 24 weeks of age, animals received subcutaneous silastic implants of 5% oestradiol and were tested for scent marking and socio/aggressive behaviour 3 and 4 weeks later, after which oestradiol implants were removed. After a washout period, animals received implants of 50% testosterone and were behaviourally tested again. Data analyses revealed that after receipt of oestrogen implants, tamoxifen-treated females showed less threat behaviour (screeching) toward stimulus males than control females (Figure 5.9) and tamoxifen-treated males showed less precopulatory behaviour and clicking vocalisation than control males. Following testosterone treatment in adulthood, sex differences but no neonatal treatment-related differences in scent marking were seen (Fadem 1995). These results indicate that blockage of oestrogen receptors by tamoxifen in grey opossums at around 16 days postnatal affects:
(a) testicular descent, phallus size and body weight;
(b) the organisation of screeching threat behaviour in females;
(c) the organisation of precopulatory and clicking behaviour in males. These findings suggest further that behaviour patterns activated by testosterone, such as scent marking, are not affected by perinatal treatment with tamoxifen and thus may be organised by gonadal hormones other than oestrogen or at times other than 16 days postnatal in grey opossums.

GENERAL CONCLUSIONS

Findings about behavioural differences between sexes in adulthood in grey opossums suggest similarities between this marsupial species and commonly studied eutherian species in sexually dimorphic behaviour, sexually dimorphic morphology and in the effects of gonadectomy and gonadal hormone replacement on behaviour. Findings in grey opossums about the developmental timetable for the appearance of neural oestrogen receptors and gonadal hormone levels in blood and development of the aromatase and luteinising hormone-releasing hormone systems in brain also follow those of eutherians of similar neural and reproductive developmental stages. Since these similarities exist in the marsupial neonate as it undergoes differentiation and development during lactation rather than gestation, these findings suggest that mechanisms controlling early hormonal events in mammals are intrinsic to the neonate rather than due to control by maternal and/or placental factors. Future studies will continue to examine the hormonal control of the development of sex differences in behaviour utilising this marsupial species.

Future directions

Examine the contributions of other possible critical periods to the organisation by gonadal hormones of sexually dimorphic behaviour in grey opossums
The findings described here suggest that 16 days postnatal is a critical age for oestrogen action on the organisation of certain sexually dimorphic behaviour patterns. Since other sexually dimorphic behaviour patterns were not affected by blockage of oestrogen receptors with tamoxifen on 16 days postnatal, in future studies, earlier (4 to 8 days postnatal) and later (30 day postnatal) possible critical periods for the organising action of oestrogen on behaviour will be examined. Comparison of the findings

reported here with those of future studies will provide information on the relative contributions of oestrogenic influences at different developmental periods on the expression of sexually dimorphic behaviour patterns in this species.

Examine the contribution of peptide neurotransmitters to the organisation of sexually dimorphic behaviour

In rodents, sex differences exist in the morphology of neuropeptide-containing neurons in brain areas important in the control of sexually dimorphic behaviour (Baum 1992). Because the neuropeptide, substance P, is associated with the expression of aggressive and sexual behaviour (Steinberg *et al.* 1992) and because sex differences in its distribution associated with gonadal hormones have been demonstrated in rodent brain during development and in adulthood (Malsbury & McKay 1987; Malsbury & Brown 1989), in collaboration with Dr Allen Siegel we will examine grey opossum brain for sex differences in anatomical distribution and concentration of substance P-like immunoreactivity in developing animals and adults. If differences are found, we will manipulate hormonal environment and observe the effects of such manipulation on the substance P system.

Examine the role of oestrogen in the sexual differentiation of the female hypothalamus as measured by physiological responsiveness to pheromones

In some eutherian species, exposure to androgen perinatally alters the physiological responsiveness of females to pheromones produced by males (Smale *et al.* 1985). Because female grey opossums enter oestrus when exposed to pheromonal stimulation provided by males, future studies will assess the effects of alteration in the availability of gonadal hormones perinatally on the responsiveness of female grey opossums to male pheromones in adulthood.

Examine the role of hormones in the organisation and activation of maternal behaviour in grey opossums.

Newly parturient eutherian species display maternal behaviour immediately because of hormonal events associated with pregnancy and parturition. However, the relative contributions of hormones associated with parturition, including progesterone, oestradiol, prolactin and other lactogenic hormones and oxytocin, have not yet been defined clearly in the induction of this behaviour in eutherian mammals (Bridges & Mann 1994). Because parturition and the initial events of maternal behaviour are temporally separated in marsupials and because we have characterised aspects of maternal behaviour (Fadem *et al.* 1986) in grey opossums, this species will be used to study the role of hormones in the organisation and display of maternal behaviour.

Acknowledgments

I would like to acknowledge and to thank my collaborators on the projects described here including (in alphabetical order) Emilie Cole, Aysha Corbett, Dianne Erianne, Greg Erianne, Anne Etgen, John Harder, Larry Karen, Doris Kraus, Neil MacLusky, Marlene Schwanzel-Fukuda, Robert Schwartz, Rachel Sheffet, Lorraine Taylor-Ali, John Tesoriero, Michael Walters and Michael Whang. I would also like to acknowledge support for these studies by the National Institutes of Health and the National Science Foundation.

References

Adkins-Regan E, Orgeur P and Signoret J P (1989). Sexual differentiation of reproductive behaviour in pigs: Defeminizing effects of prepubertal estradiol. *Hormones and Behaviour*, 23, 290–303.

Arnold A P and Breedlove S M (1985). Organizational and activational effects of sex steroid hormones on vertebrate behaviour: A re-analysis. *Hormones and Behaviour*, 19, 479–498.

Baker P J, Moore H D M, Penfold L M, Burgess A M C and Mittwoch U (1990). Gonadal sex differentiation in the neonatal marsupial. *Monodelphis domestica. Development*, 109, 699–704.

Baum M J (1984). Differentiation of coital behaviour in mammals: A comparative analysis. *Neuroscience and Biobehavioural Reviews*, 3, 265–284.

Baum M J (1992). Neuroendocrinology of sexual behaviour in the male. In *Behavioural Endocrinology*. Eds J B Becker, S M Breedlove and D Crews. Massachusetts Institute of Technology, Boston. pp 97–130.

Baum M J and Erskine, M S (1984). Effects of neonatal gonadectomy and administration of testosterone on coital masculinization in the ferret. *Endocrinology* , 115, 2440–2444.

Biggins, J G (1984). Communication in possums: a review. In *Possums and Gliders*. Eds A Smith and I Hume. Surrey Beatty & Sons Ltd, Sydney. pp 35–57.

Bridges R E and Mann P (1994). Hormone-brain interactions in maternal behaviour. *Psychoneuroendocrinology*, 19, 611–622.

Burns R K (1961). Hormones in the differentiation of sex. In *Sex and Internal Secretions*. Ed Young. Tindall and Cox, London. pp 76–160.

Call R N, Catling P C and Janssens P A (1980). Development of the adrenal gland in the tammar wallaby (Macropus eugenii). *Australian Journal of Zoology*, 72, 551–558.

Davidson I A, Scott J E, Longstaff A, Abramovich D R and Pearson C K (1983). Estrogen conjugation in mid-trimester human fetal brain: a possible role in its sexual differentiation. *Journal of Steroid Biochemistry*, 18, 525–529.

Dunlap J L, Gerall A A and Carlton S F (1978). Evaluation of prenatal androgen and ovarian secretions on receptivity in female and male rats. *Journal of Comparative and Physiological Psychology*, 92, 280–288.

Eisenberg J F and Golani I (1977). Communication in metatheria. In *How Animals Communicate*. Ed J Sebeok. Indiana University Press, Bloomington. pp 575–599.

Etgen A M and Fadem B H (1987). Estrogen receptors in hypothalamus-preoptic area of male and female gray short-tailed opossums *Monodelphis domestica. General and Comparative Endocrinology*, 66, 441–446.

Etgen A E and Fadem B H (1989). Ontogeny of estrogen binding sites in the brain of gray short-tailed opossums. *Developmental Brain Research*, 49 131–133.

Fadem B H (1985). Evidence for the activation of female reproduction by males in a marsupial, the gray short-tailed opossum *Monodelphis domestica. Biology of Reproduction*, 33 112–116.

Fadem B H (1986). Chemical Communication in gray short-tailed opossums (*Monodelphis domestica*) with comparisons to other marsupials and with reference to monotremes. In *Chemical Signals in Vertebrates IV, Ecology, Evolution and Comparative Biology*. Eds D Duvall, D Muller-Schwarze and R M Silverstein. Plenum Press, New York. pp 587–607.

Fadem B H (1987). Activation of estrus by pheromones in a marsupial: Stimulus control and endocrine factors. *Biology of Reproduction*, 36, 328–332.

Fadem B H (1989a). The effects of pheromonal stimuli on estrus and peripheral plasma estradiol in female gray short-tailed opossums *Monodelphis domestica. Biology of Reproduction*, 41, 213–217.

Fadem B H (1989b). Sex differences in aggressive behaviour in a marsupial, the gray short-tailed opossum *Monodelphis domestica. Aggressive Behaviour*, 15, 435–441.

Fadem B H (1990). The effects of gonadal hormones on scent marking and related behaviour and morphology in female gray short-tailed opossums *Monodelphis domestica. Hormones and Behaviour*, 24, 459–469.

Fadem B H (1995). The effects of neonatal treatment with tamoxifen on sexually dimorphic behaviour and morphology in gray short-tailed opossums (*Monodelphis domestica*). *Hormones and Behaviour*, 29, 296–311.

Fadem B H and Barfield R J (1981). Neonatal hormonal influences on the development of proceptive and receptive feminine sexual behaviour in rats. Hormones and Behaviour, 15, 282–288.

Fadem B H and Cole E A (1985). Scent-marking in the grey short-tailed opossum *Monodelphis domestica*. *Animal Behaviour*, 33, 730–738.

Fadem B H and Corbett A (1993). Sex differences and the role of aromatization in the control of sexually dimorphic behaviour in gray short-tailed opossums *Monodelphis domestica*. *Hormones and Behaviour*, 27, 366–379.

Fadem B H and Corbett A (1997). Sex differences and the development of social behavior in a marsupial, the gray short-tailed opossum (*Monodelphis domestica*). *Physiology and Behavior*. (in press).

Fadem B H, Erianne G and Karen L (1989). The hormonal control of scent marking and precopulatory behaviour in male gray short-tailed opossums *Monodelphis domestica*. *Hormones and Behaviour*, 23, 381–392.

Fadem B H and Harder J D (1992a). Evidence for high levels of plasma androgen during postnatal development in a marsupial: The gray short-tailed opossum *Monodelphis domestica*. *Biology of Reproduction*, 46 105–108.

Fadem B H and Harder J D (1992b). Estrogen in peripheral plasma during postnatal development in gray short-tailed opossums. *Physiology and Behaviour*, 52, 613–616.

Fadem B H, Kraus D B and Sheffet R H (1986). Nest-building in gray short-tailed opossums: Temperature effects and sex differences. *Physiology and Behaviour*, 36, 667–670.

Fadem B H, MacLusky N and Walters M (1993). Neural aromatase activity in a marsupial, the gray short-tailed opossum *Monodelphis domestic*: Ontogeny during postnatal development and androgen regulation in adulthood. *Developmental Brain Research*, 74, 199–205.

Fadem B H and Schwanzel-Fukuda M (1990). The luteinizing hormone-releasing hormone (LH-RH) system in neonatally estradiol-feminized male gray short-tailed opossums (*Monodelphis domestica*). *Developmental Brain Research*, 53, 116–120.

Fadem B H and Schwartz R (1986). A sexually dimorphic suprasternal gland in the gray short-tailed opossum, *Monodelphis domestica*. *Journal of Mammalogy*, 67, 205–208.

Fadem B H, Taylor-Ali L and Erianne D C (1996). The hormonal induction of mating behavior in female gray short-tailed opossums (*Monodelphis domestica*). *Hormones and Behavior*, 30, 40–44.

Fadem B H and Tesoriero J V (1986). Inhibition of testicular development and feminization of the male genitalia by neonatal estrogen treatment in a marsupial. *Biology of Reproduction*, 34, 771–776.

Fadem B H, Tesoriero J V and Whang M (1992). Early differentiation of the gonads in the gray short-tailed opossum *Monodelphis domestica*. *Biology of the Neonate*, 61, 131–136.

Feder H H (1981). Hormonal actions on the sexual differentiation of the genitalia and the gonadotropin-regulating systems. In *Neuroendocrinology of Reproduction*. Ed NT Adler. New York, Plenum Press. pp 89–126,

Gerall A A, Hendricks S E, Johnson L L and Bounds T W (1967). Effects of early castration in male rats on adult sexual behaviour. Journal of *Comparative and Physiological Psychology*, 64 (2), 206–212.

Goy R W and McEwen B S (1980). *Sexual Differentiation of the Brain*. Massachusetts Institute of Technology, Boston.

Goy R W, Phoenix C H and Young WC (1962). A critical period for the suppression of behavioural receptivity in adult female rats by early treatment with androgen. *Anatomical Record*, 142, 307.

Hendricks S E and Weltin M (1976). Effects of estrogen given during various periods of prepubertal life on the sexual behaviour of rats. *Physiological Psychology*, 4, 105–110.

Hinds L A, Reader M, Wernberg-Moller S and Saunders N R (1992). Hormonal evidence for induced ovulation in *Monodelphis domestica*. *Journal of Reproduction and Fertility*, 95, 303–312.

Hunsaker D and Shupe D (1977). Behaviour of New World marsupials. In *The Biology of Marsupials*. Ed D Hunsaker. Academic Press, New York. pp 279–344.

Kraus D and Fadem B H (1987). Laboratory management of the gray short-tailed opossum II. Reproduction, development and physiology. *Laboratory Animal Science*, 37, 478–482.

Malsbury C W and Brown S A (1989). Testosterone masculinizes the development of the substance P innervation of the medial nucleus of the amygdala and bed nucleus of the stria terminalis. *Society for Neuroscience Abstracts*, 39, 5.

Malsbury C W and McKay K (1987). A sex difference in the pattern of substance P-like immunoreactivity in the bed nucleus of the stria terminalis. *Brain Research*, 420, 365–370.

McCrady E (1938). The embryology of the opossum. *American Anatomical Memoirs*, 40, 544–546.

McEwen B S, Lieberburg I, Chaptal C and Krey L C (1977). Aromatization: Important for sexual differentiation of the neonatal rat brain. *Hormones and Behaviour*, 9, 249–263.

Moore H D M and Thurstan S M (1990). Sexual differentiation in the grey short-tailed opossum, *Monodelphis domestica*, and the effects of oestradiol benzoate on development in the male. *Journal of Zoology*, 221, 639–658.

Perakis A and Stylianopoulou F (1986). Effects of a prenatal androgen peak on rat brain sexual differentiation. *Journal of Endocrinology*, 108, 281–285.

Russell E M (1970). Observation on the behaviour of the red kangaroo (Megaleia rufa) in captivity. *Zeitschritt Fur Tierpsychologie*, Tierennahrung, 27, 385–404.

Saunders N R, Adam E, Reader M and MØllgärd K (1989). *Monodelphis domestica* (grey short-tailed opossum): An accessible model for studies of early neocortical development. *Anatomical Embryology*, 180, 227–236.

Schwanzel-Fukuda M, Fadem B H, Garcia M S and Pfaff D W (1988). The immunocytochemical localization of luteinizing hormone-releasing hormone (LHRH) in the brain and nervus terminalis of the adult and early neonatal gray short-tailed opossum *Monodelphis domestica*. *Journal of Comparative Neurology*, 276, 44–23.

Sharman G B (1970). Reproductive physiology of marsupials. *Science* , 167, 1221–1228.

Smale L, Nelson N J and Zucker I (1985). Neuroendocrine responsiveness to estradiol and male urine in neonatally androgenized prairie voles. *Journal of Reproduction and Fertility*, 74, 491–496.

Steinberg A, Shaikh M B and Siegel A (1992). The role of substance P in medial amygdaloid facilitation of defensive rage behavior in the cat. *Society for Neuroscience Abstracts*, 152, 3.

Stewart J and Cygnan D (1980). Ovarian hormones act early in development to feminize adult open-field behaviour in the rat. *Hormones and Behaviour*, 14, 20–32.

Stonerook M J and Harder J D (1992). Sexual maturation in female gray short-tailed opossums, *Monodelphis domestica*, is dependent upon male stimuli. *Biology of Reproduction*, 46, 290–294.

Toran-Allerand C D (1984). Gonadal hormones and brain development: Implications for the genesis of sexual differentiation. *Annals of the New York Academy of Science*, 435 101–111.

Tyndale-Biscoe C H and Janssens P A (1988). The *Developing Marsupial. Models for Biomedical Research*. Springer-Verlag, Berlin.

VandeBerg J L (1983). The gray short-tailed opossum: a new laboratory animal. *International Laboratory Animal Research News*, 29, 9.

vom Saal FS, Grant W, McMullin C and Laves K (1983). High fetal estrogen concentrations: correlation with increased adult sexual performance and decreased aggression in male mice. *Science*, 220 1306–1308.

Wade G N, Gray J M and Blaustein J D (1993). Effects of estrogens and antiestrogens on eating behaviour, metabolism and energy balance. *Oncology* (Supplement), 7(11), 61–68.

Wemmer C and Collins L (1978). Communication patterns in two phalangerid marsupials, the Grey Cuscus (*Phalanger gymnotis*) and the Brush Possum (*Trichosurus vulpecula*). *Saugetierkundliche Mitteilunge*, 3, 161–172.

Wilson J, George F and Griffin J (1981). The hormonal control of sexual development. *Science*, 211, 1278–1284.

Yalom J D, Green R and Fisk N (1973). Prenatal exposure to female hormones. *Archives of General Psychiatry*, 28, 554–561.

CHAPTER 6

THE USE OF *MONODELPHIS DOMESTICA* AS A MODEL FOR MUSCULO-SKELETAL DEVELOPMENT

Charles W Archer, Heather Morrison, Michael T Bayliss and Mark W J Ferguson

INTRODUCTION

A variety of animal species has been used over the last century for studying skeletogenesis (for review, Murray 1985) although the chick embryo has attracted the most attention over the recent past. Primarily this is because the developing limb of the chicken embryo has been used extensively for the study of pattern formation with whole mount-stained preparations of the cartilaginous elements being used as convenient indicators of changes in skeletal pattern (Tickle & Eichele 1994). Similarly, the chick has also been used to study early events in chondrogenic differentiation (Solursh 1986). Because of the fundamental differences in the composition of avian cartilage (Craig *et al.* 1987) and the general biology of avian bone when compared to mammalian species, the chick is an inappropriate model for later stages of differentiation and morphogenesis. For example, when one considers the development of articular cartilage, the chick has a fibrous articular surface, rich in collagen type I (Craig *et al.* 1987) but no collagen type II, thus contrasting with mammals where collagen type II is predominant throughout the tissue.

More recently, the mouse and rat have been used more widely for addressing fundamental questions of skeletal biology, though the use of eutherian mammals has several limitations. For example, mothers have to be sacrificed in order to harvest the embryos and even determining the stage of gestation can be unreliable. In terms of experimental studies, manipulating embryos *in utero* is technically difficult. Also, compared with marsupials, eutherian mammals are born at later developmental stages hence limiting the usefulness of litters of neonates.

We have circumvented the above problems by utilising the South American grey short-tailed opossum, *Monodelphis domestica*, as a model system for studies of joint and articular cartilage development. It has several advantages for the study of skeletogenesis. First, it is a non-pouched marsupial, the young suspending themselves from the mothers' teats. Second, whilst at birth the forelimbs are precociously developed, the hindlimbs are at an early stage of development, being at the point of overt chondrogenesis (equivalent to Hamburger & Hamilton stage 25/26 in the chick; Hamburger and Hamilton 1951). Third, the hindlimbs 'catch up' with the forelimbs by 30 days post-partum thereby making them a good model for the study of differential growth.

We have been using *Monodelphis* to study the development of knee articular cartilage. The rationale is that if we know how articular cartilage develops, then we might gain insights into how it behaves during degenerative joint disease and we may be better placed to develop strategies for stimulating or augmenting both extrinsic and intrinsic repair.

Accordingly, we have carried out extensive immunohistochemical mapping analyses of the patterns of extracellular matrix component distribution in the epiphyseal and subsequent articular cartilage during development. In addition, we have immunolocalised the distribution of a number of growth factors that are likely to be central to the regulation of cartilage development and growth, some of which is discussed in this chapter.

GENERAL GROWTH CHARACTERISTICS OF *MONODELPHIS* LIMBS

At parturition (embryonic day 14), it can be seen from alcian blue/alizarin red stained whole mounts that the forelimbs are well developed and possess claws whilst the hindlimbs are still buds at the point of initial chondrification (Figure 6.1). Subsequent overall growth of the limbs expressed as increase in length from the day of parturition to 30 days postnatal is shown in Figure 6.2. It can be seen that it is not until between 20 and 30 days that there is a convergence in the two plots and that at 30 days , the hindlimb has overtaken the forelimb in overall length. A very similar pattern is seen if one plots the increase in humerus length against femur length over the same time scale (Figure 6.3). On an expanded scale between parturition and 15 days (Figure 6.4), it can be seen that although a great disparity exists between the overall lengths of the forelimbs and hindlimbs (and the constituent proximal elements), the pattern of length increase is very similar. These data suggest that the elements are responding to common circulating growth factors. The 'catch up' between femur and humerus occurs between 15 and 26 days. Although preliminary, the

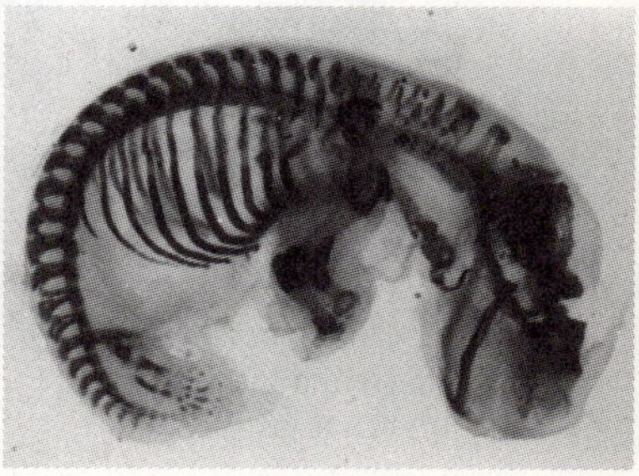

Figure 6.1 Whole-mount embryo preparation of *Monodelphis domestica* at parturition stained with alcian blue and alizarin red. It can be seen that the forelimbs are well developed with the proximal elements possessing ossified bony diaphyseal collars. In contrast, the hindlimbs are very much smaller and the elements are in the early stages of chondrogenesis.
Crown-rump length 10mm.

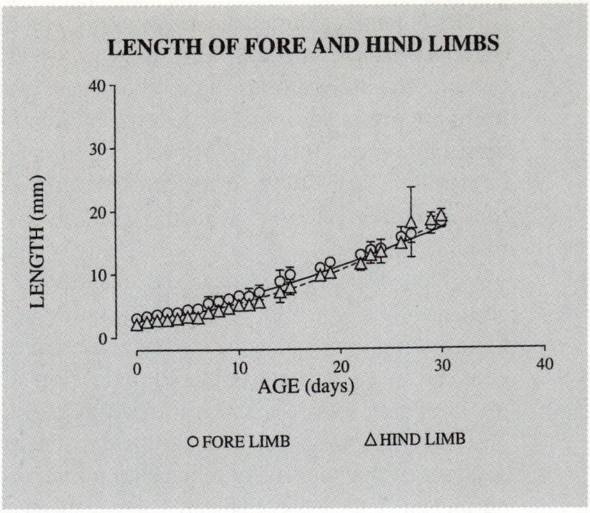

Figure 6.2 Plot of the increase in length of *Monodelphis* fore-limbs and hindlimbs from parturition to 30 days post-partum. Note: The data points represent the mean limb lengths of 5 animals from each developmental stage; the bars represent standard deviation of the means.

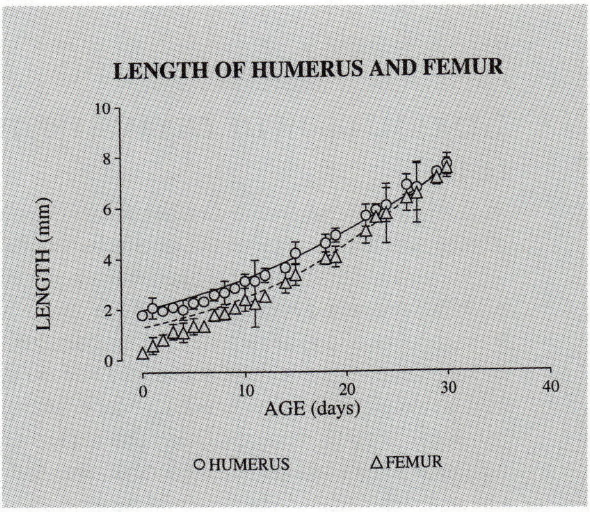

Figure 6.3 Plot of the increase in length of the humerus and femur in *Monodelphis* from parturition to 30 days post-partum.
Notes:
(1) A similar plot is obtained to the whole limb lengths in Figure 6.2.
(2) The femur 'catches-up' with the humerus at 26 days.
(3) Bars represent standard deviation of the means.

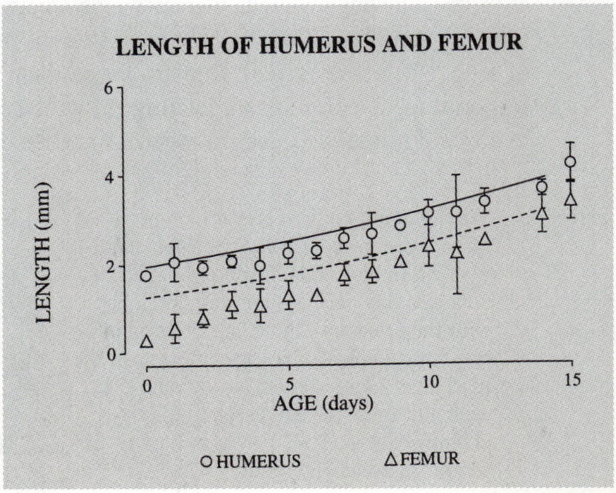

Figure 6.4 Detailed plot of the increase in length of humerus and femur between parturition and 15 days post-partum.
Notes:
(1) It can be seen that the growth rates are very similar indicating that 'catch-up' occurs between 15 and 26 days post-partum.
(2) Bars represent standard deviation of the means.

data also indicate that *Monodelphis* would provide an interesting model for the study of skeletal growth mechanisms, particularly in relation to differential growth. In addition, the accessibility of the hindlimbs for experimental manipulation at early developmental stages adds considerably to the versatility of this model system.

The distribution of glycosaminoglycans and proteoglycans in developing articular cartilage

We have utilised a panel of monoclonal and polyclonal antibodies, which recognise a number of epitopes on both proteoglycan core proteins and glycosaminoglycan chains. Monoclonal antibody (Mab) 3B3 recognise stubs of chondroitin-6-sulphate after chondroitinase ABC digestion whilst antibody 2B6 recognises stubs of chondroitin-4-sulphate after chondroitinase ABC digestion. The latter digestion procedure is used since if chondroitinase AC is used, Mab 2B6 will identify the stubs carrying the neoepitopes of dermatan sulphate also (that is chondroitin-B in old nomenclature). Mab 5D4 recognises epitopes on keratan sulphate in its native form without predigestion. (Caterson *et al.* 1985; Mabs were a kind gift of Professor B Caterson.) Polyclonal antibodies raised against fibromodulin, decorin and biglycan were used to detect these small leucine-rich proteoglycans (kind gifts of Professor D Heinegard.) Immunolocalisation procedures were as described by Archer *et al.* (1994) and employed both fluorescent and silver-enhanced secondary detection systems. Briefly, fluorescent procedures utilised an indirect fluorescein-conjugated species and isotype specific secondary antibody system. The silver system utilised an indirect gold particle-conjugated secondary antibody which was subsequently enhanced.

It is useful to consider the distribution patterns of the components individually prior to comparative discussion.

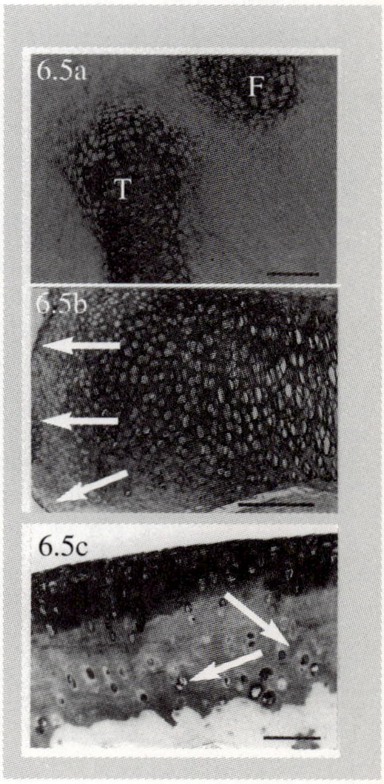

Figure 6.5 Localisation of chondroitin-6-sulphate using antibody 3B3 after chondroitinase ABC predigestion.
(a) Longitudinal section of a developing knee joint of *Monodelphis* at 1 day. Chondroitin-6-sulphate can be detected throughout the epiphyses of both the femoral (F) and tibial (T) anlagen.
(b) At 10 days postpartum, the presumptive articular cartilage of the tibial condyle is very weakly labelled (arrows) contrasting the intense label observed in the epiphyseal round cell and flattened cell zones.
(c) Adult articular cartilage showing intense labelling in the upper third of the cartilage depth. Some cellular labelling is seen in deeper regions of the tissue (arrows).
Scale bars = 100 μm.

Chondroitin-6-sulphate

At parturition, the entire chondrifying femoral and tibial/fibial anlagen was uniformly immunoreactive (Figure 6.5a). During early development (up to 11 days postnatal), the pattern of localisation became more concentrated in the interterritorial matrix, but there was a paucity in binding particularly at the epiphyseal ends, in the region of presumptive articular cartilage (Figure 6.5b). Subsequent development of the cartilage showed that the entire articular cartilage was immunoreactive, but binding intensity was greatly increased in the outer third of the tissue depth (Figure 6.5c), a pattern which persisted into adulthood.

Chondroitin–4-sulphate

At parturition, the entire analagen matrix was immunoreactive although it was concentrated pericellularly and in the diaphyseal hypertrophic zone. Least labelling was seen at the epiphyseal ends of the element (Figure 6.6a). Subsequently, the epiphyseal cartilage became strongly immunopositive (by 9 days). As with chondroitin-6-sulphate, the presumptive articular cartilage was not strongly labelled (Figure 6.6b). After the formation of the secondary centre of ossification at 2 months (when the articular cartilage is physically separated from the epiphyseal growth plate cartilage), the pattern remained rather constant until adulthood in that immunopositive labelling was again more pronounced in the outer half of the articular tissue depth, especially pericellularly (Figure 6.6c).

Keratan sulphate

Using Mab 5D4, keratan sulphate was detected in the diaphyseal and metaphyseal regions of the cartilage anlagen at parturition and was particularly marked pericellularly. As with the chondroitin sulphates, there was a paucity in labelling in the presumptive articular cartilage (Figure 6.7a). By 11 days the pericellular pattern was still prevalent, although the developing articular surfaces were only slightly immunoreactive (Figure 6.7b). During further development, keratan sulphate labelling remained weakly through the matrix, but strong pericellularly. At two months, keratan sulphate was localised through the articular cartilage with

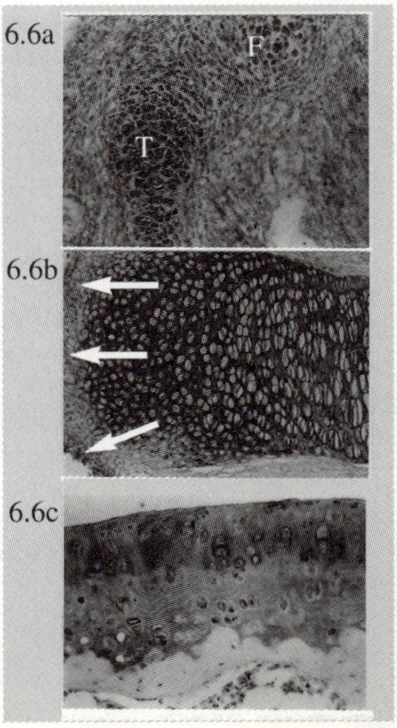

Figure 6.6 Localisation of chondroitin-4-sulphate using antibody 2B6 after chondroitinase ABC predigestion.
(a) At day 1, C-4-S is detected pericellularly throughout the tibial (T) and femoral (F) anlagen.
(b) By 10 days postpartum, the localisation now extends throughout the matrix, but like C-6-S, the presumptive articular cartilage labels weakly (arrows).
(c) In the adult, C-4-S is localised preferentially to the upper half of the tissue depth with more intense pericellular label. Scale as in Figure 5.

strong pericellular labelling around the basal cells. However, by three months interterritorial labelling became prominent in the basal region whilst intense labelling was detected through the outer third of the cartilage depth. The latter pattern was maintained into adulthood (Figure 6.7c).

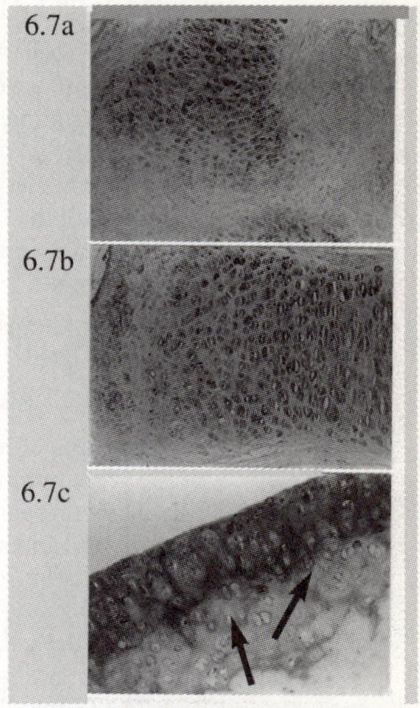

Figure 6.7 Localisation of keratan sulphate using antibody 5D4.
(a) At parturition, labelling is mainly cellular and pericellular and is stronger in the metaphyseal and diaphyseal regions of the femur.
(b) By 10 days postpartum, the labelling remains around the cells with little evidence of extensive labelling within the epiphysis.
(c) In the adult, most of the label is restricted to the upper half of the tissue depth with strong interterritorial labelling which is also evident in the basal half of the tissue (arrows).
Scale as in Figure 5.

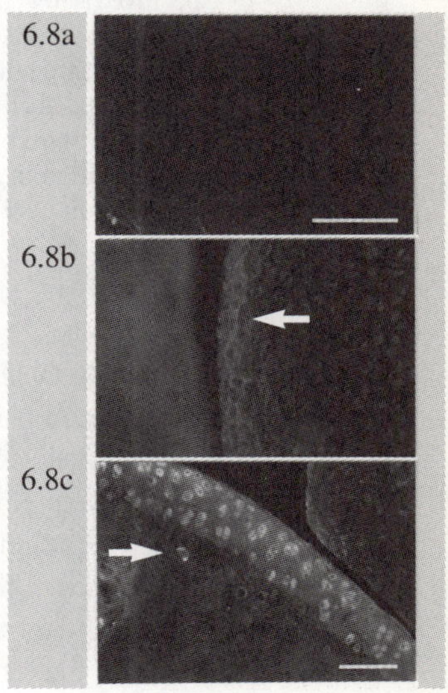

Figure 6.8 Localisation of fibromodulin.
(a) At parturition, fibromodulin was not detected in the developing epiphyses.
(b) After 8 days fibromodulin was detected intercellularly between the articular surface three or four cells (arrowed).
(c) In the adult both cellular and matrix labelling was evident through the upper half to two-thirds of the tissue depth. Some cellular fluorescence was observed in the more basal regions (arrowed).
Scale bars = 100 μm; (a) and (b) are at the same magnification.

Fibromodulin

Fibromodulin was not detected in the cartilage matrix at parturition (Figure 6.8a). By 8 days it was detected in the presumptive articular cartilage occupying 3 or 4 cells of the surface depth (Figure 6.8b). By 27 days, fibromodulin could also be detected in the matrix of the hypertrophic zone, and by two months when the secondary centre of ossification develops, the entire articular cartilage is immunopositive. Within the adult cartilage, labelling is observed in all regions except the most basal regions where occasional cellular fluorescence was observed (Figure 6.8c).

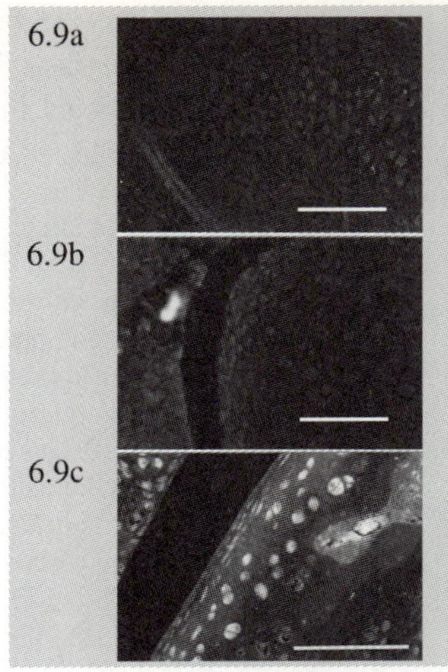

Figure 6.9 Localisation of decorin.
(a) At parturition, decorin was weakly detectable in the epiphyseal cartilage, but more strongly so in the cells of the metaphysis and diaphysis.
(b) At 6 days decorin became more intensely labelled in the developing articular surface cells although still detectable in the rest of the epiphyses.
(c) In the adult decorin was localised within both the cells and the matrix but predominantly in the upper third of the tissue depth.
Scale bars = 100 µm.

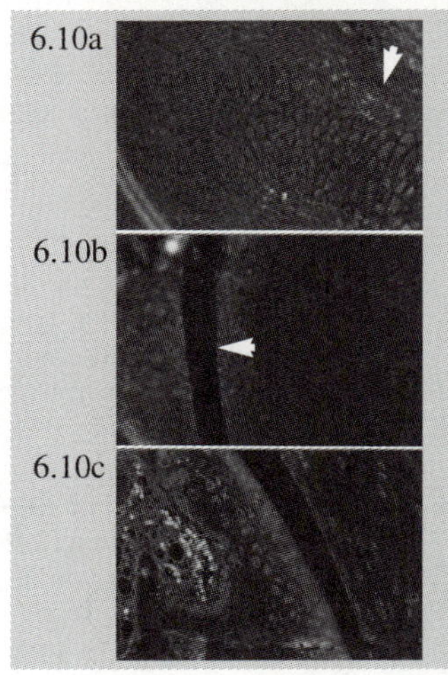

Figure 6.10 Localisation of biglycan.
(a) At parturition, biglycan was present cellularly throughout the cartilage anlage whilst matrix immunopositivity was present in the metaphyseal region (arrowed).
(b) At 10 days, the matrix surrounding the uppermost 2-3 cell layers were positive (arrowed).
(c) In the adult cellular localisation was apparent throughout the tissue depth, and matrix labelling was also detected in the upper third of the tissue depth.
Magnification in a) and b) as for Figure 6.8a; and in c) as for Figure 6.8c.

Decorin

Decorin was first detectable in cartilage prior to parturition within the diaphyseal chondrocytes. At parturition, immunolabelling was present throughout the anlagen both cellularly and extracellularly though more intense at the developing metaphyses (Figure 6.9a). From 6 days, immunopositivity became intensified to the articular epiphyses (Figure 6.9b) and in later devolopment, to the surface of the articular cartilage. However, from the establishment of the secondary centre of ossification to adulthood, decorin was detected throughout the cartilage matrix though stronger in the outer third of the tissue depth. Strong immunopositivity was also localised within the chondrocytes again in the upper half of the tissue depth (Figure 6.9c).

Biglycan

Like decorin, biglycan was present throughout the cartilage anlagen at parturition and particularly intracellularly (Figure 6.10a). However, by 2 days postpartum, its localisation was restricted to to the presumptive articular surfaces and the perichondrium and by 10 days, it was also evident in the extracellular matrix of the hypertrophic zone (Figure 6.10b). At 2 months to adulthood, the pattern of binding was very similar to decorin eventually being present through the tissue depth, but preferentially localised to the outer third of the matrix depth (Fig 6.10c).

CARTILAGE GROWTH

Cartilage grows by a combination of matrix secretion, cell hypertrophy and cell proliferation (Rooney *et al.* 1984). In articular cartilage, matrix secretion accounts for the the greatest proportion of volume increase, particularly since cell density decreases during growth (Stockwell 1979). Cell division, on the other hand, accounts for the least proportion of volume increase, although it is essential for sustained growth. An important question in relation to growth in cartilage is to what extent is it appositional, as opposed to interstitial? A previous report (Mankin 1962) using the immature rabbit as a model, described two bands of proliferating cells (as detected by autoradiography of tritiated thymidine labelled chondrocytes) within the tissue, one band above the secondary centre of ossification and the other near the articular surface. However, the study had the drawback of a limited number of developmental stages, which in part, reflects the limitation in the use of eutherian mammals for such studies. Nevertheless, it was evident that growth was a combination of appositional and interstitial mechanisms.

We have extended these studies using *Monodelphis*. Proliferating chondrocytes were localised using an antibody (PC10 anti-PCNA) which recognises an acidic 36kDa non-histone nuclear protein which is an auxillary protein for DNA polymerase whose level peaks during S-phase of the cell cycle. Insulin-like growth factors (IGF) are known to play a central role in cartilage matrix metabolism and are mitogenic for chondrocytes (reviewed by Hill & Logan, 1992). Consequently, we have used monoclonal antibodies to detect IGF1 and IGF2 in the developing distal femoral cartilage from parturition to the adult. In terms of the development of the articular cartilage *per se,* we have focused on the patterns of distribution of the above antibodies between 2 months (when the secondary centres of ossification is developing and the articular cartilage is physically delineated from growth plate cartilage) and at 3, 4, 6 and finally at 24 months, which is the mature adult condition. The distribution patterns of IGF1, IGF2 and PCNA, were very similar and showed the same developmental trend namely a restriction towards the articular surface during growth. Thus, at two months, the upper (outer) third of the

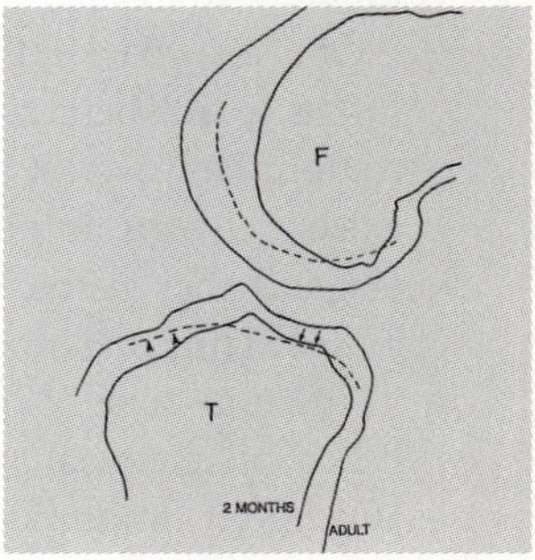

Figure 6.11 Camera lucida drawings of longitudinal sections of *Monodelphis* knees at 2 months and 8 months (sexually mature adult) superimposed. The diagram shows that daughter cells originating in the surface zone at two months may occupy basal regions of the articular cartilage at 8 months (arrows) or be replaced altogether through endochondral ossification of the subchondral plate during growth (arrowheads).
Key: — Level of subchondral bone in the adult.
T: tibia F: femur

tissue depth contains immunopositive cells which gradually become restricted to the 2 or 3 layers of chondrocytes at the articular surface.

Figure 6.11 shows camera lucida drawings taken from longitudinal sections of knees between 2 and 8 months which have been superimposed using the top of the epiphyseal growth plate as a reference point. From these drawings it can be seen that if one considers the tibial plateau, the position of the articular surface at 2 months corresponds to the basal zone at 8 months, thus representing the growth during that time. Therefore, if one considers a dividing chondrocyte at the articular surface at 2 months, then the progeny of that division at eight months would occupy a basal position (see figure legend for further explanation).

DISCUSSION

The use of marsupials in biomedical research is becoming more widespread. However, a main limitation to their use as a 'model' is the extent to which they 'mimic' the eutherian counterpart and ultimately man. From an embryological perspective, because marsupials are born at developmentally earlier stages than eutherian mammals, they clearly have a number of advantages over standard laboratory animals such as rats or mice. *Monodelphis,* in particular, has proved useful, since it is small (being the size of a rat), and breeds fairly well in captivity.

Studies of the development of articular cartilage have not been carried out in any detail in other mammals so we do not know how reliable *Monodelphis* is as a model for eutherians. However, there are differences in the distribution of certain extracellular matrix components between *Monodelphis* and other mammals. Perhaps the most striking is keratan sulphate which, in pig (Zanetti *et al.* 1985), human (Archer *et al.* 1990) and rabbit (personal observations, C W Archer) is not found in the surface articular zone, but increases with depth from the articular surface. However, in *Monodelphis*, intense labelling is found in the surface zone and more restricted labelling in the deeper zone. Whilst such a species difference may be considered a disadvantage, it questions some of the ideas which have been generated concerning the functional significance of the patterns of extracellular matrix component distribution. For example, it has been suggested that the substitution of keratan sulphate chains on to aggrecan was epigenetically controlled and occurred in regions of low oxygen tension, that is, in the deeper zone of the cartilage farthest away from the source of nutrient (Scott & Haigh 1988). Articular cartilage derives its nutrient by diffusion from the articular surface with little or none from the underlying subchondral bone. However, unless the entire physiology of the joint is different in *Monodelphis*, such a notion must be questioned.

One drawback of immunohistochemistry is that it is not stochiometric, intensity of labelling often reflecting epitope accessibility rather than concentration. However, it has been shown for Mabs 3B3 and 2B6 that a correlation exists between immunolabelling density at the ultrastructural level using silver and the chemical concentration of chondroitin sulphate (Ratcliffe *et al.* 1984). Thus, maximal labelling density for chondroitin-6-sulphate was found in the mid-zone of human articular cartilage which correlates well with published biochemical analyses (Bayliss *et al.* 1983). Our data (which also utilises silver/gold immunocytochemistry) suggest that in *Monodelphis*, maximal concentration of chondroitin-6-sulphate occurs in the upper third of the tissue. Again, our data contrasts that reported for other mammalian cartilages, but would further emphasise the points raised in relation to keratan sulphate.

The distributions of the small leucine-rich proteoglycans during early joint formation are particularly interesting for a number of reasons. First, it appears that fibromodulin delineates presumptive articular cartilage from 8 days when it is first detected in the limb skeleton. In contrast, both decorin and biglycan are present within the embryonic cartilage albeit at different locations (in the hypertrophic and entire cartilage matrix respectively). In this context, the current data, in part, confirm earlier reports which have localised dermatan sulphate proteoglycan II in bovine tissue (Poole *et al.* 1986) and the distributions of both biglycan and decorin in human skeletal tissue albeit at a single developmental stage (Bianco *et al.* 1990). Consequently, this is a first report which has documented (in comparative terms) the appearance and spatio-temporal distributions of these proteoglycans in a defined developmental sequence. It can be seen that the three molecular species occupy differing spatial distributions at a variety of stages and that these distributions can change markedly within a short time. For example, biglycan was present throughout the matrix at parturition, but was restricted to the presumptive articular cartilage by 2 days. We do not know the significance of these observations, nor indeed, how they are controlled. However, it would suggest that these small proteoglycans play multifunctional roles throughout the developmental process. The role of decorin and, to a lesser extent, biglycan in sequestering TGFß within the extracellular matrix must be noted with interest particularly since both become intensely immunopositive in the upper third of the tissue depth in later development.

The restriction of proliferative cells towards the articular surface between two and eight months (and similarly the localisation of mitogenic growth factor) suggest that articular cartilage grows appositionally that is, from the articular surface and further suggest that there exists a chondrogenic 'stem' cell within the surface of cartilage whose progeny will occupy more basal positions. Consequently, the phenotype that a chondrocyte adopts is positionally specified. Those daughter cells which originate at two months will, in the adult tissue, become basal cells, perhaps in the calcified cartilage layer above the bony subchondral plate. Cells which originate later will occupy more superior positions within the tissue and thus possess a different phenotype. Such a concept may provide insights into why articular cartilage has a limited ability to repair since, during the onset of osteoarthritis, it is the surface cells that are either lost or die. It has yet to be established whether *Monodelphis* represents a more global growth pattern for mammals since our data partly contrast that of Mankin who observed a second band of proliferating cells above the subchondral plate in neonate rabbits (Mankin 1962). However, morphometric data emerging from other work also suggest that the articular cartilage of rabbits grows by apposition (E Hunziker, personal communication).

In conclusion, *Monodelphis* has provided many new insights into the development of articular cartilage. The added advantage of young being born at an immature stage compared with newborn eutherian, with hindlimbs that are retarded in their stage of growth compared with their forelimbs, facilitates the study of earlier stages of skeletogenesis in a mammalian species. It would also appear to hold considerable scope for the study of differential growth of the appendicular skeleton.

Acknowledgments

Charles W Archer, Heather Morrison, Michael T Bayliss and Mark W J Ferguson would like to thank the Arthritis and Rheumatism Council of Great Britain for financial support.

References

Archer C W, McDowell J, Bayliss M T and Bentley G (1990). Phenotypic modulation in sub-populations of human articular chondrocytes *in vitro*. *Journal of Cell Science*, 97, 361–371.

Archer C W, Morrison E H and Pitsillides A A (1994). Cellular aspects of the development of diarthrodial joints and articular cartilage. *Journal of Anatomy*, 184, 447–456.

Bayliss M, Venn M, Maroudas A and Ali S Y (1983). Structure of proteoglycans from different layers of human articular cartilage. *Biochemical Journal*, 209, 387–400.

Bianco P, Fisher L W, Young M F, Termine J D and Gehron Robey P (1990). Expression and localisation of the two small proteoglycans biglycan and decorin in developing human skeletal and non-skeletal tissues. *Journal of Histochemistry and Cytochemistry*, 38, 1549–1563.

Caterson B, Christner J E, Baker J R and Couchman J R (1985). Production and characterisation of monoclonal antibodies directed against connective tissue proteoglycans. *Federation of American Societies for Experimental Biology Journal*, 44, 386–393.

Craig F M, Bentley G and Archer C W (1987). The spatial and temporal pattern of collagens I and II and keratan sulphate in the developing chick metatarso phalangeal joint. *Development*, 99, 383–391.

Hamburger V and Hamilton H (1951). A series of normal stages in the development of the chick embryo. *Journal of Morphology*, 88, 49–92.

Hill D J and Logan A (1992). Peptide growth factors and their interactions during chondrogenesis. *Progress in Growth Factor Research*, 4, 45–68.

Murray P D F (1985). *Bones*. Cambridge University Press, Cambridge. pp 1–203.

Mankin H (1962). Localisation of tritiated thymidine in articular cartilage of rabbits. I. Growth in immature cartilage. *Journal of Bone and Joint Surgery*, 44A, 682–688.

Poole A R, Webber C, Pidoux I, Choi H and Rosenburg C (1986). Localisation of a dermatan sulphate proteoglycan (DS-PGII) in cartilage and the presence of an immunologically related species in other tissues. *Journal of Histochemistry and Cytochemistry*, 34, 619–625.

Ratcliffe A, Fryer P R and Hardingham T E (1984). The distribution of aggregating proteoglycans in articular cartilage: comparison of quantitative immunoelectron microscopy with radioimmunassay and biochemical analysis. *Journal of Histochemistry and Cytochemistry*, 32, 193–201.

Rooney P, Archer C and Wolpert L (1984). Morphogenesis of cartilaginous long bone rudiments. In *The Role of the Extracellular Matrix in Development*. Ed R.Trelstad, Alan R Liss, New York. pp 305–322.

Scott J E and Haigh M (1988). Keratan sulphate and the ultrastructure of cornea and cartilage: a 'stand in' for chondroitin sulphate in conditions of oxygen lack. *Journal of Anatomy*, 158, 95–108.

Solursh M (1986). Environmental regulation of limb chondrogenesis. In *Articular Cartilage Biochemistry*. Eds K Kuettner, R Schleyerbach and V Hascall, Raven Press, New York. pp. 145–161.

Stockwell R A (1979). *Biology of Cartilage Cells*. Cambridge University Press. Cambridge. pp. 1–329.

Tickle C and Eichele G (1994). Vertebrate limb development. *Annual Review of Cell Biology*, 10, 121–152.

Zanetti M, Ratcliffe A and Watt F M (1985). Two sub-populations of differentiated chondrocytes identified with a monoclonal antibody to keratan sulphate. *Journal of Cell Biology*, 101, 53–59

SECTION 2

GENETICS

CHAPTER 7

GENETICS OF TAMMAR WALLABIES

Desmond W Cooper and Louise M McKenzie

INTRODUCTION

Geneticists have always concentrated on a few organisms — *Escherichia coli*, *Neurospora crassa*, *Drosophila melanogaster*, *Caenorhbaditis elegans* and *Mus musculus*, for example. An in-depth knowledge of a single species is assumed to be easily generalised to other species within the group of organisms to which it belongs. The organisms chosen for study are those that can be bred quickly, cheaply and in large numbers. It must then be asked why would one want to use the tammar, *Macropus eugenii* for genetic studies. The tammar is large, breeds only once a year under ordinary circumstances, and is a mammal with the same general characteristics as many other better studied mammals.

The answer lies in its being a marsupial and in the value of comparative biological studies. Subjects such as X-inactivation, sex differentiation, sex differences in recombination and genetic variability for the major histocompatibility complex (MHC) region have given a different slant when studied in a marsupial as opposed to a eutherian. A full understanding of the evolution of reproduction in mammals is not possible without consideration of marsupials and genetic analysis is a powerful tool for such studies. The low chromosome number of marsupials is of interest in comparative gene mapping and could help establish the linkage relationships of ancestral mammals.

Three marsupial species have been bred in captivity with the explicit intention of doing genetic work. Two, *Sminthopsis crassicaudata* and *Monodelphis domestica* can be bred in larger numbers than the tammar. However, we have recently shown that a wide subspecies cross is possible in tammars which, to some extent, compensates for its slow rate of reproduction. In addition, manipulation of its reproduction allows 4 to 5 offspring per year. Most of these must be sacrificed for tissue samples since they are too immature for independent existence, although a small number can be fostered into the pouches of other females. This, together with the tammar's possession of the phenomenon of embryonic diapause influenced by both light and lactation, allow it to be used for studies not possible in other mammals.

In this chapter we summarise the results obtained since the mid-1980s.

DISTRIBUTION AND GENETIC RELATIONSHIPS BETWEEN POPULATIONS

In 1629 Francisco Pelsaert, captain of the Dutch East India Company ship *Batavia*, was shipwrecked on the Abrolhos Island Group off the Western Australian coast. It was here that he discovered the tammar. His description is the first known of an Australian macropodid. While common on some islands, tammars are considered vulnerable to extinction (Kennedy 1992). Their original range included semi-arid areas of the south-west of Western Australia (WA) and South Australia (SA), five WA islands and four SA islands (Table 7.1). Tammars remain on all five WA islands (East and West Wallabi [or Houtman and Abrolhos respectively], Garden, Middle and North Twin Peaks) and are abundant on Kangaroo Island (KI) in SA. There are remnant populations in south-west WA, but tammars are probably extinct on mainland SA and three SA islands (Flinders, St Peter and Thistle). Tammars have been introduced to Granite, Greenly, Wardang and Boston islands in SA. In the 1870s, tammars were taken to New Zealand by Sir George Grey and are now found on Kawau Island and a large area around Rotorua (Table 7.1).

Table 7.1 Former and current distribution of populations of tammars.

POPULATION	AREA OF ISLANDS (HA)	SUBSPECIES	COMMENTS
Flinders	4000–5000	SA	extinct
St. Peter	approx. 4000	SA	extinct
Thistle	4113	SA	extinct
St. Francis*	809	SA	fossil remains only
Reevseny*	approx. 500	SA	fossil remains only
Wedge*	967	SA	fossil remains only
Boston	967	SA	introduced 1971
Granite"	32	SA	introduced 1970's
Greenly	141	SA	introduced 1905
Kangaroo	450 000	SA	extant
Mainland SA	na	SA	probably extinct
Mainland WA	na	SA	extant
East and West Wallabi^	894	WA	extant
Garden	1054	WA	extant
Middle	1036	WA	extant
North Twin Peaks	272	WA	extant
Kawau	2075	SA	introduced in 1870s
Rotorua environs	na	SA	introduced in 20th century

Key:
SA=South Australia; WA=Western Australia; na = not applicable.
* It is unclear whether these were naturally occurring populations or the result of introductions by sailors in the eighteenth and nineteenth centuries.
" This population has been removed (SA Tourist Bureau, personal communication)
^ East and West Wallabi Islands, part of the Abrolhos Group, are connected at low tide. Tammars move between the two (W E Poole, personal communication), and as such are considered one area for the purposes of this discussion.
Note: Tammars have also been introduced to Wardang Island in Spencer's Gulf (Hugh Tyndale-Biscoe, personal communication).Source: Adapted from Kennedy (1992).

The island populations have probably been isolated from the mainland for about 10 000 years after the rise in sea level caused by the ending of the Pleistocene ice age (Kennedy 1992). The SA and WA populations have been isolated from each other for at least 50 000 years (Oliver *et al.* 1979). The subspecies can be distinguished by cranial morphometrics (Poole *et al.* 1991) and by allozymes (van Oorschot & Cooper 1988a). The genetic distance between KI and Garden Island (GI) tammars, calculated using allozymes and serum protein markers, suggests that they are distantly related subspecies or closely related species (Cooper and Bell, unpublished observations). Results of attempts at interbreeding WA and SA tammars indicate that production of F1 hybrids is approximately 50% as successful as producing pure within-subspecies progeny (McKenzie and Cooper, unpublished observations).

The biology of the KI population of tammar wallabies has been studied intensively since the late 1970s and, as such, is now considered a reference model for macropodid species (Hinds *et al.* 1990). Most advances in understanding have pertained to its reproductive cycle and genetics. While there is an increasing body of literature for this, the only SA population of tammars, relatively little is known of the WA populations.

That individual tammar populations can be distinguished morphometrically, are genetically distinct, and are beginning to show reproductive isolation suggests that they may be on the verge of speciation. This clearly has important implications for formulation of conservation policy and management strategies. Whether they should be regarded as widely distant subspecies or very closely related species would reflect the definition of 'a species' preferred by the person making the judgment. The need for further studies of the taxonomy, behaviour and genetics of tammar wallaby populations is clear. Such studies could lead to the relisting of some populations from 'vulnerable to extinction' to 'endangered' or 'extinct'. The genetic distinctiveness between the KI and GI populations would indicate that any programs designed to reintroduce tammars to historical ranges, or to increase numbers in dwindling mainland populations, must take into account the genetic diversity of what is now, perhaps incorrectly, considered one species.

GENETIC VARIATION IN TAMMARS

Australia's 160 offshore islands have proved to be immensely valuable in providing refuge for some critically threatened species. Thirty-one species listed as endangered, vulnerable, or potentially vulnerable to extinction, occur on offshore islands, including the tammar. The tammar is now restricted to offshore islands and to the Perup region of south-west WA , the area with the greatest concentration of endangered or vulnerable marsupial species on the continent. Along with many other species from this region, the numbers of 'Perup' tammars have declined dramatically in recent years. Australia's offshore islands will probably prove to be the last refuge for the tammar wallaby in its native habitat.

Small, isolated populations of mammals, like those confined to islands, are particularly vulnerable to loss of genetic variability due to random genetic drift as well as other stochastic events such as disease, fire and predation (George & Brown 1992). As tammar numbers decline on the mainland, it is of great importance to assess the genetic variability of extant island populations.

Kangaroo Island is 450 000 ha and hence it is reasonable to assume that this population is not under the pressures of a 'typical' island situation. Garden Island (1054 ha) is approximately 430 times smaller than KI and as such would be under the pressures of isolation and small population size. Population genetics theory predicts that small, isolated populations will undergo a loss of heterozygosity due to genetic drift, in comparison to a larger population. The level of genetic variation in these two populations has been assessed using allozyme, serum protein and DNA markers.

An unexpected finding was made. There is no significant difference in the level of genetic variability in GI tammars compared with KI tammars. This result is seemingly incongruent with the small size of Garden Island and the time since its isolation from the mainland. This is even more surprising considering that humans and sheep have levels of heterozygosity not significantly different from the tammar (McKenzie and Cooper, unpublished observations).

There are several possible explanations for this result:

(a) The effective population size of GI could approximate that of KI. This seems unlikely considering the large difference in size between the two islands.

(b) Selection could be acting. While this possibility must be considered, the number and randomness of loci examined suggests that this is an unlikely explanation.

(c) GI may not have been isolated from the mainland for 10 000 years as predicted by the known rise in sea level after the Pleistocene Ice Age.

(d) Translocations of animals from mainland populations by Aborigines could have occurred.

A survey of the populations and reproductive behaviour of tammars on islands will enable a more stringent interpretation of the surprising result that there is no significant difference between the level of genetic variability in KI and GI tammar wallabies. The various sizes of island populations of tammars and the growing body of literature about the genetics of some of these populations provides a unique opportunity to study the effects of isolation of small populations of macropodids.

GENERATION OF A MARSUPIAL GENETIC LINKAGE MAP USING TAMMAR WALLABY SUBSPECIES CROSSES

Comparisons between genetic maps of different species has increased understanding of mammalian genome evolution. To date, the majority of available maps are for eutherian species. As marsupials and eutherians diverged at between 80 to 170 million years ago (Hope *et al.* 1990; Novacek 1992), comparisons between marsupial and eutherian gene maps may, in general terms, help elucidate the evolution of genetic linkage groups and to identify ancient therian chromosome structure and organisation. Specifically, comparisons between marsupials and eutherians may provide further understanding of gene regulation and expression in mammals. Additionally, there are several cytogenetic characteristics of marsupials which make them especially useful for such studies. Marsupials are characterised by much lower chromosome numbers than eutherians; the marsupial range is 2n=14 to 32 with two strong modes of 2n=14 and 16 (Hayman 1990) compared to the eutherian range of 2n=6 to 84 (reviewed in White 1973). Also, a considerable proportion of marsupial karyotypes appear to be conserved over wide evolutionary distances (Hayman *et al.* 1987; Hayman 1990), suggesting that the rate of karyotypic evolution in marsupials may be slower than in some eutherian lineages.

Compared to the wealth of data available for eutherian species, few loci have been mapped in marsupials (Graves *et al.* 1993). This paucity of genetic linkage information results largely from the difficulty of identifying a marsupial species which firstly exhibits sufficient genetic variation and secondly produces enough offspring for linkage analysis.

These difficulties have been overcome by crossing the KI and GI subspecies of tammars. F1s of both sexes are fertile and can be readily backcrossed to the parental strains. The KI and GI populations are sufficiently genetically distinct (van Oorschot & Cooper 1988a) for efficient identification of allelic differences. Identification of large gene frequency or fixed differences between KI and GI tammars has become substantially more efficient with the use of DNA probes as opposed to the more traditionally used allozyme and protein markers. Seventy percent of DNA probes detected fixed differences between the two populations compared to 13% of biochemical markers (Table 7.2). An additional, but not trivial, advantage of this system is that the reproduction of female tammars can be manipulated so that each female can, theoretically, produce 4 to 5 offspring per year (Hinds *et al.* 1990).

Matings between F1 hybrid males and KI females have produced more than 100 backcross progeny. Similarly, matings between F1 hybrid females and KI or GI males have generated over 100 test-cross progeny. These two types of matings allow for detection of linkage based on male and female recombination frequencies respectively.

The tammar system described here is the most powerful available for marsupial gene mapping through recombination values. Previous to this work only five linkage groups had been identified in marsupials. These are ADA-GPI-PI and PGD-SOD-TRF in the Australian species *S. crassicaudata*, (the fat-tailed dunnart) (Bennett *et al.* 1986); AK1-PI and C6-C7-GPT in the distantly related American marsupial, *M. domestica* (the South American grey short-tailed opossum) (van Oorschot *et al.* 1992; van Oorschot personal communication); and GPI-PI in the tammar (van Oorschot & Cooper 1990). Fifteen markers have been identified to date which show fixed or large gene frequency differences between KI and GI tammars (Table 7.2). Seven of these belong to linkage groups — two autosomal (McKenzie *et al.* 1993) and one X-linked (McKenzie *et al.* 1996). Clearly, as the body of information for additional markers increases, so will the probability of detecting linkage between loci.

SEX DIFFERENCES IN RECOMBINATION RATES

In general, higher levels of recombination rates are seen in the heterogametic sex in comparison to the homogametic sex. This has been shown in such disparate species as *Bombyx*, *Drosophila* and eutherian mammals. In striking contrast, significantly lower female than male recombination rates occur in two distantly related marsupials species

Table 7.2 Comparison of the number of fixed and large gene differences detected between Garden Island (GI) and Kangaroo Island (KI) tammar wallabies using restriction fragment length polymorphisms (RFLP), allozyme and protein markers.

DIFFERENT?	RFLP	ALLOZYME	PROTEIN
Yes	15	2	2
No	7	26	2
Total	22	28	4

Note: In addition a total of 21 polymorphisms within KI and 14 within GI were described.

S. crassicaudata and *M. domestica*. In parallel, females of these species have significantly lower numbers of chiasmata per cell than do males, particularly in interstitial regions (Bennett *et al.* 1986; Hayman *et al.* 1988). This dramatic difference in chiasmata frequency is not observed in two other marsupial species *Bettongia penicillata* (Hayman *et al.* 1990) and *Trichosurus vulpecula* (Hayman & Rodger 1990) suggesting that, since the likely level of recombination can be inferred from chiasmata frequencies, severely reduced female recombination might not be a general characteristic of marsupials.

The system of using subspecies crosses to generate a linkage map of a marsupial has now enabled a comparison to be made between male and female recombination rates in a macropodid. In contrast to *S. crassicaudata* and *M. domestica*, there is no significant sex difference or higher female than male recombination in the tammar (McKenzie *et al.* 1995), similar to observations obtained in eutherian species (Dunn & Bennett 1967).

That a didelphid (*M. domestica*) and a dasyurid (*S. crassicaudata*) have lower female than male recombination rates; a phalangerid (*T. vulpecula*) has lower numbers of chiasmata in females than males, but the difference is not as great as that seen in *S. crassicaudata* and *M. domestica*; a potoroid (*B. penicillata*) shows no sex differences in chiasmata frequency or distribution; and a macropodid (*M. eugenii*) shows no significant difference or higher female than male recombination rates suggests that recombination frequencies have altered throughout mammalian evolution.

STUDIES OF THE LEVEL OF GENETIC VARIATION AT MHC LOCI IN THE TAMMAR

The MHC is a group of closely linked loci that code for proteins involved in immune responses. This group of loci can be divided into several functional classes including genes involved in graft rejection (Class I) and genes involved in antigen recognition (Class II). Class I and Class IIb loci have been shown to exhibit extensive levels of genetic polymorphism in the majority of eutherian species studied to date. We have recently published a paper on MHC variability of the tammar, from which this account is taken (McKenzie & Cooper 1994). References to earlier work can be found in Deane & Cooper (1988).

There is little information on the marsupial MHC. Class I MHC responses, that is, responses to skin grafts or injected white cells, have been identified in the tammar wallaby (Rodger *et al.* 1985; van Oorschot & Cooper, unpublished data). Similar responses have also been shown in *M. domestica* (Infante *et al.* 1991; Stone *et al.* 1987), *Setonix brachyurus* (Yadav *et al.* 1974) and *Didelphis virginiana* (Rowlands 1976). While characterisation of MHC Class I antigens and genes in these species is very limited, the frequency with which transplantation responses and leucocyte antibodies are generated after immunisations suggests that the level of variability for Class I genes is comparable with that in eutherian species. In contrast, weak or non-existent mixed lymphocyte culture responses have been reported in outbred populations of several species; *M. domestica* (Infante *et al.* 1991; Fox *et al.* 1976), *Phascolarctos cinereus* (Wilkonson *et al.* 1992) and *D. virginiana* (Rowlands 1976). Since mixed lymphocyte culture responses are the result of polymorphism at Class II loci, these data suggest that marsupials, unlike eutherians, have a Class II system which shows limited variation.

There have been suggestions that genetic variability at MHC loci plays a role in successful reproduction in eutherians. The most striking difference between the divergent marsupial and eutherian mammals is in their mode of reproduction. Eutherian pregnancies are of a relatively long duration and involve intimate connections between fetal placental and maternal uterine tissue. Pregnancy in marsupials is short, young are born in an immature state and the majority of their development occurs in the pouch. With the exception of bandicoots (*Peramelidae*), nothing approaching the eutherian invasive placenta is found in marsupials (Hughes 1974; Padykula & Taylor 1977). Nonetheless, it is worth noting, that all marsupials have a placenta, which makes the use of the term 'placental' to differentiate them from eutherians inappropriate.

The role, if any, of the MHC in marsupial reproduction has been the subject of a small number of investigations. Hughes (1974) classified the trophoblast of the tammar yolk sac as being slightly invasive, that is, having rudimentary implantation. Several writers have suggested that marsupials have a short gestation period because, unlike eutherians, they have not evolved a trophoblast capable of successfully resisting an immunological attack upon the embryo by the mother. In particular, Moors (1974) postulated that an allograft reaction was initiated by the presence of the embryo in the uterus, and that this led to birth. Tyndale-Biscoe (1973) and Lillegraven (1975, 1979) have also raised the possibility that the length of marsupial gestation might be limited immunologically, that is, that birth might be initiated by an immunological response of mother to fetus. These speculations are not supported by the little data subsequently collected. Walker & Tyndale-Biscoe (1978) and Rodger *et al.* (1985) showed that grafting of the sire's skin onto mothers did not impair reproduction, while van Oorschot and Cooper (1988b) could not detect any complement-dependent cytotoxic antibodies to paternally derived antigens in multiparous tammars.

The idea that there should be a relationship between reproduction in eutherians and the level of MHC polymorphism goes back to Medawar's influential 1953 paper on the 'immunological paradox of viviparity', from which emerged the term 'fetal allograft'. Partly through the stimulus of this idea and intense medical interest in the causes of reproductive difficulties in humans, a large amount of data has accumulated in the intervening years (reviewed in Chaouat 1993). However, no clear principles have emerged.

Genetic polymorphism at MHC Class II loci has been investigated in the tammar wallaby at the DNA level. In contrast to eutherians, there are extremely low levels of variability at these loci. Similar observations have been made in *M. domestica* (W H Stone, personal communication). Comparison of data from the tammar with data from a number of eutherian species shows that the level of MHC variation in this marsupial species falls between that found in cheetahs and a population of lions, both of which have a restricted genetic base. There is no significant difference between the average heterozygosity found in tammars compared with sheep and humans. Hence, low MHC Class II variation in tammars is not a reflection of an overall decrease in genetic polymorphism in this species.

It is tempting to speculate that the differential MHC Class II variation in marsupials and eutherians is connected to their respective modes of reproduction, although it is difficult, in the present state of knowledge, to propose a specific mechanism.

Slade *et al.* (1994) have characterised a Class II gene, homologous to human HLA-DNA, in the tammar. *In situ* hybridisation places it on chromosome 1.

MARSUPIAL MILK PROTEINS AND THE EVOLUTION OF MAMMALIAN REPRODUCTION

The mammary gland with its milk is the eponymous mammalian structure. It is presumed to have evolved once. The proteins which make up milk have few recognisable homologues in non-mammalian vertebrates. One protein in particular seems to be marsupial specific, namely late lactation protein. This protein was first recognised by Bailey & Lemon (1966) and ultimately characterised by Nicholas (1988).

Marsupial lactation can be divided into three phases: phase I is lactogenesis, when the mammary is prepared for lactation by the appropriate antenatal endocrine environment, phase II involves early milk production and phase III late milk production. Early and late milk are very different in their composition, even though both can be produced simultaneously by the one female from different glands (Green & Merchant 1988; Tyndale-Biscoe & Janssens 1988). In eutherians, there is only one postnatal phase of lactation.

The distinction between the two phases is defined by the appearance of late lactation protein. It appears in the tammar at about the same time as the young first leaves the pouch (Nicholas 1988). A large part of the increase in protein which occurs at this time is attributable to its appearance. Whether it has a more precise physiological function remains to be determined.

Its evolutionary origin was not known until recently. In collaboration with Dr Chris Collet, we have shown that it is related to β-lactoglobulin, a protein found in the milk of many, but not all, marsupials studied to date (Woodlee *et al.* 1993). The evidence is two-fold. Both cDNA and inferred amino acid sequence data indicated a distant relationship to β-lactoglobulin, with preservation of five amino acids in positions which define the lipocalin group of proteins. Linkage and linkage disequilibrium data indicate that the two genes are closely linked.

An estimate of the time of divergence of the two genes of 1.5×10^8 was obtained. This estimate did not involve any 'gap' penalty and also assumed no change in rate of evolution consequent upon duplication and change of function. Thus the estimate is conservative. It is therefore reasonable to presume that the duplication took place before eutherians and marsupials diverged. This argues that the ancestral therian had late lactation protein, and hence a marsupial type of lactation which thus implies altricial young. The old hypothesis, which has been attributed to Huxley (Tyndale-Biscoe 1973), that eutherians went through a marsupial-like phase in their evolution is supported by our interpretation of the late lactation protein/β-lactoglobulin relationship.

A more detailed examination of β-lactoglobulin and late lactation protein in marsupials and, if possible, in monotremes, would be most interesting.

SEX DETERMINATION AND DIFFERENTIATION

Sex determination in eutherian mammals has been extensively studied. Maleness is dominant, being determined by the *SRY* gene carried on the Y chromosome. Absence of *SRY* leads to femaleness. In principle a simple sequence of events is involved, summarised by genetic sex (*SRY*) → male gonad → biochemical sex → physiological sex → morphological and behavioural sex.

Much less information is available for marsupials. The cloning of the Y-linked *SRY* gene from tammars and *Sminthopsis* has been accomplished (Foster *et al.* 1992). Data on the karyotypes of intersexual marsupials accumulated by Sharman and Hughes over a twenty year period indicated that the Y, and presumably the *SRY* gene, is testis-determining (Sharman *et al.* 1970, 1990; Hughes *et al.* 1993). We have continued to collect these data, and information is now available for about 25 animals across a wide range of marsupial species, including several tammars (Cooper *et al.* unpublished). Over half of these animals are karyotypically abnormal. At least 4 main types are evident

(a) Males with no scrotum, a pouch and mammary glands

(b) Females with an empty scrotum but no pouch or mammary glands

(c) Females with a hemiscrotum on one side and a hemipouch with half the usual number of mammaries on the other

(d) A single male with a hemiscrotum containing a testis on one side and a hemipouch on the other

Animals entirely lacking pouch, mammary glands or scrotum have recently been found.

While the data are not completely clear-cut, it seems that presence of the pouch is associated with two X chromosomes and presence of the scrotum with only one. These data are compatible with two hypotheses: the pouch plus mammary gland versus scrotum dichotomy is controlled by X chromosome dosage. Alternatively, it could be that presence of the pouch plus mammary glands is dependent upon the presence of a paternally derived X, since normal males are X^MY and normal females are X^MX^F. It is clear that X-inactivation must play a role in sex differentiation in marsupials and this theme is developed further in Cooper *et al.* (1993).

Renfree (1992) and Shaw *et al.* (Chapter 8 in this volume) have carried out parallel studies on normal sexual development in tammars and also conclude that the process is more complex than in eutherians.

FUTURE DIRECTIONS

Interest in studying mammalian genome organisation and evolution has gained considerable momentum over the last decade. There are now genetic linkage maps available for many species (O'Brien 1993), the majority of which are eutherian. A map of a marsupial genome would increase substantially our understanding of these important biological questions.

The system described here, of crossing genetically distinct populations of tammar wallabies, is the most efficient available for generating a marsupial genetic map through recombination. Already comparisons can be made between the organisation of the tammar genome and those of 4 eutherian species. As the tammar map grows, so will its potential for studying specific aspects of genome organisation and evolution and gene expression, such as:

(a) further detailed comparisons between marsupial and eutherian linkage groups

(b) understanding the distribution and significance of differential recombination rates in male and female tammar wallabies

(c) the underlying mechanisms of X chromosome inactivation

(d) the origins and maintenance of sex differentiation and determination in marsupials and their evolution in mammals.

As the tammar becomes increasingly important as a model or reference species, it is clear that detailed investigations comparing the different populations are warranted. This is important both from a conservation perspective and also from a scientific perspective. Further investigations into the effects of isolation on small populations of tammars will be of great importance in formulating conservation policy about endangered macropodid species.

Acknowledgments

We thank Hugh Tyndale-Biscoe, Geoff Sharman, Lyn Hinds, Bill Poole, Marilyn Renfree and Leon Hughes for discussion. Our work on tammars has been supported by the Australian Research Council, the Australian Nature Conservation Agency and Macquarie University.

References

Bailey L F and Lemon M (1966). Specific milk proteins associated with resumption and development by the quiescent blastocyst of the lactating Red Kangaroo. *Journal of Reproduction and Fertility*, 11, 473–475.

Bennett J H, Hayman D L and Hope R M (1986). Novel sex differences in linkage values and meiotic chromosome behaviour in a marsupial. *Nature*, 323, 59–60.

Chaouat G (1993). *Immunology of Pregnancy*. CRC Press, Ann Arbor.

Cooper D W, Johnston P G, Watson J M and Graves J A M (1993). X-inactivation in marsupials and monotremes. *Seminars in Developmental Biology*, 4, 117–128.

Deane E M and Cooper D W (1988). Immunological development in pouch young marsupials. In *The Developing Marsupial. Models for Biomedical Research*. Eds C H Tyndale-Biscoe, P A Janssens. Springer-Verlag, Berlin. pp 190–199.

Dunn L C and Bennett D (1967). Sex differences in recombination of linked genes in animals. *Genetical Research*, 9, 211–220.

Foster J W, Brennan F E, Hampikian G K, Goodfellow P N, Sinclair A H, Lovell-Badge R, Selwood L, Renfree M B, Cooper D W and Marshall Graves J A (1992). Evolution of sex determination and the Y chromosome: SRY-related sequences in marsupials. *Nature*, 359, 531–533.

Fox D H, Rowlands D T and Wilson D B (1976). Proliferative reactivity of opossum peripheral blood leukocytes to allogeneic cells, mitogens and specific antigens. *Transplantation*, 21, 164–168.

George G G and Brown P R (1992). Conservation management of small populations. In *Australasian Marsupials and Monotremes: an action plan for their conservation*. Ed M Kennedy. International Union for Conservation of Nature and Natural Resources, Gland, Switzerland. pp 38–42.

Graves J A M, Cooper D W, McKenzie L M, Hope R M and Watson J M (1993). Marsupials and Monotremes. In *Genetic Maps: locus maps of complex genomes*. Ed S J O'Brien, Book 4, 6th edn. Cold Spring Harbor Laboratory Press, Laboratory of Viral Carcinogenesis, National Cancer Institute. pp 282–289.

Green B and Merchant J C (1988). The composition of marsupial milk. In *The Developing Marsupial. Models for Biomedical Research*. Eds C H Tyndale-Biscoe and P A Janssens Springer-Verlag, Berlin. pp 41–54.

Hayman D L (1990). Marsupial Cytogenetics. In *Mammals from Pouches and Eggs: Genetics, Breeding and Evolution of Marsupials and Monotremes*, Eds J A M Graves, R M Hope and Cooper D W. CSIRO, Melbourne. pp 189–207.

Hayman D L and Rodger J C (1990). Meiosis in male and female *Trichosurus vulpecula* (Marsupialia). *Heredity*, 64, 251–254.

Hayman D L, Smith M J and Rodger J C (1990). A comparative study of chiasmata in male and female *Bettongia penicillata* (Marsupialia). *Genetica* 83, 45–49.

Hayman D L, Moore H D M and Evans E P (1988). Further evidence of novel sex differences in chiasma distribution in marsupials. *Heredity*, 61, 455–458.

Hayman D L, Rofe R H and Sharp P J (1987).

Chromosome evolution in mammals. *Chromosomes Today*, 9, 91–102.

Hinds L A, Poole W E, Tyndale-Biscoe C H, van Oorschot R A H and Cooper D W (1990). Reproductive biology and the potential for genetic studies in the tammar wallaby (*Macropus eugenii*). *Australian Journal of Zoology*, 37, 223–234.

Hope R M, Cooper S and Wainwright B (1990). Globin macromolecular sequences in marsupials and monotremes. In *Mammals from Pouches and Eggs: Genetics, Breeding and Evolution of Marsupials and Monotremes*. Eds J A M Graves, R M Hope and D W Cooper. CSIRO, Melbourne. pp 147–171.

Hughes R L (1974). Morphological studies on implantation in marsupials. *Journal of Reproduction and Fertility*, 39, 173–186.

Hughes R L, Pearse A M, Cooper D W, Joss J M P, Johnston P G and Jones M K (1993). In *Sex Chromosomes and Sex-determining Genes,* Eds K C Reed and J A Marshall Graves. Harwood Academic Publishers, Switzerland. pp 17–48.

Infante A J, Samples N K, Croix D A, Redding T S, VandeBerg J L and Stone W H (1991). Cellular immune response of a marsupial, *Monodelphis domestica. Developmental and Comparative Immunology*, 15, 189–199.

Kennedy M (1992). *Australasian Marsupials and Monotremes: an action plan for their conservation*. International Union for Conservation of Nature and Natural Resources, Gland, Switzerland.

Lillegraven J A (1979). Reproduction in Mesozoic mammals. In *Mesozoic Mammals: The Fist Two-thirds of Mammalian History* Eds J A Lillegraven, Z Kielan- Jawcrowska and W A Clemens. University of California Press, Berkeley. Chapter 13, pp 259–276.

Lillegraven J A (1975). Biological considerations of the marsupial-placental dichotomy. *Evolution*, 29, 707–722.

Medawar P B (1953). Some immunological and endocrinological problems raised by the evolution of viviparity in vertebrates. *Symposium of the Society for Experimental Biology*, 7, 320–338.

McKenzie L M, Poole W E, Collet C and Cooper D W (1995). Higher female than male recombination rates in a marsupial mammal, the tammar wallaby (*Macropus eugenii*). *Cytogenetics and Cell Genetics*, 68, 64–66.

McKenzie L M and Cooper D W (1994). Low MHC class II variability in a marsupial. *Reproduction Fertility and Development*, 6, 721–726.

McKenzie L M, Collet C and Cooper D W (1993). Use of a subspecies cross for efficient development of a linkage map for a marsupial mammal, the tammar wallaby (*Macropus eugenii*). *Cytogenetics and Cell Genetics*, 64, 264–267.

McKenzie LM, Johnston P G, Eldridge M D B and Cooper D W (1996). Close linkage between R N R and G P D genes on the tammar wallaby (*Macropus eugenii*) X chromosome. *Cytogenetics and Cell Genetics*, 72, 197–199.

Moors P J (1974). The feto-maternal relationship and its significance in marsupial reproduction: A unifying hypothesis. *Australian Mammalogy*, 1, 263–266.

Nicholas K R (1988). Control of milk protein synthesis in the marsupial *Macropus eugenii*: a model system to study prolactin-dependent development. In *The Developing Marsupial. Models for Biomedical Research*. Eds C H Tyndale-Biscoe and P A Janssens. Springer-Verlag, Berlin. pp 68–85.

Novacek M J (1992). Mammalian phylogeny: shaking the tree. *Nature*, 356, 121–125.

O'Brien S J (1993). *Genetic Maps: locus maps of complex genomes*, Book 4, 6th edn Cold Spring Harbor Laboratory Press, Laboratory of Viral Carcinogenesis, National Cancer Institute. Cold Spring Harbor, New York.

Oliver A J, King D R and Mead R J (1979). Fluoroacetate tolerance, a genetic marker in some Australian marsupials. *Australian Journal of Zoology*, 27, 363–372.

Padykula H A and Taylor J M (1977). Uniqueness of the bandicoot chorioallantoic placenta (Marsupialia: Peramelidae) Cytological and evolutionary interpretations. In *Reproduction and Evolution*. Eds J H Calaby and C H Tyndale-Biscoe. Australian Academy of Science, Canberra. pp 303–323.

Poole W E, Wood J T and Simms N G (1991). Distribution of the tammar, *Macropus eugenii*, and the relationships of populations as determined by cranial morphometrics. *Wildlife Research*, 18, 625–639.

Renfree M B (1992). The role of genes and hormones in marsupial sexual differentiation. *Journal of Zoology*, 226,165–173.

Rodger J C, Fletcher T P and Tyndale-Biscoe C H

(1985). Active antipaternal immunization does not effect the success of marsupial pregnancy. *Journal of Reproductive Immunology*, 8, 249–256.

Rowlands, D T (1976). Immunity in prototherian and metatherian mammals. In *Comparative Immunology*, Ed J J Marchalonis. Blackwell Scientific, London. pp 367–389.

Sharman G B, Robinson E S, Walton S M and Berger P J (1970). Sex chromosomes and reproductive anatomy of some intersexual marsupials. *Journal of Reproduction and Fertility*, 21, 57–68.

Sharman G B, Hughes R L and Cooper D W(1990). The chromosome basis of sex differentiation in marsupials. *Australian Journal of Zoology*, 37, 451–466.

Slade R W, Hale P T, Francis D I, Marshall Graves J A and Sturn R A (1994). The marsupial MHC: the tammar wallaby, *Macropus eugenii*, contains an expressed *DNA*-like gene on chromosome 1. *Journal of Molecular Evolution*, 38, 496–505.

Stone W H, Samples N K, Kraig E and VandeBerg J L (1987). Definition of class I antigens of the MHC of a marsupial (*Monodelphis domestica*). *Archivos de Zootecnia*, 36, 219–235.

Tyndale-Biscoe C H (1973). *Life of Marsupials*. Edward Arnold, London.

Tyndale-Biscoe C H and Janssens P A (1988). *The Developing Marsupial. Models for Biomedical Research*, Springer-Verlag, Berlin.

Tyndale-Biscoe C H and Hinds L A. (1989). Influence if the immature testis on sexual differentiation in the tammar wallaby, *Macropus eugenii* (Macropodidae: Marsupalia. *Reproduction Fertility and Development*, 1, 243–254.

van Oorschot R A H, Porter P A, Kammerer C M and VandeBerg J L (1992). Severely reduced recombination in females of the South American marsupial *Monodelphis domestica*. *Cytogenetics and Cell Genetics*, 60, 64–67.

van Oorschot R A H and Cooper D W (1990). A conserved linkage between PI and GPI in a marsupial mammal, the tammar wallaby (*Macropus eugenii*). *Cytogenetics and Cell Genetics*, 51, 1097.

van Oorschot R A H and Cooper D W (1988a). Limited allozymic variation in a marsupial, the tammar wallaby (*Macropus eugenii*). *Biochemical Genetics*, 26, 481–4

van Oorschot R A H and Cooper D W (1988b). Lack of evidence for complement-dependent cytotoxic antibodies to fetal paternally derived antigens in the marsupial *Macropus eugenii* (Tammar Wallaby). *American Journal of Reproductive Immunology and Microbiology*, 17, 145–148.

Walker K Z and Tyndale-Biscoe C H (1978). Immunological aspects of gestation in the tammar wallaby, *Macropus eugenii*. *Australian Journal of Biological Sciences*, 31, 173–182.

White M J D (1973). *Animal Cytology and Evolution*. Cambridge University Press, Cambridge. p 391.

Wilkonson R, Kotlarski I and Barton M (1992). Koala lymphoid cells: analysis of antigen-specific responses. *Veterinary Immunology and Immunopathology*, 33, 237–247.

Woodlee G L, Gooley A A, Collet C and Cooper D W (1993). Origin of late lactation protein from b-lactoglobulin in the tammar wallaby. *Journal of Heredity*, 84, 460–465.

Yadav M, Waring H and Stanley N F (1974). Effect of thymectomy on skin allograft survival in a macropod marsupial *Setonix brachyurus*. *Transplantation*, 17, 30–36.

CHAPTER 8

SEXUAL DETERMINATION AND DIFFERENTIATION IN THE MARSUPIAL *MACROPUS EUGENII*

Geoffrey Shaw, Jenny L Harry, Deanne J Whitworth and Marilyn B Renfree

INTRODUCTION

Sexual development in mammals involves sequential genetic and hormonal cascades. Initially male and female develop alike, with indifferent gonadal ridges medial to the mesonephros. Paired Wolffian and Müllerian ducts on the side of the mesonephros will ultimately form the male and female reproductive duct systems respectively (Jost 1970). Activation of a Y-linked gene, *SRY*, in the gonadal ridge acts to transform the indifferent gonad of the male into a testis (Berta *et al.* 1990; Jager *et al.* 1990). Testicular production of the hormones Müllerian inhibiting substance and testosterone masculinises the internal and external genitalia. Although marsupial and eutherian mammals follow a similar pattern of sexual differentiation there are some major differences (Burns 1961; Shaw *et al.* 1990; Renfree 1992, 1994). In contrast to eutherians, where differentiation of the scrotum and mammary glands is hormone-dependent, the differentiation of these structures in the tammar, *Macropus eugenii* appears to be determined by a genetic mechanism. The scrotum of male marsupials develops anterior to the penis and therefore cannot be homologous to that of eutherians. Also, the scrotum, mammary primordia, gubernaculum and processus vaginalis become sexually dimorphic before gonadal differentiation and before the onset of testosterone and production of Müllerian inhibiting substance by the developing testes (O *et al.* 1988) (Figure 8.1). Treatment of female neonates with testosterone, or male neonates with oestrogens, has no effect on the normal development of the scrotum, mammary gland or pouch, even though these hormones have dramatic effects on the internal genitalia (Shaw *et al.* 1988; Lucas *et al.* 1995) (Figure 8.2). The phenotype is also unaffected in females carrying testicular transplants (Tyndale-Biscoe & Hinds 1989). In natural 'intersexes' a scrotum can occur in females with ovaries and a female internal phenotype. Further,

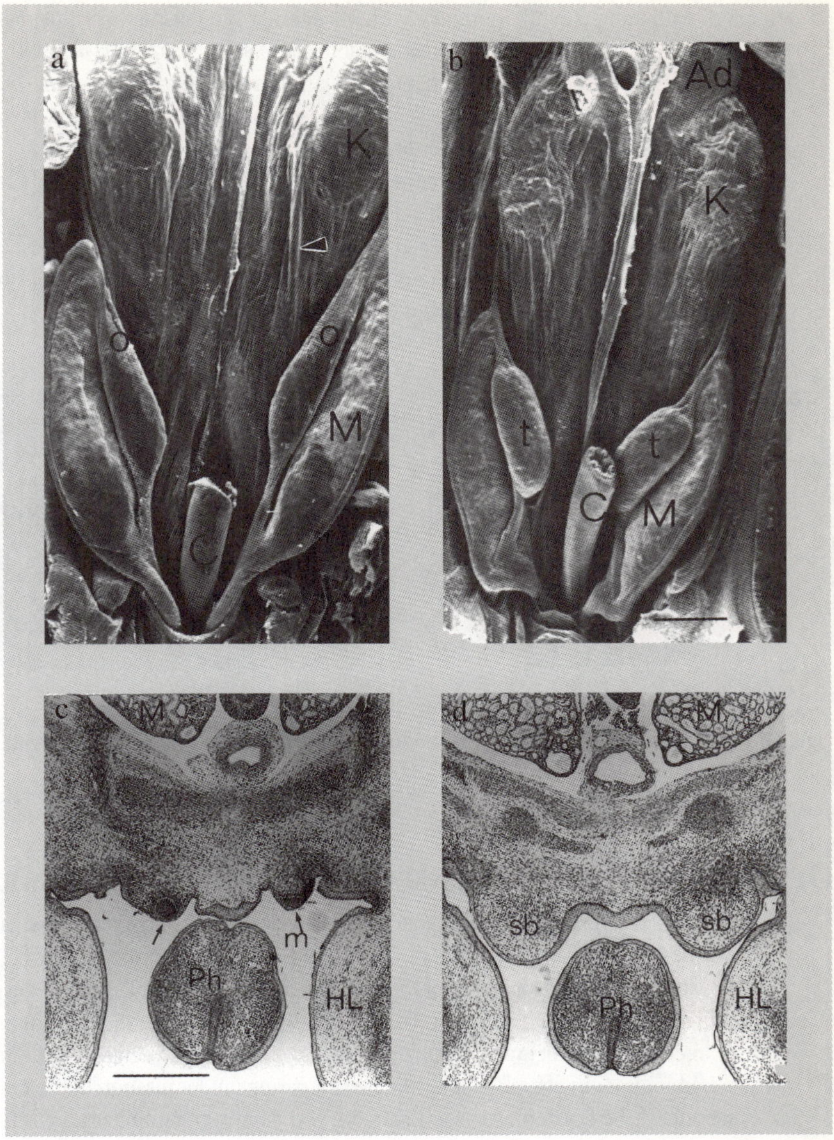

Figure 8.1 Anatomy of the developing tammar urogenital system. Scanning electron micrographs of dissections of a 10-day-old pouch young: (a) female; (b) male. The developing ovary (o) and testis (t) lie on the medial side of the mesonephros (M). The ridge containing the Wolffian and Müllerian ducts is evident lateral to the mesonephros, running down to the urogenital sinus. The arrowhead shows the ureters. The metanephros (K) and adrenal (Ad) are also visible. At this stage the well-developed testis tubules are present, and the testis has become very rounded compared to the still poorly differentiated, spindle-shaped, ovary. In females (c) the mammary primordia (m) are well developed by day 25 of gestation, while in males of the same age (d) the scrotal bulges (sb), though still separate, are well developed. The phallus (Ph) is not yet sexually dimorphic. The colon (C) and hindlimbs (HL) are also labelled.
[Scale bars a) and b) = 1.0mm, c) and d) = 0.2 mm.]

males with an apparently normally virilised internal genitalia and phallus may lack a scrotum, and instead have a pouch and mammary glands (Sharman *et al.* 1990; Cooper 1993; Cooper & McKenzie, see chapter 7 this volume). These data suggest that development of mammary gland and scrotum is controlled by an X-linked mechanism which is discussed in detail by Cooper (1993).

Sexual determination and differentiation in marsupials has been extensively reviewed recently (Tyndale-Biscoe & Renfree 1987; Renfree & Short 1988; Shaw *et al.* 1990; Sharman *et al.* 1990; Renfree 1992, 1994; Cooper 1993; Hughes *et al.* 1993), so this paper focuses specifically on the tammar where studies on the genetic and hormonal controls of sexual determination and differentiation are ongoing.

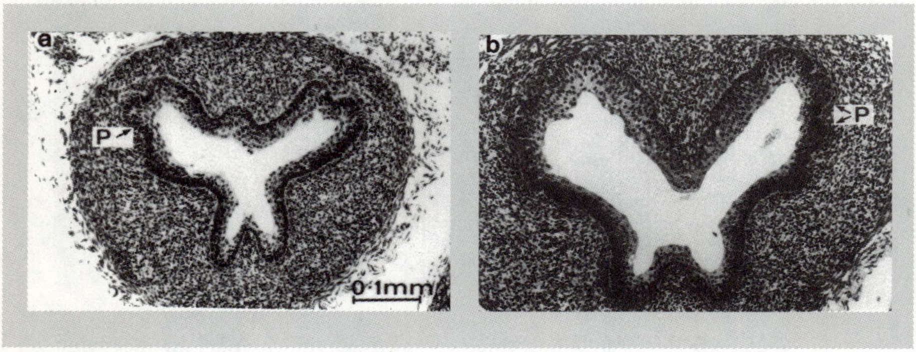

Figure 8.2 (a) Prostate buds (P) start to form in the urogenital sinus of males about day 25 in male pouch young.
(b) Testosterone treatment of females can also induce the start of prostate development.

DEVELOPMENT OF THE GENITAL SYSTEM IN THE TAMMAR

In tammars the gonad first differentiates as an elongate thickening on the medial border of the mesonephros at about day 21 of the 26.5-day gestation period (Figure 8.1). Over the next few days the gonads of both male and female grow similarly and develop an undifferentiated core of blastemal cells. The first signs of testicular differentiation can be seen in electron micrographs of fetal gonadal ridges on day 25 of gestation. There is a localised increase in cytoplasmic volume and a peripheral localisation of the nucleus in some cells of the developing testis (Renfree 1994). However, seminiferous cords are not obvious at the light microscopic level until day 2 after birth (O *et al.* 1988) (Figure 8.3). In females, ovarian differentiation is defined by the appearance of distinct cortical and medullary regions which are not seen until about 7 days after birth.

The Wolffian duct forms as the mesonephric duct and is patent to the urogenital sinus about day 21 of gestation. By day 25 of gestation the Müllerian duct has formed in both males and females as an invagination on the cranio-lateral border of the mesonephros. It grows caudally, lateral to the Wolffian duct then passes ventro-medially, eventually fusing with the Wolffian duct at its junction with the urogenital sinus by day 10 post-partum (Figure 8.1). In males the Wolffian duct develops into the epididymis and vas deferens. In females the Müllerian duct develops to form the fallopian tubes, uterus and lateral vaginae.

The scrotum forms anterior to the urogenital tubercle and thus is not homologous to the eutherian scrotum which forms posterior to the phallus from the labioscrotal folds. In females the same area of abdominal skin develops mammary primordia (Renfree *et al.* 1990; Robinson *et al.* 1991).

Testis determination

SRY is believed to be the testis-determining gene on the mammalian Y chromosome. Mutation analysis of the human *SRY* gene (Berta *et al.* 1990; Jager *et al.* 1990) and the production of XX male mice transgenic for *SRY* (Koopman *et al.* 1991) have provided direct evidence that this gene is required for testis development in eutherians. Homologues of the human *SRY* gene have been localised to the Y chromosome of marsupial mammals (Foster *et al.* 1992), suggesting that its testis-determining function has been highly conserved in mammals.

In the fetal mouse, *SRY* is expressed transiently from the appearance of the genital ridge at 10.5 days until 12.5 days *post-coitum* when the Sertoli cells align into seminiferous tubules (Koopman *et al.* 1991). As Sertoli cells are the first cell type to differentiate in the testis and as they appear during this period of *SRY* expression, it is assumed that *SRY* directs the differentiation of Sertoli cells from the supporting cell lineage (Lovell-Badge 1993). The equivalent stages of gonadal morphogenesis are observed over a 7 day period in the tammar (Figure 8.3), making it possible to test critically the correlation between *SRY* expression and testis development.

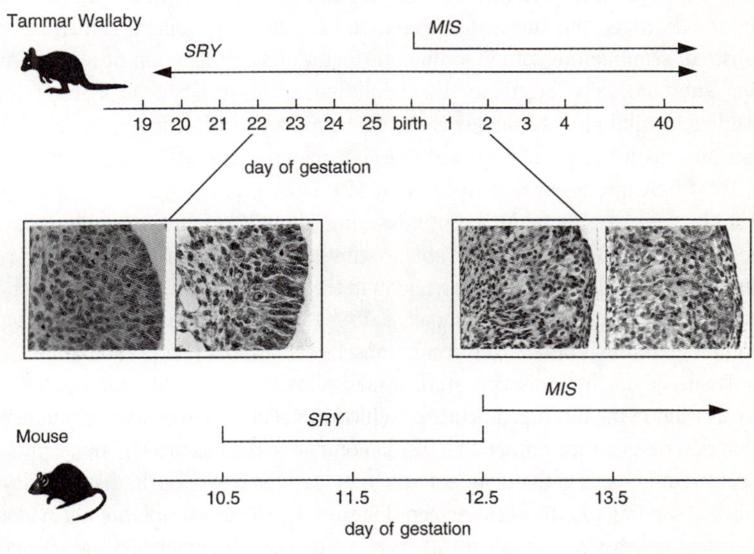

Figure 8.3 Timing of expression of *SRY* and *MIS* genes in the tammar compared to the mouse. Gonadal development of day 10.5 fetal mice is roughly equivalent to day 21 fetal tammars, and day 12.5 fetal mice correspond to tammars on day 2 after birth. Insets show micrographs of male and female tammar gonads at day 22 of gestation and at day 2 postpartum. Note the long period of expression of SRY in the tammar, continuing well after testicular differentiation. Müllerian inhibiting substance production starts as the testis tubules begin to form.

Examination of *SRY* expression in tammar fetuses and neonates, using reverse transcription and the polymerase chain reaction, provided some surprising results (Harry *et al.* 1995). The *SRY* gene is not only transcribed in the male gonad throughout this entire 7-day period of testis differentiation, but *SRY* transcripts are detected at all developmental stages examined, from day 19 of gestation (early head-fold; 20 somites) until 40 days after birth (Figure 8.3). *SRY* mRNA is also found in many male fetal and pouch young tissues and a variety of extra-gonadal tissues in the adult male (Harry *et al.* 1995). These data are strikingly different to the narrow window of *SRY* expression for the mouse, although there are reports of transcripts detected in preimplantation stages (Zwingman *et al.* 1993; Boyer *et al.* 1994). The only *SRY* mRNA detected in the adult mouse is a circular, non-polyadenylated (and presumably inactive) transcript in the testis.

Our observations however, do appear to parallel the pattern of *SRY* expression reported for humans. Comparative developmental studies are not possible in humans, but *SRY* transcripts are found in a variety of non-gonadal male embryonic and adult tissues (Clepet *et al.* 1993). While the widespread distribution of *SRY* transcripts in marsupials and humans suggests a more general role for this gene, the normal female phenotype of human XY individuals with mutations in *SRY* affirms that the action of this gene is limited to testis differentiation. To understand the significance of our findings, it is necessary to establish if, and when, marsupial transcripts are translated to produce functional protein.

MÜLLERIAN INHIBITING SUBSTANCE IN EARLY MALE DEVELOPMENT

Müllerian inhibiting substance is produced by the Sertoli cells of the developing testis and is the factor responsible for the regression of the Müllerian duct. In the male mouse fetus *MIS* transcripts are first detected at 12.5 days *post-coitum*, coinciding with the onset of seminiferous tubule formation (Figure 8.3). Production of Müllerian inhibiting substance also occurs in the adult mouse testis albeit at reduced levels. As Müllerian inhibiting substance is the first product to be secreted by the developing Sertoli cells it has been suggested that *SRY* may regulate *MIS* expression. Haqq *et al.* (1993) have proposed that there is an SRY DNA binding site in the *MIS* promoter. However, expression of *MIS* continues long after *SRY* expression ends in the mouse testis (Figure 8.3). Whilst *SRY* is able to stimulate *MIS* expressionin a gonadal ridge cell line, mutation of the SRY-binding region in the *MIS* promoter did not diminish the *MIS* response to *SRY* expression (Haqq *et al.* 1994). Thus, *SRY* must initiate *MIS* expression indirectly through other intervening transcription factors (Haqq *et al.* 1994).

Tissue *in situ* hybridisation studies have shown that the temporal pattern of *MIS* expression in the marsupial neonate (Whitworth *et al.* 1994) is markedly different from that described for the eutherian fetus. In contrast to the mouse, *MIS* transcripts are present in Sertoli cells of the tammar testis by at least the day of birth, that is, 2 days before distinct seminiferous tubules are seen (Figure 8.3). *MIS* transcripts are not evident in the developing ovary at equivalent stages. The presence of tammar *MIS* transcripts is coincident with the period of *SRY* transcription (Figure 8.3). While there is no direct evidence that SRY regulates the expression of *MIS* in the tammar, the significant overlap in expression profiles suggests that there may be some interaction between *SRY* and *MIS*.

Bioassays have established that functional Müllerian inhibiting substance is produced by the testis, from at least 2 to 80 days post-partum (Hutson *et al.* 1988). In the mouse, *MIS* is transcribed only in the Sertoli cells of the developing testis from day

12.5 *post-coitum*, and at a lower level in the granulosa cells of the post-pubertal ovary. In contrast, tammar *MIS* is expressed in a variety of tissues and for a longer period. Western blot analysis of total protein extracts from both male and female tammar pouch young tissues, using an anti-bovine Müllerian inhibiting substance monoclonal antibody, show that Müllerian inhibiting substance (or an Müllerian inhibiting substance-like peptide) is produced not only by the developing testis, but also by the mesonephros, liver, kidney and adrenal of both sexes (Whitworth *et al.* 1994). This antibody does not recognise any protein bands in tammar pouch young ovary.

Marsupial young are ideal models for tissue transplantation experiments because of their highly altricial state at birth. Undifferentiated gonads from female tammar neonates can be transplanted under the skin of slightly older male hosts to determine the effects of male hormones, including Müllerian inhibiting substance, on the developing ovary. Such transplants develop seminiferous tubule-like structures reminiscent of those observed in the bovine freemartin (Whitworth *et al* 1996) (Figure 8.4). Similarly, Müllerian inhibiting substance has a masculinising effect on the tammar ovary, since undifferentiated ovaries cultured for 4 days in medium containing recombinant human Müllerian inhibiting substance also developed seminiferous tubule-like structures (Whitworth *et al* 1996). Müllerian inhibiting substance, therefore, appears to be capable of inducing the masculinisation of tammar ovaries *in vitro* and is likely to be the factor responsible for the transdifferentiation of the female gonads transplanted into male hosts.

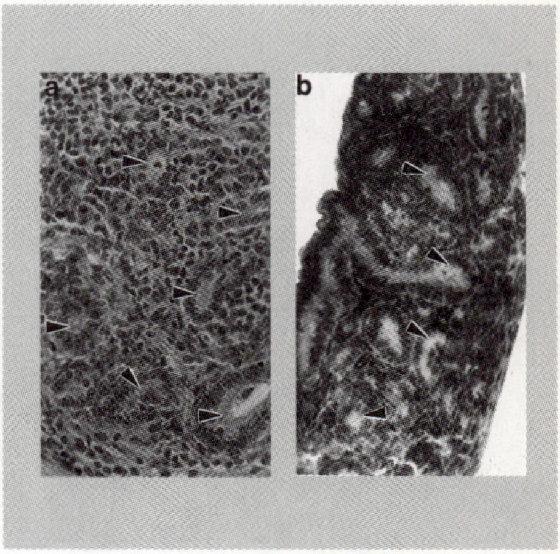

Figure 8.4 Induced gonadal sex reversal in tammars. (a) Twenty-five days after transplanting a neonatal ovary into a male host, the tissue contains obvious tubular structures (arrowheads) quite uncharacteristic of a normal ovary. (b) culture of neonatal ovaries in the presence of Müllerian inhibiting substance also develop seminiferous tubule-like structures.

ANDROGENS

As in developing eutherian mammals, androgens are essential for normal male tammar development. The main gonadal androgen produced during sexual differentiation is testosterone, with negligible amounts of 5α-dihydrotestosterone (Renfree *et al.* 1992). Gonadal testosterone levels remain low in male and female gonads during the first few days of pouch life (Renfree *et al.* 1992) (Figure 8.5). After the formation of seminiferous tubules in testes, between 2 days and 10 days, testosterone production increases (Figure 8.5). Testosterone levels remain high in the testis until after 40 days by which time sexual differentiation of the internal genitalia is essentially complete. In females the testosterone content of ovaries remains minimal.

Androgens directly stimulate the differentiation of the Wolffian duct. In females the Wolffian duct normally begins to regress by 7 days, but if female neonates are treated daily with testosterone from birth until 25 days, the Wolffian duct grows and differentiates (Shaw *et al.* 1988). This treatment also induces the onset of prostatic development in the tissues of the urogenital sinus (Figure 8.2). Prostatic tissue normally does not form in females, but is present as small buds by 25 days in both control males and those females treated with testosterone. Treatment of male pouch young with the androgen receptor blocker flutamide from 0 to 40 days also dramatically reduced prostatic development compared to control treated males (Butler *et al.* 1995). The phallus is also

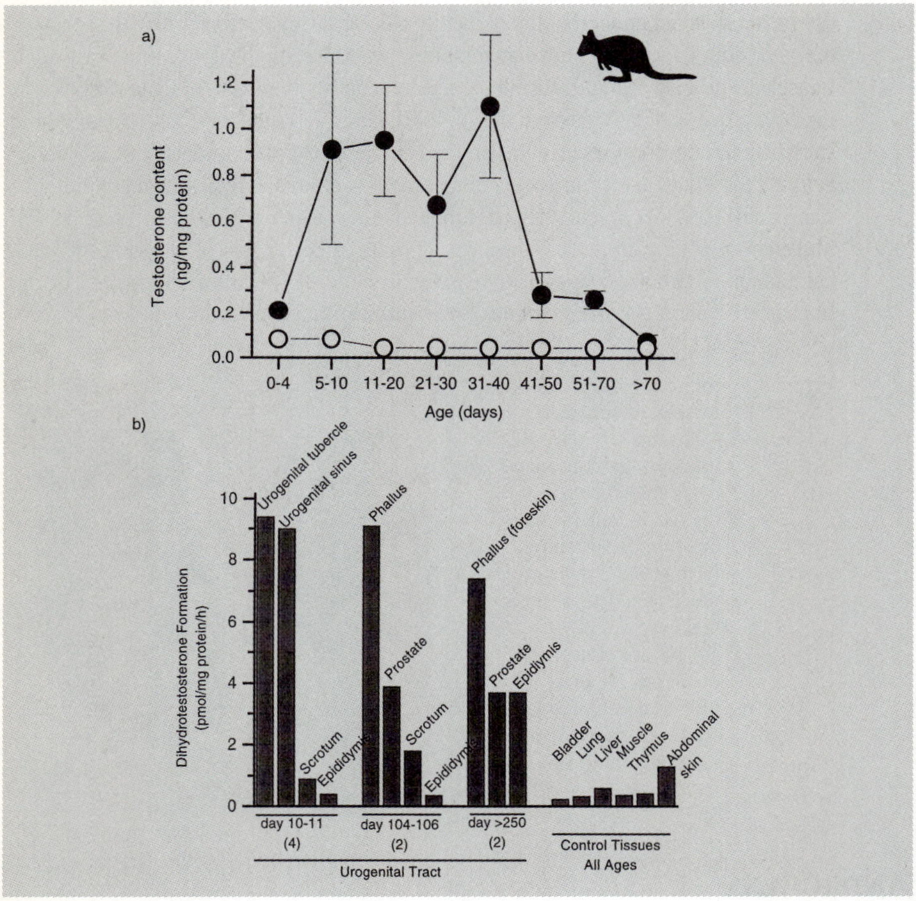

Figure 8.5
Androgens in developing tammars.(a) Gonadal testosterone content in developing tammars. Pooled gonads from males (closed circles) and females (open circles) were assayed for testosterone using a sensitive radioimmunoassay. The number of individual pools assayed at each point is indicated on the graph. Testosterone content is significantly higher in testes than ovaries between days 11 and 70. (b) Activity of 5α-reductase enzyme in tissues from male tammar pouch young. Fresh tissue was incubated with [1,2,6,7-^{3}H]-testosterone, and conversion to 5α-dihydrotestosterone was assessed. High activity is seen in urogenital sinus and urogenital tubercle, and their derivatives, but not in scrotum, epididymis, or control tissues. Data are averages from 2 pools (10 to 11 days) or tissue from 2 pouch young (later ages), with control tissues showing the average of 6 replicates. Source: Renfree *et al.* (1992).

androgen-dependent. Tyndale-Biscoe & Hinds (1989) removed testes from 10-day male pouch young and grafted these into similar aged female pouch young. The castrated males had no development of the phallus beyond that of normal females, whereas the female pouch young with testicular grafts developed a male-type phallus. The urogenital sinus and phallus both contain 5α-reductase (Renfree *et al.* 1992) which is present in high levels by 10 to 11 days (Figure 8.5). These observations suggest 5α-dihydrotestosterone is the active androgen in these tissues.

The timing of virilisation well after the increase in androgen levels poses an interesting question. Although gonadal testosterone increases markedly before 10 days, and high levels of 5α-reductase are already present in the male tract, the prostate does not start to differentiate until about 25 days. Even very high doses of exogenous testosterone do not accelerate the timing of prostate development. The phallus does not become sexually dimorphic until after 60 days, by which stage testicular testosterone concentrations have fallen (Figure 8.5). Clearly low androgen levels cannot account for this delay in virilisation, and it is possible that these events are controlled by expression or activation of the androgen receptor.

CONCLUSIONS

The tammar has a homologue of the eutherian testis determining gene, *SRY* which is transcribed from before genital ridge formation until at least 40 days after birth, unlike the brief (2-day) period of expression in the mouse. Marsupial *SRY* transcripts are also detected in a variety of extra-gonadal tissues in the developing young and adult, as in humans, suggesting that *SRY* may have multiple roles in addition to a function in testis determination.

Differentiation of the testis and internal genitalia is controlled by the hormones, Müllerian inhibiting substance and testosterone produced by testes as the seminiferous tubules start to differentiate, but not by ovaries. Surprisingly, *MIS* is also expressed in non-gonadal tissue unlike in eutherians. Müllerian inhibiting substance induces Müllerian duct regression and may also induce seminiferous tubule formation, as ovaries cultured with Müllerian inhibiting substance or transplanted into male recipient pouch young develop tubular structures. Testosterone production by testes increases dramatically over the first 10 days after birth, whilst production by ovaries is minimal. Testosterone stimulates Wolffian duct development to form the epididymis and vas deferens in males. Virilisation of the prostate and phallus is also androgen dependent: it can be prevented in males by administration of androgen receptor inhibitors, and stimulated in females treated with testosterone. Virilisation may depend on local conversion of testosterone to the more potent androgen 5α-dihydrotestosterone since the urogenital sinus and the urogenital tubercle contain high levels of 5α-reductase activity by day 10 after birth. Since virilisation does not begin until 3 to 4 weeks after birth there may be a delay in the expression of androgen receptors in these tissues. Testicular hormones, however, do not control the sexual dimorphisms of the mammary glands, scrotum, gubernaculum and processus vaginalis. These appear to be determined by an X-linked genetic mechanism.

FUTURE DIRECTIONS

Despite the basic similarities between marsupials and eutherians, there are important differences in timing and location of *SRY* and *MIS* expression, the potential for gonadal sex reversal, and the role of the androgen receptor in virilisation. A number of important questions remain to be answered:

- Why is *SRY* expressed in extragonadal tissues?
- Does *SRY* have functions other than inducing Sertoli cell formation?
- What interactions are there between *SRY* and *MIS*?
- Does Müllerian inhibiting substance have a role in testicular morphogenesis?
- Is the extra-gonadal Müllerian inhibiting substance active, and if so, does it have a function?

Is the androgen receptor controlling the timing of virilisation?

The differences in sexual differentiation between marsupial and eutherian mammals will help clarify the underlying molecular and endocrine pathways of sexual determination and differentiation in all mammals.

References

Berta P, Hawkins J R, Sinclair A H, Taylor A, Griffiths B L, Goodfellow P N and Fellous M (1990). Genetic evidence equating SRY and the testis-determining factor. *Nature*, 348, 448–450.

Boyer T R and Erickson R P (1994). Detection of circular and linear transcripts of Sry in pre–implantation mouse embryos: differences in requirement for reverse transcriptase. *Biochemica and Biophysica Research Communications*, 198, 492–496.

Burns, R K (1961). Role of hormones in the differentiation of sex. In *Sex and Internal Secretions.* Ed W C Young. Williams and Wilkins, Baltimore. *Vol 1. 3rd edn.* pp 76–158.

Butler C M, Shaw G and Renfree M B (1995). Delayed virilization of the reproductive system of the male tammar wallaby. *Journal of Reproduction and Fertility Abstract Series*, 15, 52.

Clepet C, Schafer A J, Sinclair A H, Palmer M S, Lovell-Badge R and Goodfellow P N (1993). The human SRY transcript. *Human Molecular Genetics,* 2, 2007–2012.

Cooper D W (1993). The evolution of sex determination, sex chromosome dimorphism, and X-inactivation in Therian mammals: A comparison of metatherians (marsupials) and eutherians ('placentals'). In *Sex Chromosomes and Sex Determining Genes.* Eds K C Reed and J A Marshall-Graves. Harwood Academic Publishers, Switzerland. pp 183–200.

Cooper D W and Hope R M (1990). Marsupial and monotreme phylogeny. *Australian Journal of Zoology,* 37, 271–272.

Foster J W, Brennan F E, Hampikian G K, Goodfellow P N, Sinclair A H, Lovell-Badge R, Selwood L, Renfree M B, Cooper D W and Marshall Graves J A (1992). Evolution of sex determination and the Y chromosome: SRY-related sequences in marsupials. *Nature,* 359, 531–533.

Haqq C M, King C-Y, Donahoe P K and Weiss M A (1993). SRY recognizes conserved DNA sites in sex-specific promoters. *Proceedings of National Academy of Sciences USA*, 90, 1097–1101.

Haqq C M, King C-Y, Ukiyama, E, Falsafi S, Haqq T N, Donahoe, P K and Weiss A (1994). Molecular basis of mammalian sexual determination: Activation of Müllerian inhibiting substance gene expression by SRY. *Science,* 266,1494–1500.

Harry J L, Koopman P, Brennan F E, Graves J A M and Renfree M B (1995). Widespread expression of the testis-determining gene, *SRY,* in a marsupial. *Nature Genetics,* 11, 347–349.

Hughes R L, Pearse A, Cooper D W, Joss J M P, Johnston P G and Jones M K (1993). The genetic basis of marsupial gonadogenesis and sexual phenotype. In: *Sex Chromosomes and Sex-Determining Genes.* Eds K C Reed and J A Marshall Graves. Harwood Academic Publishers, Switzerland. pp 17–48.

Hutson J M, Shaw G, Wai-Sum O, Short R V and Renfree M B (1988). The ontogeny of

Mullerian inhibiting substance production and testicular differentiation, migration and descent in the pouch young of a marsupial. *Development,* 104, 549–556.

Jager R J, Anvret M, Hall K and Scherer G (1990). A human XY female with a frame shift mutation in the candidate testis-determining gene SRY. *Nature,* 348, 452–454.

Jost A (1970). Hormonal factors in the sex differentiation of the mammalian foetus. *Philosophical Transactions of the Royal Society of London,* B 259, 119–130.

Koopman P, Gubbay J, Vivian N, Goodfellow P and Lovell-Badge R (1991). Male development of chromosomally female mice transgenic for Sry. *Nature,* 351, 117–121.

Lovell-Badge R (1993). Sex determining gene expression during embryogenesis. *Philosophical Transactions of the Royal Society of London,* B 339, 159–164.

O W-S, Short R V, Renfree M B and Shaw G (1988). Primary genetic control of somatic sexual differentiation in a mammal. *Nature,* 331, 716–717.

Renfree M B, Robinson E S, Short R V and VandeBerg J L (1990). Mammary glands in male marsupials. 1. Primordia in neonatal opossums, *Didelphis virginiana* and *Monodelphis domestica*. *Development,* 110, 385–390.

Renfree M B (1992). The role of genes and hormones in marsupial sexual differentiation. *Journal of Zoology,* 226, 165–173.

Renfree M B (1994). Sexual dimorphisms in the gonads and reproductive tract of marsupial mammals. In *The Differences between the Sexes*. Eds R V Short and E Balaban. Cambridge University Press, Cambridge. pp 213–230.

Renfree M B and Short R V (1988). Sex determination in marsupials: evidence for a marsupial–eutherian dichotomy. *Philosophical Transactions of the Royal Society of London,* B 322, 41–54.

Renfree M B, Wilson D, Short R V, Shaw G and George F (1992). Steroid hormone content of the gonads of the tammar wallaby during sexual differentiation. *Biology of Reproduction,* 47, 644–647.

Robinson E S, Renfree M B, Short R V and VandeBerg J L (1991). Mammary glands in male marsupials. 2. Development and regression of mammary primordia in *Monodelphis domestica* and *Didelphis virginiana*. *Reproduction, Fertility and Development,* 3, 295–301.

Sharman G B, Hughes R L and Cooper D W (1990). The chromosomal basis of sex differentiation in marsupials. *Australian Journal of Zoology,* 37, 451–466.

Shaw G, Renfree M B, Short R V and O W-S (1988). Experimental manipulation of sexual differentiation in wallaby pouch young treated with exogenous steroids. *Development,* 104, 689–701

Shaw G, Renfree M B and Short R V (1990). Primary genetic control of sexual differentiation in marsupials. *Australian Journal of Zoology,* 37, 443–450.

Tyndale-Biscoe C H and Renfree M B (1987). *Reproductive Physiology of Marsupials*. Cambridge University Press, Cambridge. 476pp.

Tyndale-Biscoe C H and Hinds L A (1989). Influence of the immature testis on sexual differentiation in the tammar wallaby *Macropus eugenii* (Macropodidae: Marsupialia). *Reproduction, Fertility and Development,* 1, 243–254.

Whitworth D J, Shaw G and Renfree M B (1996). Gonadal sex reversal of the developing marsupial ovary *in vivo* and *in vitro*. *Development,* 122, 4057–4063

Whitworth D J, Harry J L, Shaw G and Renfree M B (1994). Müllerian inhibiting substance gene expression in the developing testis of the tammar wallaby (*Macropus eugenii*). *Proceedings of the Australian Society for Reproductive Biology,* 68.

Zwingman T, Erickson R P, Boyer T, Ao A (1993). Transcription of the sex-determining region genes Sry and Zfy in the mouse preimplantation embryo. *Proceedings of National Academy of Sciences USA,* 90, 814–817.

SECTION 3

ECOLOGY

CHAPTER 9

FIELD ENERGETICS AND WATER FLUXES IN MARSUPIALS

Brian Green

INTRODUCTION

Water and energy are two of the most important resources that animals must obtain from their environment in order to survive and reproduce. The advent of isotope turnover techniques in the past 30 years has enabled the water and food/energy requirements of animals to be determined under totally free-living conditions. The most important contribution to field studies of water and energy turnover was that provided by Lifson and his co-workers (Lifson *et al.* 1949; Lifson & McClintock 1966); that is, the doubly-labelled water turnover technique. Here, the hydrogen atoms in a water molecule are labelled either as deuterium (^{2}H) or tritium (^{3}H), while the oxygen atom is labelled as ^{18}O. When this isotope mixture is injected into the body of an animal the isotopes become completely mixed into the body water pool within a few hours. Subsequently, the isotopes leave the body; hydrogen and oxygen as water by way of excretion and evaporation, and some oxygen in the form of respiratory CO_2. Both isotopes are replaced in the body when water and oxygen are taken up from the environment by way of feeding, drinking, metabolic water formation and respiration. Both isotopes are diluted in an exponential way, the difference between total ^{18}O loss and that via water representing CO_2 production. The basic calculations of CO_2 production presented by Lifson & McClintock (1966), have been modified to some extent by Nagy (1980) and Speakman *et al.* (1993).

When the chemical composition of the diet and assimilation efficiencies are known it is possible to calculate the amount of energy metabolised and subsequently the amount of food intake required to achieve energy balance. In situations where food provides all of the water intake (by way of free and metabolic water), water influx estimates can provide alternative measures of food intake.

^{22}Sodium has also been used to provide a direct estimate of food intake in field energetics studies, particularly for carnivores (Green 1978; Green & Eberhard 1983); sodium influx rates can be converted to food consumption rates if the sodium content of the diet is known. When animals are in energy balance, the amount of energy metabolised can be derived from food intake estimates if the metabolisable energy efficiency is known. Sodium turnover can also be used to measure the milk consumption rates of sucklings (Green & Newgrain 1979; Green *et al.* 1988).

During the first 15 to 20 years of field isotope studies, most of the data pertained to a single season for a given species. However, over the past 10 years or so increasing attention has been paid to the influences of seasonal climatic variations, diet, reproduction and body size on water and energy metabolism.

The purpose of this review is to assess the relative importance of these influences on the energy and water use by free-living marsupials, and to indicate some possible future directions in this field of research.

MAJOR INFLUENCES

Diet

The diets of carnivores and herbivores are substantially different with respect to chemical composition and degree of variation. The food of carnivores is characterised by its relative uniformity between taxa (Table 9.1a) and the absence of major seasonal variation, being restricted to comparatively minor changes in body fat content. Prey also exhibit high free water contents, a high digestibility of body solids, high available nitrogen and a minimal provision of carbohydrate, the bulk of energy being provided by protein and fat.

Table 9.1a Composition of prey

SPECIES	WATER (% OF FRESH MASS)	ENERGY (KJ G^{-1} FRESH MASS)
Vertebrates		
Leiolopisma sp.	73.6	5.16
Gehyra variegata	74.0	5.31
Mus domesticus	65.9	7.57
Macropus eugenii	69.6	6.45
Invertebrates		
insect larvae	79.1	5.02
isopods	64.9	5.05
beetles	65.4	8.61
earthworms	75.1	3.53
scorpions	68.9	7.54
cockroaches	70.4	5.67

Source: King & Green (1993).

In contrast, the composition of plant materials utilised by herbivores can vary substantially between taxa and season (Table 9.1b). The water contents of plants, particularly grasses and forbs, can range between 10 and 90% of fresh mass, and are usually inversely related to the fibre and available nitrogen contents. Complex carbohydrates form the bulk of plant materials ingested by herbivores and provide the main source of energy. The efficient utilisation of such diets is associated with the development of complex digestive tracts in herbivores.

Table 9.1b Composition of plant leaves in wet (w) and dry (d) seasons

PLANTS	SEASON	WATER (% FRESH MASS)	ENERGY (KJ G^{-1}FRESH)
Grasses[1] (mixed)	w	84.0	3.03
	d	14.0	16.28
Forbs[2]			
Malhania	w	66.1	6.14
	d	6.1	16.90
Tephrosia	w	67.5	6.34
	d	15.7	16.31
Clerodendrum	w	65.9	6.65
	d	16.3	
Trees			
Leptospermum[3] *laevigatum*	w	51.1	11.24
	d	50.7	11.34
Casuarina sticta[3]	w	50.6	10.42
	d	53.8	9.75
Acacia sophorae[3]	w	62.5	7.61
	d	64.0	7.31
Callitris rhomboidea[3]	w	60.7	8.76
	d	55.0	10.04
Eucalyptus creba[4]	w	44.1	13.02
	d	43.7	13.12
Ficus[5]	w	69.7	6.18
	d	61.6	7.80

Sources: [1]King & Green (1993); [2,5]Merchant *et al.* (unpublished observations); [3]Munks (1990); [4]Ellis *et al.* (1995).

In the absence of drinking water, an adequate free water content of plant food is crucial to the proper functioning of the herbivore digestive tract. For example, when the free water content of herbage falls below 55% and no drinking water is available, rabbits (*Oryctolagus cuniculus*) cease feeding (Cooke 1982), and water restriction leads to reduced food intake in *Macropus eugenii* (Barker *et al.* 1970) and *Macropus robustus* (Ealey *et al.* 1965). However, when drinking water is available, herbivores are able to digest extremely dry, fibrous diets.

The prey of carnivores is generally mobile and widely dispersed, thereby imposing major energetic expenditure on active hunters. In contrast, the food of herbivores is immobile and generally more readily available than the food of carnivores, obviating the need for large energy expenditure in foraging for food. Also, folivory is associated with depressed basal metabolic rates (McNab 1986). Consequently, it is to be expected that water and energy turnover in carnivores and herbivores will be dictated by the composition and availability of their food, and wide divergence between them is also to be expected.

Seasonality

Carnivores: Seasonal water flux and field metabolic rate (FMR) data are available for 7 dasyurid species inhabiting temperate environments (Table 9.2). Water fluxes tend to be higher in winter than in summer in the smaller species, but not in the largest species,

Table 9.2 Rates of water influx and field metabolic rate of free-living marsupial carnivores

SPECIES	SEASON	MASS (G)	WATER mL DAY^{-1}	ENERGY kJ DAY^{-1}
Marmosa robinsonii [1]	sp	28	9.4	53
Sminthopsis crassicaudata [2,3]	s	20	22.4	
	w	13	19.4	
	sp	17	13.4	69
Antechinus swainsonii [4]	s	43	23.4	74
	w	49	24.4	162
	a	43	23.4	130
	sp(e)	57	27.9	222
	sp(l)	60	52.9	175
Antechinus stuartii [5,6]	w	26	16.4	72
	s	16	8.9	45
	w	25	14.6	82
	sp(l)	30	26.6	160
Phascogale calura [7]	a	36	6.7	43
	w	34	11.2	110
	sp	35	14.0	70
	s	30	6.8	54
Parantechinus apicalis [8]	s	51	9.4	64
	w	47	15.2	92
	a	59	9.1	127
	sp	49	16.4	94
Dasyurus viverrinus [9]	s	1137	205	*846
	w	1019	289	*1075
	s	1029	148	793
	w	1102	262	1169
Sarcophilus harrisii [10]	s	7900	724	*2591
	w	7100	743	*2890

Key: s=summer a=autumn w=winter sp=spring (e)=early (l)=late *=^{22}sodium derived
Sources: [1]Green, King & Pefaur (unpublished observations); [2]Morton (1980); [3]Nagy *et al.* (1988); [4]Green & Crowley (1989); [5]Nagy *et al.* (1978); [6]Green *et al.* (1991); [7]Green *et al.* (1989); [8]Green, King & Dickman (unpublished observations); [9]Green & Eberhard (1983); [10]Green (unpublished observations).

the Tasmanian devil (*Sarcophilus harrisii*). However, the water flux rates for two *Antechinus* species in sub-nivian conditions were not significantly higher in summer (Green & Crowley, 1989), nor were there significant differences in summer and winter water fluxes for *Antechinus stuartii* in a temperate coastal habitat (Green *et al.* 1991).

There is a more pronounced trend in the relationship of FMRs measured in summer and in winter, the smaller species showing significantly higher rates in winter than in summer (Table 9.3), whereas in *S. harrisii*, the rates in winter and summer do not differ significantly.

The relatively small seasonal influence on mass specific rates of water and energy turnover in *A. stuartii* inhabiting a mild, coastal environment (Green *et al.* 1991), are complicated by the fact that the rates measured in summer pertain to small juveniles while those in winter relate to adults; that is, the summer rates are elevated due to the smaller size of animals (see below) and the higher metabolic demands of growth.

Herbivores: Seasonal FMR data are available for only 6 herbivorous species (Tables 9.4 and 9.5); 4 macropodoids (*Bettongia penicillata, B. gaimardi, Potorous tridactylus* and *Aepyprymnus rufescens*) and 2 arboreal folivores (*Pseudocheirus peregrinus* and *Phascolarctos cinereus*). Three of these species show significantly higher FMRs in winter; 30% above summer levels in both *B. penicillata* and *P. tridactylus*, and 12% in *P. cinereus*. However, such seasonal effects on FMR are not evident in *B. gaimardi* and *A. rufescens* which occupy climatically similar habitats to *B. penicillata* and *P. tridactylus*. Ellis *et al.* (1995) suggest that the small, but significant, seasonal difference in FMR of *P. cinereus* (Table 9.5) is explained by reduced activity in summer.

Macropodoids exhibit strong seasonal variations in water flux rates (Green 1989) that are dictated by the seasonal water status of grasses and forbs (Table 9.1b) that predominate in the diet of most species. However, Wallis & Green (1992) found no seasonal differences in the water fluxes of *A. rufescens* inhabiting a temperate habitat, reflecting the uniform seasonal water status of tubers which constitute a major dietary component in this species.

The southern hairy-nosed wombat, *Lasiorhinus latifrons*, is a grazing herbivore that shows a seasonal pattern of water use (Wells 1973), similar to that of most macropodoids.

The field water fluxes of 2 arboreal folivores (*P. peregrinus* and *P. cinereus*) do not show the extreme seasonal variations seen in grazing species, again reflecting the uniform seasonal water status of the leaves of trees (Table 9.1 b). The water fluxes of *P. cinereus* in summer are only slightly elevated, probably due to increased thermoregulatory evaporation (Ellis *et al.* 1995), but there are no significant seasonal variations in the water influxes of *P. peregrinus* (Munks & Green 1995).

The absence of seasonal variation in the FMRs of *P. peregrinus* probably reflects the relatively benign coastal environment where the study was carried out (Munks & Green, 1995).

Omnivores: Seasonal FMR data are available for only 3 terrestrial omnivores: the bandicoots *Isoodon auratus, I. obesulus* and *Macrotis lagotis* (Table 9.5). The major seasonal variation of FMR in *I. auratus* was related to change in the water status of the environment (Bradshaw *et al.* 1994) rather than any change in thermal conditions. Similarly, the FMRs of *I. obesulus* were not significantly different in summer and late autumn/early winter, but FMRs were significantly higher in late spring (Green, unpublished observations). No significant seasonal differences were found in the FMRs and water fluxes of *M. lagotis* inhabiting arid, central Australia (R. Southgate *et al.*, unpublished observations). The data base for terrestrial omnivores is too limited to detect any obvious trends in seasonal or reproductive influences on water and energy use.

Table 9.3 Seasonal field metabolic rates (FMR) and water influx in temperate marsupial carnivores

SPECIES	MASS (G)	WATER INFLUX (mL kg^{-1} day^{-1}) W	S	W/S	FMR (kJ kg^{-1} day^{-1}) W	S	W/S
Sminthopsis[1] *crassicaudata*	17	1490	1098	1.36			
Antechinus stuartii[2]	22	584	559	1.04	3 286	2803	1.17
A. stuartii[3]		548	509	1.08	4 218	2526	1.67
A. swainsonii[3]	42	610	541	1.13	4 046	1719	2.35
Phascogale calura	35	326	188	1.73	3 210	1212	2.65
Parantechinus apicalis[4]	9	324	185	1.75	1 964	1245	1.58
Dasyurus viverrinus[5]	1 100	284	180	1.56	1 055	733	1.44
Sarcophilus harrisii[6]	7 500	105	92	1.14	407	328	1.24

Key: w = winter s = summer

Sources: [1]Morton (1980); [2]Green *et al.* (1991); [3]Green & Crowley (1989); [4]Green, King & Dickman (unpublished observations); [5]Green & Eberhard (1983); [6]Green (unpublished observations).

Table 9.4 Rates of water influx and field metabolic rate of free-living macropodoids

SPECIES	SEASON	MASS (G)	WATER mL DAY⁻¹	ENERGY kJ DAY⁻¹
Potorous tridactylus[1]	s(d)	816	131	454
	sp(w)	784	174	589
Bettongia penicillata[2]	s(d)	1100	67	524
	a	1100	99	570
	w(w)	1100	196	695
B. gaimardi[3]	s	1700	118	892
	s	1700	277	856
Setonix brachyurus[4, 5, 6]	s(d)	1900	90	548
	s(d)	2600	130	
	w(w)	2880	314	
Lagorchestes hirsutus[7]	w(d)	1453	148	661
	s(w)	1351	247	532
L. conspicillatus[8]	sp(d)	2230	91	
Aepyprymnus rufescens[9]	s(d)	2850	398	1422
	w(w)	2900	374	1491
Petrogale inornata[10]	Annual	3180	257	
P. rothschildi[11]	sp(d)	3350	149	
	s(w)	2740	365	
Macropus eugenii[11]	w	4863	826	
	sp	5530	412	
Petrogale xanthopus[12]	S(d)	8900	475	2209
	w(d)		497	
	w(w)		1304	
Thylogale billardierii[4]	s(d)	5980	646	1630
Macropus giganteus[4]	s(d)	43900	2634	8170
M. rufus[13]	s(d)	21800	861	
M. robustus[13]	s(d)	28100	1107	

Key: s=summer a=autumn w=winter sp=spring (d)=dry (w)=wet

Sources: [1]Green & Wallis (unpublished observations); [2]Green, Best & Turner (unpublished observations); [3]Johnson et al. (unpublished observations); [4]Nagy *et al.* (1990); [5]Rice (1976); [6]Horwitz (1980); [7]Lundie-Jenkins (unpublished observations); [8]Bakker & Bradshaw (1989); [9]Wallis & Green (1992); [10]Kennedy & Heinsohn (1974); [11]Bradshaw (unpublished observations); [12]Green & Robinson (unpublished observations); [13]Dawson *et al.* (1975).

Breeding

There are few data about the energy costs associated with mating behaviour in marsupials. Many small dasyurid species exhibit an intense period of mating which is followed by the death of most or all males (Lee & Cockburn 1985). Green *et al.* (1991) found that in the mating period immediately before the male die-off in *A. stuartii*, the FMRs of males were significantly reduced, body masses declined by about 20% and body fat reserves became depleted. During the mating period males spend much of their time in leks (Lazenby-Cohen & Cockburn 1988), which may explain the reduced FMR in males at this time. Further, there may be a reduction in the time available for foraging, with concomitant reduction in food and energy acquisition and the decline in body mass and fat reserves. It is not known how representative this pattern of energy use in breeding males is among small dasyurids.

Reproduction in marsupials is characterised by:
(a) a relatively brief gestation, the birth of extremely small neonates and a relatively long period of lactation (Russell 1982; Tyndale-Biscoe & Janssens 1988);
(b) mammary glands that increase substantially in size throughout lactation, with associated increases in milk production (Green *et al.* 1988; Munks & Green 1995; Merchant *et al.* 1994); and
(c) substantial changes in the composition of milk (reviewed by Green 1984 and Green & Merchant, 1988).

In the early stages of lactation, marsupial milk is quite dilute, daily production is low, and consequently, the commitment of energy to milk is low (Table 9.6). In contrast, at later stages of lactation the milk is concentrated, milk production is high and the commitment of energy to milk is substantial (Table 9.6).

Water is the major component of milk at all stages of lactation in marsupials, with the greatest demand on maternal body water being at peak lactation (Table 9.6 and Figure 9.1). Thus, in the later stages of lactation, young marsupials exert substantial demands on maternal energy and water reserves by way of milk provision. These demands are further amplified by the metabolic costs associated with obtaining additional food and water from the environment, as well as the direct metabolic costs involved in milk synthesis.

There are no detectable increases in the FMRs of marsupials during the early stages of lactation when the young are small. However, the FMRs and food consumption rates of late-stage lactating marsupials are significantly higher than those of non-lactating animals in most field studies; for example, *Antechinus* (Green *et al.* 1991), *Dasyurus* (Green & Eberhard 1983), *Sarcophilus* (B Green, unpublished observations), *Pseudocheirus* (Munks & Green 1995), but Wallis & Green (1992) were unable to detect any significant difference in the mean FMRs of lactating and non-lactating *Aepyprymnus rufescens*.

The presence or absence of an increase in FMR in the later stages of lactation reflects differences in the number of young being suckled, milk composition and the rates of milk production (Table 9.7), with polytocous species being subjected to much greater lactational energy demands in the late stages of lactation. For *Dasyurus viverrinus* the total energy expenditures of lactating females are about double those of non-lactating animals.

A similar pattern is seen with respect to water flux rates of lactating females; there are no detectable differences between early and non-lactating animals, whereas at peak lactation water flux rates are increased by between 30 and 60 % (Figure 9.1). However, there are no detectable differences in water fluxes of late and non-lactating *A. rufescens* (Wallis & Green 1992), nor have they been found in *B. penicillata* or *M. eugenii* (B Green, unpublished observations). Again, this is probably because macropodoids usually have a single suckling and the relatively low rates of milk production impose a comparatively small amount of additional water loss on the female (for example, *B. penicillata* in Table 9.6); an amount that lies within the coefficient of variation observed in field water flux measurements.

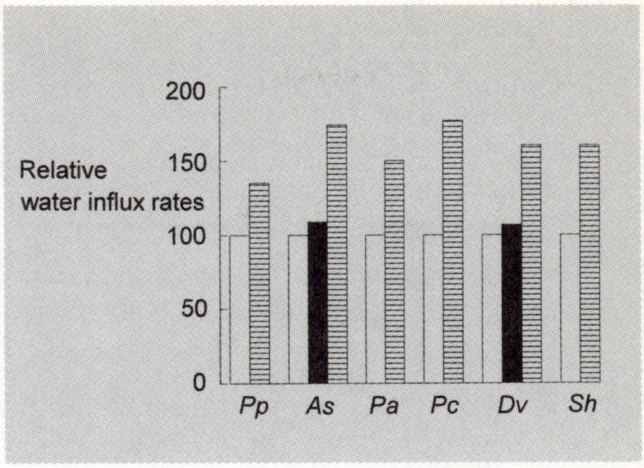

Figure 9.1 Relative water influx rates during lactation in marsupials.
Key: [open bar] non lactating [solid bar] early lactation [line bar] late lactation
Pp: *Pseudocheirus peregrinus*; *As*: *Antechinus stuartii*; *Pa*: *Parantechinus apicalis*; *Pc*: *Phascogale calura*; *Dv*: *Dasyurus viverrinus* *Sh*: *Sarcophilus harrisii*

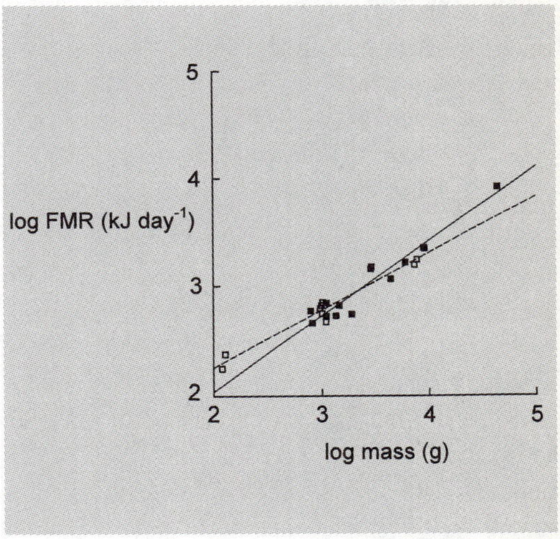

Figure 9.2 Field metabolic rate (FMR) of macropodoid ■ and non-macropodoid □ herbivorous marsupials.
Notes:
Macropodoid;
kJ d^{-1} =.664(.202)mass $(g)^{0.687(.058)}$
Non-macropodoid;
kJ d^{-1} =1.185(.106) $mass(g)^{0.523(.034)}$
(S.E.)

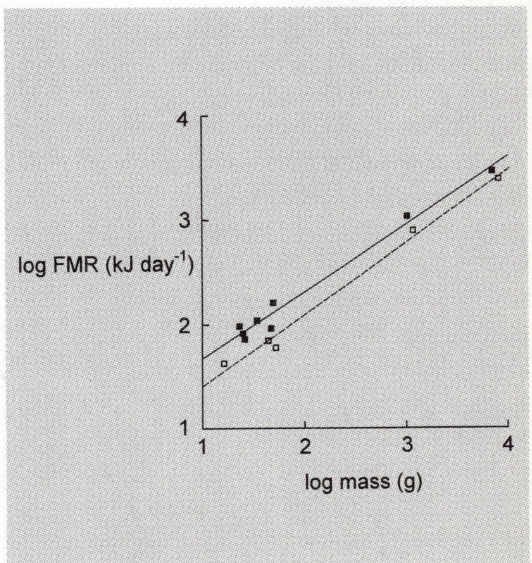

Figure 9.3 Field metabolic rate (FMR) of marsupial carnivores in winter ■ and summer □ .
Note:
Winter:
kJ d^{-1}=1.029(0.080) $mass(g)^{0.642(.037)}$
Summer:
kJ d^{-1} = 0.742(.102) $mass(g)^{0.691(0.041)}$
(S.E.)

Table 9.5 Rates of water influx and field metabolic rate of free-living non-macropodoid herbivores and omnivores

SPECIES	SEASON	MASS (g)	WATER mL DAY^{-1}	ENERGY kJ DAY^{-1}
Petaurus brevicauda [1]	sp	121	25	169
Gymnobelideus leadbeateri [2]	sp	129	35	226
	w	129	48	
Pseudocheirus herbertensis [3]	sp	1103	140	446
Hemibelideus lemuroides [3]	sp	1026	155	675
Petauroides volans [3]	w	1018	88	520
Pseudocheirus peregrinus [5]	s	1002	114	631
	w	980	118	591
Trichosurus vulpecula [6]	An	1520	136	
Phascolarctos cinereus [7, 8, 9]	s	7400	296	1480
	w	7800	265	1654
	w	9200	414	2088
	s	7286	321	
Lasiorhinus latifrons [10]	w	22200	777	
	s	22200	555	
Bandicoots				
Isoodon obesulus [11, 12]	a	1230	109	644
	s/a	486	38	366
	sp	517	40	459
I. auratus [13]	(d)	307	27	76
	w	333	46	257
Isoodon macrourus [14]	w	1456	341	
	w	873	103	
	s	1120	152	
	s	1388	355	
Macrotis lagotis [15]	w	1208	56	626
	s	1132	68	455

Key:s=summer a=autumn w=winter sp=spring (w)=wet (d)=dry (an)=Annual

Data from: [1]Nagy & Suckling (1985); [2]Smith *et al* (1982); [3]Goudberg (1988); [4]Foley *et al.* (1990); [5]Munks & Green (1995); [66]Kennedy & Heinsohn (1974); [7]Ellis *et al.* (1995); [8]Nagy & Martin (1985); [9]Eberhard (1972); [10]Wells (1973); [11]Nagy *et al.* (19910; [12]Green *et al.* (unpublished observations); [13]Bradshaw *et al.* (1994); [14]Hulbert & Gordon (1972); [15]Southgate *et al.* (unpublished observations).

Body size

The impact of body mass on FMRs of 13 marsupial species has been assessed by Nagy (1987): 3 carnivore, 4 macropodoids and 6 other herbivore/omnivore species. However, the regression analyses provided are flawed due to the multiple representation of some species, seasonal factors are ignored, and species are only characterised as herbivores or non-herbivores. This analysis indicates that FMRs of marsupials are related to body mass in the following ways:

herbivore FMR (kJ day^{-1}) = 6.37 $mass(g)^{0.64}$
non-herbivore FMR (kJ day^{-1}) = 9.51 $mass(g)^{0.64}$

These body mass exponents are not significantly different, but marsupial non-herbivore FMRs are significantly higher than those of marsupial herbivores.

Another review by Nagy (1994) covers 18 species (4 carnivores, 8 macropodoids and 6 others) and does not use multiple data for any species. However, the mean FMR value derived for some species is derived from different seasons and cohorts. This analysis shows that in general the FMRs of marsupials scale to $mass^{0.58}$, significantly lower than in eutherians ($mass^{0.76}$).

There are now data on FMRs for 28 marsupial species; 8 carnivores, 10 macropodoids and 10 others (Tables 9.2, 9.4 and 9.5). No significant seasonal influences are apparent in the FMR data for macropodoids and non-macropodoid herbivores (petaurids and the single phascolarctid), FMRs being described by the expressions:

macropodoid FMR (kJ day^{-1}) = 4.61 mass $(g)^{0.69}$
non-macropodoid FMR (kJ day^{-1}) = 15.3 mass $(g)^{0.52}$

The difference in these slope exponents is significant ($p<0.02$) as shown in Figure 9.2.

In contrast, there is a significant seasonal influence on FMRs of carnivorous marsupials:

winter FMR (kJ day^{-1}) = 10.69 mass $(g)^{0.64}$
summer FMR (kJ day^{-1}) = 5.52 mass $(g)^{0.69}$

There is no significant seasonal difference in carnivore slope exponents, but the intercept value is significantly higher in winter ($p<0.01$) (Figure 9.3). There are no significant differences between the slope exponents for macropodoids and carnivores, but the intercept values for carnivores are higher. The slope exponents of carnivores and non-macropodoid herbivores are significantly different ($p<0.05$).

The influence of body mass on the field water fluxes of 18 marsupial species has been reviewed by Nagy & Peterson (1988); 4 carnivores, 6 macropodoids, and 8 non-macropodoid herbivores/omnivores. Nagy & Peterson (1988) provided two regressions of water influx on body mass:

herbivores; influx (ml day^{-1}) = 0.874 body mass $(g)^{0.71}$
carnivores; influx (ml day^{-1}) = 1.862 body mass $(g)^{0.71}$

However, in these analyses there is multiple representation of some species, especially carnivores, and data are from all seasons. In addition, the term 'herbivore' is used to cover grazers, arboreal folivores and omnivores.

There is now a much larger data set available for assessing the influence of body mass on field water fluxes of marsupials; 8 carnivores, 15 macropodoids, 9 non-macropodoid herbivores, and 4 omnivores (Tables 9.2, 9.4 and 9.5). In addition, seasonal effects can be clearly discerned.

There is a substantial seasonal influence on field water fluxes in macropodoid herbivores; the rates measured in wet seasons are significantly higher ($p< 0.001$) than

Table 9.6 Milk production and transfer of water and energy to young in four similarly sized (1 kg) marsupials

	Pouch young		Milk composition		Milk in	Pouch Young intake		Litter		Dam provision	
Species	Age (days)	Mass (g)	kJ mL^{-1}	mL H_2O mL^{-1}	mL day^{-1}	kJ day^{-1}	mL day^{-1}	n	Mass (g)	kJ day^{-1}	mL H_2O day^{-1}
Bettongia[1] *penicillata*	34	9.0	5.3	0.8	1.0	5.3	0.8	1	9.0	5.3	0.8
	64	35.6	6.0	0.75	4.2	25.4	3.2		35.6	25.4	3.2
	94	183	7.7	0.70	22.8	175	16		183	175	16.0
Pseudocheirus[2] *peregrinus*	70	19.8	3.5	0.85	2.4	8.4	2.0	2	40	17	4.0
	110	65.0	6.0	0.75	10.4	60.1	7.8		130	120	15.6
	160	211	4.3	0.85	18.3	77.8	15.6		420	155	31.2
Isoodon[3] *macrourus*	23	8.5	5.6	0.8	2.3	13.1	1.9	3	26	39	6
	38	35.2	8.8	0.7	5.7	50.4	4.0		105	151	12
	52	112	11.6	0.6	17.2	200	10.3		336	600	31
Dasyurus[4] *viverrinus*	10	13	2.2	0.9	2.5	5.5	2.3	6	78	33	14
	63	40	5.9	0.8	8.0	47.2	6.4		240	283	38
	105	190	8.8	0.7	20.0	176	14.0		1140	1056	84

Sources: [1]Merchant *et al.* (1994); [2]Munks & Green (1995); [3]Merchant *et al.* (1996); [4]Green *et al.* (unpublished observations)

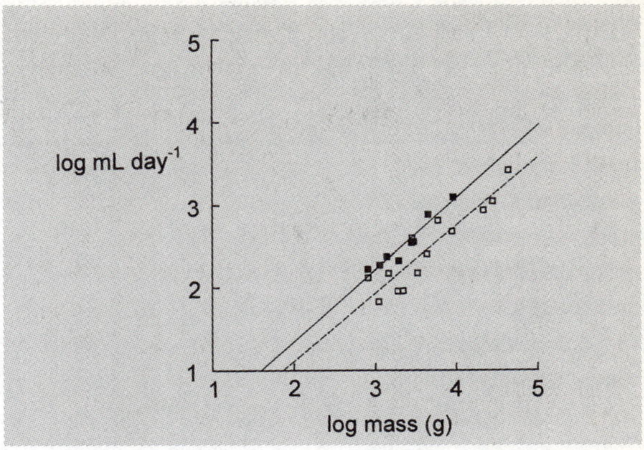

Figure 9.4 Water influx rates in macropodoids in wet (closed symbol) and dry seasons (open symbol).
Note:
Wet:
ml d^{-1} =0.372(0.344) mass(g)$^{0.872(.102)}$
Dry:
ml d^{-1} = -.0530(0.382) mass(g)$^{0.820(.103)}$
(S.E.)

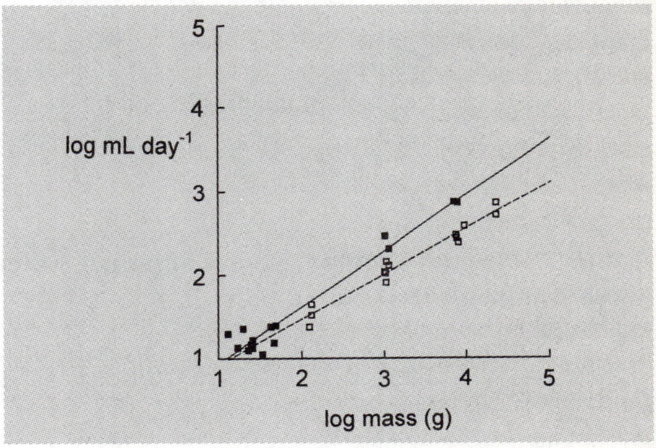

Figure 9.5: Water influx rates in carnivores (closed symbol) and non-macropodoid herbivores (open symbol).
Note:
Carnivores:
ml d^{-1} = 0.133(0.110) mass$^{0.706(.054)}$
Non-macropodoid:
ml d^{-1} = 0.404(0.107) mass$^{0.549(0.032)}$
(S.E.)

during dry seasons (Fig 9.4), however the slope exponents are not significantly different. The regressions take the form:

wet season; ml day^{-1} = 0.425 body mass $(g)^{0.87}$

dry season; ml day^{-1} = 0.295 body mass $(g)^{0.82}$

There are no significant seasonal influences on the field water fluxes of non-macropodoid herbivores or carnivores (Figure 9.3), regressions taking the form:

non-macropodoids; ml day^{-1} = 2.53 body mass$(g)^{0.55}$

carnivores; ml day^{-1} = 1.35 body mass$(g)^{0.71}$

The slope exponent for carnivores is not significantly different to that for macropodoids in either wet or dry seasons, nor is the intercept value for the wet season different. However, the intercept value for macropodoids in the dry season is significantly lower than that for carnivores (p= 0.013). The slope exponent for field water fluxes of non-macropodoid herbivores is significantly lower than those for both carnivores and macropodoids (p= 0.032) (Figure 9.5). Even if the data for *Gymnobelideus* and *Petauroides* (smaller arboreal species that include nectar, plant exudates and insects in their diets) are excluded from the non-macropodoid herbivore analysis, there are no significant changes to the regression exponents.

Insufficient data are currently available for the omnivorous bandicoots to assess the influence of body mass on FMRs and water flux rates in this group.

A general comparison of relative field energy and water use by three similar sized

Table 9.7 Peak lactation energetics; milk energy transfer and field metabolic rate (FMR) in three similarly sized marsupials

SPECIES	PY(n)	FMR (kJ DAY^{-1})			MILK	TOTAL kJ
		L	NL	L/NL	kJ DAY^{-1}	L/NL
Dasyurus viverrinus[1]	6	1240	950	1.30	640	1.98
Pseudocheirus[2] *peregrinus*	2	770	580	1.33	156	1.60
Bettongia penicillata,[3]	1	585	585	1.00	175	1.30

Key: PY=pouch young L=lactating NL=non-lactating
Sources: [1]Green *et al.* (unpublished observations); [2]Munks & Green (1995); [3]Green *et al.* (unpublished observations); Merchant *et al.* (1994).

Table 9.8 General comparison of seasonal water and energy turnover in three similarly sized marsupials

SPECIES	SEASON	H_2O IN (mL KG^{-1} DAY^{-1})	FMR (kJ KG^{-1} DAY^{-1})
D. viverrinus[1]	w	284 ± 44	1120
	s	180 ± 42	770
P. peregrinus[2]	w + s	120 ± 20	610
B. penicillata[3]	w	140 ± 25	790
	s	61 ± 14	490

w=winter, s=summer.
Sources: [1]Green & Eberhard (1983); [2]Munks & Green (1995); [3]Green *et al.* (unpublished observations)

(1 kg) non-breeding marsupials is presented in Table 9.8. The representative carnivore utilises water and energy at the highest rates, especially in winter, while the macropodoid herbivore exhibits the lowest FMRs and water fluxes in the dry season. During wet periods the water fluxes of a macropodoid are equal to those of a carnivore, while the rates for the arboreal folivore are intermediate in every respect.

DISCUSSION

Body mass is the major determinant of energy and water use in endotherms (Nagy 1987, 1994; Nagy & Peterson 1988), accounting for over 90% of the variation in these parameters in mammals. However, while the allometric analyses of marsupials and eutherians provided by these authors are of general value, they are of limited predictive value as they include data from a variety of seasons, breeding conditions, and activity regimes (for example, active, fasting, torpid). Also, they only delineated herbivorous and non-herbivorous/carnivorous categories in their analyses. The enlarged data base now available allows the more detailed analysis of field metabolism and water use by marsupials presented here.

The regression exponents for macropodoid FMRs are not significantly different to those for 'marsupial herbivores' (that is, macropodoids and non-macropodoids) provided by Nagy (1987), however the slope exponent for non-macropodoid herbivores is significantly lower than that for macropodoids and 'marsupial herbivores'. While the slope exponents for carnivorous marsupial FMRs are not significantly different from that for 'non-herbivores' (Nagy 1987), the distinct seasonal influence on the intercept values was not distinguished by Nagy (1987).

The regression exponents for macropodoid FMRs and water fluxes, in both wet and dry conditions, are not significantly different to those derived by Green (1989) from a smaller data base.

Even when the regression descriptors for both FMR and water flux in the present analysis are not significantly different to those described for marsupials by Nagy (1987), Nagy & Peterson (1988) and Green (1989), the larger data set utilised here should provide more accurate predictive estimates of field water and energy use in marsupials.

Although body size is the most important proximate determinant of water and energy turnover, it is clear that season, activity patterns, and breeding, can exert significant influences also. While seasonal climatic effects on FMR are most obvious in carnivorous marsupials inhabiting cooler, more mesic environments, the effects are more pronounced in small species, and not detectable in the largest species, *S. harrisii*. Unfortunately there are few seasonal data for macropodoids, particularly for larger species, however, it is likely that seasonal climatic influences on FMR will also be restricted to smaller species (probably those weighing less than 2 kg).

While there do not appear to be major seasonal influences on water flux rates in carnivorous and non-macropodoid folivorous marsupials, clear seasonal differences are shown by macropodoid herbivores that relate to the water status of herbage and the general environment. The variations in water flux are strictly seasonal in habitats where rainfall patterns are relatively uniform and reliable, for example, wet/dry tropics, mesic mediterranean regions. However, in xeric regions where rainfall is unreliable and irregular, water flux rates in grazing herbivores follow the local rainfall pattern, and are not strictly seasonal.

There is a fundamental difference between marsupials and eutherians with respect to their relative emphasis on lactation and gestation in the growth and development of young; in marsupials gestation is relatively brief while lactational duration is extensive (Lee & Cockburn 1985; Tyndale-Biscoe & Janssens 1988). Marsupial neonates are extremely small and the early stages of lactation are associated predominantly with organ differentiation, the daily amounts of energy and water required for growth in body mass being miniscule compared to the total daily turnover of energy and water by lactating females. However, with the onset of endothermy in marsupial young, (suggested by Russell (1982) as being equivalent to birth in eutherians), there is a major increase in milk provision that allows an acceleration in the growth of young (Green *et al.* 1988; Dove & Cork 1989). Thus, in the latter stages of lactation the provisioning of young places a significant burden on the energy and water resources of marsupial females, especially polytocous species.

The apparent absence of detectable increases in the FMRs and water expenditures of monotocous macropodoids suggests that the lactational pattern of this group (that is, a long period of low level investment in a single young) is well suited to exploiting habitats where water and food resources may be constrained. This pattern of maternal investment is carried to even further extremes by the larger arid/semi-arid kangaroos, which can 'abort' small pouch young before any significant energy investment has occurred, replacing the young with a neonate derived from a dormant blastocyst, then repeating the process if there is no improvement in habitat resources (Tyndale-Biscoe 1973).

The comparatively slow growth of most marsupial young (Lee & Cockburn 1985; Cockburn & Johnson 1988), reflects maternal regulation of energy supplied in the milk (Green & Merchant 1988), and allows marsupials in general to spread the costs of reproduction in habitats that may be impoverished or unreliable in providing adequate energy and water resources.

Some appreciation of the relative magnitude of the factors that influence energy and water turnover can be gained by equating seasonal and breeding increases in these parameters to the notional body mass of a marsupial using similar amounts of energy and water. For example, a 1 kg macropodoid increases its water influx from 85 to 173 ml day^{-1} in passing from dry to wet conditions, the latter value being equivalent to the water flux of a 2.4 kg macropodoid in dry conditions. Similarly, a 1 kg marsupial carnivore increases its FMR from 650 to 890 kJ day^{-1} in passing from summer to winter conditions, the winter FMR being equivalent to that of a carnivore weighing 1.6 kg in summer. The total metabolic expenditure of a late-stage lactating *D. viverrinus* (1880 kJ day^{-1}) is equivalent to a carnivorous marsupial weighing 4.7 kg in summer. Clearly, factors other than body mass can be extremely important in determining the energy and water expenditures of free-living animals, and it is important that these factors are considered in generating reliable allometric regressions.

FUTURE DIRECTIONS

The macropodoids and dasyurids represent two excellent taxonomic groups for concentrating efforts to clarify the influence of body size on field energetics and water fluxes; they both offer a wide size range, individuals within each group are similar with respect to diet, body form and locomotion, and breeding characteristics. There is a particular need to increase the data available for larger macropodoids, as well as the smallest species (*Hypsiprymnodon moschatus*), and for intermediate-sized dasyurids (for example, *Dasyuroides* and *D. maculatus*), as well as the smallest forms such as *Planigale* and *Ningaui*. Such data should greatly enhance the allometric descriptors for FMR and water flux in these two marsupial groups, and further elucidate the relative impacts of seasonality and breeding.

There is a dearth of information on seasonal FMRs and water fluxes in arboreal folivorous marsupials, especially in smaller forms that are physiologically incapable of utilising a totally folivorous diet (Cork & Foley 1991) and must resort to including nectar, plant exudates and invertebrates in their diet to satisfy energy requirements.

FMR and water flux estimates are often associated with coefficients of variation that exceed 25% of mean values. It is likely that a major part of this variation is due to different levels of activity shown by individuals. The deployment of activity loggers in conjunction with doubly labelled water experiments should help resolve the extent of the influence of activity patterns on energy and water use in the field.

Although there is a reasonable amount of published information on changes in milk composition in marsupials, milk consumption rates have only been established for a few species. In view of the comparative ease of measuring sodium turnover in pouch young and the sodium content of milk, the estimation of milk intake in free-living marsupials is also worth further research attention.

Acknowledgments

I wish to thank my research collaborators for permission to use unpublished data in this review. I thank Keith Newgrain for his able assistance in isotope analyses and for preparing the figures.

References

Bakker H R and Bradshaw S D (1989). Rate of water turnover and electrolyte balance of an arid-zone marsupial, the spectacled hare wallaby (*Lagorchestes conspicillatus*) on Barrow Island. *Comparative Biochemistry and Physiology*, 92A, 521–529.

Bakker H R, Bradshaw S D and Main A R (1982). Water and electrolyte metabolism of the tammar wallaby *Macropus eugenii*. *Physiological Zoology*, 55, 209–219.

Barker S, Lintern S M and Murphy C (1970). The effects of water restriction on urea retention and nitrogen excretion in the Kangaroo Island Wallaby *Protemnodon eugenii* (Desmarest). *Comparative Biochemistry and Physiology*, 34, 883–893.

Bradshaw S D (1988). *Garden Island tammar research programme: an investigation into the physiological well-being of the tammar wallaby, Macropus eugenii on Garden Island, in Western Australia*. Department of Zoology, University of Western Australia, Perth.

Bradshaw S D, Morris K D, Dickman C R, Withers P C and Murphy D (1994). Field metabolism and turnover in the golden bandicoot (*Isoodon auratus*) and other small mammals from Barrow Island, Western Australia. *Australian Journal of Zoology*, 42, 29–41.

Cockburn A and Johnson C N (1988). Patterns of growth. In *The Developing Marsupial; Models for Biomedical Research*. Eds C H Tyndale-Biscoe and P A Janssens. Springer-Verlag, Berlin. pp 28–40.

Cooke B D (1982). Reduction of food intake and other physiological responses to a restriction of drinking water in captive wild rabbits, *Oryctolagus cuniculus* (L.). *Australian Wildlife Research*, 9, 247–252.

Cork S J and Foley W J (1991). Digestive and metabolic strategies of arboreal mammalian folivores in relation to chemical defences in temperate and tropical forests. In *Plant defences against mammalian herbivory*. Eds R T Palo and C T Robbins. CRC Press, Boca Raton, Florida. pp 133–166.

Dawson T J, Denny M J S, Russell E M and Ellis B (1975). Water usage and diet preferences of free-ranging kangaroos, sheep and feral goats in the Australian arid zone during summer. *Journal of Zoology*, 177, 1–23.

Dove H and Cork S J (1989). Lactation in the tammar wallaby (*Macropus eugenii*). 1. Milk consumption and the algebraic description of the lactation curve. *Journal of Zoology*, 219, 385–397.

Eberhard I H (1972). Ecology of the koala, *Phascolarctos cinereus* (Goldfuss), on Flinders Chase, Kangaroo Island. PhD thesis, University of Adelaide, Adelaide.

Ealey E H M, Bentley P J and Main A R (1965). Studies on water metabolism of the hill kangaroo, *Macropus robustus* (Gould) in north-west Australia. *Ecology*, 46, 473–479.

Ellis W, Melzer A, Green B, Newgrain K, Hindell M and Carrick F (1995). Seasonal variation in water flux, field metabolic rate and food consumption of free-ranging koalas (*Phascolarctos cinereus*). *Australian Journal of Zoology*, 43, 59–68.

Foley W J, Kehl J C, Nagy K A, Kaplan I R and Borsboom A C (1990). Energy and water metabolism in free-living greater gliders, *Petauroides volans*. *Australian Journal of Zoology*, 38, 1–9.

Goudberg N J (1988). The feeding ecology of three species of north Queensland upland rainforest ringtail possums, *Hemibelideus lemuroides*, *Pseudocheirus herbertensis* and *Pseudocheirus archeri* (Marsupialia: Petauridae). PhD thesis, James Cook University of North Queensland, Townsville.

Green B (1978). Estimation of food consumption in the dingo, *Canis familiaris dingo*, by means of ^{22}Na turnover. *Ecology*, 59, 207–210.

Green B (1984). Composition of milk and energetics of growth in marsupials. *Zoological Society of London Symposia*, 51, 369–387.

Green B (1989). Water and energy turnover in free-living Macropodoids. In *Kangaroos, Wallabies and Rat-Kangaroos*. Eds G C Grigg, I D Hume and P Jarman. Surrey-Beatty & Sons Pty Ltd, Sydney. pp 223–229.

Green B and Eberhard I (1983). Water and sodium intake and estimated food consumption in free-living eastern quolls, *Dasyurus viverrinus*. *Australian Journal of Zoology*, 31, 871–880.

Green B, King D and Bradley A (1989). Water and energy metabolism and estimated food consumption rates of free-living wambengers, *Phascogale calura* (Marsupialia: Dasyuridae). *Australian Wildlife Research*, 16, 501–508.

Green B and Merchant J (1988). The composition

of marsupial milk. In: *The developing marsupial; models for biomedical research*. Eds C H Tyndale-Biscoe and P A Janssens. Springer-Verlag, Berlin. pp 41–54.

Green B, Merchant J and Newgrain K (1988). Milk consumption and energetics of growth in pouch young of the tammar wallaby, *Macropus eugenii*. *Australian Journal of Zoology*, 36, 217–227.

Green B and Newgrain K (1979). Estimation of the milk intake of sucklings by means of ^{22}Na. *Journal of Mammalogy*, 60, 556–559.

Green B, Newgrain K, Catling P and Turner G (1991). Patterns of prey consumption and energy use in a small carnivorous marsupial, *Antechinus stuartii*. *Australian Journal of Zoology*, 39, 539–547.

Green, K and Crowley, H (1989). Energetics and behaviour of active subnivean insectivores *Antechinus swainsonii* and *A. stuartii* (Marsupialia: Dasyuridae). *Australian Wildlife Research*, 16, 509–516.

Horwitz P (1980). Condition, total body water and aspects of metabolism of quokkas (*Setonix brachyurus*) and tammars (*Macropus eugenii*). In *Electrolyte metabolism in the quokka (Setonix brachyurus, Quoy and Gaimard) and the tammar (Macropus eugenii, Desmarest)*. Eds S Ambrose, S Bunn, J Dolra, B Fergusson, P Horwitz, P and Kelly. Department of Zoology, University of Western Australia, Perth.

Hulbert A J and Gordon G (1972). Water metabolism of the bandicoot *Isoodon macrourus* Gould in the wild. *Comparative Biochemistry and Physiology*, 41A, 27–34.

Inns R W (1980). Ecology of the Kangaroo Island wallaby, *Macropus eugenii* (Desmarest) in Flinders Chase National Park, Kangaroo Island. PhD thesis, University of Adelaide, Adelaide.

Kennedy P M and Heinsohn G E (1974). Water metabolism of two marsupials — the brush-tailed possum, *Trichosurus vulpecula* and the rock-wallaby, *Petrogale inornata* in the wild. *Comparative Biochemistry and Physiology*, 47A, 829–834.

King D and Green B. (1993). *Goanna: The Biology of Varanid Lizards*. University of New South Wales Press, Sydney.

Lazenby-Cohen K A and Cockburn A (1988). Lek promiscuity in a semelparous mammal, *Antechinus stuartii* (Marsupialia: Dasyuridae). *Behavioural Ecology and Sociobiology*, 22, 195–202.

Lee A K and Cockburn A (1985). *The evolutionary ecology of marsupials*. Cambridge University Press, Cambridge.

Lifson N, Gordon G B, Visscher M B and Nier A O (1949). The fate of utilised molecular oxygen and the source of the oxygen of respiratory carbon dioxide, studied with the aid of heavy oxygen. *Journal of Biological Chemistry*, 180, 803–811.

Lifson N and McClintock R (1966). Theory of use of the turnover rates of body water for measuring energy and material balance. *Journal of Theoretical Biology*, 12, 46–74.

McNab B K (1986). Food habits, energetics, and the reproduction of marsupials. *Journal of Zoology*, 208, 595–614.

Merchant J C, Libke J A and Newgrain K (1996). Milk consumption and estimates of growth energetics in pouch young of the northern brown bandicoot, *Isoodon macrourus* (Peramelidae, Marsupialia), in captivity. *Journal of Zoology*, 238, 483–494.

Merchant J C, Libke J A and Smith M J (1994). Lactation and Energetics of growth in the Brush-tailed Bettong, *Bettongia penicillata* (Marsupialia: Potoroidae) in Captivity. *Australian Journal of Zoology*, 42, 267–277.

Morton S R (1980). Field and laboratory studies of water metabolism in *Sminthopsis crassicaudata* (Marsupialia: Dasyuridae). *Australian Journal of Zoology*, 28, 213–217.

Munks S A (1990). Ecological energetics and reproduction in the common ringtail possum, *Pseudocheirus peregrinus*. PhD thesis, University of Tasmania, Hobart.

Munks S and Green B (1995). Energy allocation for reproduction in a marsupial arboreal folivore, the common ringtail possum (*Pseudocheirus peregrinus*). *Oecologia*, 101, 94–104.

Nagy K A (1980). CO_2 production in animals: analysis of potential errors in the doubly labelled water method. *American Journal of Physiology*, 238, R466–473.

Nagy K A (1987). Field metabolic rate and food requirement scaling in mammals and birds. *Ecological Monographs*, 57, 111–128.

Nagy K A (1994). Seasonal water, energy and food use by free-living, arid-habitat mammals. *Australian Journal of Zoology*, 42, 55–63.

Nagy K A, Bradley A J and Morris K D (1990). Field metabolic rates, water fluxes, and feeding rates of quokkas, *Setonix brachyurus*, and tammars, *Macropus eugenii*, in Western Australia. *Australian Journal of Zoology*, 37, 553–560.

Nagy K A, Bradshaw S D and Clay B T (1991). Field metabolic rate, water flux, and food requirements of short-nosed bandicoots, *Isoodon obesulus* (Marsupialia: Peramelidae). *Australian Journal of Zoology*, 39, 299–305.

Nagy K A, Lee A K, Martin R W and Fleming M R (1988). Field metabolic rate and food requirement of a small dasyurid marsupial, *Sminthopsis crassicaudata*. *Australian Journal of Zoology*, 36, 293–299.

Nagy K A and Martin R W (1985). Field metabolic rate, water flux, food consumption and time budgets of koalas, *Phascolarctos cinereus* (Marupialia: Phascolarctidae) in Victoria. *Australian Journal of Zoology*, 83, 655–665.

Nagy K A and Peterson C C (1988). *Scaling of water flux rate in animals*. University of California Publications in Zoology, University of California Press, Berkeley. Vol. 120.

Nagy K A, Sanson G D and Jacobsen N K (1990). Comparative field energetics of two macropod marsupials and a ruminant. *Australian Wildlife Research*, 17, 591–599.

Nagy K A, Seymour R S, Lee A K and Braithwaite R (1978). Energy and water budgets in free–living *Antechinus stuartii* (Marsupialia: Dasyuridae). *Journal of Mammalogy*, 59, 60–68.

Nagy K A and Suckling G C (1985). Field energetics and water balance of sugar gliders, *Petaurus breviceps* (Marsupialia: Petauridae). *Australian Journal of Zoology*, 33, 683–691.

Rice G E (1976). Water metabolism of the quokka. In *Salt appetite and condition in the quokka*. Eds R Homes, P Prideaux, S Halse, J Trendall, D Harrington, O Nichols and G Rice. Department of Zoology, University of Western Australia, Perth.

Russell E M (1982). Patterns of parental care and parental investment in marsupials. *Biological Reviews* 57, 423–486.

Smith A P, Nagy K A, Fleming M R and Green B (1982). Energy requirements and water turnover in free-living Leadbeaters possums, *Gymnobelideus leadbeateri* (Marsupialia: Petauridae). *Australian Journal of Zoology*, 30, 737–749.

Speakman J R, Sreekumaran N and Goran M I (1993). Revised equations for calculating CO_2 production from doubly labeled water in humans. *American Journal of Physiology*, 264, E912–917.

Tyndale-Biscoe C H (1973). *Life of Marsupials*. Edward Arnold, London.

Tyndale-Biscoe C H, and Janssens P A (1988), *The Developing Marsupial. Models for Biomedical Research*. Springer-Verlag, Berlin. pp 1–7.

Wallis I R and Green B (1992). Seasonal field energetics of the rufous rat-kangaroo (*Aepyprymnus rufescens*). *Australian Journal Zoology*, 40, 279–290.

Wells R T (1973). Physiological and behavioural adaptations of the hairy-nosed wombat (*Lasiorhinus latifrons* Owen) to its arid environment. PhD thesis, University of Adelaide, Adelaide.

CHAPTER 10

LIVING SLOW AND DYING YOUNG: SENESCENCE IN MARSUPIALS

Andrew Cockburn

INTRODUCTION

In humans, a decline in reproductive ability and age-specific performance with age is reasonably obvious. This deterioration is collectively called senescence. Although it was once widely believed that senescence is only visible in populations coddled from extrinsic mortality agents, it is increasingly clear that senescence is a feature of many free-living populations, including all well-studied mammalian populations.

I argue here that senescence in small- to medium-sized marsupials differs to a great extent from that observed in eutherians and other vertebrates, and that this difference represents a major unexplained challenge for evolutionary biologists and marsupial physiologists.

AGE-SPECIFIC VARIATION IN REPRODUCTION AND SURVIVAL IN EUTHERIANS

Comparing the reproductive performance of individuals of different ages is usually done cross-sectionally, by comparing the performance of individuals of different ages within a sample. This approach is severely flawed, as the older individuals in a population may be a non-random sample of the younger generation from which they emerged. For example, performance might appear to improve with age because superior individuals both reproduce and survive better, though the age-specific performance of any particular individual might be constant. For this reason, a longitudinal approach is necessary to interpret age-specific variation reliably.

Such longitudinal studies of the reproductive performance of eutherian mammals and birds reveal an almost ubiquitous pattern. At least some aspects of reproduction initially improve with each successive reproductive bout or season (Clutton-Brock 1988; Partridge 1989). After a plateau of variable duration, reproductive performance starts to decline. In the case of the menopause observed in humans and whales, reproduction may cease altogether (Kasuya & Marsh 1984; Marsh & Kasuya 1984). This decline in reproductive ability can occur even when reproductive effort increases (Clutton-Brock 1984).

In general, patterns of survival in eutherian mammals and birds follow a similar pattern (for example, Finch 1990; Promislow 1991). Juveniles are usually vulnerable, but once reproduction commences, survival rates decline at an increasing rate, which can be statistically approximated using a Gompertz function (Finch *et al.* 1990). It was once widely believed to be only visible in those populations coddled from the mortality agents prevalent in free-living populations, such as predation and disease. Old and infirm animals in free-living populations were supposed to be eliminated by such agents before they were observed. However, it is now clear that senescence is very common in free-living populations (Finch 1990) and such conspicuous changes in age-specific performance demand some sort of explanation, be it mechanistic or adaptive.

There is now universal acceptance that both mechanistic and adaptive explanations are necessary (see Rose 1991; Finch 1990 for excellent reviews). 'Medical' models of senescence which attribute importance to proximate factors, such as radiation damage to nucleic acids, or the highly reactive nature of free radicals produced as a by-product of cellular metabolism, obviously contribute to 'wear and tear' which help explain the deterioration in organisms as they age. However, it is clear that 'wear-and-tear' models do not predict why species differ in their susceptibility to wear and tear. Further, the inevitability of senescence is thrown into doubt by straightforward selection experiments which show that it is possible to extend the lifespan of organisms subject to identical environmental hazards to unselected control lines (Rose 1991). In both *Drosophila* and *Caenorhabditis* it is possible to generate very long-lived strains in only a small number of generations of selection, and in the latter species the genes causing senescence have been identified (Kenyon *et al.* 1993). Two adaptive explanations have gained currency. The first suggests that senescence arises because it is caused by genes which have favourable effects early in life, but deleterious effects late in life (antagonistic pleiotropy hypothesis; Williams 1957). Because there is stronger selection on early-acting genes, the balance of selection will favour early-acting genes despite their deleterious effects. The second hypothesis also depends on the declining force of selection as organisms grow older. If a deleterious mutation only manifests itself at a particular age, it will be selected against less strongly at advanced ages (mutation-selection hypothesis, Hamilton 1966). There is strong experimental support for both hypotheses (Service *et al.* 1988; Finch 1990; Rose 1991).

VARIATION IN EUTHERIAN LIFESPAN

One avenue to understanding senescence is to ask why some species rarely exceed 1 year in life, while others reach extremely advanced ages, apparently immune to the hazards imposed by the environment or by the by-products of metabolism. Comparative analyses have shed considerable light on the diversity of life histories in eutherian mammals. Early attempts to understand this diversity have focused on the positive relation between lifespan and body size. Initial attempts to explain this relation suggested an effect of metabolic rate on life history variables, citing as support the common allometric exponent (0.25 and 0.75) for life history and energetic variables (for example, McNab 1980; Calder 1984). These initial studies used comparative methods which either ignored statistics or which were inappropriate because they ignored phylogenetic history in the formulation of the comparisons (Read & Harvey

1989). For example, whole clusters of closely related species were often treated as independent units in correlation analyses, grossly distorting the importance of these groups through pseudoreplication (Harvey & Pagel 1991). Recent analyses using procedures designed to overcome this phylogenetic pseudoreplication lend little support to these simple metabolic models of life history (for birds, Trevelyan *et al.* 1990; for mammals, Elgar & Harvey 1987; Read & Harvey 1989; Harvey *et al.* 1991), but suggest instead that eutherians can be arranged along a gradient of species from those that 'live fast' to those that 'live slow', even when allometric factors are considered explicitly. High speed life histories are characterised by fast time to maturity, and comparatively short periods of reproduction as adults. Promislow and Harvey (1990) characterised this life history as 'living fast and dying young', in contrast to the obvious alternative — living slow and dying old.

An explicit model synthesising relations among life history variables in eutherian mammals has been popularised by Ric Charnov and his co-workers (Charnov 1991, 1993; Charnov & Berrigan 1993; Berrigan *et al.* 1993). According to this model, mammals have a production energy which can be channelled into growth or reproduction, and which increases with body size. Until puberty, energy is devoted exclusively to growth, but thereafter, at least for females, it is dominated by reproduction. A mother can therefore devote energy to offspring at an equivalent rate as she did when she herself was growing. Slow-growing groups, such as primates, should therefore have little energy to devote to offspring, and will take a long time to raise them to independence, as on average young are weaned at a similar weight as is true for other mammals (Charnov & Berrigan 1993). Thus slow growth, low offspring number, and prolonged parental investment are explicitly related. The model can be extended to allow the age at maturity (where reproduction takes over from growth) to respond to externally imposed mortality. There is a trade-off between growing to a greater size and increasing offspring production, and the higher pre-reproductive mortality which results from the greater time required to reach a greater size. Natural selection should favour the size at maturity which maximises lifetime production of offspring. It is comparatively simple to derive the prediction that slow growth rates should be associated with prolonged adult lifespans, which will only be true when mortality rates are low (Charnov 1993). In general this is true for both birds and eutherians. Slow growth is associated with longer lifespans (Finch 1990; Charnov 1991, 1993).

How are marsupials different?

It has been realised for some time that marsupials grow more slowly than their eutherian analogues (Russell 1982; Lee & Cockburn 1985a; Zullinger *et al.* 1984; Cockburn & Johnson 1988). Such prolongation of early development might therefore be predicted to lead to an extended post-reproductive life. However, only a few studies have persisted sufficiently long to characterise lifespan in the wild accurately. These studies present a paradox. While some marsupials enjoy the predicted extended lifespan (notably the burramyids, Cockburn *et al.* 1990), others exhibit the shortest reproductive spans known for mammals. The most abrupt is the obligate male die-off found in some species of dasyurid, of which the antechinuses are most famous (Lee & Cockburn 1985a). Die-off is characterised by the complete mortality of all males at the conclusion of the rut, and is now well characterised physiologically (reviewed by Lee & Cockburn 1985a).

Apart from naive group selectionist hypotheses, there are only two published adaptive explanations for the evolution of die-off. Dickman (1993) suggested that die-off is a direct consequence of male attempts to maximise the success of their sperm in competition with those of other males. Because of the short rut found in *Antechinus*, the existence of prolonged sperm storage, and because males lose sperm throughout the mating season via spermatorrhoea, he suggested that males need to produce large numbers of ejaculates early in the mating season, and suffer mortality as a consequence. This hypothesis is unlikely to be valid for at least two reasons. First, the problems of loss of sperm via spermatorrhoea arise because spermatogenesis ceases well before the rut (Kerr & Hedger 1983; Taggart & Temple-Smith 1990). However, spermatogenic failure is likely to be a consequence of, rather than a cause of, die-off, resulting from the failure of selection for its maintenance once die-off has evolved. Second, comparative studies of dasyurids do not lend support to the idea that male reproductive anatomy and female sperm storage are as diverse as Dickman (1993) implied (Taggart & Temple-Smith 1994).

Braithwaite and Lee (1979) suggested that die-off was a consequence of the prolonged period marsupials required to raise an offspring from conception to weaning. The long period required to produce a litter means that it is unlikely that marsupials living in environments where resources for reproduction are available only seasonally could complete two reproductive bouts within a single season. Because there are allometric constraints on lifespan, and hence a low probability of surviving an additional year, it may pay individuals to invest extremely heavily in the first bout of reproduction, even if this reduces future reproductive possibilities. Despite the potential circularity in this argument, and the long lifespan of some small marsupials (Cockburn *et al.* 1990), comparative analyses support this argument.

Braithwaite and Lee's (1979) argument has been extended to provide an explicit physiological model for die-off in *Antechinus* (Lee & Cockburn 1985a, b; Lazenby-Cohen & Cockburn 1988; Cockburn 1992). According to this view, synchronised female reproduction and prolonged female investment mean that females are only sexually receptive during late winter when it is comparatively difficult for males to procure food. At least in *Antechinus stuartii* this problem is exacerbated because mating takes place at, or near, communal aggregations which are often considerable distances from male foraging ranges (Lazenby-Cohen & Cockburn 1988, 1991; Cockburn & Lazenby-Cohen 1992). Males therefore risk losing mating opportunities if they leave the mating aggregations to feed in areas with which they are familiar. Because males are small and are likely to consume easily metabolised energy reserves quickly they need to exploit protein energy reserves stored in tissues in order to sustain attendance at mating arenas.

This exploitation of protein reserves is achieved through the use of physiological stress to promote gluconeogenic mobilisation of body tissues (Lee & Cockburn 1985a, b). Gluconeogenesis is promoted by sustained high levels of active corticosteroids, which, in turn, results from three events. First, there is a sharp increase in testosterone levels before the rut. This causes a decrease in corticosteroid binding globulin (Bradley *et al.* 1980). Second, there is also competition between the dominant corticosteroids (cortisol and corticosterone) for the diminishing number of corticosteroid binding globulin sites, leading to a dramatic increase in free cortisol (25-fold in the best studied species, *Phascogale calura* — Bradley 1987), which is much more physiologically potent than corticosterone. Third, in most mammals sustained high levels of corticosteroids are prevented by negative feedback on adrenocorticotropin secretion. In *Antechinus* this negative feedback is suppressed (McDonald *et al.* 1986).

Early attempts to interpret die-off against a phylogenetic background suggested that it had evolved only once, in the clade of mammals that diverged to form the modern genera *Antechinus* and *Phascogale*. However, it is now clear that die-off occurs in at least some populations of three other dasyurid genera. As for *Phascogale* and *Antechinus*, die-off appears to be obligate in *Dasykaluta rosamondae* (Woolley 1991a), but facultative in *Parantechinus apicalis* (compare Dickman & Braithwaite 1992 with Woolley 1991b), and *Dasyurus hallucatus* (Dickman & Braithwaite 1992; M. Oakwood, personal communication). More remarkably, die-off has been inferred for two species of didelphids, *Marmosa incana* (Lorini *et al.* 1994) and *Monodelphis dimidiata* (Pine *et al.* 1985). Set against current interpretations of the phylogeny of these groups (for example, Baverstock *et al.* 1990; Kirsch *et al.* 1990), die-off therefore appears to have at least five independent origins, even assuming that *Dasykaluta* is closely related to one of the other clades, which is currently most uncertain.

This extraordinary repeated evolution of a trait which is otherwise unknown among mammals provides some support for the Braithwaite-Lee model. Semelparous Australian marsupials tend to be confined to the coastal fringe (Lee & Cockburn 1985a). In many of these environments predictable seasonal fluctuations occur, possibly confining species to a limited window of opportunity in which to reproduce (Lee & Cockburn 1985a), though this argument is a little difficult to apply to *Dasykaluta* (Woolley 1991a). The two semelparous didelphids are essentially sympatric and coastal (Streilein 1982). However, the model seems only weakly applicable to species with larger body weights, such as *Dasyurus hallucatus,* which presumably have a greater capacity to store energy reserves in rapidly metabolisable forms such as fat. Further, the mating system described by Lazenby-Cohen and Cockburn (1988; 1991) for *Antechinus* seems to be inapplicable to other species with die-off (T Soderquist and M Oakwood, personal communication), and may not apply to all *Antechinus* (A Watt, personal communication).

SENESCENCE IN FEMALES

A further complication in the generality of these explanations arises from the increasingly frequent observation that female dasyurids and didelphids are also subject to exaggerated senescence, to the point that many species defy an otherwise sound generalisation for all mammals and birds. In these marsupials, individuals reproduce best at their first opportunity and thereafter suffer marked declines in reproductive performance. Two examples illustrate this phenomenon well.

Antechinus stuartii females in the second year (and hence rearing their second litter):

(a) produce fewer embryos (Selwood 1983);

(b) are more likely to reduce their broods, which they do by killing sons (Cockburn 1994);

(c) despite an apparent preference for investing in daughters, produce daughters which are themselves unlikely to survive (Cockburn 1994); and

(d) are more susceptible to mortality, during both reproduction (Figure 10.1) (Cockburn 1994) and outside the breeding season, when they suffer severe infestation with parasites (Cockburn, Cooper, Promislow & Lazenby-Cohen, unpublished observations).

The reproductive performance of free-living *Didelphis* in their first season of reproduction is superior to that in the second season of reproduction (for example, Reynolds 1952). Austad (1993) has related these observations explicitly to evolutionary models of senescence by comparing reproduction and survivorship of *D. virginiana* living on an island where predators were absent, with those living on the mainland, and hence subject to an important additional source of intrinsic mortality. Opossums on the mainland had shorter lifespans (Figure 10.2), but larger litters (7.6 young on average) than opossums on the island (5.7 young). As a general indicator of ageing, Austad (1993) used the fibre breaking time of tail tendon fibres, which grow stiffer as collagen is subject to increased intra- and intermolecular crosslinking. The fibres of island animals stiffened much more slowly with age in the island opossums. Capacity to invest in young also differed with age between the two populations. Although one-year-old mainland opossums produced a greater total litter mass in their first year of life (as would be expected from the increased litter size), litters produced in the second year of life grow very slowly. There is no evidence of reproductive senescence over the first two seasons in island opossums.

WHY ARE DIDELPHIDS AND DASYURIDS DIFFERENT?

This deterioration in females with age cannot be explained in terms of rapid mobilisation of body tissues for a very short period, which potentially provides an explanation for male die-off (see above). Therefore a general explanation is required for the general tendency of female didelphids and dasyurids to senesce immediately reproduction commences, in contrast to the usual pattern of a period of improvement in reproduction followed by a period of decline.

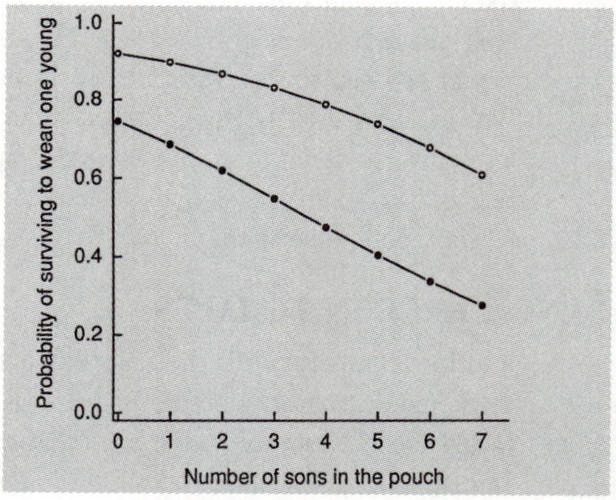

Figure 10.1
The effect of rearing sons on the probability that one-year-old (o) and two-year-old (●) female *Antechinus stuartii* will die during lactation.
Note: Because teats are usually saturated with young, one more son usually means one less daughter. The lines depict predictions of a logistic regression model.
Source: Modified from Cockburn (1994).

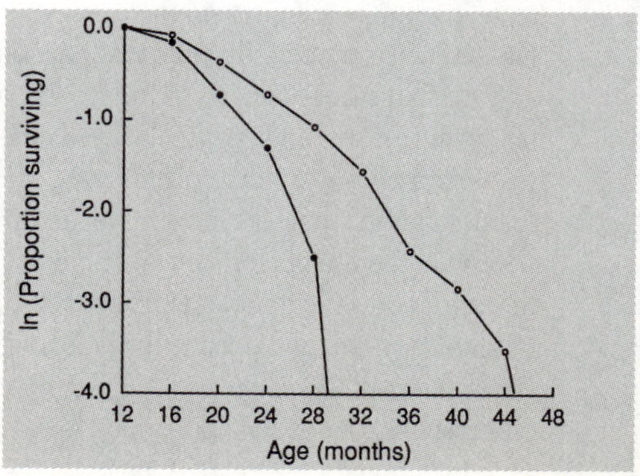

Figure 10.2
Proportion of *Didelphis virginiana* surviving for successive periods past a baseline of 12 months in an island population without predators (o), and in a mainland population subject to predation (●).
Source: Modified from Austad (1993).

One way to tackle this question is to compare the marsupial clades in which there is an acute tendency to develop exaggerated senescence with those in which senescence is less obvious, or follows the general pattern. Because rates of reproduction and development are likely to be at least partly sensitive to diet, in this discussion I concentrate on the groups which share a similar diet to the didelphids and dasyurids. All these clades are usually at least partly insectivorous, though fungus and nectar may play an important role in the diet of all these groups (Lee & Cockburn 1985a).

Peramelids (bandicoots and bilbies) differ from other marsupials in extreme acceleration of growth and development, both before and after maternal care ceases (Cockburn 1989, 1990). Although the means by which the acceleration of development arises remains unclear, bandicoots differ from other marsupials through the possession of a chorioallantoic placenta in which the chorionic villi are highly invasive, and the persistence of the corpora lutea of pregnancy through much of lactation.

The enigma of embryonic diapause

The three diprotodont groups with similar diets (acrobatids, burramyids and potoroines) at first seem to have little in common. However, it has recently been recognised that all three groups are united in the possession of embryonic diapause (potoroines — Tyndale-Biscoe & Renfree 1987; burramyids — Ward 1990a, b; acrobatids — Renfree 1980; Ward & Renfree 1988). The free-living longevity of some of these species is high for their size (Cockburn *et al.* 1990; Ward 1990c). As for senescence, cladistic and comparative evidence suggest that diapause has probably evolved separately in all three groups (Baverstock *et al.* 1990). By contrast, full diapause in didelphids and dasyurids appear to be confined to the prolonged periods of stasis well developed in at least two genera of dasyurids (*Antechinus* — Selwood 1980, 1981; *Dasykaluta* — Woolley 1991a).

Despite this repeated evolution, diapause has defied adaptive explanation. While there are several obvious advantages of prolonged diapause, such as reducing the interval between litters by allowing the rapid initiation of a new pregnancy (Thompson 1987; Lee & Ward 1989), these advantages only arise once diapause is well developed. As diapause appears to be a derived trait that has arisen on several occasions, rather than some bizarre evolutionary accident, such *post hoc* explanations are clearly unsatisfactory. Tyndale-Biscoe (1989) provides the only hypothesis that might satisfactorily explain the *origin* of diapause which he views as a means to accomplish prolongation of gestation, by extending the luteal phase of the cycle into the follicular phase. He regarded the principal benefit from prolongation of gestation to be the production of more advanced young at birth, influenced by Hughes and Hall's (1988) demonstration of variation in the state of development of marsupial pouch young, with macropodoids conspicuously more developed at birth. There are several difficulties with this argument. First, diapause occurs in several groups where it does not lead to any more pronounced development of the young at birth. This is most notable in *Tarsipes*, the neonates of which are the smallest free-living mammals, sometimes weighing less than 5 mg (Renfree *et al.* 1984). Even in dasyurids, specialised apomorphies (notably a temporary cervical bladder which straightens the fetus from a closely contorted position at birth; Hughes & Hall 1988) suggest that their extremely small size at birth is an apomorphic trait (Cockburn 1989).

I propose as an alternative that the prolonged organogenesis of the young of macropodoids not only increases their size and development, but also slows down the pace of growth (Figure 10.3) (Hughes *et al.* 1989). We can therefore recognise three developmental tactics among marsupials: a) peramelids, where a very brief period of organogenesis produces a well-developed neonate because of enhanced embryonic performance, leading to a fast pace of life; b) the species where diapause allows more leisurely development. In this respect it is interesting that in burramyids the blastocyst continues to grow slowly during diapause (Ward 1990b, c) the species without any special early developmental mechanisms. It is in these latter groups that senescence occurs most dramatically. I therefore agree with the general argument advanced by Tyndale-Biscoe (1989), but suggest that embryonic diapause arises as a way of influencing the pace of early development. This, in turn, may alter the pace of life in marsupials in general, and impact on the expectation of future life once reproduction commences. Ecological circumstances will then dictate whether selection will favour rapid early development or prolongation of reproductive life.

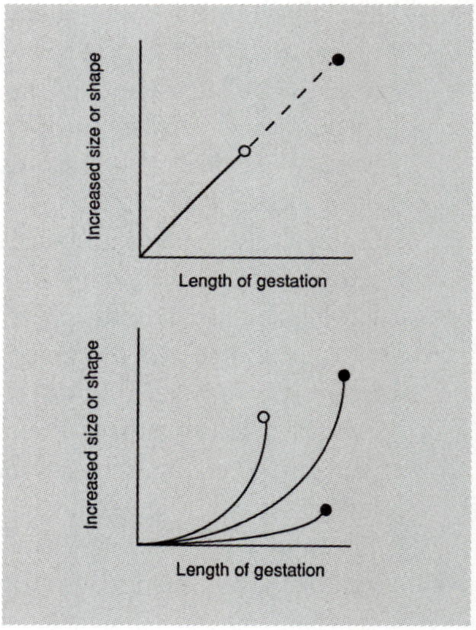

Figure 10.3 Alternative ways of viewing the developmental trajectories of marsupials during gestation. (a) Increase in size and shape associated with prolonged gestation. (b) Variation in shape and size at birth dependent on both rate of development and when birth takes place. See also Hughes *et al.* (1989).

FUTURE DIRECTIONS

In summary, dasyurid and didelphid marsupials have repeatedly evolved a pattern of senescence and post-reproductive die-off not found elsewhere among mammals. Study of this phenomenon in one genus (*Antechinus*) has helped clarify the role of stress in free-living populations, and illuminated the intimate relationship between stress, reproduction, and the immune system. It is likely that this approach could be profitably extended, in particular, by taking advantage of the explosion of our understanding of mammalian immunology. However, the frequent evolution of the phenomenon of acute senescence cautions against evolutionary explanations based on the idiosyncrasies of a single well-studied genus. Instead, I have offered a speculative hypothesis relating two traits which have evolved repeatedly in marsupials but which have defied adaptive explanation (exaggerated senescence, embryonic diapause), to the most fundamental aspect of marsupialness: the production of a tiny neonate which while completing most of its development during lactation, must still be sufficiently developed at the time of birth to exist independently. The correlation of the rates of development of marsupial embryos with aspects of their adult life history is potentially a rewarding topic for further investigation by both evolutionary ecologists and developmental physiologists.

Acknowledgments

My work on *Antechinus* life histories was supported by Australian Research Council grants. Most of my thoughts on antechinuses and senescence have been shaped by conversations with Wendy Cooper, Katie Lazenby-Cohen, Tony Lee, Ian MacDonald, and Daniel Promislow. Meri Oakwood, Todd Soderquist, and Anne-Marie Watt, have successfully eroded my confidence that my *Antechinus* escapades had revealed a general explanation for marsupial senescence. I am grateful to Lyn Hinds for comments on the manuscript and Hugh Tyndale-Biscoe for persistent scepticism.

References

Austad S N (1993). Retarded senescence in an insular population of virginia opossums (*Didelphis virginiana). Journal of Zoology*, 229, 695–708.

Baverstock P R, Krieg M and Birrell J (1990). Evolutionary relationships of Australian marsupials as assessed by albumin immunology. *Australian Journal of Zoology*, 37, 273–287.

Berrigan D, Charnov E L, Purvis A and Harvey P H (1993). Phylogenetic contrasts and the evolution of mammalian life histories. *Evolutionary Ecology*, 7, 270–278.

Bradley A J (1987). Stress and mortality in the red-tailed phascogale, *Phascogale calura* (Marsupialia: Dasyuridae). *General and Comparative Endocrinology*, 67, 85–100.

Bradley A J, McDonald I R and Lee A K (1980). Stress and mortality in a small marsupial (*Antechinus stuartii* Macleay). *General and Comparative Endocrinology*, 40, 188–200.

Braithwaite R W and Lee A K (1979). A mammalian example of semelparity. *American Naturalist*, 113, 151–156.

Calder W A (1984). *Size, Function and Life History*. Harvard University Press, Cambridge.

Charnov E L (1991). Evolution of life history variation among female mammals. *Proceedings of the National Academy of Sciences USA*, 88, 1134–1137.

Charnov E L (1993). *Life history invariants: some explorations of symmetry in evolutionary ecology*. Oxford University Press, Oxford.

Charnov E L and Berrigan D (1993). Why do female primates have such long lifespans and so few babies? or life in the slow lane. *Evolutionary Anthropology*, 1, 191–194.

Clutton-Brock T H (1984). Reproductive effort and terminal investment in iteroparous animals. *American Naturalist*, 123, 212–229.

Clutton-Brock T H (1988). Reproductive success. In *Reproductive success*. Ed T H Clutton-Brock. University of Chicago Press, Chicago. pp 472–485.

Cockburn A (1989). Adaptive patterns in marsupial reproduction. *Trends in Ecology and Evolution*, 4, 126–131.

Cockburn A (1990). Life history of the bandicoots: developmental rigidity and phenotypic plasticity. In *Bandicoots and Bilbies*. Eds J H Seebeck, P R Brown, R L Wallis and C M Kemper. Surrey Beatty & Sons Pty Ltd, Sydney. pp 285–292.

Cockburn A (1992). Evolutionary ecology of the immune system: why does the thymus involute? *Functional Ecology*, 6, 364–370.

Cockburn A (1994). Adaptive sex allocation by brood reduction in antechinuses. *Behavioral Ecology and Sociobiology*, 35, 53–62.

Cockburn A and Johnson C N (1988). Patterns of growth. In *The Developing Maruspial. Models for Biomedical Research*. Eds C H Tyndale-Biscoe and P A Janssens. Springer-Verlag, Berlin. pp 28–40.

Cockburn A and Lazenby-Cohen K A (1992). Use of nest trees by *Antechinus stuartii*, a semelparous lekking marsupial. *Journal of Zoology*, 226, 657–680.

Cockburn A, Mansergh I, Broome L S and Ward S (1990). Molecular clocks and generation time in burramyid marsupials. *Molecular Biology and Evolution*, 7, 283–285.

Dickman C R (1993). Evolution of semelparity in male dasyurid marsupials: a critique and an hypothesis of sperm competition. In *The biology and management of Australasian carnivorous marsupials*. Eds M Roberts, J Carnio, G Crawshaw and M Hutchins. Metropolitan Toronto Zoo, Toronto. pp 25–38.

Dickman C R and Braithwaite R W (1992). Postmating mortality in the dasyurid marsupials, *Dasyurus* and *Parantechinus*. *Journal of Mammalogy*, 73, 143–147.

Elgar M A and Harvey P H (1987). Basal metabolic rates in mammals: allometry, phylogeny and ecology. *Functional Ecology*, 1, 25–36.

Finch C E (1990). *Longevity, Senescence and the Genome*. Chicago University Press, Chicago.

Finch C E, Pike M C and Witten M (1990). Slow mortality rate accelerations during aging approximate that of humans. *Science*, 249, 902–906.

Hamilton W D (1966). The molding of senescence by natural selection. *Journal of Theoretical Biology*, 12, 12–45.

Harvey P H and Pagel M D (1991). *The Comparative Method in Evolutionary Biology*. Oxford University Press, Oxford.

Harvey P H, Pagel M D and Rees J A (1991). Mammalian metabolism and life histories. *American Naturalist*, 137, 556–566.

Hughes R L and Hall L S (1988). Structural adaptations of the newborn marsupial. In *The Developing Marsupial. Models for Biomedical Research*. Eds C H Tyndale-Biscoe and P A Janssens. Springer-Verlag, Berlin. pp 8–27.

Hughes R L, Hall L S, Tyndale-Biscoe C H and Hinds L A (1989). Evolutionary implication of macropodid organogenesis. In *Kangaroos, Wallabies and Rat-Kangaroos*, Eds G Grigg, P Jarman and I Hume. Surrey Beatty & Sons Pty Ltd, Sydney. pp 377–405.

Kasuya T and Marsh H (1984). Life history and reproductive biology of the short-finned pilot whale, *Globicephala macrohynchus*, off the Pacific coast of Japan. *Report of the International Whaling Commission Special*, 6, 259–310.

Kenyon C, Chang, J, Gensch E, Rudner A and Tabtiang R (1993). A *C. elegans* mutant that lives twice as long as wild type. *Nature*, 366, 461–464.

Kerr J B and Hedger M P (1983). Spontaneous spermatogenic failure in the marsupial mouse *Antechinus stuartii* Macleay (Dasyuridae: Marsupialia). *Australian Journal of Zoology*, 31, 445–466.

Kirsch J A W, Krajewski C, Springer M S and Archer M (1990). DNA-DNA hybridisation studies of carnivorous marsupials. II. Relationships among dasyurids (Marsupialia: Dasyuridae). *Australian Journal of Zoology*, 38, 673–696.

Lazenby-Cohen K A and Cockburn A (1988). Lek promiscuity in a semelparous mammal, *Antechinus stuartii* (Marsupialia: Dasyuridae)? *Behavioral Ecology and Sociobiology*, 22, 195–202.

Lazenby-Cohen K A and Cockburn A (1991). Foraging and social components of the home range in *Antechinus stuartii*. *Australian Journal of Zoology*, 16, 301–307.

Lee A K and Cockburn A (1985a). *Evolutionary Ecology of Marsupials*. Cambridge University Press, Cambridge.

Lee A K and Cockburn A (1985b). Spring declines in small mammal populations. *Acta Zoologica Fennica*, 173, 75–76.

Lee A K and Ward S J (1989). Life histories of macropodoid marsupials. In *Kangaroos, Wallabies and Rat-Kangaroos*. Eds G Grigg, P Jarman and I Hume. Surrey Beatty & Sons, Sydney. pp 105–115.

Lorini M L, De Oliveira J A and Persson V G (1994). Annual age structure and reproductive patterns in *Marmosa incana* (Lund, 1841) (Didelphidae, Marsupialia). *Zeitschrift fur Saugetierkunde-International Journal of Mammalian Biology*, 59, 65–73.

Marsh H and Kaysuya T (1984). Changes in the ovaries of the short-finned pilot whale, *Globicephala macrorhyncus*, with age and reproductive activity. *Report of the International Whaling Commission Special*, 6, 311–335.

McDonald I R, Lee, A K, Than K A and Martin R W (1986). Failure of glucocorticoid feedback in males of a population of small marsupials (*Antechinus swainsonii*) during the period of mating. *Journal of Endocrinology*, 108, 63–68.

McNab B K (1980). Food habits, energetics, and the population biology of mammals. *American Naturalist*, 116, 106–124.

Partridge L (1989). Lifetime reproductive success and life-history evolution. In *Lifetime Reproduction in Birds*. Ed I Newton. Academic Press, London. pp 421–440.

Pine R H, Dalby P L and Matson J O (1985). Ecology, postnatal development, morphometrics, and taxonomic status of the short-tailed opossum, *Monodelphis dimidiata*, an apparently semelparous annual marsupial. *Annals of the Carnegie Museum*, 54 (6), 195–231.

Promislow D E L (1991). Senescence in natural populations of mammals: a comparative study. *Evolution*, 45, 1869–1887.

Promislow D E L and Harvey P H (1990). Living fast and dying young: a comparative analysis of life-history variation among mammals. *Journal of Zoology*, 220, 417–437.

Read A F and Harvey P H (1989). Life history differences among the eutherian radiations. *Journal of Zoology*, 219, 329–353.

Renfree M B (1980). Embryonic diapause in the honey possum *Tarsipes spenserae*. *Search*, 11, 81.

Renfree M B, Russell E M and Wooller R D (1984). Reproduction and life history of the honey possum *Tarsipes rostratus*. In *Possums and Gliders*. Eds A P Smith and I D Hume. Australian Mammal Society, Sydney. pp 427–437.

Reynolds H C (1952). Studies on reproduction in the opossum (*Didelphis virginiana virginiana*). *University of California Publications in Zoology*, 52, 223–284.

Rose M R (1991). *Evolutionary Biology of Aging*. Oxford University Press, Oxford.

Russell E R (1982). Patterns of parental care and parental investment in marsupials. *Biological*

Reviews of the Cambridge Philosophical Society, 57, 423–486.

Selwood L (1980). A timetable of embryonic development of the dasyurid marsupial *Antechinus stuartii* (Macleay). *Australian Journal of Zoology*, 28, 649–668.

Selwood L (1981). Delayed embryonic developent in the dasyurid marsupial, *Antechinus stuartii. Journal of Reproduction and Fertility Supplement,* 29, 79–87.

Selwood L (1983). Factors influencing pre-natal fertility in the brown marsupial mouse, *Antechinus stuartii. Journal of Reproduction and Fertility,* 68, 317–324.

Service P M, Hutchinson E W and Rose M R (1988). Multiple genetic mechanisms for the evolution of senescence in *Drosophila melanogaster*. *Evolution,* 42, 708–716.

Streilein K E (1982). Behavior, ecology, and distribution of South American marsupials. In *Mammalian Biology in South America*. Eds M A Mares and H H Genoways. Pymatuning Symposia in Ecology, Pittsburg, PA. Vol 6, pp 231–250.

Taggart D A and Temple-Smith P D (1990). Effect of breeding season and mating on total number and distribution of spermatozoa in the epididymis of the brown marsupial mouse, *Antechinus stuartii. Journal of Reproduction and Fertility,* 88, 81–91.

Taggart D A and Temple-Smith P D (1994). Comparative studies of epididymal morphology and sperm distribution in dasyurid marsupials during the breeding season. *Journal of Zoology,* 232, 365–381.

Thompson S D (1987). Body size, duration of parental care, and the intrinsic rate of natural increase in eutherian and metatherian mammals. *Oecologisa (Berlin)*, 71, 201–209.

Trevelyan R, Harvey P H and Pagel M D (1990). Metabolic rates and life histories in birds. *Functional Ecology*, 4, 135–141.

Tyndale-Biscoe C H (1989). The adaptiveness of reproductive process. In *Kangaroos, Wallabies and Rat-Kangaroos*. Eds G Grigg, P Jarman and I Hume. Surrey Beatty & Sons Pty Ltd, Sydney. pp 277–285.

Tyndale-Biscoe C H and Renfree, M (1987). *Reproductive Physiology of Marsupials*. Cambridge University Press, Cambridge.

Ward S J (1990a). Life history of the eastern pygmy-possum, *Cercartetus nanus* (Burramyidae: Marsupialia), in south-eastern Australia. *Australian Journal of Zoology*, 38, 287–304.

Ward S J (1990b). Reproduction in the western pygmy-possum, *Cercartetus concinnus* (Marsupialia: Burramyidae), with notes on reproduction of some other small possum species. *Australian Journal of Zoology*, 38, 423–438.

Ward S J (1990c). Life history of the feathertail glider, *Acrobates pygmaeus* (Acrobatidae: Marsupialia), in south-eastern Australia. *Australian Journal of Zoology*, 38, 503–517.

Ward S J and Renfree M B (1988). Reproduction in females of the feathertail glider *Acrobates pygmaeus* (Marsupialia). *Journal of Zoology,* 216, 225–239.

Williams G C (1957). Pleiotropy, natural selection, and the evolution of senescence. *Evolution,* 11, 398–411.

Woolley P A (1991a). Reproduction in *Dasykaluta rosamondae* (Marsupialia: Dasyuridae): field and laboratory observations. *Australian Journal of Zoology*, 39, 549–568.

Woolley P A (1991b). Reproductive patterns of captive Boullanger Island dibblers, *Parantechinus apicalis* (Marsupialia: Dasyuridae). *Wildlife Research*, 18, 157–163.

Zullinger E M, Ricklefs R E, Redford K H and Mace G (1984). Fitting sigmoidal equations to mammalian growth curves. *Journal of Mammalogy,* 65, 607–636.

CHAPTER 11

REPRODUCTIVE ECOLOGY OF NEW WORLD MARSUPIALS

John D Harder and David W Fleck

INTRODUCTION

New World marsupials have much to offer in providing a full appreciation of the evolution and diversity of marsupials. Both North and South America have been regarded as the place of origin for marsupials. The oldest undisputed fossil marsupials are didelphids of the mid-Cretaceous deposits of North America and late Cretaceous deposits of South America (80 million years BP) (Marshall 1982), which existed long before the oldest (Oligocene) marsupials of Australia. The evidence appears to favour North America as a place of origin, but with North and South America sharing a diverse marsupial fauna by the late Cretaceous. During the late Cretaceous, South America, Antarctica, and Australia lay in close proximity, such that dispersal of both plant and animal species was possible. The discovery of a marsupial fossil in Antarctica (Woodburne & Zinmeister 1982) strongly supports the notion of a southern dispersal route of South American marsupials into Antarctica and then into Australia before the Eocene separation of these continents (Marshall *et al.* 1990).

Modern marsupials represent an important segment of the mammalian fauna of the Neotropics. Didelphimorphia is the third largest order of South American mammals, following rodents and bats, which together comprise about 75% of mammalian diversity on the continent. Although only 3 marsupial families are found in South America, together they represent about 30% of the world's total of 250 species.

The nomenclature in this chapter is based on the classification of Gardner (1993), which recognises three orders of New World marsupials: Didelphimorphia (63 species), Paucituberculata (5 species), and Microbiotheria (1 species) (Table 11.1). Several new genera are listed and *Marmosa* now includes just 9 species in contrast to more than 50 species listed in previous classifications (Kirsch & Calaby 1977). *Monodelphis* includes 15 species, making it the largest genus of New World marsupials. It should be noted that the taxonomy of many South American genera is poorly developed and, therefore, the reader should expect discrepancies between the nomenclature of Gardner (1993) and species cited in this text.

Table 11.1. Classification of New World marsupials according to Gardner (1993) with common names, distributions, and whole body masses

SPECIES AND COMMON NAMES[1]	DISTRIBUTION[2]	BODY MASS(g)[3]
ORDER DIDELPHIMORPHIA		
Family Didelphidae		
Subfamily Caluromyinae		
Caluromys derbianus Central American Woolly Opossum	Mexico, Central America, west Colombia, west Ecuador	245–370
Caluromys lanatus Western Woolly Opossum	Colombia, Venezuela, east Ecuador, east Peru east Bolivia, north Argentina, Brazil	310–410
Caluromys philander Bare-tailed Woolly Opossum	Venezuela, Trinidad and Tobago, Guyana, Surinam, French Guiana, Brazil	140–270
Caluromysiops irrupta Black-shouldered Opossum	south-east Peru, west Brazil	
Glironia venusta Bushy-tailed Opossum	Amazonian Brazil, Ecuador, Peru, Bolivia	
Subfamily Didelphinae		
Chironectes minimus Water Opossum	south Mexico, Central America, Colombia, Ecuador, Brazil, Peru, Venezuela, the Guianas, Paraguay, and north-east Argentina	590–700
Didelphis albiventris White-eared Opossum	Venezuela, Surinam, Brazil, Colombia, Ecuador, Peru, Brazil, Bolivia Paraguay, Uruguay, north Argentina	500–2000
Didelphis aurita	east Brazil, south-east Paraguay, north-east Argentina	
Didelphis marsupialis Common Opossum	south Mexico, central and South America south to Peru, Bolivia and Brazil	565–1610
Didelphis virginiana Virginia Opossum	south Canada, USA, Mexico, Central America into North Costa Rica	500–2300
Gracilinanus acermarcae	La Paz, Bolivia	
Gracilinanus agilis Agile Gracile Mouse Opossum	Brazil, east Peru, east Bolivia, Paraguay, Uruguay, north-east Argentina	20–30
Gracilinanus dryas	Andes west of Venezuela	18

Table 11.1. (*Continued*)

SPECIES AND COMMON NAMES[1]	DISTRIBUTION[2]	BODY MASS(g)[3]
Gracilinanus emiliae Emilia's Gracile Mouse Opossum	north-east Brazil	
Gracilinanus marica Northern Gracile Mouse Opossum	north Colombia, Venezuela	
Gracilinanus microtarsus Brazilian Gracile Mouse Opossum	south-east Brazil	19–29
Lestodelphys halli	east and south Argentina	
Lutreolina crassicaudata Thick-tailed Opossum	Colombia, Venezuela, Guyana; Bolivia south-east Brazil, Paraguay, Uruguay, north Argentina	540–812
Marmosa andersoni	Cuzco, Peru	
Marmosa canescens	Mexico from south Sonora to Oaxaca, Yucatán, and Tres Mar'as Islands	
Marmosa lepida Little Rufous Mouse Opossum	Surinam, east Colombia, Ecuador, Peru, Bolivia, probably Brazil	10
Marmosa mexicana Mexican Mouse Opossum	south Mexico to west Panama	40–100
Marmosa murina Murine Mouse Opossum	Colombia, Venezuela, the Guianas, Brazil, east Ecuador, east Peru, east Bolivia	43–60
Marmosa robinsoni Robinson's Mouse Opossum	Belize, Honduras, Panama, Colombia, west Ecuador, north-west Peru, north Venezuela	36–132
Marmosa rubra Red Mouse Opossum	east Ecuador, Peru	59–67
Marmosa tyleriana	Guayanan Highland tepuis of Venezuela	
Marmosa xerophila	north-east Colombia, north-west Venezuela	
Marmosops cracens	Falcón, Venezuela	
Marmosops dorothea	Bolivia	
Marmosops fuscatus Grey-bellied Sender Mouse Opossum	east Andes of Colombia, north Venezuela	41–85
Marmosops handleyi	Antioquia, Colombia	
Marmosops impavidus	west Panama to Venezuela and Colombia, Ecuador, Peru	36–45

Table 11.1. (*Continued*)

SPECIES AND COMMON NAMES[1]	DISTRIBUTION[2]	BODY MASS(g)[3]
Marmosops incanus Gray Slender Mouse Opossum	east Brazil	54
Marmosops invictus Slaty Slender Mouse Opossum	Panama	
Marmosops noctivagus White-bellied Slender Mouse Opossum	Amazonian Brazil, Ecuador, Peru, Bolivia	55–85
Marmosops parvidens Delicate Slender Mouse Opossum	Colombia, Peru, Venezuela, Guyana, Surinam, Brazil	15–27
Metachirus nudicaudatus Brown Four-eyed Opossum	Nicaragua to Paraguay and north Argentina	300–450
Micoureus alstoni	Central America from Belize to Panama and adjacent Caribbean Islands	
Micoureus constantiae Woolly Mouse Opossum	east Bolivia and adjacent Brazil south into north Argentina	
Micoureus demerarae	Colombia, Venezuela, French Guiana, Guyana, Surinam, Brazil, east Paraguay	80–152
Micoureus regina	Colombia, Ecuador, Peru, Bolivia	
Monodelphis adusta Sepia Short-tailed Opossum	east Panama, Colombia, Ecuador, Peru	15–35
Monodelphis americana Three-striped Short-tailed Opossum	east Brazil	23–35
Monodelphis brevicaudata Red-legged Short-tailed Opossum	Venezuela, Surinam, French Guiana, Amazon Basin of Brazil and Bolivia	46–150
Monodelphis dimidiata	Uruguay, south-east Brazil, north-east Argentina	
Monodelphis domestica Grey Short-tailed Opossum	Brazil, Bolivia, Paraguay	36–98
Monodelphis emiliae Emilia's Short-tailed Opossum	Amazon Basin of Peru and Brazil	60
Monodelphis iheringi	south-east Brazil	
Monodelphis kunsi Pygmy Short-tailed Opossum	Bolivia, Brazil	
Monodelphis maraxina	Pará and Marajo Island, Brazil	
Monodelphis osgoodi	south-east Peru, central Bolivia	

Table 11.1. (*Continued*)

SPECIES AND COMMON NAMES[1]	DISTRIBUTION[2]	BODY MASS(g)[3]
Monodelphis rubida Chestnut Striped Short-tailed Opossum	east Brazil	45–46
Monodelphis scalops Long-nosed Short-tailed Opossum	south-east Brazil,	
Monodelphis sorex Shrewish Short-tailed Opossum	south-east Brazil, south Paraguay, north-east Argentina	48
Monodelphis theresa	east Brazil, Peruvian Andes	
Monodelphis unistriata One-striped Short-tailed Opossum	south-east Brazil	
Philander andersoni Anderson's Grey Four-eyed Opossum	south Venezuela, west Brazil, east Colombia, Ecuador, Peru	225–425
Philander opossum Common Grey four-eyed Opossum	south Mexico, through Central and South America to Paraguay and north-east Argentina	200–660
Thylamys elegans	south Peru, east Chile, Bolivia, east Argentina	
Thylamys macrura	Paraguay, south Brazil	
Thylamys pallidior	Argentina, east and south Bolivia	
Thylamys pusilla	Brazil, Paraguay, south -east Bolivia, north Argentina	
Thylamys velutinus	south-east Brazil	
ORDER PAUCITUBERCULATA **Family Caenolestidae**		
Caenolestes caniventer	south-west Ecuador, north-west Peru	
Caenolestes convelatus	west Colombia, north-west Ecuador	
Caenolestes fuliginosus	Colombia, Ecuador, north-west Venezuela	
Lestoros inca	south Andean Peru	
Rhyncholestes raphanurus	south-central Chile, Chiloé Island	
ORDER MICROBIOTHERIA **Family Microbiotheriidae**		
Dromiciops gliroides	Chile and east-central Argentina	

Sources:
[1]. Common names: Emmons (1990) & Eisenberg (1989).
[2]. Distributions: Gardner (1993).
[3]. Body weights: Emmons (1990); Tyndale-Biscoe & Renfree (1987).

New World marsupials show relatively little variation in general body plan and morphology. Only a few show specialisations for feeding and locomotion, the most remarkable being the water opossum or yapok (*Chironectes minimus*). This picivore has webbed hind feet and a waterproof pouch which protects the young when the mother dives for food (Marshall 1982). Although rainforest marsupial communities exhibit considerable vertical stratification, specialisations, such as prehensile tails, are found in terrestrial species as well as arboreal species.

Radiation of didelphid marsupials is most apparent in body mass, which ranges from 10 g (*Marmosa lepida*) to more than 3 kg in some populations of the Virginia opossum, *Didelphis virginiana*, in northern USA and southern Canada (Table 11.1). Pouch development and morphology is highly variable among marsupials. Some marsupials, such as *Didelphis*, have a deep, well-developed pouch that completely covers and protects a large litter (6 to 12) for up to two-thirds of the lactation period. Others have only lateral skin folds that partially enclose small nursing young during the early stages of lactation. In fact, about half of all marsupial species lack a pouch or skin folds of any kind (Vaughan 1986) and among New World marsupials that proportion is higher; 52 of 69 species are pouchless (Harder 1992).

The pouchless condition is probably a primitive state in marsupials (Tyndale-Biscoe & Renfree 1987). It is also strongly correlated with body mass; only the larger didelphids (> 250 g) possess a well-developed pouch. Persistence of the pouchless condition in small didelphids probably indicates strong selection for leaving young in the nest at an early age (Russell 1982; Harder 1992).

Marsupials are limited to the Australasian and Neotropical Regions with one species in the Nearctic. Many hypotheses have been advanced to explain this limited distribution, most often centring on arguments that marsupials are disadvantaged when competing with eutherians. That debate is beyond the scope of this chapter. However, among the New World marsupials, distributions are restricted at high latitudes (McNab 1986; 1989). The Virginia opossum, *D. virginiana*, the single marsupial in North America, is found only as far north as southern Ontario, Canada and northern New York (about 44°N) (Gardner 1982) and, although representatives of all three neotropical marsupial families can be found in Patagonia and southern Chile to about 46°S (Table 11.1), most neotropical marsupials are found in tropical regions. Only about 16% (11 out of 68) would be considered temperate; that is, those having all or most of their known distribution south of the Tropic of Capricorn (23°S) (Harder 1992).

This predominance of tropical species might be expected, given the fact that most (about 79%) of South America lies above 23°S. Nevertheless, the largest, most diverse genera (*Marmosa, Marmosops*, and *Monodelphis*) are tropical and, certainly, local species richness of neotropical marsupials is highest in tropical rainforests. Whether this reflects the general diversity of tropical rainforest communities, or physiological limits on marsupial distribution, is not clear. McNab (1986) argues that marsupials are generally ill-suited for cold, temperate regions because of their low metabolic rates and related low reproductive rates. This argument would appear to be more valid for *D. virginiana* near the northern limits of its distribution in North America than for didelphids in South America. Biogeographic factors related to the small land mass below 45°S might be a more important factor in limiting marsupial distributions than physiological limits *per se*.

METABOLISM AND REPRODUCTIVE ENERGETICS

The allocation of energy to maintenance, growth, and reproduction in marsupials reflects a mode of reproduction that is characterised by a brief period of gestation, low caloric investment in neonates, and an extended period of lactation. Thus, questions regarding the energetic costs and efficiencies of gestation and lactation in marsupials are long standing and have been rendered even more intriguing by early reports of low metabolic rates in this group. The basal metabolic rate (BMR) of marsupials is approximately 30% lower than that of eutherian mammals of equivalent body mass (Dawson & Hulbert 1970; Lustick & Lustick 1972; McNab 1978). BMR (that is the minimum required for maintenance of non-reproductive animals held in thermoneutrality) of eutherians generally follows the three-fourths power of body mass relationship (heat production in kcal day^{-1} = 70 x (body mass, $kg^{-.75}$, Kleiber 1961). Deviations of observed BMR from the Kleiber-predicted BMR (hereinafter, %BMR) in eutherians is correlated with diet (McNab 1980, 1986). However, interspecific variation in BMR in marsupials does not appear to be related to diet (McNab 1980). Instead, low BMR is reflected in lower body temperature (Dawson & Hulbert 1970). Body temperature (T_b), measured at basal metabolism, is typically between 33 to 35^oC in marsupials (Dawson & Wolfers 1978; McNab 1978) compared to 36 to 38^oC in eutherians.

Metabolic rate has been measured in a number of didelphid marsupials (McNab 1978), but most of the work on reproductive energetics has focused on 2 species: the Virginia opossum, *D. virginiana*, and the South American grey short-tailed opossum, *Monodelphis domestica*. Fortunately, these 2 species represent quite well the diverse array of didelphid marsupials, which can be placed in two basic groups of contrasting body size and pouch development. *Chironectes*, *Didelphis*, and *Philander* comprise a group of 8 species ranging in body size from 250 g to 3 kg. All have a well-developed pouch in which they carry their young for at least the first half of the lactation period. They tend to be more terrestrial, far-ranging, and less specialised in feeding habits than the smaller opossums. *D. virginiana*, *D. marsupialis* and *Philander opossum* are the best-known members of this group. Most of the remaining didelphids, about 55 species, belong to the second group, which are small in body size (less than 150 g) and pouchless. Included are members of three large genera: *Marmosa*, *Marmosops*, and *Monodelphis*. They tend to be scansorial and probably leave their young in a nest at an early age when foraging.

A small South American opossum: Monodelphis domestica

Since its introduction as a laboratory animal, little more than a decade ago (VandeBerg 1983; VandeBerg & Robinson, chapter 14 in this volume, Saunders, chapter 22 in this volume), the grey short-tailed opossum has become one of the most widely studied of all marsupials. Unfortunately, relatively little is known of the natural history or ecology of this species. This small (60 to 100 g) marsupial is one of 15 species of short-tailed opossums (*Monodelphis*) that are distributed throughout much of tropical South America, but only about half inhabit rainforests. *M. domestica* is distributed in the southern rim of the Amazon Basin from Bolivia to south-eastern Brazil. It is a scrub species that has accompanied people into cleared rainforest regions (Emmons 1990). *M. domestica* is best known from the Caatinga region of south-eastern Brazil, a semi-arid area vegetated with scattered small trees and shrubs, where it is most often trapped around mesic rock outcroppings (Streilein 1982). *M. domestica* is omnivorous, feeding on fruits, insects, and small vertebrates.

The BMR of *M. domestica* is typical of most marsupials, that is it is 55 to 64% of the Kleiber-predicted value for eutherians (Thompson & Nicoll 1986;l Nicoll & Thompson 1987; Dawson & Olson 1988; Harder *et al.* 1996) and about 10% lower than that measured in some Australian marsupials (Dawson & Hulbert 1970; MacMillen & Nelson 1969). The average T_b of this species, within thermoneutrality, is 32.6°C (Dawson & Olson 1988), typical of neotropical marsupials (McNab 1978).

Thompson and Nicoll (1986) reported a 12.5% increase in %BMR in *M. domestica* during gestation and a 52.3% increase during lactation. Values obtained by Harder *et al.* (1996) were higher (28.9%) for pregnant animals but very similar (57.2%) for lactating females. These increases in metabolic rate during lactation are substantial, particularly during the 2 weeks before weaning, but still modest in comparison to the sustained (summit) metabolism of 8 to 9 times BMR measured in response to low (5°C) ambient temperatures in this species (Dawson & Olson 1988). The metabolic rate of opossums increases during lactation but T_b increases as well (Thompson & Nicoll 1986; Harder *et al.* 1996). The average T_b in non-reproductive (dioestrous) opossums (32.1°C) increased to 35°C during late lactation (Harder *et al.* 1996). Facultative reduction in T_b has also been observed in opossums; for example, *M. brevicaudata* becomes torpid (T_b = 27°C) following food deprivation at 24°C (McNab 1978). Even small reductions in T_b (1°C or 2°C in sleep) are associated with detectable decreases in metabolic rate (Walker *et al.* 1979) and, thus, heterothermy and associated variations in metabolic rate might reflect a general pattern and strategy for energy conservation in small didelphids.

A large North American opossum: Didelphis virginiana

The Virginia opossum, *D. virginiana*, is, by far, the best known and longest studied of the New World marsupials. In fact, the pioneering work of Carl Hartman (1916; 1921; 1923; 1925; 1928) established *D. virginiana* as the model of marsupial reproductive physiology decades before comparable studies were accomplished on Australian marsupials. Interest in this species has widened during the intervening years, particularly among ecologists, in that *D. virginiana* provides the only opportunity in all of North America to study a marsupial in the wild.

D. virginiana is a relatively new species, having evolved from an isolated stock of *D. marsupialis* that invaded temperate North America during the Pleistocene (Gardner 1973). The distribution of *D. virginiana* extends north as far as southern Ontario and northern New York to approximately 44°N latitude (Gardner 1982). These recent distribution records are somewhat beyond the -7°C January isotherm, suggested by Tyndale-Biscoe (1973) as the northern limit of *D. virginiana*. However, additional range extensions northward are unlikely because *D. virginiana* is poorly adapted for winter conditions.

D. virginiana does not hibernate in winter. However, snow cover and frozen ground sharply limit foraging activities and opossums seldom leave their dens at ambient temperatures below -7°C (Wiseman & Hendrickson 1950). Thus, extended periods of severe winter weather require individuals to remain inactive and rely on metabolism of body stores. Perhaps up to one-third of the winter energy requirements of *D. virginiana* are met through catabolism of body tissues (Brocke 1970). In fact, changes in body mass of *D. virginiana* held captive under ambient winter conditions reveal a seasonal pattern of weight gain and loss that appears to be independent of food supplies. Females fed a near *ad libitum* diet lost 27% of their body mass from autumn (3.74 ± 0.44 kg, mean ± sem) to spring (2.74 ± 0.33 kg) (Hsu *et al.* 1988).

The intrinsic pattern of weight loss in *D. virginiana*, which would accommodate a reduction in food intake during unfavourable conditions, appears to be the major physiological adjustment for low ambient temperatures in winter. Measurements of resting metabolic rate (RMR) and T_b revealed no thermoregulatory adjustments in opossums acclimatised to the four seasons in Ohio, including winter when the mean monthly ambient temperature = -1°C and minimum = -17°C. RMR measured at an ambient temperature of 0°C did not vary with season and averaged 0.66 ± 0.1 ml O^2/g/h, which suggests that *D. virginiana* has not evolved mechanisms for reducing heat loss in winter (Hsu *et al.* 1988). Although pelage density increased 25% from summer to winter, thermal conductance at 0°C did not decline in winter, indicating little or no increase in insulation from fur or fat.

The reproductive energetics of *D. virginiana* has received relatively little attention. In fact, measurements of metabolic rate during reproduction appear to be limited to one study of 8 lactating females (5 to 8 young per litter) in which the cumulative resting metabolic rate (RMR) was 92% higher than that of nonreproductive females. Total cost in RMR to raise an average opossum nursling from birth to weaning was 400 kcal (Fleming *et al.* 1981).

Metabolic rate and reproductive performance

McNab (1989) proposed that the selective persistence of mammalian taxa in South America during the Tertiary Period is related to metabolic rates and food habits. Mammals that rely on fruit, leaves of woody plants, or invertebrates have low basal metabolic rates, presumably because such foods are seasonally unavailable, indigestible, or loaded with toxins (McNab 1986; Cork & Foley, chapter 12 in this volume). This relationship of low BMR and diet is most apparent in sedentary, arboreal species. Feeding on herbs and vertebrates is associated with high BMR in eutherian mammals but not in marsupials (McNab 1986). McNab argues that these foods permit high BMR and are exploited by eutherians because high BMR leads to high reproductive rates, expressed in high fecundity, rapid postnatal development, and short generation times. Conversely, marsupials show limited correlation of BMR and reproductive performance (McNab 1980, 1986).

Although interspecific variation in BMR among eutherians may be positively correlated with reproductive rates, it is also true that life history traits, BMR, and reproductive rates are correlated with body mass. Moreover, the analysis of Thompson (1992) suggests that BMR might be more closely related to reproductive effort (total daily energy used during reproduction) than to speed of reproduction *per se*. If so, reproductive performance might be constrained by morphophysiological limits on food acquisition and energy assimilation. Additional work on a wider array of species will be required to identify the reproductive consequences of taxonomic variation in BMR.

McNab (1986) has noted that, in contrast to eutherians, marsupials show little or no correlation between BMR and reproductive traits, such as fecundity or the time from conception to weaning. The latter is due in part to a narrow range of variation in basal rates and slow postnatal growth in mass (M) in marsupials. In fact, Lee & Cockburn (1985) have developed an alternative perspective on marsupial-eutherian differences that relates to pace of growth (α) where:

$$\alpha_{eutherian} = 1.76M^{-0.061}\ \alpha_{marsupial}$$

This indicates that the difference is most pronounced for small body masses, such that a 10 g eutherian grows about 2 times faster than a marsupial of similar mass while a 10 kg eutherian grows only 50% faster than a comparable marsupial. Lee & Cockburn (1985) noted a close similarity in their expression relating the pace of growth between eutherians and marsupials to one derived for energy expenditure during locomotion, which provides support for the notion that the pace of life is slower in marsupials than eutherians.

Discussions of marsupial energetics that focus on low BMR, low T_b, and slow pace of growth tend to obscure important questions about metabolic rates during times of high energy demand, for example, for thermoregulation at low ambient temperature or during late lactation. This aerobic scope or summit metabolism during cold stress is similar among eutherians and marsupials, about 3 to 7 times BMR (Dawson & Dawson 1982; Hinds & MacMillen 1984). In fact sustained summit metabolism in *Monodelphis domestica* was 8 to 9 times BMR. Clearly, the elevation in RMR during late lactation in this species (52% of %BMR, Thompson & Nicoll 1986) falls well below summit metabolism. Thus, it seems that reproductive rate (litter production) in small marsupials might not be constrained by metabolic capacity, but by limits set by food availability and quality (Green, chapter 9 in this volume) and the ability to process energy during a prolonged period of lactation (Thompson 1992). Such limits are accommodated in a mode of reproduction that features a slow pace of growth and low energy investment per unit time.

ECOLOGY AND REPRODUCTION

Previous discussions of neotropical marsupials have often noted the dearth of ecological information relative to the size and biological complexity of the region. Fortunately, during the last decade several excellent studies have been published that not only clarify patterns of diversity and abundance of marsupials in tropical rainforests but also reveal important relationships between their food supply and reproduction. This review focuses on tropical species, primarily from rainforests, and concludes with a comparison to the nearctic representative, *Didelphis virginiana*.

Marsupial species diversity and abundance

Mammal surveys have revealed high species richness of marsupials at study sites throughout Amazonia. The number of species per locality varies from 5 to 8 (Table 11.2). However, in many cases the trapping and sampling procedures employed probably fail to detect all species living in a given study area. For example, traps set on the ground or in the understorey (2 to 3 m height) are unlikely to detect canopy species such as *Caluromys philander* and will certainly underestimate their abundance (Malcolm 1991; McClearn *et al.* 1994). A recent survey revealed the presence of 17 marsupials in a single river valley in the Manu Biosphere Reserve in south-eastern Peru (Pacheco *et al.* 1993), which is close to the total number recorded at a number of sites throughout Amazonia. Species–richness of marsupials in northern neotropical locations (4 to 5 species, O'Connell 1989; Rabinowitz & Nottingham 1989; Fleming 1973) and in the coastal shrubby vegetation of Brazil (3 species, Cerqueira *et al.* 1993) appears to be lower than for localities in Amazonia.

Meaningful comparison of data from separate studies, using different procedures, is difficult and, therefore, the work of a single investigator, Louise Emmons (1984) at 7 different Amazonian locations in Brazil, Ecuador, and Peru has been widely cited. Relative abundance of small species varied markedly among localities as did species-richness. For example, the generally lower abundance of mammals in the 'Guianan' region north of Manaus, Brazil, is reflected in lower species richness than in the western Amazonian region of south-eastern Peru (68 species of non-flying mammals versus 86 species in total).

The composition of marsupial communities at several sites throughout Amazonia (Table 11.2) generally confirms the prediction for didelphid marsupials in South American lowland rainforests (Emmons 1990, p.11): '.. one to two species of *Marmosa*, one each of *Micoureus* and *Marmosops*, occur together with one each of *Didelphis*, *Caluromys*, *Metachirus*, and *Philander*, and sometimes *Monodelphis*.' The predictable nature of the species composition of marsupial assemblages suggests the existence of structure and resource partitioning in marsupial fauna in rainforest communities.

Emmons (1984) concluded that soil fertility and undergrowth density were positively correlated with mammalian abundance and diversity. However, other factors, such as body size and dietary specialisations, are also important correlates of population density (Robinson & Redford 1986). Fleck & Harder (1995) documented marked differences in marsupial species richness (8 versus 3) and abundance (6.9 versus 0.5 captures per 100 trapnights) in two primary rainforests separated by only 100 km in north-eastern Peru. Habitat characteristics, including vegetative density, were not sufficiently different ($p>0.05$) to account for the difference in marsupials between the 2 sites. Other factors including agricultural activity near Jenaro Herrera (the richer site) and annual flooding at the Yarapa River site (the poorer location) were probably more important.

Emmons (1984) observed that numbers of small mammals varied markedly from site to site while those of larger species were more constant. She postulated that large mammals experience higher survival in unfavourable habitats in Amazonia compared to related species of smaller body size. The data of Fleck & Harder (1995) seem to support that notion. Three of 8 opossum species captured at Jenaro Herrera were small (<100 g), but none of the species captured at the poorer site (Yarapa) were small; only the 3 largest species (> 350 g) of opossums (*D. marsupialis*, *M. nudicaudatus*, and *P. andersoni*) were captured there. What are the ecological factors that allow larger species to persist in unfavourable environments where smaller species are rare or drop out? The answer is not clear, but Emmons (1984) suggests that differences in diet, home range size, and competition are factors worthy of future study.

Habitat use and foods of rain forest marsupials

Explanations for high species-richness of marsupials in many rainforest communities are not readily apparent. A certain amount of differentiation is probably related to body size but, excluding 1 or 2 of the largest or smallest opossums in most communities, the range in body mass is less than an order of magnitude (for example, 40 to 350 g). With few exceptions (for example, *Chironectes* and *Lutreolina*), neotropical marsupials are similar in body plan and relatively unspecialised in feeding habits. The recent studies of Malcolm (1991) and McClearn *et al.* (1994) have confirmed earlier reports of considerable vertical stratification in use of forests by opossums. But here again, behavioural flexibility and overlap in habitat use is more the rule than the exception. Only a few opossums (for example, *Caluromys* and *Metachirus*) appear to be substrate specialists. Progress in understanding the structure of marsupial rainforest faunas requires more detailed information on the diets and use of terrestrial and arboreal habitats by opossums, particularly as they relate to the seasonal and spacial patterns of precipitation, flooding and food availability.

Table 11.2. Relative abundance (within localities) of marsupials in rainforests in South America. H = high; M = medium; L = low abundance. Species are listed in order of decreasing body mass

SPECIES	COCHA CASHU[1]	JENARO HERRERA[2]	MANUAS BRAZIL[3]	FRENCH GUIANA[4]	GUATOPO VENEZUELA[5]
Didelphis marsupialis	H	H	M	H	H
Philander opossum	M	M		H	
Philander andersoni		H			
Metachirus nudicaudatus	M	L	H		
Caluromys lanatus	L	M			
Caluromys philander			H	H	M
Caluromysiops irrupta	M				
Marmosa cinerea (=*Micoureus cinereus*)	M	L	H	L	M
Monodelphis brevicaudata			M		M
Marmosops noctivagus	M	H			
Marmosa fucata (=*Marmosops fuscatus*)					H
Marmosa murina		L	L	M	
Marmosa parvidens (=*Marmosops parvidens*)			M		

Sources:
[1] Janson and Emmons (1990).
[2] Fleck and Harder (1995).
[3] Malcolm (1991).
[4] Charles-Dominique (1983).
[5] O'Connell (1989).

A general, but rather complete, picture of diet and substrate use by rainforest opossums is now available from the literature (Table 11.3). Information is also available on seasonal patterns of precipitation and food availability (primarily fruitfall and insect abundance). What is missing is an understanding of the interaction between these factors in the dynamic and heterogeneous landscape of a rainforest. In this section we turn to the people who live in the forest and whose livelihood depends upon an understanding of the factors that determine the availability of game. We seek to compare what is known of the biology of rainforest marsupials with the related knowledge of the Matses Indians of north-eastern Peru. In so doing we hope to advance the pace of investigation by generating hypotheses that might not emerge from scientific investigation alone.

ETHNOBIOLOGY OF A PERUVIAN RAINFOREST

Ethnobiology is the study of the knowledge and uses of living organisms by different human societies. Traditionally, the most extensive applications have been made in anthropology and pharmacology, but during the last 15 years ethnobiology has found application in basic ecological studies as well. For example, interviews with Aguaruna Indians in Peru provided insights regarding mammalian classification (Patton *et al.* 1982). Similarly, the work of Parker *et al.* (1983) with *caboclos* (local residents) of Lake Coari, Brazil revealed an ecological classification based on the floral and faunal composition of resource units. Hypotheses generated from interviews of Kayapó— Indians in Brazil formed the basis of Posey's (1986) research on stingless bees in Amazonia. Newsome (Chapter 13 in this volume) has found remarkable congruence between sites in Australian Aboriginal legends of the ancestral red kangaroo and drainage lines that represent the original preferred habitat for this species.

Information presented in the following section was obtained by David Fleck in his studies with Matses Indians in March through June 1994. The Matses (also known as Mayoruna) are a relatively unassimilated upland (*terra firme*) rainforest Amerindian group in the Pano language family living in north-eastern Peru and Brazil along the Javari River and its tributaries. These indigenous Amazonians, through their cultural heritage and lifetimes of wildlife observation, have acquired an encyclopaedic knowledge of natural history and a remarkable appreciation of long-term ecological patterns.

Hunters were interviewed with a standard list of questions regarding the natural history of each species of local mammal (or genus in the case of marsupials, bats, and small rodents). Matses were also questioned about general climatic and ecological trends of the region, including long-term yearly rainfall patterns, seasonal river level fluctuations, and the seasonality of fruit production. Five Matses hunters, who know Spanish, were selected as principal informants; the older men who speak only Matses were also interviewed, with the aid of a translator. Several informants were interviewed in order to acquire comparable information on each species. Matses were accompanied on hunting trips several times per week to document hunter information first-hand. When Matses hunters brought a kill to the village, the hunter was interviewed about details of the hunt, including the habitat where the animal was found, its behaviour, and the size of its herd or troop.

Monthly rainfall patterns (periods of highest and lowest rainfall and duration of wet and dry periods) were plotted on a graph in consultation with Matses informants. Subsequently, these relative values were set to a scale of known monthly average, maximum, and minimum rainfall levels (Figure 11.1a). This plot of Matses-perceived seasonal variation in rainfall was remarkably similar to a plot of monthly average rainfall at the nearest weather station in Iquitos, 200 km to the north (Figure 11.1b). The Matses-perceived seasonal patterns of water level fluctuation of the Galvez River, the tributary of the Javari River where most Matses live, showed a wide flux of about 10 m between the highest and lowest water levels (Figure 11.1c).

Matses knowledge revealed asynchrony in seasonal production of different types of fruit, including a period from September to November when production of all fruit types is low (Figure 11.1d). The Matses claim that during this period of fruit scarcity, frugivores must turn to a wide range of alternative foraging methods and food types. Variation in foraging strategies during the period of food scarcity may provide a basis for niche separation and partially explain the large number of sympatric frugivorous species within families of rainforest mammals.

Matses knowledge is most extensive for game animals such as ungulates and large monkeys and relatively weak for small, nocturnal mammals such as opossums, small rodents, and bats. Nevertheless, Matses knowledge of basic marsupial natural history corresponded surprisingly well with published information. For example, their information on diet and substrate use was almost identical to that described in Emmons (1990), (Table 11.3).

Matses recognise and have names for at least 41 habitat types within the rainforest, subdivisions of what scientists often recognise as only *várzea* (flooded forest) and *terra firme* (upland forest). Names applied to habitats within the *várzea* (Figures 11.2 and 11.3) have meaning relative to seasonal changes in water levels and use by game animals. This terminology is the cultural heritage of hunters and, therefore, it conveys potentially significant information regarding seasonal changes in the distribution and abundance of animals.

According to several Matses informants, opossums and many other mammals exhibit predictable, seasonal changes in habitat use. During the dry season (June to September) opossums concentrate in swamp palm, *Mauritia flexuosa*, forests (Figures 11.2 and 11.3) and many even make their nests there. When the swamp palms stop

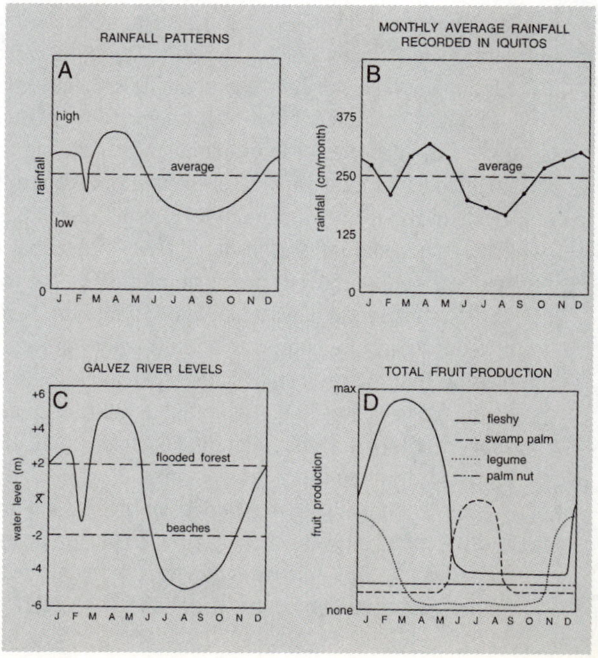

Figure 11.1
Seasonal variation in precipitation, water levels and fruit production as perceived by Matses Indians living in an Amazonian rainforest on the Galvez River (a tributary of the Yaveri River) in north-eastern Peru
(a) Annual rainfall patterns reported by Matses Indians.
(b) Average monthly precipitation recorded at a weather station 200 km north in Iquitos, Peru.
(c) Seasonal fluctuation in water levels of the Galvez River.
(d) Seasonality of fruit production of the 4 main types of fruit-trees. Drawn by David Fleck.

Table 11.3 Habitat, substrate use, and known foods of representative species of the genera in the Order Didelphimorphia (from Eisenberg 1989 & Emmons 1990)

SPECIES	HABITAT	SUBSTRATE	FOODS
Caluromys philander	rainforest, viney midstorey and canopy	arboreal	fruits, invertebrates, nectar
Caluromysiops irrupta	rainforest canopy	arboreal	fruits, flower nectar
Glirona venusta	rainforest, midstorey	arboreal	insects, fruit, seeds
Didelphis marsupialis	humid secondary and gallery forests, gardens	scansorial	small vertebrates, inverte brates, fruit, nectar
Philander opossum	rainforests, gallery forests, and gardens	scansorial	small vertebrates, invertebrates, fruit
Metachirus nudicaudatus	rainforests, gallery forests	terrestrial	termites, other invertebrates, eggs, fruit
Lutreolina crassicaudata	drier ground habitats	terrestrial	small vertebrates, fish, insects plants, seeds
Chironectes minimus	near fast-flowing rivers in tropical forests	semiaquatic	fish, crustaceans, other invertebrates
Micoureus demerarae	evergreen rainforests, midstorey and canopy	arboreal	small vertebrates, invertebrates, fruit, nectar
Marmosa murina	secondary lowland forest viney understory, gardens	scansorial	small vertebrates, invertebrates, fruit
Marmosa robinsoni	diverse forest habitats	scansorial	insects, fruit
Marmosops noctivagus	rainforest understory	scansorial	insects, fruit
Gracilinanus agilis	rainforest to Caatinga	arboreal	
Monodelphis domestica	deciduous forests and caatinga, houses	terrestrial	small vertebrates, invertebrates, fruit
Monodelphis brevicaudata	rainforests, gardens	terrestrial	insects

producing fruit in October, opossums move to riverside and lakeside habitats to prey on the nests of waterside-nesting birds. They remain in waterside habitats through November and add to their diets the pods of leguminous trees, many of which start fruiting in November along bodies of water (Figure 11.1d). From December to May (rainy season) terrestrial species must move to upland forests to avoid inundation, while the arboreal species can remain in the flooded forest. During the period of high fruit productivity (December to May), terrestrial opossums are located in *terra firme*

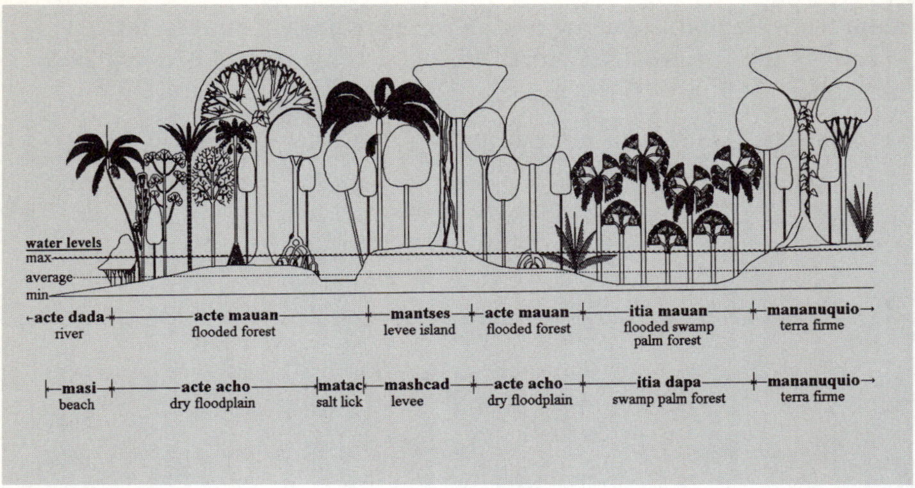

Figure 11.2 Cross-section of the *várzea* (seasonally inundated rainforest) along the Galvez River in north-eastern Peru showing maximum, average and minimum annual river water levels and terminology for the major *várzea* habitats recognised by the Matses Indians.
Note: The upper line provides the Matses' names (with English equivalents) for habitats recognised during the wet season (maximum water levels). Note that names for the same areas are different during the dry season (lower line) when water levels are minimum.
Drawn by David Fleck.

Figure 11.3 Cover map of the Galvez River *várzea* rainforest and surrounding *terra firme* (upland) rainforest showing distribution of habitat types and the location (bar) of the area represented in the cross-section of forest in Figure 11.2.
Drawn by David Fleck.

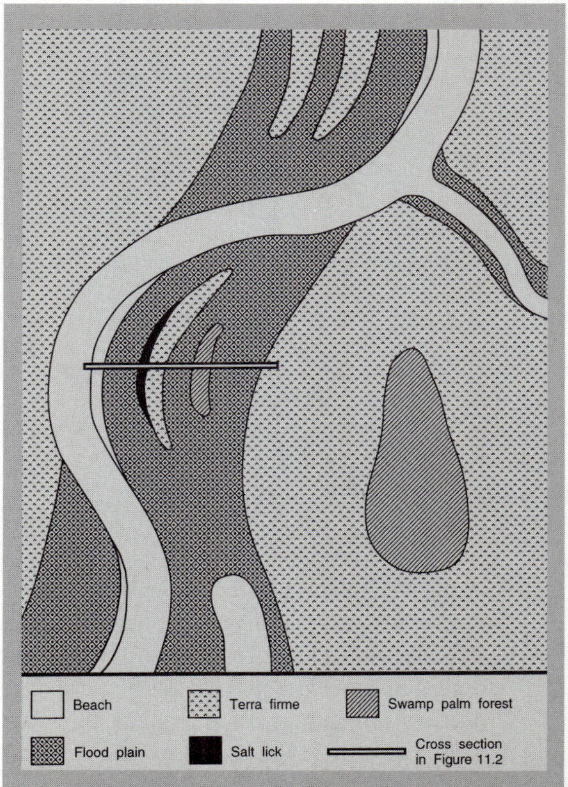

rain forest feeding under fruiting trees, while arboreal species can be found in both flooded and upland sites feeding on fruit hanging in trees. In June these animals return to the swamp palm forests and resume their cycle of seasonal micro-migrations among habitats (Figures 11.2 and 11.3).

Although this information from the Matses on seasonal movements of opossums awaits confirmation with scientific studies, it may explain, in part, the wide temporal variation in opossum abundance recorded in some studies. Clearly, opossum densities do vary considerably over time. However, if surveys are not conducted in the same location at the same time of the year, wide variation in trapping success should be expected, if the Matses are correct about seasonal movements of rainforest marsupials.

Table 11.4. Litter sizes and temporal features (in days) of the reproductive cycles of selected New World marsupials

SPECIES	LITTER SIZE	GESTATION LENGTH	OESTROUS CYCLE LENGTH	WEANING AGE	REFERENCES[1]
Caluromys philander	4	20+	39	120 (80)[2]	1, 2, 3
Chironectes minimus	2–3				4, 5
Didelphis albiventris	4–5				6,7
Didelphis marsupialis	4–9			90–100	3, 6, 8
Didelphis virginiana	6–9	13	28	100 (70)	9, 10
Marmosa robinsoni	10–14	14	26	60–70 (20)	8, 11, 12, 13
Monodelphis brevicaudata	7–8				13
Monodelphis dimidiata	8–16			56 (14)	14
Monodelphis domestica	7	15	0	56 (14)	15, 16, 17
Philander opossum	4.5			90	3
Caenolestes fuliginosus =*C. obscurus*	3.4				18, 19, 20

[1]. Numbers refer to the following references:
1 Atramentowicz (1986), 2 Perret & Atramentowicz (1989), 3 Julien-Laferrier & Atramentowicz (1990), 4 Enders (1966), 5 Marshall (1978), 6 Tyndale-Biscoe & Mackenzie (1976), 7 Cerqueira (1984), 8 Fleming (1973), 9 Hartman (1921), 10 Reynolds (1952), 11 Barnes & Barthold (1969), 12 Godfrey (1975), 13 O'Connell (1979), 14 Pine *et al.* (1985), 15 VandeBerg (1983), 16 Kraus & Fadem (1987), 17 Harder *et al.* (1993), 18 Osgood (1921), 19 Kirsch & Waller (1979), 20 Tyndale-Biscoe (1980).

[2]. Numbers in parentheses indicate the approximate age at which young are left in the nest; that is, pouch exit or the end of the obligatory teat attachment phase in pouchless species.

REPRODUCTIVE ECOLOGY OF DIDELPHID MARSUPIALS

Although much has been learned about the reproductive biology of New World marsupials in the 20 years that have passed since the publication of Hugh Tyndale-Biscoe's (1973) *Life of Marsupials*, the list of well-known species remains remarkably short (Table 11.4). With the exception of litter size, basic reproductive features (for example, gestation length, litters per year, and duration of lactation) have been published for fewer than 10 of the 69 New World marsupials. On the other hand, the last decade has seen considerable progress in our understanding of the ecological constraints on seasonal breeding, fecundity, and juvenile survival. For purposes of discussion, species covered in this literature are grouped by body size, life history traits, and geographic distribution.

Monodelphis domestica and other small, pouchless species

Most New World opossums are small (30 to 150 g), scansorial, tropical species of *Marmosa, Marmosops*, and *Monodelphis*. Perhaps the best known of this group, relative to reproductive ecology, is Robinson's mouse opossum (*Marmosa robinsoni*). This attractive opossum is recognised by its dark eye patches, cinnamon-brown dorsal pelage with pale orange-yellow underparts (Emmons 1990). Males are larger than females and can exceed 350 mm in total length. *M. robinsoni* occupies a wide range of habitats in the northern Neotropics, including evergreen, deciduous and thorn forest, secondary growth, disturbed areas and seasonally dry habitats, where it forages for invertebrates and fruit in vines, trees and undergrowth (Eisenberg 1989; Emmons 1990).

M. robinsoni is polyoestrous with a cycle length of 26 days; the luteal phase of the cycle is the same length as gestation, about 14 days (Godfrey 1975). Young are firmly and continuously attached to the mother's teats during the obligate teat attachment phase, which lasts only for the first 20 days of the 70-day lactation period (Table 11.4). During this time the young travel with the mother without the protection of a pouch. Young older than 20 days of age are left in a nest while the mother forages for food (Eisenberg & Maliniak 1967; Barnes & Barthold 1969).

M. robinsoni is reproductively inactive from November through March in Panama (Fleming 1973); in this area one large litter of 10 might be most common (Enders 1966). Lactating females were observed from May through October in the *llanos* of northern Venezuela with no clear peaks in activity during this period. Females apparently produce 1 or 2 large litters of 14 during a single breeding season in Venezuela (O'Connell 1979). In the premontane humid forests of northern Venezuela, *M. robinsoni* and *M. brevicaudata* breed early in the wet season and weaning of young coincides with the rains (O'Connell 1989). Juvenile survival of didelphids was low (<50%) on the forest site and fell to near zero for all opossums on the *llanos* at the end of the dry season.

Monodelphis dimidiata is an interesting example of a small didelphid of the south temperate region (Pine *et al.* 1985). It inhabits the grasslands of northern Argentina where it feeds primarily on insects, being most active in late afternoon. Four litters were observed during the summer months (December to January) the largest being comprised of 16 young. Pine *et al.* (1985) concluded, based on trapping records, that adults do not survive into the winter and that females are semelparous, having one large litter during their lifetime. A similar reproductive pattern has been inferred from museum records of *M. incana* (Lorini *et al.* 1994).

The breeding season of *Monodelphis domestica* in the semi-arid Caatinga of north-eastern Brazil (8°S to 12°S) appears to be controlled by photoperiod such that pregnancy rates are lowest in July through November, which are also the driest months (Bergallo & Cerqueira 1994). However, females living around mesic rock outcroppings may breed all year round (Streilein 1982). Although little more is known of the reproductive ecology of *M. domestica*, the reproductive physiology of this species is better known than for any other neotropical marsupial, particularly among the diverse array of small, pouchless opossums. Therefore, a brief review of the reproductive features of relevance to the natural history and ecology of this group seems appropriate.

Perhaps the most noteworthy reproductive feature of *M. domestica* (Table 11.4) is that females do not exhibit a true oestrous cycle (Fadem 1985; Fadem & Rayve 1985) or attain puberty unless they are exposed to male pheromones (Stonerook & Harder 1992). Isolated females remain anoestrus, but, within 4 to 11 days after exposure to only male pheromones, can be induced to oestrus and will mate (Fadem 1987). Changes associated with oestrus occur in the cells of the urogenital sinus an average of 5.3 days after male-female cage switching. If paired at this time, copulation occurs 2.0 days later and ovulation 24 hours after copulation (Harder *et al.* 1993; Baggott *et al.* 1987; Baggott & Moore 1990). Plasma luteinising hormone levels peak in females (presumably reflecting a preovulatory luteinising hormone surge) 6 to 8 days after pairing with males (Hinds *et al.* 1992). Ovulation in *M. domestica* is not spontaneous; isolated oestrous females do not ovulate (Stonerook & Harder 1992; Hinds *et al.* 1992). However, proestrous females will ovulate when caged with a male but prevented from copulating (Baggott *et al.* 1987). The ovulation rate (average corpora lutea per pregnant female = 11.9, maximum = 16) does not greatly exceed the number of embryos (9.8) or the average mid-lactational litter size (7.1) (Harder *et al.* 1993). These values from laboratory studies are similar to an average of 7.9 embryos (range: 2 to 16) observed in *M. domestica* taken from the Caatinga of north-eastern Brazil (Bergallo & Cerqueira 1994).

M. domestica does not have an oestrous cycle or a luteal phase, except during gestation. Plasma progesterone is elevated for 16 days in pregnant females, falling sharply at parturition (Hinds *et al.* 1992). However, based on time-lapse video recordings, the interval from copulation to birth is only 15.2 days (Harder *et al.* 1993) with an interval of 13.5 to 14 days from fertilization to parturition (Moore 1992). Thus, it would appear that progesterone might be elevated in oestrous females 1 or 2 days before copulation and ovulation.

Although true (obligatory) male-induced oestrus has not been reported in didelphids or New World marsupials other than *M. domestica*, male urine is known to have a stimulatory effect on the occurrence of oestrus in woolly opossums (*Philander opossum*) (Perret & M'Barek 1991). Also, induced oestrus and ovulation have been observed in at least one Australian marsupial, the brush-tailed bettong (*Bettongia penicillata*, Potoroidae) (Smith 1992; Hinds & Smith 1992). Thus, a systematic examination of pheromonal responses in didelphid species with different life histories and habitat affiliations would be worthwhile.

Caluromys: The arboreal specialists

Woolly opossums represent, with the two other monotypic genera (Table 11.1), the Subfamily Caluromyinae. Caluromyids are the most highly arboreal of New World opossums; they seldom forage on the ground, depending instead upon the insects, nectar, and fruits in the canopy of rainforests (Eisenberg 1989; Emmons 1990). The bare-tailed opossum, *C. philander*, is the most thoroughly studied marsupial in the Neotropics, particularly in regard to reproductive physiology and ecology. It is somewhat smaller (140 to 270 g) than *C. derbianus* and *C. lanatus* and is widely distributed in northern and eastern regions of South America. *C. philander* nests in tree cavities and feeds in the canopy of multistratal forest (Charles-Dominique 1983; Eisenberg 1989; Emmons 1990), where they run along branches and vines, bounding from limb to limb in a manner reminiscent of lemurs (Hunsaker & Shupe 1977).

The oestrous cycle of *C. philander* varies from 28 to 45 days in length with an average 38.6 days. The luteal phase of the cycle is of longer duration (25 days) and is marked by higher levels of plasma progesterone (30 to 40 ng/ml) (Perret & Atramentowicz 1989) than in *D. virginiana* (10 to 15 ng/ml) (Harder & Fleming 1981). Gestation in captive females lasts 20 to 28 days (Perret & Atramentowicz 1989), the longest known for any American marsupial (Table 11.4).

The period of postnatal care in *C. philander* is relatively long and the average litter size of 4 (Charles-Dominique *et al.* 1981; Atramentowicz 1986) is small compared to other didelphids of similar size. The pouch is developed only when the female is carrying young and regresses to lateral folds at other times (Perret & Atramentowicz 1989; Emmons 1990). Young leave the pouch at about 80 days of age but continue to nurse in the nest for an additional 40 days (Julien-Laferriere & Atramentowicz 1990). Juveniles disperse from the natal area at about 130 days of age (Eisenberg 1989).

Progress in understanding the complex relationships between food availability, reproduction and survival of *C. philander* and other didelphids in the rainforests of French Guiana represents one of the most significant advances in New World marsupial biology during the last decade. Charles-Dominique *et al.* (1981), Atramentowicz (1982, 1986) and Perret & Atramentowicz (1989) have shown a positive relationship between the number of fruiting trees and the percentage of females that are reproductively active, that is, lactating or in oestrus. Young born near the end of the fruiting season do not survive to weaning. Food intake in lactating *C. philander* is correlated with litter mass at weaning but not litter size and, thus, it appears that food resources have a greater impact on survival of litters than on regulation of litter size (Atramentowicz 1992).

C. philander and *P. opossum* breed all year long in French Guiana, but populations in secondary forests experience lower survival rates of pouch young during the period of fruit scarcity (July to November) than those in mature forests, where more supplemental food, such as flower nectar, is available at this time. Conversely, reproduction in *D. marsupialis* is highly seasonal but successful (low pouch mortality) in both primary and secondary forests and appears to be less directly affected by variation in environmental conditions (Julien-Laferriere & Atramentowicz 1990).

Caluromys is long-lived in captivity and more apt to breed in more than one season in the wild than *Marmosa* or *Monodelphis* (O'Connell 1979; Charles-Dominique 1983). These features, along with small litter size, an extended lactation period, and relatively high encephalisation quotients, were seen by Eisenberg & Wilson (1981) as evidence of an adaptive syndrome that might show convergence with nocturnal prosimians. However, Lee & Cockburn (1985) note that *Caluromys* is more reliant on plant material than other didelphids (Charles-Dominique *et al.* 1981; Atramentowicz 1992) and that herbivory in mammals is generally associated with prolonged development. They argue that the dietary restrictions of herbivory in *Caluromys* are more likely to be the cause of longer generation time rather than an adaptive syndrome associated with increased encephalisation and social organisation. The studies of Atramentowicz and colleagues are important in this regard because they focus attention on nutrition and community interactions as determinants of reproductive strategies in rainforest marsupials.

Didelphis: The New World habitat generalist

The four species of *Didelphis* collectively range from northern USA (44°N) to southern Argentina (46°S), making *Didelphis* the most widely distributed marsupial genus in the world (Table 11.1). The common opossum (*D. marsupialis*) is widely distributed and abundant in a variety of habitats, particularly rainforests. It is a tropical species, however, and its geographical limits correspond closely to 23°N and 23°S. The white-eared opossum (*D. albiventris*) is found throughout much of South America, except in the Amazon Basin, from the higher elevations of Colombia to approximately 39°S in southern Argentina. It apparently favours cooler climates and is widely distributed in temperate regions of South America (Streilein 1982).

In higher latitudes of North America, food shortages and low winter temperatures ultimately restrict breeding activity in *D. virginiana* to the months of February through July. Less is known of these seasonal factors in temperate regions of South America, but even in subtropical regions in south-eastern Brazil, increasing daylength appears to be an important cue for the onset of seasonal breeding as in *P. opossum* (Cerqueira *et al.* 1993). Accessory glands of male *D. albiventris* increase in weight during the first part of the breeding period, June to January (Nogueira 1988). Females are in oestrus during the dry season (July to November) and pouch young are observed from August to March (Mascarenhas 1980; Rigueira *et al.* 1987, cited in Nogueira 1988). Breeding in *D. albiventris* begins in the dry season and young are subsequently weaned during the rainy season (September to March) when food is abundant (Nogueira 1988). Seasonal breeding is more closely related to precipitation in north-east Brazil, where annual variation in daylength is lower (Cerqueira 1984). *D. marsupialis* starts breeding in late January in Panama and weans 2 litters during the period of fruit abundance (May to September) (Fleming 1973).

The reproductive physiology of *Didelphis*, as represented by *D. virginiana*, is well known and has been reviewed in Tyndale-Biscoe and Renfree (1987) and Harder (1992). However, a summary is included here for comparison with descriptions of other didelphids discussed in this chapter. *D. virginiana* is seasonally polyoestrous with a cycle length of 28 days and a gestation period of 13 days (Table 11.4). In the absence of conception, females cycle and ovulate 6 or 7 times per year, but oestrous cycles are restricted to January through July in northern populations. The luteal phase of the oestrous cycle (defined by elevated progesterone levels) is 11 days in duration and is hormonally equivalent to gestation (Harder & Fleming 1981). Ovulation rate is high (60 ova/cycle)(Fleming & Harder 1983) but the number of embryos *in utero* (22) is much lower (Hartman 1923; Reynolds 1952).

Average litter size in *D. virginiana* varies from 9 in the north (Hamilton 1958; Hossler *et al.* 1994), where most females have only one litter per year, to about 6 in Texas and Florida (Hartman 1928; Burns & Burns 1957), where nearly all females raise 2 litters per year. The generality of such latitudinal clines in litter size within *Didelphis* has been recognised for some time (Tyndale-Biscoe 1973). For example, litter sizes in *D. marsupialis* increase from 6.5 at 5°N in French Guiana (Charles-Dominique 1983) and 4.5 at 3°N in western Colombia, where most females have 3 litters per year (Tyndale-Biscoe & MacKenzie 1976), to averages of 7.1 to 8.5 (Hill 1918; Davis 1945) at 23°S in south-eastern Brazil. Thus, *Didelphis* females apparently invest in larger litters when opportunities for second or third litters are reduced by cold weather, dry conditions, or when high neonatal mortality might be expected (Hunsaker 1977; Gardner 1982; Lee & Cockburn 1985; Eisenberg 1988).

The success of *D. virginiana* in North America, particularly its northern range extensions of the nineteenth century (Gardner 1982), is generally attributed to a high reproductive rate (judged from litter size), high omnivory, and adaptability as a habitat generalist. However, reproduction in populations existing in far northern areas is constrained by a short growing season and an inherently long interval (3.5 months) from conception to weaning of young. Moreover, opossums have a brief reproductive life in the north. They do not breed in their first year (8 to 12 months) and reproductive capacity declines sharply in their third year (Reynolds 1952; Jurgelski & Porter 1974).

D. virginiana seldom lives beyond 2 years in the wild (Petrides 1949; Fitch & Sandidge 1953; Sanderson 1961; Sunquist *et al.* 1987). Thus, northern opossums have but one season to achieve reproductive success. Under these extreme conditions, *D. virginiana* is effectively semelparous; the size and survival of a single litter are the near-sole determinants of a female's lifetime reproductive success. Cockburn (chapter 10 in this volume) notes that accelerated senescence and semelparity observed in some didelphids is unexpected in mammals with slow growth and slow pace of life. Austad (1993) sought an explanation for this enigma in senescence theory, which predicts the evolution of accelerated senescence in populations living in hostile environments. He compared age-specific mortality and reproduction in a mainland (South Carolina) population of *D. virginiana* with an isolated population on Sapelo Island, Georgia, which has a 4000 year history of reduced exposure to predators. Under these conditions, several indicators of senescence, including an age-specific reduction in reproductive performance, were retarded in the island population.

Litter size in *Didelphis* is related to maternal nutritional condition as shown in body mass (Motta *et al.* 1983; Austad & Sunquist 1986; Eisenberg 1988). Maximum litter size also correlates with body mass in *D. albiventris*, ranging from 5 in the smallest mothers to 10 in the largest (Cerqueira 1984). It appears that nutritional factors can also influence the sex ratio and survival of offspring. Free-ranging *D. marsupialis* that received supplemental food produced litters with male-biased sex ratios, higher body weights, and better survival than control females (Austad & Sunquist 1986).

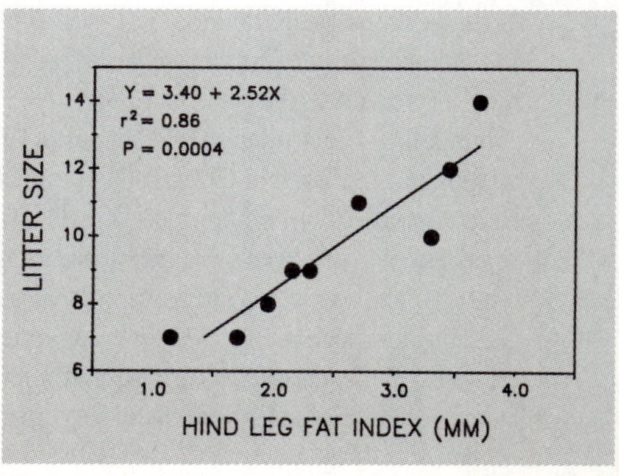

Figure 11.4 Linear relationship between the hindleg fat index and litter size in the North American opossum (*Didelphis virginiana*) in south-eastern New York.
Source: Hossler *et al.* (1994).

D. virginiana depends upon body reserves to survive northern winters, wherein perhaps one-third of winter energy requirements are met through catabolism of fat and muscle (Brocke 1970). In a study conducted just 65 km south of the January-7°C isotherm in New York, Hossler *et al.* (1994) found that many *D. virginiana* captured in late winter (March) showed signs of muscle atrophy and weight loss. Thus, many females probably enter the breeding season in February in poor condition and the extent to which they can obtain food and gain weight might very well determine their reproductive success. Indeed, females gained an average of 11% in body weight during lactation and litter size (mean = 9.4) was closely correlated with a hindleg fat index ($r^2 = 0.86$) measured in early lactation (Figure 11.4) (Hossler *et al.* 1994).

Juvenile survival is the second major determinant of reproductive success in northern *D. virginiana*. Young are left in the den at 70 days of age and after weaning (93 to 106 days, Gardner 1982), they forage outside the den where they are vulnerable to predation. Juveniles stay near their natal area during the first two or three months of independence (Gillette 1980; Wright 1989) where more than 60% might die during the first month (Wright 1989). Thus, the quality of habitat (protective cover and food) near the weaning den might well determine juvenile survival and, if so, maternal denning behaviour should reflect this reality. Hossler *et al.* (1994) found support for this hypothesis in that den fidelity was higher during the weaning period (young are 70 to 103 days old) and weaning dens were located in more densely vegetated habitats than non-weaning dens. Nevertheless, juvenile survival before dispersal was only 22.3%.

REPRODUCTIVE STRATEGIES OF DIDELPHID MARSUPIALS

The picture that emerges from this review of recent literature on the reproductive ecology of New World marsupials reinforces and extends earlier conclusions regarding the importance of energetics and nutrition. That is, the latitudinal clines to larger but fewer litters in northern populations of *Didelphis* have long been seen to reflect the thermoregulatory and energetic constraints of a brief growing season. It now seems that this pattern should also be viewed in the context of accelerated senescence and semelparity (Cockburn, chapter 10 in this volume), which might represent in *D. virginiana* a response to high mortality due to predation (Austad 1993).

Reproductive success of females in northern populations of *D. virginiana* pivots on their nutritional status following overwinter deprivation and the extent to which body reserves are replenished early in the breeding season. Climatic factors, such as the timing of the spring thaw and subsequent growing conditions and related foraging opportunities, would appear to be critical determinants of not only reproductive performance but of maternal survival as well. A female that is able to gain weight early in the season might wean a large litter by July, which would provide her young with the opportunity to survive, if weaned in favorable habitat, and gain body mass sufficient to overwinter and breed the following year.

The reproductive strategies of tropical rainforest didelphids are most certainly more diverse, if not more complex, than for opossums at high latitudes. The well-documented synchrony of reproduction with precipitation and seasonal variation in food resources is a common feature that overlies more complex community interactions. The interesting studies of *C. philander* and associated didelphids in the rainforests of French Guiana (for example, Julien-Laferriere & Atramentowicz 1990) have revealed a degree of differentiation in reproductive strategies that might not be expected from differences in body size, substrate uses, or even diet. It is now clear that detailed knowledge of reproductive strategies offer a promising avenue for understanding niche differentiation and marsupial community structure in Neotropical rainforests.

FUTURE DIRECTIONS

We believe that the coming decade will see progress in three areas of reproductive ecology of New World marsupials. First, studies of energy assimilation and respirometry will be applied to the study of reproduction in a greater variety of marsupials. As knowledge of natural systems becomes more complete and hypotheses are generated relative to reproductive strategies, laboratory studies of energetics will naturally follow (for example, Atramentowicz 1992). Energetic considerations appear dominant in the reproductive strategy of the *D. virginiana* and yet very little is known of the metabolism of reproduction in this species. This deficiency should be addressed in future studies. Techniques suitable for both laboratory and field studies are available and the questions are important, at the ecological level as well as for phylogenetic considerations.

Basic reproductive ecology of marsupials will develop as a second area because it offers an opportunity to understand resource partitioning and structure in complex tropical communities. The species richness of marsupials in these areas can be understood best with detailed information on natural history and knowledge of nutritional and reproductive strategies of each species as they relate to the dynamic and heterogeneous environments of tropical forests. The knowledge of natural history of rainforest animals possessed by unassimilated Amerindians is a valuable resource that should be used whenever possible.

Surveys of biodiversity in the Neotropics offer a third opportunity for progress in reproductive biology of marsupials. The growing interest in conservation biology and systematics should place increasing numbers of biologists in the field. In particular, we might hope that the confused state of didelphid taxonomy will attract mammalogists, equipped with molecular techniques, to marsupial systematics. If so, the associated fieldwork will yield basic reproductive information on a wide variety of New World marsupials.

Acknowledgments

Our work was supported by the Institute for Ecosystems Studies, Instituto de Investigaciones de la Amazonia Peruana, Ohio Board of Regents, and The Ohio State University. We are grateful for the hospitality and assistance of Peruvians, particularly Contralmirante Raul Zuazo Tolmos, Pablo Reategui Rodriquez, and Antonio Jimenez Tafur, our Matses host and principal informant in our ethnobiological studies. Dave Dennis assisted with the graphics.

References

Atramentowicz M (1982). Influence du milieu sur l'activite locomotrice et la reproduction de *Caluromys philander*. *Revue d'Ecologie la Terre et la Vie*, 36, 373–395.

Atramentowicz M (1986). Dynamique de population chez trois marsupiaux Didelphides de Guyane. *Biotropica*, 18, 136–149.

Atramentowicz,M (1992). Optimal litter size: does it cost more to raise a large litter in *Caluromys philander*? *Canadian Journal of Zoology*, 70, 1511–1515.

Austad S N and Sunquist M E (1986). Sex-ratio manipulation in the common opossum. *Nature*, 324, 58–60.

Austad S N (1993). Retarded senescence in an insular population of virginia opossums (*Didelphis virginiana*). *Journal of Zoology*, 229, 695–708.

Baggott L M, Davis-Butler S and Moore H D M (1987). Characterization of oestrus and timed collection of oocytes in the grey short-tailed opossum, *Monodelphis domestica*. *Journal of Reproduction and Fertility*, 79, 105–114.

Baggott L M and Moore H D M (1990). Early embryonic development of the gray short-tailed opossum, *Monodelphis domestica*, *in vivo* and *in vitro*. *Journal of Zoology*, 222, 623–639.

Barnes R D and Barthold S W (1969). Reproduction and breeding behaviour in an experimental colony of *Marmosa mitis* Bangs Didelphidae. *Journal of Reproduction and Fertility Supplement*, 6, 477–492.

Bergallo H G and Cerqueira R (1994). Reproduction and growth of the opossum *Monodelphis domestica* (Mammalia: Didelphidae) in northeastern Brazil. *Journal of Zoology*, 232, 551–563.

Brocke R H (1970). *The Winter Ecology and Bioenergetics of the Opossum, Didelphis marsupialis, as Distributional Factors in Michigan*. Ph.D. dissertation, Michigan State University, East Lansing.

Burns R K and Burns L M (1957). Observations on the breeding of the American opossum in Florida. *Revue Suisse de Zoologie*, 64, 595–605.

Cerqueira R (1984). Reproduction de *Didelphis albiventris* dans le nord-est du Bresil Polyprotodontia, Didelphidae. *Mammalia*, 48, 95–104.

Cerqueira R, Gentile R, Fernandez F A S and D'Andrea P S (1993). A five-year population study of an assemblage of small mammals in southeastern Brazil. *Mammalia*, 57, 507–517.

Charles-Dominique P (1983). Ecology and social adaptations in didelphid marsupials: comparisons with eutherians of similar ecology. In *Advances in the Study of Mammalian Behaviour*. Eds J F Eisenberg and D G Klieman. American Society of Mammalogists, Philadelphia. pp 395–422.

Charles-Dominique P, Atramentowicz M, Charles-Dominique M, Gerald H, Hladik A, Hladik C M and Prevost M F (1981). Les mammiferes frugivores arboricoles nocturnes d'une foret guyanaise: Interrelations plantes-animaux. *Revue d'Ecologie la Terre et la Vie*, 35, 341–435.

Davis D E (1945). The annual cycle of plants, mosquitoes, birds and mammals in two Brazilian forests. *Ecological Monographs*, 15, 245–295.

Dawson T J and Dawson W R (1982). Metabolic scope and conductance in response to cold of some dasyurid marsupials and Australian rodents. *Comparative Biochemistry and Physiology*, 71(A), 59–64.

Dawson T J and Hulbert A J (1970). Standard metabolism, body temperature, and surface areas of Australian marsupials. *American Journal of Physiology*, 218, 1233–1238.

Dawson T J and Olson J M (1988). Thermogenic capabilities of the opossum *Monodelphis domestica* when warm and cold acclimated: similarities between American and Australian marsupials. *Comparative Biochemistry and Physiology*, 89A, 85–91.

Dawson T J and Wolfers J M (1978). Metabolism, thermoregulation and torpor in shrew sized marsupials of the genus *Planigale*. *Comparative Biochemistry and Physiology*, 59A, 305–309.

Eisenberg J F (1988). Reproduction in polyprotodont marsupials and similar-sized eutherians with a speculation concerning the evolution of litter size in mammals. In *Evolution of Life Histories of Mammals*. Ed M S Boyce. Yale University Press, New Haven. pp 291–311.

Eisenberg J F (1989). *Mammals of the Neotropics: The northern neotropics*. The University of Chicago Press, Chicago. 449 pp.

Eisenberg J F and Maliniak E (1967). Breeding

the murine opossum *Marmosa* in captivity. *International Zoo Yearbook*, 7, 78–79.

Eisenberg J F and Wilson D E (1981). Relative brain size and demographic strategies in didelphid marsupials. *American Naturalist*, 118, 1–15.

Emmons L H (1984). Geographic variation in densities and diversities of non-flying mammals in Amazonia. *Biotropica*, 16, 210–222.

Emmons L H (1990). *Neotropical Rainforest Mammals, a Field Guide*. The University of Chicago Press, Chicago. 281pp.

Enders R K (1966). Attachment, nursing, and survival of young in some Didelphids. In *Comparative Biology of Reproduction in Mammals*. Ed J W Rowlands. Academic Press, New York. pp 195–203.

Fadem B H (1985). Evidence for the activation of female reproduction by males in a marsupial, the gray short-tailed opossum (*Monodelphis domestica)*. *Biology of Reproduction*, 33, 112–116.

Fadem B H (1987). Activation of estrus by pheromones in a marsupial: stimulus control and endocrine factors. *Biology of Reproduction*, 36, 328–332.

Fadem B H and Rayve R S (1985). Characteristics of the oestrus cycle and influence of social factors in grey short-tailed opossums, *Monodelphis domestica*. *Journal of Reproduction and Fertility*, 73, 337–342.

Fitch H S and Sandidge L L (1953). Ecology of the opossum on a natural area in Northeastern Kansas. *University of Kansas Museum of Natural History*, 7, 305–338.

Fleck D F and Harder J D (1995). Ecology of marsupials in two Amazonian rain forests in northeastern Peru. *Journal of Mammalogy*, 76, 809–818.

Fleming M W and Harder J D (1983). Luteal and follicular populations in the ovary of the opossum *Didelphis virginiana* after ovulation. *Journal of Reproduction and Fertility*, 67, 29–34.

Fleming M W, Harder, J D and Wukie J J (1981). Reproductive energetics of the Virginia opossum compared with some eutherians. *Comparative Biochemistry and Physiology*, 70B, 645–648.

Fleming T H (1973). The reproductive cycles of three species of opossums and other mammals in the Panama canal zone. *Journal of Mammalogy*, 54, 439–453.

Gardner, A L (1973). The systematics of the genus *Didelphis marsupialia*: Didelphidae in North and Middle America. *Special Publication of the Museum, Texas Technical University*, 4, 1–81.

Gardner A L (1982). Virginia opossum. In *Wild Mammals of North America: Biology, Economics, and Management*. Eds J A Chapman and G A Feldhamer. The John Hopkins University Press, Baltimore. pp 3–36.

Gardner A L (1993). Order Didelphimorphia. In *Mammal Species of the World: a Taxonomic and Geographic Reference*. Eds D E Wilson and D M Reeder. Smithsonian Institution Press, Washington D C. pp 15–23.

Gillette L N (1980). Movement patterns of radio-tagged opossums in Wisconsin. *The American Midland Naturalist*, 104, 1–12.

Godfrey G K (1975). A study of oestrus and fecundity in a laboratory colony of mouse opossums *Marmosa robinsoni*. *Journal of Zoology*, 175, 541–555.

Hamilton W J (1958). Life history and economic relations of the opossum, *Didelphis marsupialis virginiana* in New York State. *Cornell University Agricultural Experiment Station Memoir*, 354, 1–48.

Harder J D (1992). Reproductive biology of South American marsupials. In *Reproduction in South American Vertebrates*. Ed W C Hamlett. Springer-Verlag, New York. pp 211–228.

Harder J D and Fleming M W (1981). Estradiol and progesterone profiles indicate a lack of endocrine recognition of pregnancy in the opossum. *Science*, 212, 1400–1402.

Harder J D, Stonerook M J and Pondy J (1993). Gestation and placentation in two new world opossums: *Didelphis virginiana* and *Monodelphis domestica*. *Journal of Experimental Zoology*, 266, 463–479.

Harder J D, Hsu M J and Garton D W (1996). Metabolic rates and body temperature of the gray short-tailed opossum (*Monodelphis domestica*) during gestation and lactation. *Physiological Zoology*, 69, 317–339.

Hartman C G (1916). Development of the opossum. *Journal of Morphology*, 27, 1–142.

Hartman C G (1921). Breeding habits, development, and birth of the opossum. *Annual Report of the Smithsonian Institution*, 1921, 347–364.

Hartman C G (1923). The oestrous cycle of the opossum. *American Journal of Anatomy*, 32, 353–421.

Hartman C G (1925). Hysterectomy and the oestrous cycle in the opossum. *American Journal of Anatomy*, 35, 25–30.

Hartman C G (1928). The breeding season of the opossum (*Didelphis virginia)* and the rate of intrauterine and postnatal development. *Journal of Morphology*, 46, 143–215.

Hill J P (1918). Some observations on the early development of *Didelphis aurita. Quarterly Journal of Microscopic Science*, 63, 91–139.

Hinds D A and MacMillen R E (1984). Energy scaling in marsupials and eutherians. *Science*, 225, 335–337.

Hinds L A, Reader M, Wernberg-Moller S and Saunders N R (1992). Hormonal evidence for induced ovulation in *Monodelphis domestica. Journal of Reproduction and Fertility,* 95, 303–302.

Hinds L A and Smith M J (1992). Evidence from plasma progesterone concentrations for male-induced ovulation in the brush-tailed bettong, *Bettongia pencicillata. Journal of Reproduction and Fertility*, 95, 291–302.

Hossler R J, McAninch J B and Harder J D (1994). Maternal denning behavior and survival of juveniles in opossums in southeastern New York. *Journal of Mammalogy*, 75, 60–70.

Hsu M, Harder J D and Lustick S I (1988). Seasonal energetics of opossums (*Didelphis virginiana)* in Ohio. *Comparative Biochemistry and Physiology*, 90A, 441–443.

Hunsaker D (1977). Ecology of new world marsupials. In *The Biology of Marsupials*. Ed D Hunsaker. Academic Press, New York. pp 95–156.

Hunsaker D and Shupe D (1977). Behavior of new world marsupials. In *The Biology of Marsupials*. Ed D Hunsaker. Academic Press, New York. pp 279–347.

Janson C H and Emmons L H (1990). Ecological structure of the nonflying mammal community at Cocha Cashu Biological Station, Manu National Park, Peru. In *Four Neotropical Rainforests*. Ed A H Gentry. Yale University Press, New Haven. pp 314–338.

Julien-Laferriere D and Atramentowicz M (1990). Feeding and reproduction of three didelphid marsupials in two neotropical forests (French Guiana). *Biotropica*, 22, 404–415.

Jurgelski W Jr and Porter M E (1974). The opossum (*Didelphis virginiana)* as a biomedical model, III. Breeding the opossum in captivity: methods. *Laboratory Animal Science*, 24, 412–425.

Kirsch J A W and Calaby J H (1977). The species of living marsupials: an annotated list. In *The Biology of Marsupials*. Eds B Stonehouse and D Gilmore. Macmillan, London. pp 9–26.

Kirsch J A W and Waller P F (1979). Notes on the trapping and behavior of the Caenolestidae. *Journal of Mammalogy*, 60, 390–395.

Kleiber M (1961). *The Fire of Life: an Introduction to Animal Energetics*. John Wiley & Son Inc, New York. 454 pp.

Kraus D B and Fadem,B H (1987). Reproduction, development and physiology of the gray short-tailed opossum (*Monodelphis domestica*). *Laboratory Animal Science*, 37, 478–482.

Lee A K and Cockburn A (1985). *Evolutionary Ecology of Marsupials*. Cambridge University Press, Cambridge. 274 pp.

Lorini, M L, DeOliveira J A and Persson V G (1994). Annual age structure and reproductive patterns in *Marmosa incana* (Lund, 1841) (Didelphidae, Marsupialia). *Zeitschrift fur Saugetierkunde*, 59, 65–73.

Lustick S and Lustick D D (1972). Energetics in the opossum, *Didelphis marsupialis virginiana. Comparative Biochemistry and Physiology*, 43A, 643–647.

MacMillen R E and Nelson J E (1969). Bioenergetics and body size in dasyurid marsupials. *American Journal of Physiology*, 217, 1246–1251.

Malcolm J R (1991). Comparative abundances of neotropical small mammals by trap height. *Journal of Mammalogy*, 72, 188–192.

Marshall L G (1978). *Chironectes minimus. Mammalian Species*, 109, 1–6.

Marshall L G (1982). Evolution of South

American Marsupialia. In *Mammalian Biology in South America*. Eds M A Mares and H H Genoways. Pymatuning Laboratory of Ecology, University of Pittsburgh, Pittsburgh. pp 251–272.

Marshall L G, Case J A and Woodburne M O (1990). Phylogenetic relationships of the families of marsupials. In *Current Mammalogy*. Ed H H Genoways. Plenum Press, New York. pp 433–481.

Mascarenhas A E A S (1980). *Histologia e inervacao do sistema genital feminino na fase de anestro no gamba Didelphis albiventris (Lund, 1841) (Didelphidae Marsupialia)*. Tese de Mestrado em Morfologia, ICB/UFMG, Belo Horizonte, Brasil.

McClearn D, Kohler J and McGowan K J (1994). Arboreal and terrestrial mammal trapping on Gigante Peninsula, Barro Colorado Nature Monument, Panama. *Biotropica*, 26, 208–213.

McNab B K (1978). The comparative energetics of neotropical marsupials. *Journal of Comparative Physiology*, 125, 115–128.

McNab B K (1980). Food habits, energetics, and the population biology of mammals. *The American Naturalist*, 116, 106–124.

McNab B K (1986). Food habits, energetics, and the reproduction of marsupials. *Journal of Zoology*, 208, 595–614.

McNab B K (1989). On the selective persistence of mammals in South America. In *Advances in Neotropical Mammology*. Eds K H Redford and J F Eisenberg. The Sandhill Crane Press, Gainsville. pp 605–614.

Moore H D M (1992). Reproduction in the gray short-tailed opossum (*Monodelphis domestica*). In *Reproduction of South American Vertebrates*. Ed W H Hamlett. Springer-Verlag, New York. pp 229–241.

Motta M, Dezonna J C and Franco A M R (1983). A note on the reproduction of *Didelphis marsupialis* in captivity. *Memorias do Instituto Oswaldo Cruz: Rio de Janeiro*, 78, 507–509.

Nicoll M E and Thompson S D (1987). Basal metabolic rates and energetics of reproduction in therian mammals: marsupials and placentals compared. *Symposium Zoological Society of London*, 57, 7–27.

Nogueira J C (1988). Anatomical aspects and biometry of the male genital system of the white-belly opossum *Didelphis albiventris* (Lund, 1841) during the annual reproductive cycle. *Mammalia*, 52, 233–242.

O'Connell M (1979). Ecology of didelphid marsupials from northern Venezuela. In *Vertebrate Ecology in the Northern Neotropics*. Ed J F Eisenberg. Smithsonian Institution Press, Washington D C. pp 73–87.

O'Connell M A (1989). Population dynamics of neotropical small mammals in seasonal habitats. *Journal of Mammalogy*, 70, 532–548.

Osgood W H (1921). A monographic study of the American marsupial, *Caenolestes. Field Museum of Natural History Zoological Series*, 14, 1–162.

Pacheco V, Patterson B D, Patton J L, Emmons L H, Solari S and Ascorra C F (1993). List of mammal species known to occur in Manu Biosphere Reserve, Peru. *Universidad Mayor de San Marcos: Publicaciones del Museo de Historia Nacional, Universidad Nacional Serie A Zoologia*, 44, 1–12.

Patton J L, Berlin B and Berlin E A (1982). Aboriginal perspectives of a mammal community in Amazonian Peru: Knowledge and utilization patterns among the Aguruna Jivaro. In *Mammalian Biology in South America*. Eds M A Mares and H H Genoways. Pymatuning Symposia in Ecology, Pittsburg. Vol. 6, pp 111–128.

Parker E, Posey D, Frenchione J and Francelino da Silva L (1983). Resource exploitation in Amazonia: ethnoecological examples from four populations. *Annals of the Carnegie Museum of Natural History*, 52, 163–203.

Perret M and Atramentowicz M (1989). Plasma concentrations of progesterone and testosterone in captive woolly opossums (*Caluromys philander*). *Journal of Reproduction and Fertility*, 85, 31–41.

Perret M and M'Barek S B (1991). Male influence on oestrous cycles in female woolly opossum (*Caluromys philander*). *Journal of Reproductive Fertility*, 91, 557–566.

Petrides G A (1949). Sex and age determination in the opossum. *Journal of Mammalogy*, 30, 364–378.

Pine R H, Dalby P L and Matson J O (1985). Ecology, postnatal development, morphometrics, and taxonomic status of the short-tailed opossum, *Monodelphis dimidiata*, an apparently semelparous annual marsupial. *Annals of Carnegie Museum*, 54, 195–231.

Posey D A (1986). Topics and issues in ethnoentomology with some suggestions for the development of hypothesis-generation and testing in ethnobiology. *Journal of Ethnobiology*, 6, 99–120.

Rabinowitz A and Nottingham B G Jr (1989). Mammal species richness and relative abundance of small mammals in a subtropical wet forest of Central America. *Mammalia*, 53, 217–226.

Reynolds H C (1952). Studies on reproduction in the opossum (*Didelphis virginiana*). *University of California Publications in Zoology*, 52, 223–284.

Rigueira S E, Valle C M C, Varejao J B M, Albuquerque P V and Nogueira J C (1987). Algumas observacoes sobre o ciclo reprodutivo anual de femeas do gamba *Didelphis albiventris* (Lund, 1841) Didelphidae, Marsupialia. Em populacoes naturais no Estado de Minas Gerais, Brasil. *Revista Brasileira de Zoologia*, 4, 129–137.

Robinson J G and Redford K H (1986). Body size, diet, and population density of neotropical forest mammals. *The American Naturalist*, 128, 665–680.

Russell E M (1982). Patterns of parental care and parental investment in marsupials. *Biological Review*, 57, 423–486.

Sanderson G C (1961). Estimating opossum populations by marking young. *Journal of Wildlife Management*, 25, 20–27.

Smith M J (1992). Evidence from the oestrous cycle for male-induced ovulation in *Bettongia penicillata*. *Journal of Reproduction and Fertility*, 95, 283–289.

Stonerook M J and Harder J D (1992). Sexual maturation in female gray short-tailed opossums, *Monodelphis domestica*, is dependent upon male stimuli. *Biology of Reproduction*, 46, 290–294.

Streilein K E (1982). Behavior, ecology, and distribution of South American marsupials. In *Mammalian Biology in South America*. Eds M A Mares and H H Genoways. Pymatuning Symposia in Ecology, Pittsburg. Vol. 6, pp 231–250.

Sunquist M E, Austad S N and Sunquist F (1987) Movement patterns and home range patterns in the common opossum (*Didelphis marsupialis*). *Journal of Mammalogy*, 68, 173–176.

Thompson S D (1992). Gestation and lactation in small mammals: basal metabolic rate and the limits of energy use. In *Mammalian Energetics: interdisciplinary views of metabolism and reproduction*. Eds T E Tomasi and T H Horton. Comstock Publishing Associates, Cornell University Press, Ithaca. pp 213–259.

Thompson S D and Nicoll M E (1986). Basal metabolic rate and energetics of reproduction in eutherian mamals. *Nature*, 321, 690–693.

Tyndale-Biscoe C H (1973). *Life of Marsupials*. American Elsevier Publishing Co., New York. 254pp.

Tyndale-Biscoe C H (1980). Observations on the biology of marsupials in Colombia and Venezuela. *National Geographic Society Research Reports*, 12, 711–720.

Tyndale-Biscoe C H and MacKenzie R B (1976). Reproduction in *Didelphis marsupialis* and *Didelphis albiventris* in Colombia. *Journal of Mammalogy*, 57, 249–265.

Tyndale-Biscoe C H and Renfree M (1987). *Reproductive physiology of marsupials*. Cambridge University Press, Cambridge. 476pp.

VandeBerg J L (1983). The gray short-tailed opossum: A new laboratory animal. *Institute of Laboratory Animal Research News*, 26, 9–12.

Vaughan T A (1986). *Mammalogy*. Saunders College Publishing, New York. 576 pp.

Walker J M, Garber A and Berger R J (1979). Sleep and estivation (shallow torpor): continuous processes of energy conservation. Science, 204, 1098–1100.

Wiseman G L and Hendrickson G O (1950). Notes on the life history and ecology of the opossum in southeast Iowa. *Journal of Mammalogy*, 31, 331–337.

Woodburne M O and Zinmeister W J (1982). Fossil land mammals from Antarctica. *Science*, 218, 284–286.

Wright, D D (1989). *Mortality and Dispersal of Juvenile Opossum, Didelphis virginiana*. MS thesis, University of Florida, Gainsville.

CHAPTER 12

DIGESTIVE AND METABOLIC ADAPTATIONS OF ARBOREAL MARSUPIALS FOR DEALING WITH PLANT ANTINUTRIENTS AND TOXINS

Steven J Cork and William J Foley

INTRODUCTION

Hugh Tyndale-Biscoe has long advocated marsupials as a group that can give us insights into physiological and ecological strategies of mammals generally. While Hugh and his colleagues have demonstrated this with respect to reproductive biology, it is just as true for other fields including nutritional physiology and ecology. In this chapter, we discuss recent research on the physiological and behavioural strategies that arboreal, leaf-eating marsupials (possums, gliders and koalas) employ to deal with the nutritional and toxicological challenges posed for them by a diet of *Eucalyptus* foliage, and what studying these animals has told us about the underlying physical and physiological constraints on herbivory in mammals and vertebrates generally. Hugh and his students were early contributors to research on both the habitat requirements and physiological adaptations of arboreal, leaf-eating marsupials (Smith 1969; Tyndale-Biscoe & Smith 1969a, b). But for us and others who have worked in this field subsequently Hugh's influence has also been one of encouraging new ideas and contributing a good dose of his own lateral thinking. It is a pleasure for us to contribute to this volume in Hugh's honour.

Mammals eating tree foliage face chemical challenges posed by the generally low protein and high fibre concentrations in this diet and the array of chemical defences employed by woody plants against herbivores (Cork & Foley 1991). The ability of arboreal species to deal with these challenges is limited, in theory, by physiological constraints associated with small body size (Foley & Cork 1992; Cork 1994) and by the capacity of mammalian metabolic pathways to cope with toxin loads (Foley 1992; Foley *et al.* 1995). The studies on arboreal marsupials reviewed here have given insights into both the theoretical limits to herbivory in small mammals and the adaptations that permit some species to offset those theoretical limits.

Arboreal marsupials are ideal experimental models for studying the problems of dealing with dietary fibre, as they are constrained to small body size by the requirements of life in forest canopies. This makes them too small, in theory, to utilise high fibre diets and yet they do remarkably well on diets that rank with some of the least digestible utilised by herbivores. Research since 1980 has revealed extreme digestive and metabolic adaptations in several of the most folivorous arboreal marsupials that, subsequently, have been shown to be characteristic of small mammals in general that 'break the rules' by utilising more fibre in their diet than predicted from theoretical considerations.

Arboreal marsupials have also proved to be useful models for understanding the role of plant secondary metabolites (PSMs) in plant-herbivore interactions. PSMs are compounds that have no apparent role in the primary metabolic processes of plants and many are suggested to defend plants against herbivores and pathogens (Harbourne 1988). Australian forests contain some of the most specialised of all mammalian herbivores living sympatrically with generalists. Both groups may face very different loads of PSMs and, yet, have many similarities in physiological strategies for dealing with the challenges posed by these loads (Cork & Foley 1991; Foley *et al.* 1995). It has been known for some time that PSMs have the potential to act as deterrents to feeding (Bryant *et al.* 1992) or disrupt nutrient absorption and utilisation (Foley & McArthur 1994). They also may have substantial impacts on the structure of plant communities and on the population ecology of herbivores over landscape scales and over long periods (Jeffries *et al.* 1994).

However, emergence of theory about the effects of PSMs that might allow us to predict their effects in other ecosystems has been slow because most of the interactions so far documented involve a single PSM and a single species of herbivore. A comparative approach involving several species that utilise a common diet differently is providing a series of unifying themes applicable to other plant-herbivore interactions (Cork & Foley 1991; Foley *et al.* 1995).

Food for folivores in *Eucalyptus* forests

The main food sources available to arboreal mammals in forests are foliage, fruits, flowers, seeds, sap, and invertebrates. Among the mammalian arboreal fauna in the temperate eucalypt forests of Australia, there is a greater proportion of leaf-eaters than in many other forests (for example, wet tropical forests), probably because there are fewer edible seeds and fruits than in other forests (Cork 1995; Smith & Ganzhorn 1995), and because the degree of leaf-eating practised by the most folivorous species is higher than among any other group of arboreal vertebrates (Cork & Foley 1991; Cork 1995).

For mammals eating the foliage of woody plants, nutritional challenges are presented by low concentrations of nutrients, especially protein, high concentrations of components that interfere with digestion, especially lignin and tannins, and a range of potentially toxic secondary metabolites (Cork & Foley 1991). Plant constituents resistant to digestion by mammalian enzymes are termed collectively 'dietary fibre' by nutritionists. Dietary fibre often is equated with plant cell walls, which consist mainly of lignin and various structural polysaccharides (cellulose and others). Lignins inhibit the digestion of cell-wall polysaccharides by the symbiotic microorganisms that herbivorous mammals rely on for energy extraction from dietary fibre (van Soest 1982). The foliage of trees and arborescent woody plants often has a lower content of cell walls than do herbaceous plants but the ratio of lignin to other cell-wall constituents is usually much higher in tree foliage (Cork 1984; see also Cork & Foley 1991). The high lignin concentration offsets the potential benefit from the lower overall cell-wall content to the extent that the range of dry-matter digestibilities (in ruminants) for browsers is similar to or lower than that for grasses (Robbins *et al.* 1987). Therefore, browsing is not necessarily an easier option nutritionally than grazing, especially where the browse is mature tree foliage.

Foliage of Australian eucalypt forests is known for being sclerophyllous and having low concentrations of nutrients, especially nitrogen and phosphorus, compared with other forests in the world (Attiwill & Leeper 1987; Specht & Rundel1990; Cork, 1996).

The concentration of cell walls is not as high as in grasses, but is high compared with foliage of shrubs and herbs and most fruits and flowers (Cork 1984; Cork & Foley 1991). The degree of lignification of the cell walls appears to be high compared with grasses, and the concentrations both of cell-wall carbohydrates and of lignin, do not vary greatly between species of eucalypt within the same age class (Cork 1984; Cork & Foley 1991) (Figure 12.1). The fruits of eucalypts are woody and there are few other plants in temperate eucalypt forests that produce large fleshy fruits (Smith & Ganzhorn 1995).

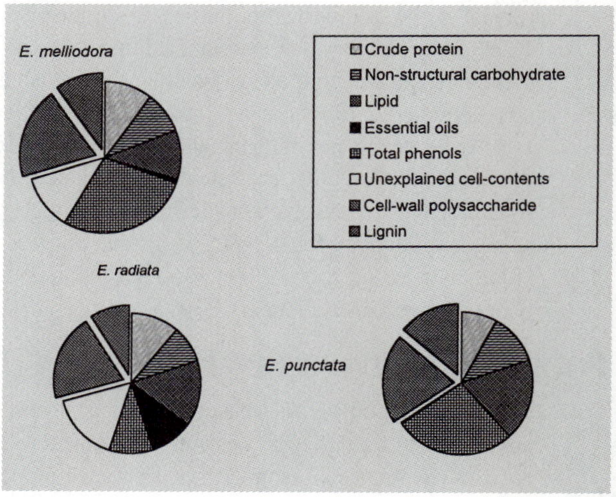

Figure 12.1
Data on 3 species of eucalypt for which comprehensive analyses of leaf chemical composition are available.
Note: Unexplained cell contents for *E. punctata* (<1% of leaf dry matter) includes essential oils. Cell-wall constituents (polysaccharides and lignin) are exploded from the pie graph.
Sources: Cork *et al.* (1983); Foley & Hume (1987a); Foley (1992).

Eucalypts and many of the understorey plants associated with them produce mainly carbon-based chemical defences, especially phenolics and terpenoids (essential oils) (Fox & Macauley 1977; Cork 1984; Foley *et al.* 1987). This is consistent with hypotheses that predict high investment in carbon-based, immobile defences in plants with long leaf-lifetimes growing on nutrient-poor soils (Coley *et al.* 1985; Cork 1992; Landsberg & Cork 1995). Phenolic secondary metabolites may comprise up to 40% of foliar dry matter (Fox & Macauley 1977; Cork 1984) (Figure 12.1) and include condensed and hydrolysable tannins, phenolic glycosides, phenolic acids and saponins (Dayal 1988). The concentrations of phenolics in eucalypt forests can be much higher than in other forests in the world and rarely are low (Cork 1996). The potential effects of tannins on mammals include decreased food intake, decreased digestibility of proteins and cell-wall carbohydrates, and metabolic or pathological effects (Mole & Waterman 1987; Robbins *et al.* 1987; Cork & Foley 1991; Hanley *et al.* 1992).

Woody plants also produce a diversity of other secondary metabolites, including alkaloids, terpenoids and cyanogenic compounds, that are absorbed from the gut and can have toxic effects once in the bloodstream (Cork & Foley 1991). Concentrations of terpenoids (including essential oils that give eucalypts their characteristic scent) range from 0 to 24% of dry matter (Morrow & Fox 1980) (Figure 12.1) and there may be upwards of 65 compounds in the oil mixtures (Boland *et al.* 1991). A number of species in the subgenus Symphomyrtus (for example, *E. ovata*, *E. yarraensis*, *E. orgadophila*, *E. polyanthemos*) contain the cyanogenic glycoside prunasin but its occurrence is polymorphic within these species and cyanogenic and acyanogenic individuals can be found growing adjacent to one another (E E Conn, unpublished observations; L I Pahl, personal communication; D M Pass & W J Foley, unpublished observations). Although no alkaloids have been found in *Eucalyptus*, several non-eucalypts that are eaten by arboreal marsupials do contain a variety of alkaloids (for example, the mangrove *Exoecaria agallocha*, *Erythophleum chlorostachys* and various species of *Solanum*) (Kerle 1985; Van Dyck 1979; Collins 1990; S A Dalla Pozza & W J Foley, unpublished observations).

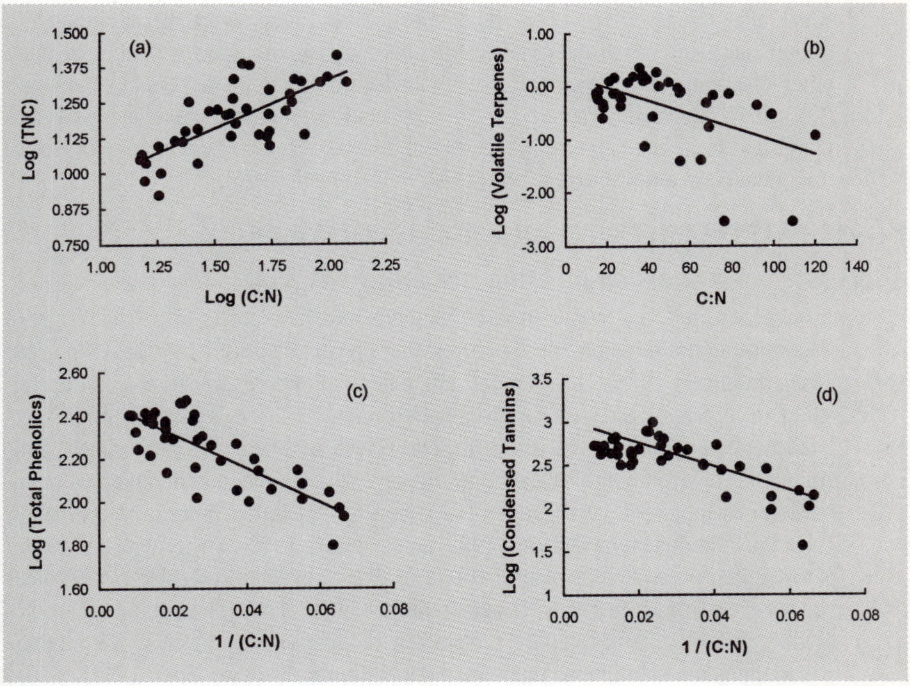

Figure 12.2 Relationship between the carbon to nitrogen ratio (C:N) of foliage of *Eucalyptus tereticornis* and the concentration of: (a) total non-structural carbohydrates (TNC); (b) volatile terpenes; (c) total phenolics; and (d) condensed tannins

Notes:

(1) All values are mg/g dry matter.

(2)The C:N ratio was varied by growing seedlings under all combinations of low and high nutrients (1/24 and 1/4 strength Hoagland's solution), low and high light (30% and full sunlight) and low and high atmospheric carbon dioxide (350 mmol.mol^{-1} and 800 mmol.mol^{-1}).(3) Regression lines shown are statistically significant (p<0.001 for plots a, b and c, and p=0.005 for plot d).

Source: Lawler (1993).

The occurrence of many PSMs in *Eucalyptus* is under genetic control (Barton *et al.* 1991) but there is some evidence that the concentration of PSMs is plastic depending on the resources available to the plant. For example, seedlings of *E. tereticornis* and *E. citriodora* showed substantial increases in the concentration of total phenols and condensed tannins when grown under enhanced CO_2 or in full light (Lawler 1993). This suggests that phenolic PSMs are repositories of excess carbon (Figure 12.2). Similarly, Cork (1992) found that there were significant inverse relationships over a landscape scale between foliage nutrients and concentrations of phenolic metabolites. In contrast there was no effect of elevated CO_2, soil nutrients or light on the concentration of volatile terpenes in either *E. tereticornis* or *E. citriodora*. This difference probably relates to the different biosynthetic pathways of these groups of compounds (Lawler 1993). Presumably lack of biosynthetic pathways and a low availability of nitrogen limits the use of nitrogen-based PSM-like alkaloids in eucalypts. Various metabolic pathways are employed by mammals to detoxify absorbed secondary metabolites, but detoxification ultimately increases the requirements of mammals for energy and nutrients, especially nitrogen (Foley 1992).

Eucalypt forests contrast with some other temperate forests and many tropical forests which offer a wider variety of nutritional possibilities for herbivores because there is a greater range in growth rates and leaf-lifetimes among tree species present in these other forests and more species produce fleshy fruits adapted for dispersal by mammals (Cork 1995; Smith & Ganzhorn 1995). These inter-forest differences in chemical challenges may explain differences in digestive and metabolic strategies employed by arboreal marsupials and primates (Cork & Foley 1991; Cork 1996).

POSTULATED LIMITS TO ARBOREAL FOLIVORY

Allometric constraints on utilising tree foliage as food

If mammals are to be mobile in forest canopies and have access to the full range of foliage, fruits and seeds produced by trees, they apparently must weigh less than about 15 kg (Eisenberg 1978). However, this size limitation imposes constraints on the ability of mammals to meet their nutritional requirements from a diet of tree foliage. Various researchers have discussed the role played by allometric (size-related) constraints in the evolution of dietary and digestive strategies among herbivorous mammals (for example, Bell 1969; Jarman 1974; Janis 1976; Clutton-Brock & Harvey 1977; Parra 1978; Demment & van Soest 1985; Foley & Cork 1992). The general conclusion is that utilisation of fibrous plant tissues is facilitated by large body size and becomes progressively less feasible as body size declines. Mammals in the mass range of arboreal folivores (less than 15 to 20 kg) are predicted to face severe problems meeting energy requirements from fibrous diets (Parra 1978; Demment & van Soest 1985).

This situation occurs because daily energy requirements among mammals scale with body mass to an exponent between 0.5 and 0.9 (Nagy 1987), yet both of the alternative digestive strategies for meeting energy requirements from fibrous diets ultimately are limited by gut capacity, which scales almost directly with body mass (Parra 1978; Demment & van Soest 1985). Energy requirements can be met from fibrous diets by either maximising retention of the diet or maximising the amount eaten (Janis 1976; Cork 1994). Maximising retention optimises the opportunity for microbial degradation of fibre and this strategy is exemplified in ruminants such as sheep and cattle. The rate of yield of energy from this strategy is the product of average daily fermentation rate, which is determined by the composition of the diet and is independent of body size, and capacity of the digestive tract (Parra 1978; Demment & van Soest 1985). Modelling of these relationships shows that an average mammal needs to have a mass of tens of kilograms to meet energy requirements from even a highly digestible browse diet using the retention-maximising strategy (Demment & van Soest 1985; Cork & Foley 1991; Cork 1994).

The yield from the intake-maximising strategy is limited by how rapidly food can pass through the gut. This strategy also appears unviable for most small mammals because of limited gut capacity relative to intake requirements, likely allometric limits on rates of passage of food through the gut, and prohibitive losses of endogenous secretions and microorganisms in faeces when high rates of throughput are maintained (Björnhag 1989; Cork 1994).

In contrast to energy, proteins are more difficult nutrients to acquire for large than small arboreal mammals. Although small mammals (<1 kg) can meet all or a large part of their protein requirements by foraging for invertebrates, this is not profitable for larger species because of the low relative biomass of invertebrates in tree canopies (Hladik 1978). Species larger than about 1 kg are forced to be at least moderately folivorous because fruit usually is a poor source of protein. As the concentration of protein in most tree foliage is close to the predicted minimum required in a fibrous diet by mammalian herbivores, folivorous species must, in theory, be highly selective for foliage with highest protein concentrations (Hladik 1978; Milton 1979).

Constraints imposed by plant chemical defences

As discussed later in this chapter, the consequences of low energy and nutrient availability in tree foliage are exacerbated by secondary metabolites because they increase requirements for various nutrients used in biotransformational pathways and for nitrogen in particular due to the acidemic effects of biotransformed secondary metabolites (Foley 1992).

BREAKING THE RULES: DIGESTIVE AND METABOLIC ADAPTATIONS OF MARSUPIAL AND OTHER MAMMALIAN ARBOREAL FOLIVORES

Feeding styles

Eisenberg (1978) estimated that only about 4% of all mammalian genera contain species that are to some extent both arboreal and herbivorous, reflecting the constraints against this combination of feeding style and small body size. Most of these species are among the Australian marsupials and the New and Old World primates.

The most folivorous arboreal mammals are among the marsupials and within this group three of the four arboreal species inhabiting *Eucalyptus* forests of south-eastern Australia are exceptional. The koala *Phascolarctos cinereus* (5 to 13 kg), appears to be the most folivorous of all arboreal mammals, eating *Eucalyptus* spp. foliage almost exclusively and including a large proportion of mature foliage when young foliage is scarce. The greater glider, *Petauroides volans* (0.8 to 1.7 kg), also is highly folivorous, but strongly selects young foliage. The common ringtail possum, *Pseudocheirus peregrinus* (0.7 to 1 kg), like *Lepilemur* among the primates, is much more folivorous than predicted from its small mass, eating mature and young eucalypt foliage and buds, fruits, and flowers of understorey species. A fourth species, the common brushtail possum, *Trichosurus vulpecula* (2 to 3 kg), provides a useful comparison with the other three because it is much less folivorous, its diet includes ground-storey plants, fruits, occasional insects and a variable amount of foliage. It differs, however, in terms of its digestive strategy (see below). Smaller marsupials inhabiting Australian eucalypt forests, like small primates, avoid foliage completely and feed largely on sap, nectar, and/or insects and invertebrates (see review by Cork & Foley 1991).

In tropical rainforests of northern Australia, ringtail possums (*Pseudocheirus* spp. and *Hemibelideus lemuroides*) practise a high degree of folivory, while tree kangaroos (*Dendrolagus* spp.) appear to be only moderately folivorous (see Cork & Foley 1991). The marsupials inhabiting South American forests are all in the range 0.04 to 1.3 kg and are either insectivore-omnivores or frugivore-omnivores (Robinson & Redford 1986; Harder & Fleck, Chapter 11 in this volume).

The forest-living primates of Africa and Asia provide informative comparisons with koalas, greater gliders and ringtail possums (Cork & Foley 1991; Cork 1996). Whereas the vast majority of arboreal primates are faunivores, frugivores and/or exudate feeders (Clutton-Brock & Harvey 1977; Chivers & Hladik 1980), evolution towards folivory has occurred among colobine monkeys (sub-family Colobinae) in Africa/Asia, two prosimian families (Lemuridae and Indriidae) in Madagascar (Hladik 1978; Ganzhorn 1988), and howler monkeys (*Alouatta* spp.) in Central and South America (Eisenberg 1978; Hladik 1978). However, only a few of the prosimians approach the degrees of folivory practised by koalas, greater gliders and ringtail possums and these also are the only primates with digestive adaptations comparable to those in the arboreal folivorous marsupials (Cork & Foley 1991; Cork 1996; see also below).

Digestive strategies

Among arboreal mammals, length, capacity and surface area of the digestive tract increase with degree of folivory, and both gut capacity and degree of specialisation are greater in folivorous species than in other herbivorous mammals of comparable body mass (Chivers & Hladik 1980; Hume 1982). Two broad categories of digestive strategy can be distinguished: species with a simple stomach and an expanded colon or caecum-colon (hindgut fermenters) and species with both a complex forestomach and a moderately expanded hindgut (foregut fermenters). Foregut-fermenting arboreal folivores are found among the colobine primates (leaf monkeys), the macropodid marsupials (tree kangaroos) and the edentates (three-toed sloths) (Langer 1988; Foley *et al.* 1995). The significance of foregut fermentation for arboreal herbivores has been discussed elsewhere in detail (Cork & Foley 1991; Cork 1996). It has been concluded that this category of digestive strategy is inefficient for dealing with highly fibrous diets like tree foliage and probably evolved under natural selection for utilising mixed fruit and leaf diets, such as are available in many tropical forests.

From the research on the arboreal, folivorous marsupials, it has been concluded that hindgut fermentation, with certain key characteristics, is the optimal digestive strategy for utilising fibrous diets in small mammals (Cork & Foley 1991; Cork 1994; Cork 1996). Among hindgut-fermenting marsupials and primates, the caecum and/or proximal colon are expanded in proportion to the degree of folivory practised (Chivers & Hladik 1980; Hume 1982). Because the caecum is offset from the main flow of digesta through the colon, digesta or digesta components can either be routed into the caecum for fermentation or passed rapidly through the colon.

Colonic separation mechanisms

Among hindgut fermenting mammals, a distinction can be drawn between 'separators'and 'non-separators' In the hindgut of separators, particulate digesta are separated by a retrograde movement of solutes so that large particles are excreted relatively rapidly, while small particles are selectively retained (Björnhag 1972, 1987; Cork & Warner 1983; Chilcott & Hume 1985; Foley & Hume 1987b; Cork & Foley 1991). The most highly folivorous hindgut-fermenting arboreal mammals (koalas, greater gliders, ringtail possums, and *Lepilemur* spp.) are all separators (Cork & Warner 1983; Chilcott & Hume 1985; Foley & Hume 1987b) and all non-separators are either non-folivorous or are much less folivorous than the separators. Because the small particles include microorganisms, epithelial debris and small, poorly lignified fragments of dietary cell walls, and because large particles principally are fragments of highly lignified fibre, these 'colonic separation mechanisms' minimise faecal losses of microbial nitrogen and retain the most digestible part of the dietary fibre while minimising retention of indigestible fibre that could limit food intake. It is postulated that hindgut fermentation without substantial colonic digesta separation and preferential excretion of large particles would not allow a sufficient food intake and rate of passage of fibre through the gut to meet energy requirements and would result in unsustainable losses of nitrogen in faeces (Cork & Foley 1991; Cork 1994). Consistent with this hypothesis, brushtail possums, which lack a substantial mechanism for selective retention of fine digesta in the hindgut, practise a much lower degree of folivory than koalas, greater gliders and ringtail possums that do have such mechanisms (Cork & Foley 1991).

Caecotrophy

Colonic separation mechanisms frequently are associated with the practice of caecotrophy in which the contents of the caecum are excreted as special faeces and ingested to provide high quality protein to the acid stomach and small intestine, the main sites of protein digestion and absorption (Björnhag 1987). This greatly reduces the requirement of small herbivores for dietary nitrogen, in general, and essential amino acids, in particular. Among arboreal mammals, ringtail possums are caecotrophic (Chilcott & Hume 1985) and sportive lemurs appear to be also (Hladik *et al.* 1971). It probably is no coincidence that these are the smallest of the folivores and, therefore, the most prone to the effects of elevated nitrogen losses due to high throughput of fibrous foods.

Digestive strategy in relation to forest type and food resources

Foregut-fermenting arboreal herbivores are found only in tropical and subtropical forests, whereas hindgut-fermenters are found in both temperate and tropical forests. The most highly folivorous arboreal marsupials and primates, including koalas, greater gliders, ringtail possums and sportive lemurs, are hindgut fermenters. On the other hand, no foregut-fermenting arboreal marsupials or primates are strict folivores and all include a highly digestible component (fruits or seeds) in their diet.

Hindgut fermentation without colonic separation of digesta is found among arboreal primates and marsupials inhabiting both temperate and tropical forests and is typical of species that include little or no mature foliage in their diets (Cork & Foley 1991; Cork 1995). The highly folivorous hindgut-fermenters all have mechanisms for selective retention of small particulate digesta (Cork & Foley 1991).

Three-toed sloths are an exception to this generalisation, being highly folivorous foregut-fermenters; however, this probably is due to the low metabolic rate of sloths (see below), which overcomes many of the physiological constraints predicted for other mammals.

Metabolic rate

Many aspects of the behaviour and life histories of arboreal folivores have been interpreted as energy-saving features, including small litter sizes, long development times, small muscle masses and low levels of activity (Eisenberg 1978; McNab 1978). Several folivorous, frugivorous and insectivorous species among the arboreal primates, edentates and marsupials have rates of metabolism and daily energy requirements less than the average for other mammals of similar body mass (see review by Cork & Foley 1991). Although there still is debate on the relative importance of food source versus phylogeny in the evolution of metabolic rate (McNab 1986; Müller 1985; Milton *et al.* 1979) it appears that most, if not all, primarily folivorous arboreal mammals benefit to some degree from low energy requirements compared with other mammals. It has been postulated that very high degrees of folivory are not possible without low rates of basal metabolism and/or low expenditure of energy on temperature regulation and activity (Cork & Foley 1991). For example, in the only two arboreal mammals that maintain almost exclusive utilisation of mature tree foliage, three-toed sloths and koalas, total energy expenditures in the field are 21% and 36%, respectively, of those predicted for placental mammals of the same body mass (Nagy & Montgomery 1980; Nagy 1987).

EFFECTS OF PLANT SECONDARY METABOLITES (PSMS) ON DIET CHOICE AND NUTRITIONAL QUALITY OF *EUCALYPTUS* FOLIAGE FOR FOLIVOROUS MARSUPIALS

Nearly all plants, and certainly all tree leaves, contain substantial concentrations of PSMs, and eucalypts conform with this generalisation (see above). Many PSMs, especially those present at high concentrations in eucalypts, have the potential to act as deterrents to feeding or disrupt nutrient absorption and utilisation (Cork & Foley 1991; Bryant *et al.* 1992; Foley & McArthur 1994). The interactions between PSMs and herbivores have been best studied in the boreal ecosystems of northern Alaska (Bryant *et al.* 1991, 1992). For example, small concentrations of a diverse range of compounds can deter voles and hares from feeding on birch trees (Bryant *et al.* 1991) and may lead to them being in an unsustainable negative sodium balance (Reichardt *et al.* 1984). PSMs may also have substantial impacts on the structure of plant communities and on the population ecology of herbivores over landscape scales and over long periods (Jeffries *et al.* 1994).

The major limitation to further studies of PSMs in animal and plant communities is that many of the interactions so far documented involve a single PSM and a single species of herbivore. Consequently it has been difficult to develop a general understanding of the effects of PSMs that might allow us to predict their effects in other ecosystems.

PSMs can have an effect on diet choice and nutritional quality of foliage eaten by marsupials in a number of ways, but it depends on the temporal scale of the analysis. They may influence acceptance/rejection of species of trees and trees within a species, both under free-ranging and captive conditions (for example, Cork & Pahl 1984; Hindell & Lee 1987; Pahl & Hume 1990). Once animals start to feed on a particular tree species PSMs can modulate the amount of food eaten (Pahl & Hume 1990). Finally the long-term survival of animals in a particular habitat depends on critical nutrient thresholds. Animals need food of sufficient nutrient concentration to be able to reproduce and PSMs may reduce the long-term nutritional quality of particular habitats.

Correlative studies of diet choice and leaf composition

To date the principal method used to study the impact of PSMs on diet selection and digestion of marsupial herbivores has been to seek correlations between the chemical composition of the foliage and leaf choice both in free-living and captive situations. These have only been partially successful.

The effect of volatile terpenes on food preferences of eucalypt herbivores serves as an example. Early studies suggested that the preferences of koalas for particular trees was related to the cooling effects of the monoterpene a-phellandrene preferred by Queensland koalas and the warming effects of 1,8 cineole preferred by Victorian koalas (Pratt 1937). More recent studies have shown that this supposed preference does not exist, nor do the compounds have any thermogenic effects (Southwell 1978).

These and other studies have failed to show any clear patterns between the concentration of volatile terpenes and the preferences of koalas for particular eucalypt species (Southwell 1978; Zoidis & Markowitz 1992). Betts (1978) claimed that the ratio of sesquiterpenes to cineole was important in food choice of koalas but only 17% of the variation was explained by this factor. Hume & Esson (1993) found that koalas would only eat leaves that contained more than a certain threshold of volatile oils and that they preferred foliage with a high proportion of monoterpenes.

Pass (1992) showed that common ringtail possums strongly preferred food that contained little or no 1,8 cineole when they were offered a choice of food containing different levels of cineole, but when the same animals were given no choice, they consumed cineole-rich food with no diminution of their normal food intake. Similarly, Krockenberger (1988) could find no effect of added 1,8 cineole on the intake and nutritional quality of artificial diets fed to common ringtail and brushtail possums.

There are three main reasons why correlative approaches to understanding diet choice have been unsuccessful. Firstly, they rely on our assessment of what factors are important in food choice by animals. For example, we can smell some terpenes in *Eucalyptus* oils very clearly and we know that, if ingested, compounds such as cineole are toxic to humans (Martindale 1989). Koalas and other folivorous marsupials carefully smell foliage before ingesting it (Pahl & Hume 1990) and so it may seem reasonable to us to measure the concentration of volatile terpenes in foliage. However, if an animal's perception of what is important differs from ours and they are more tolerant of 'toxic' compounds then we may not have measured the correct factor in our analyses.

Similarly we have incomplete knowledge of the composition of complex foods. Many PSMs that have significant biological activity occur in very low concentrations (<0.1%) in foliage (Harborne 1991; Bryant *et al.* 1992). Given the poor knowledge of the phytochemistry of the Australian flora (Collins 1990), it is quite possible that significant compounds have been omitted from even the most comprehensive analyses.

A second problem concerns our ability to make reliable measurements of plant constituents. For example, although we have been aware of the effects of plant phenolics on the measurement of plant fibre (Cork & Pahl 1984; McArthur 1988), we have only recently devised effective, alternative analytical methods (S J Cork, unpublished observations). In the same vein, plant phenolics themselves have probably been incompletely extracted from eucalypts in previous studies and analysis procedures have used inappropriate standards (Cork & Krockenberger 1991).

Finally, the importance of PSMs and nutrients varies depending on the scale of food choice being examined, and in the past the resolution of the analyses have not always matched the resolution of the feeding data. For example, it would be surprising indeed if small differences in soluble sugars or protein explained an animal's complete refusal to feed on leaves from particular trees (Geritz 1987). In contrast, measuring components such as these might explain differences in the amounts animals eat of different acceptable leaves. Clearly, it is important to specify the scale of food selection being examined.

Correlative approaches alone can never provide definitive reasons for the acceptance/rejection of particular foliages. An alternative approach involves chemical fractionation and bioassay and allows an animal's choice to be continuously tested against defined plant fractions. This approach is extremely labour-intensive and time-consuming. Future studies of the chemical basis of leaf choice in marsupials (and any other folivore) need to combine initial bioassays with targeted correlative studies.

Antifeedant compounds in Eucalyptus*: Bioassay-guided fractionation*

Folivorous marsupials prefer to eat some individual trees more than others. A number of studies of koalas (Hindell & Lee 1987) common brushtail possums (Freeland & Winter 1976) and common ringtail possums (Pahl 1984) have all noted these marked intraspecific preferences. The best-documented case is of common ringtail possums and *Eucalyptus ovata* (Geritz 1987; Pahl 1984; Pass 1992). Many *E. ovata* trees in forests south-east of Melbourne have been killed by repeated browsing by common ringtails whereas adjacent trees of the same species are flourishing untouched.

These strong feeding preferences are not related to traditional measures of nutritional quality such as leaf nitrogen, fibre, amino acids, tannins or volatile oils (Geritz 1987). Further chemical examination showed that unpalatable trees contained significantly greater concentrations of the cyanogenic glycoside prunasin and the oil fraction was dominated by cineole (Pass 1992). However, choice tests showed that neither of these fractions was responsible for the deterrent effect of methanol extracts of the unpalatable plant (Pass 1992).

Nonetheless, a comprehensive bioassay-guided fractionation has revealed the presence of a hitherto unknown group of lipid-soluble phenolic compounds and it is these compounds that deter ringtail possums from feeding (D M Pass, W J Foley & B Bowden, unpublished observations). The structures of these compounds are currently being fully characterised, but they have also been detected in other eucalypts raising the possibility that there is a common defence in eucalypts against marsupial browsers. It will be interesting to determine why these compounds occur in some eucalypts and not others and whether they are equally active against koalas, greater gliders and brushtail possums. These newly discovered antifeedant compounds have several similarities to the antifeedants found in boreal ecosystems. For example, compounds which deter feeding by snowshoe hares are all small lipid-soluble compounds and some have phenolic groups (Reichardt *et al.* 1985). It may be that lipid-soluble moieties are needed to carry reactive phenolics across the gut membrane and this is the basis of antifeedant actions. Research in progress should answer some of these questions.

Role of phenolic PSMs in regulating the daily intake of foliage

Once animals start to feed on particular leaves the rate and ultimate level of intake is affected by a complex series of interactions involving the physical, chemical and sensory characteristics of the plant, along with the animals' digestive physiology and hormonal state. Our understanding of the limits to daily intake is incomplete in all plant-herbivore systems, but there is some evidence that PSMs in *Eucalyptus* can limit the amount of particular foliages that are consumed and lower the nutritional quality of the diet.

In early studies of plant-herbivore interactions, tannins were perceived to be a general plant defence that would have predictable effects on herbivores and several general theories of plant defence incorporated this concept (Feeny 1976; Rhoades & Cates 1976). Many tannins have a strong affinity for proteins and so it was believed that they could bind with dietary proteins and reduce the amount of protein available for the animal. Studies over the past decade have shown that this simplistic view of tannins is not correct (for example, Mole & Waterman 1987; Foley & Hume 1987a; McArthur & Sanson 1993). In much of this work, studies with marsupials have allowed the development of new paradigms of plant-herbivore interactions.

Underpinning this new understanding was the realisation/demonstration that the degree of tannin-protein binding depended critically on the structure of both the tannin and the protein involved (Hagerman & Butler 1981; Zucker 1983). This complicates quantitative predictions of effects on nitrogen balance from analysis of 'tannins' as a broad group.

Natural plant tannins have been demonstrated to negatively affect the digestibility of both eucalypt (*E. melliodora*) and mangrove (*Lumnitzera racemosa*) foliage by the common brushtail possum (Foley & Hume 1987a; S A Dalla Pozza & W J Foley, unpublished observations). In both cases the effect of tannins was blocked with polyethylene glycol-4000. *Eucalyptus* tannins have a greater affinity for polyethylene glycol than for either dietary proteins (McArthur & Sanson 1993) or for other proteins such as digestive enzymes (Foley 1984) and so the addition of polyethylene glycol to a foliage diet allows the effect of tannins on intake and digestibility to be measured.

Common brushtail possums fed *E. melliodora* leaves plus polyethylene glycol ate significantly more dry matter than controls (Foley & Hume 1987a). Although the overall digestibility of the diet did not change, the animals fed polyethylene glycol digested substantially greater amounts of the cell-wall constituents. It was argued that this was due to the tannins complexing with microbial cellulases and hence reducing the ability of the animals to digest fibre (Foley & Hume 1987a). Brushtail possums do not separate coarse and fine digesta in the caecum and so lack the capacity to quickly eliminate fibrous plant particles from the gut (Foley & Hume 1987b). In contrast to the ringtail possums, greater gliders and the koala, brushtail possums must rely on chemical digestion to rid the gut of bulky fibre (Cork & Foley 1991).

This scenario predicts that polyethylene glycol should have little impact on food intake and fibre digestion in those species that can separate digesta in the hindgut and experiments with common ringtail possums have borne this out (McArthur & Sanson 1991). A necessary caveat is that the measurements of fibre digestion in the studies of both brushtail and ringtail possums were subject to interference from phenolics.

A further complication to a simple interpretation of the effects of tannins on folivorous marsupials is the occurrence in many animals of salivary proteins with an especially strong affinity for plant tannins. The importance of these tannin-binding salivary proteins has only recently been appreciated (Mehanso *et al.* 1987; McArthur *et al.* 1995). Many mammals that feed on tannin-rich foods produce a suite of tannin-binding salivary proteins in the saliva. The most common feature of these proteins is that they are rich in proline and it is the lack of an extensive tertiary structure of proline-rich proteins that makes them so effective at binding tannins. This open structure exposes hydrophobic residues in polar solutions, whereas these residues are hidden by folding in most proteins (McArthur *et al.* 1995). Proline-rich proteins have an important role in oral homeostasis (Bennick 1982; McArthur *et al.* 1995), but it is widely believed that an equally important function is to bind tannins that are released from vacuoles during mastication and so avoid the tannins binding to other plant constituents (Austin *et al.* 1989)

Studies in marsupials have caused substantial revision of these ideas. Firstly, McArthur *et al.* (1995) have shown that cross-sectional measures of proline-rich proteins are a poor indicator of their effectiveness as tannin-binding agents. It is essential to measure the maximal saliva flow rate and maximum protein concentration in the saliva before any evaluation can be made of the overall production of tannin-binding salivary proteins. Protein secretion rates vary considerably but the swamp wallaby (a browser) secretes parotid saliva that contains large concentrations of proteins that have high affinity for tannins. In contrast, grazing parma wallabies secrete little protein (McArthur *et al.* 1995). Koalas, and folivorous possums all produce proline-rich proteins, but salivary flow rates are much slower in the least folivorous species than the more folivorous species (for example, 0.087 ml min^{-1} in brushtail possums compared with 1.14 ml min^{-1} in koalas) (McArthur *et al.* 1995). However the proline-rich proteins in all of these species have little affinity for tannins compared with saliva from other species (Mole *et al.* 1990) and thus appear functionally unimportant. Research in progress (C McArthur personal communication) aims to quantify the importance of tannin-binding salivary proteins in marsupial herbivores, but at this stage it seems clear that these are not a simple generalised defence against dietary tannins.

One major advantage of tannin-binding salivary proteins for animals is that salivary proteins are relatively 'cheap' since they consist of non-essential amino acids and so can spare more valuable proteins from loss or degradation (Foley & McArthur 1994). Fermentation in the hindgut may recover part of the nitrogen (as ammonium) but it may also allow the phenolic constituents of the tannin to be absorbed from the gut. The form in which these are absorbed, and their physiological effects on the animal, depend on the action of gut microorganisms (see below).

Recent studies in koalas have identified several bacteria able to degrade tannin-protein complexes. Osawa and Mitsuoka (1990) have identified a strain of *Streptococcus bovis* ('*S. bovis* Biotype 1') and also an enterobacterium (Osawa 1992) which is capable of degrading some tannin-protein complexes. This exciting work has provided evidence to support earlier microscopic observations (O'Brien *et al.* 1986) that gut microflora may play a role in alleviating the effects of tannin-rich diets on herbivores. Both bacteria appear to make available to the animal ammonium derived from protein bound in the complexes.

However, it is still not clear to what extent the tannin molecules themselves are degraded. This is important because liberation of reactive phenolics in the hindgut has unknown effects. Few tannins can be recovered from the faeces of animals fed tannin-rich diets (Cork *et al.* 1983; McArthur & Sanson 1991) and one explanation is that the tannins have been degraded by microbial action (Clausen *et al.* 1990). Very recent studies (Terrill *et al.* 1994) support the widely held view that condensed tannins are resistant to microbial degradation, but more studies in a range of animal species are required.

Metabolism of PSMs in mammals

Although some PSMs are avoided during feeding, any animal that eats tree leaves cannot avoid ingesting substantial amounts of terpenoids and phenolic compounds. These may have no direct impact on diet choice (or at least none that we can detect) but substantial concentrations are absorbed from the gut. Absorption of PSMs from the gut of folivorous marsupials lessens the likelihood of interactions with gut microbes, but the compounds probably contribute nothing to overall metabolisable energy intake (Foley & McArthur 1994). Consequently, absorbed PSMs must be transformed to non-toxic products and rapidly be excreted in the urine or bile. The processes by which this is achieved are basically similar in all mammals and so may provide the basis of a common understanding of the effects of PSMs on animals (Foley 1992; Foley *et al.* 1995).

Acidosis and the cost of biotransformation

The biotransformation and excretion of dietary PSMs depends on two inter-related processes involving a group of enzymes called mixed function oxidases or polysubstrate membrane oxidases.

These enzyme systems serve to convert lipid-soluble, potentially toxic compounds into water-soluble non-toxic compounds which can be excreted in the urine and bile. However, they also convert compounds which are neutral, or only weakly acidic, into strong organic acids that are completely dissociated at physiological pH (Caldwell 1982). Therefore the processes of biotransformation and excretion of absorbed PSMs can lead to the formation of a substantial quantity of organic acids (Robinson *et al.* 1953).

Evidence for production of organic acids in detoxification comes from studies of common ringtail possums fed *Eucalyptus radiata* foliage (Foley 1992). When the diet of the animals was switched from a palatable artificial diet to *Eucalyptus radiata* leaf, urine pH dropped from 8.5 to 5.7 within 5 hours of leaf being presented. Over the next 13 days, the excretion of titratable acid, ammonium and glucuronic acid increased, mirroring changes in food intake. This showed that acid excretion was clearly diet-related and further experiments showed that there was a variable pattern of acid excretion among diets containing different species of *Eucalyptus* leaf (Foley 1992).

Since the organic acids will be totally dissociated at physiological pH, excretion of the hydrogen ions and the anion may occur independently (Foley *et al.* 1995). H^+ ions can be excreted via the kidney after binding to dibasic phosphate but the primary route for disposal is by titration with body bicarbonate to form CO_2 (Figure 12.3). Generation of new bicarbonate occurs by the catabolism of glutamine and alanine to form a-ketoglutarate and ammonium. If the ammonium is excreted in the urine then an equimolar amount of 'new' bicarbonate can be formed. Packaging the ammonium into a urea molecule, however, consumes this bicarbonate (Halperin *et al.* 1992).

These shifts in urinary nitrogen excretion may be an important consequence of the excretion of PSMs. For example, the reduction in urea synthesis during acidemia means that significantly less urea is recycled back to the gut to be incorporated into microbial protein (Foley 1992). Studies are currently in progress to assess the extent to which marsupials catabolise muscle proteins to provide the glutamine for generation of 'new' bicarbonate.

Terpene metabolites as pheromones

All mammals that feed on *Eucalyptus* foliage ingest substantial amounts of terpenes. The metabolism of various dietary terpenes has now been studied in koalas, common

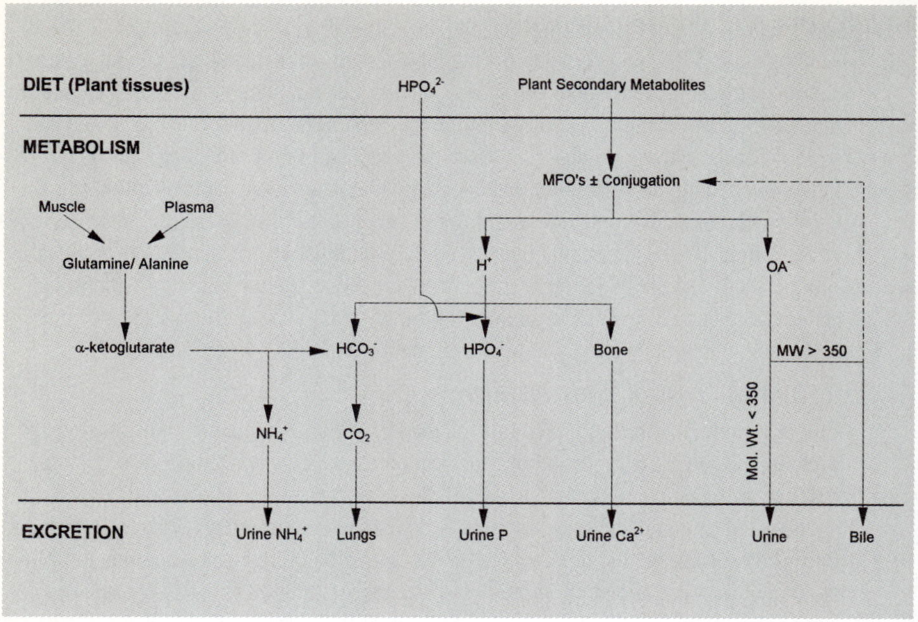

Figure 12.3 Some pathways of disposal of organic acids produced during the biotransformation of phenols and terpenes in mammals. Plant secondary metabolites are transformed by mixed function oxidases (MFO) and may be conjugated. The resulting strong organic acids (H+OA-) are completely dissociated at physiological pH. The hydrogen ion can be either buffered by bone minerals, excreted in urine as monobasic phosphate after combining with dibasic phosphate or reacted with bicarbonate and removed as CO_2. 'New' bicarbonate can be formed by the complete oxidation of a-ketoglutarate if the ammonium ion is excreted in the urine. The organic acid anions (OA-) may be actively secreted into the urine, excreted in the bile or circulated enterohepatically (dotted line).

ringtails, brushtails and greater gliders. Not all compounds have been studied in all species, but it seems clear that those species with a greater reliance on *Eucalyptus* leaf as a diet show more extensive oxidative activity (Southwell *et al.* 1980; McLean *et al.* 1993). It used to be thought that the vast majority of metabolites were excreted conjugated with glucuronic acid or glycine, but it is clear, at least in common ringtails (McLean *et al.* 1993) that polyoxygenation to form highly polar, acidic molecules that are readily excreted is the major way that terpenes are excreted from the body.

One of the major terpenes of *Eucalyptus* foliage which is eaten by folivorous marsupials is 1,8 cineole. The major metabolites of cineole in common brushtail possums are hydroxy cineole and cineolic acid (Flynn & Southwell 1979; Southwell *et al.* 1980), but Carman and Klika (1992) have shown that these are partial racemates. This means that the metabolites are a mixture of optically active isomers. Interestingly, male and female possums produce different mixtures of the two enantiomers (Carman & Klika 1992) and the smell and identical volatility of these partially racemic compounds suggest that they could act as pheromones.

HABITAT SELECTION BY ARBOREAL FOLIVOROUS MARSUPIALS IN RELATION TO THE CHEMISTRY OF EUCALYPT FOLIAGE

Various studies have shown a link between abundance of arboreal, leaf-eating marsupials and the nutrient status of eucalypt forests, as reflected in foliar concentrations of nitrogen, phosphorus and potassium and/or soil type (Braithwaite *et al.* 1983; 1984). It has been argued that this trend reflects the inverse correlation between nutrient status and concentrations of carbon-based chemical defences (for example, phenolics) predicted by the resource-availability hypothesis (Cork 1992). These studies suggest that a 'toxicity threshold' exists below which abundance of leaf-eating fauna is consistently low. It is postulated that this toxicity threshold is set by the ratio of nutrients to chemical anti-herbivore defences, especially phenolics, rather than the absolute concentration of defences (Cork & Foley 1991; Cork 1992, 1996) because of the reported effects of phenolics on acid-base balance in arboreal marsupials and the involvement of nitrogen and phosphorus in alleviating these effects (see sections above). Eucalypt communities favoured by arboreal, leaf-eating marsupials have higher ratios of nitrogen and phosphorus to phenolics than unfavoured communities (Cork 1992). Recent analyses of data on both the chemistry of foliage and climatic and structural attributes of eucalypt forests in relation to the distribution of greater gliders (Pausas *et al.* 1995; S J Cork, N Nicholls & L W Braithwaite, unpublished observations) suggest that above the nutrient/phenolic threshold, concentrations of nutrients and phenolics still explain a significant amount of the variation in probability of finding folivores, but structural and climatic variables are equally important.

FUTURE DIRECTIONS

There remain a number of aspects of mammalian nutritional physiology and ecology that can be explored profitably using arboreal marsupials as models.

The research summarised here on gut function in relation to diet choice has expanded knowledge of the processes occurring in the mammalian gut and further investigation of the physiological control of these processes could not only give insights into the evolution of digestive function in vertebrates, but also could provide valuable comparative data for interpreting the role of the hindgut in humans and domestic mammals and the consequences of malfunctioning of this system.

The studies performed so far have been largely descriptive and, although they have been valuable in development of hypotheses about limits to gut function, they have not tested any of these hypotheses directly. Such tests are difficult to perform as most animals are reluctant to eat either amounts or types of food that push them close to their physiological limits. Nevertheless, the critical experiments might be possible if arboreal marsupials can be induced to eat artificial diets ranging widely in fibre content and/or will tolerate force-feeding of diets expected to exceed their ability to process fibre and stay in energy equilibrium.

As we improve our understanding of how arboreal marsupials meet the physiological challenges posed by fibre and phenolics in eucalypt foliage, a further priority for research will be to predict how leaf chemical composition and the performance of leaf-eating animals will be affected by global climate change. For example, combining data on effects of elevated CO_2 on nitrogen:phenolic ratios in eucalypt saplings (Lawler 1993) with the observed distributions of arboreal marsupials in relation to leaf composition over a broad geographic scale (Braithwaite *et al.* 1983; Cork 1992; Pausas *et al.* 1995) predicts that large areas of present eucalypt forests could become uninhabitable for leaf-eating arboreal marsupials if the concentration of atmospheric CO_2 rises to the level predicted in the next 50 years (Figure 12.4).

An equally pressing challenge is to understand the effects of other changes to forests induced by the activity of humans, including farming, forestry, urbanisation and tourism. Each of these activities has effects on the dynamics of nutrients, water and light in forests. Our research to date suggests that if these effects can be predicted then, in turn, we can predict their consequences for food quality for arboreal marsupials. This is a good example of how physiological and ecological research are coming together to address one of the fundamental issues raised by Hugh Tyndale-Biscoe early in his career; the effects of habitat disturbance on forest-living marsupials.

With regard to PSMs, the identification of specific antifeedants in *Eucalyptus* foliage that deter common ringtails will allow us to ask much more focused questions. For example, if we can now measure the concentration of specific phenolic antifeedant compounds, rather than relying on cruder measures of total plant phenolics, we stand a much better chance of understanding and quantifying the selective effects of animal browsing on individual plants and communities. We can measure the costs of specific defence against herbivores in terms of biosynthetic and maintenance costs of antifeedant compounds and we can determine how the same compounds affect different marsupial species. As yet, we still cannot satisfactorily predict the nutritional quality of *Eucalyptus* foliage from chemical analyses. Inclusion of measures of specific antifeedants should enhance these analyses and will provide a substantial advantage to managers of forest habitats. Antifeedant compounds may also provide a practical means of reducing the damage caused by browsing mammals to native forest plantations.

We noted earlier the difficulty of making broad generalisations about the effects of PSMs on mammals because most of the interactions documented relate to single species. Clearly the dominance of *Eucalyptus* in Australian forests means that there is a greater chance of demonstrating general effects from the intensive study of a few species of eucalypts and so gain an insight into community wide processes. This is a unique opportunity for Australian ecologists.

Studies of the effects of elevated carbon dioxide on the composition of *Eucalyptus* foliage have revealed substantial shifts in parameters impacting on the overall nutritional quality of the leaf. These studies have been made with seedlings grown under glasshouse conditions and it is clearly important that these studies be extended to plants grown under field conditions. Factors such as the extent of mycorrhizal infection may be significant modulators of chemical composition under elevated CO_2.

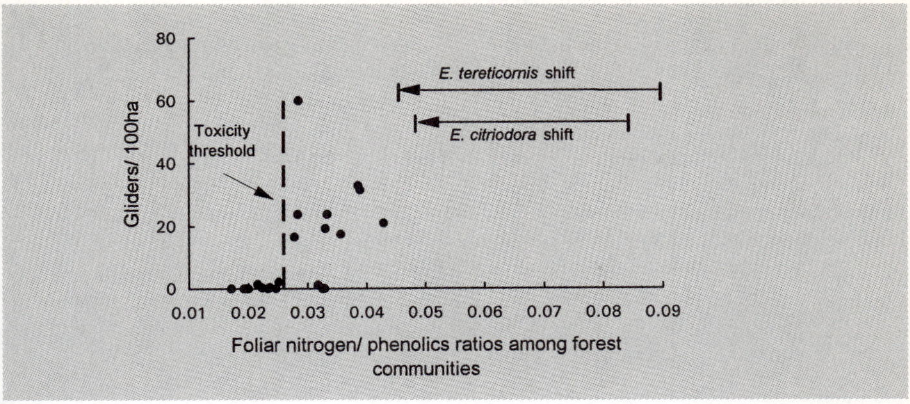

Figure 12.4 The possible effect of elevated atmospheric CO_2 on habitat quality for arboreal, leaf-eating marsupials.
Notes:
(1) The horizontal arrows indicate the average shift in the ratio of nitrogen to phenolics in foliage of eucalypt seedlings (2 species), grown under a range of nutrient and light conditions, when CO_2 is increased from 350 to 800 ppm (Lawler 1993).
(2) The plotted data are nitrogen:phenolic ratios for eucalypt communities in south-eastern New South Wales with a postulated 'toxicity threshold' indicated, below which sightings of greater gliders are rare and diet quality is postulated to be inadequate for establishment of persistent populations (Cork 1992).(3) A shift of even 20% of the magnitude seen in seedlings would cause a large proportion of the communities presently suitable for leaf-eating fauna to move to the unfavourable side of the toxicity threshold.

References

Attiwill P M and Leeper G W (1987). *Forest Soils and Nutrient Cycles*. Melbourne University Press, Melbourne.

Austin P J, Suchar L A, Robbins C T and Hagerman A E (1989). Tannin-binding proteins in saliva of deer and their absence in saliva of sheep and cattle. *Journal of Chemical Ecology,* 15, 1335–1347.

Barton A F M, Cotterill P P and Brooker M I H. (1991). Heritability of cineole yield in *Eucalyptus kochii*. *Silvae Genetica,* 40, 37–38.

Bell R H V. (1969). The use of the herb layer by grazing ungulates in the Serengeti. In *Animal Populations in Relation to their Food Resources*. Ed A Watson, Blackwell, Oxford. pp 111–123.

Bennick A (1982). Salivary proline-rich proteins. *Molecular and Cellular Biochemistry,* 45, 83–99.

Betts T J (1978). Koala acceptance of *Eucalyptus globulus* Labill as food in relation to the proportion of sesquiterpenoids in the leaves. In *The Koala*. Ed T J Bergin, Zoological Parks Board of NSW, Sydney. pp 75–85.

Bjornhag G (1987). Comparative aspects of digestion in the hindgut of mammals. The colonic separation mechanism (CSM)(a review). *Deutsche Tierarztliche Wochenschrift,* 94, 33–36.

Bjornhag G (1972). Separation and delay of contents in the rabbit colon. *Swedish Journal of Agricultural Research,* 2, 125–136.

Bjornhag G (1989). Sufficient fermentation and rapid passage of digesta. A problem of adaptation in the hindgut. *Acta Veterinaria Scandinavia,* Supplement 86, 204–211.

Boland D J, Brophy J J and House A P N. (1991). *Eucalyptus Leaf Oils: Use, Chemistry, Distillation and Marketing*. Inkata Press, Melbourne.

Braithwaite L W, Dudzinski M L and Turner J (1983). Studies of the arboreal marsupial fauna of eucalypt forests being harvested for woodpulp at Eden, New South Wales. II. Relationship between the fauna density, richness and diversity, and measured variables of the habitat. *Australian Wildlife Research,* 10, 231–247.

Braithwaite L W, Turner J and Kelly J (1984). Studies of the arboreal marsupial fauna of eucalypt forests being harvested for woodpulp at Eden, New South Wales. III. Relationships between faunal densities, eucalypt occurrence and foliage nutrients and soil parent materials. *Australian Wildlife Research,* 11, 41–48.

Bryant J P, Kuropat P J, Reichardt P B and Clausen T P (1991). Controls over the allocation of resources by woody plants to chemical antiherbivore defense. In *Plant Defenses Against Mammalian Herbivory.* Eds R T Palo and C T Robbins. CRC Press Inc. Boca Raton, Florida. pp 83–102.

Bryant J P, Reichardt P B, Clausen T P, Provenza F D and Kuropat P J (1992). Woody plant-mammal interactions. In *Herbivores: Their Interactions with Secondary Plant Metabolites. Second Edition Volume II: Ecological and Evolutionary Processes.* Eds G A Rosenthal and M R Berenbaum, Academic Press, New York. pp 344–371.

Caldwell J (1982). The conjugation reactions in foreign compound metabolism: definition, consequences and species variation. *Drug Metabolism Review,* 13, 745–778.

Carman R M and Klika K D (1992). Partially racemic compounds as brushtail possum urinary metabolites. *Australian Journal of Chemistry,* 45, 651–657.

Chilcott M J and Hume I D (1985). Coprophagy and the selective retention of fluid digesta: their role in the nutrition of the common ringtail possum (*Pseudocheirus peregrinus*) fed *Eucalyptus andrewsii* foliage. *Australian Journal of Zoology,* 33, 1–15.

Chivers D J and Hladik C M (1980). Morphology of the gastrointestinal tract in primates: Comparisons with other mammals in relation to diet. *Journal of Morphology,* 166, 337–386.

Clausen T P, Provenza F D, Burritt E A, Reichardt P B and Bryant J P (1990). Ecological implication of condensed tannin structure: a case study. *Journal of Chemical Ecology,* 16, 2381–2391.

Clutton-Brock T H and Harvey P H (1977). Species differences in feeding and ranging behaviour in primates. In *Primate Ecology.* Ed T H Clutton-Brock, Academic Press, London. pp 557–579.

Coley P D, Bryant J P and Chapin III F S (1985). Resource availability and plant antiherbivore defense. *Science,* 230, 895–899.

Collins D J (1990). *Plants for medicines: a chemical and pharmacological survey of plants in the Australian region.* CSIRO, East Melbourne.

Cork S J (1994). Digestive constraints on dietary scope in small and moderately-small mammals: How much do we really understand? In *The Digestive System in Mammals: Form, Food and Function.* Eds D J Chivers and P Langer, Cambridge University Press. Cambridge. pp 337–369.

Cork S J (1996). Optimal digestive strategies for arboreal herbivorous mammals in constrasting forest types: Why koalas an colobines are different. *Australian Journal of Ecology,* 21, 10–20.

Cork S J (1992). Polyphenols and the distribution of arboreal, folivorous marsupials in *Eucalyptus* forests of Australia. In *Plant Polyphenols: Synthesis, Properties, Significance.* Ed. R W Hemingway, Plenum Press, New York. pp 653–663..

Cork S J (1984). Utilization of *Eucalyptus* foliage by arboreal marsupials. *Proceedings of the Nutrition Society of Australia,* 9, 88–97.

Cork S J and Foley W J (1991). Digestive and metabolic strategies of arboreal mammalian folivores in relation to chemical defenses in temperate and tropical forests. In *Plant Defenses Against Mammalian Herbivory.* Eds R T Palo and C T Robbins, CRC Press Boca Raton. pp 133–166.

Cork S J, Hume I D and Dawson T J (1983). Digestion and metabolism of a natural foliar diet (*Eucalyptus punctata*) by an arboreal marsupial, the koala (*Phascolarctos cinereus*). *Journal of Comparative Physiology,* 153, 181–190.

Cork S J and Krockenberger A K (1991). Methods and pitfalls of extracting condensed tannins and other phenolics from plants: Insights from investigations on *Eucalyptus* leaves. *Journal of Chemical Ecology*, 17, 123–134.

Cork S J and Pahl L (1984). The possible influence of nutritional factors on diet and habitat selection by the ringtail possum. In *Possums and Gliders*. Eds A P Smith and I D Hume, Australian Mammal Society, Sydney. pp 269–276.

Cork S J and Warner A C I. (1983). The passage of digesta markers through the gut of a folivorous marsupial, the koala *Phascolarctos cinereus*. *Journal of Comparative Physiology*, 152, 43–51.

Dayal R (1988). Phytochemical examination of the eucalypts — a review. *Journal of Scientific and Industrial Research India*, 47, 215–220.

Demment M W and van Soest P J (1985). A nutritional explanation for body-size patterns of ruminant and nonruminant herbivores. *American Naturalist*, 125, 641–672.

Eisenberg J F (1978). The evolution of arboreal folivores in the Class Mammalia. In *The Ecology of Arboreal Folivores*. Ed. G G Montgomery, Smithsonian Institution Press, Washington, D C. pp 135–152.

Feeny P P (1976). Plant apparency and chemical defense. *Recent Advances in Phytochemistry*, 10, 1–40.

Flynn T M and Southwell I A (1979). 1,3-dimethyl-2-oxabicyclo[2,2,2]-octane-3-methanol,1,3-dimethyl-2-oxabicyclo[2,2,2]-octane-3-carboxylic acid; urinary metabolites of 1,8 cineole. *Australian Journal of Chemistry*, 32, 2093–2095.

Foley W J (1992). Nitrogen and energy retention and acid-base status in the common ringtail possum (*Pseudocheirus peregrinus*) — Evidence of the effects of absorbed allelochemicals. *Physiological Zoology*, 65, 403–421.

Foley W J (1984). *The Utilization of Eucalyptus Foliage by the Greater Glider (Petauroides volans) and the Brushtail Possum (Trichosurus vulpecula)*. PhD thesis, University of New England, Armidale.

Foley W J and Cork S J (1992 Use of fibrous diets by small herbivores: How far can the rules be 'bent'? *Trends in Ecology and Evolution*, 7, 159–162.

Foley W J, Engelhardt W v and Charles-Dominique P (1995). Digesta passage, particle size,in vitro fermentation rate in the three-toed sloth *Bradypus tridactylus* (Edentata: Bradypodidae). *Journal of Zoology*, 236, 681–696.

Foley W J and Hume I D (1987a). Digestion and metabolism of high-tannin *Eucalyptus* foliage by the brushtail possum (*Trichosurus vulpecula*) (Marsupialia: Phalangeridae). *Journal of Comparative Physiology*, 157b, 67–76.

Foley W J and Hume I D (1987b). Passage of digesta markers in two species of arboreal folivorous marsupials — the greater glider (*Petauroides volans*) and the brushtail possum (*Trichosurus vulpecula*). *Physiological Zoology*, 60, 103–113.

Foley W J, Lassak E V and Brophy J (1987). Digestion and absorption of *Eucalyptus* essential oils in greater glider (*Petauroides volans*) and brushtail possum (*Trichosurus vulpecula*). *Journal of Chemical Ecology*, 13, 2115–2130.

Foley W J and McArthur C (1994). The effects and costs of allelochemicals in mammals: an ecological perspective. In *The Digestive System in Mammals: Food, Form and Function*. Eds D J Chivers and P Langer, Cambridge University Press, Cambridge. pp 370–391.

Foley W J, McLean S and Cork S J (1995). Consequences of biotransformation of plant secondary metabolites on acid-base metabolism in mammals — a final common pathway. *Journal of Chemical Ecology*, 21, 721–743.

Fox L R and Macauley B J (1977). Insect grazing on *Eucalyptus* in response to variation in leaf tannins and nitrogen. *Oecologia*, 29, 145–162.

Freeland W J and Winter J W (1976). Evolutionary consequences of eating: *Trichosurus vulpecula* and the genus *Eucalyptus*. *Journal of Chemical Ecology*, 1, 439–455.

Ganzhorn J U (1988). Food partitioning among Malagasy primates. *Oecologia*, 75, 436–450.

Geritz R (1987). *Chemical Composition of Eucalyptus ovata Foliage browsed by the Common Ringtail Possum (Pseudocheirus peregrinus)*. Hons thesis, Monash University, Melbourne.

Hagerman A E and Butler L G (1981). The specificity of proanthocyanidin-protein interactions. *Journal of Biological Chemistry,* 256, 4494–4497.

Halperin M L, Goldstein M B, Stinebaugh B J and Jungas R L (1992). Biochemistry and physiology of ammonium excretion. In *The Kidney: Physiology and Pathophysiology.* Eds D W Seldin and G Giebisch, Raven Press, New York. pp 1471–1490.

Hanley T A, Robbins C T, Hagerman A E and McArthur C (1992). Predicting digestibe protein and digestible dry matter in tannin-containing forages consumed by ruminants. *Ecology,* 73, 537–541.

Harborne J B (1991). The chemical basis of plant defense. In *Plant Defenses Against Mammalian Herbivory.* Eds R T Palo and C T Robbins, CRC Press, Boca Raton, Florida. pp 45–59.

Harbourne J B (1988). *Introduction to Ecological Biochemistry*. Academic Press, London.

Hindell M and Lee A K (1987). Habitat use,tree preferences of koalas in a mixed eucalypt forest. *Australian Wildlife Research,* 14, 349–360.

Hladik C M (1978). Adapative strategies of primates in relation to leaf eating. In *The Ecology of Arboreal Folivores.* Ed G G Montgomery, Smithsonian Institution Press, Washington D C. pp 373–395.

Hladik C M, Charles-Dominique P, Valdebouze P, Delort-Laval J and Flanzy J (1971). La caecotrophie chez un primate phyllophage du genre *Lepilemur* et des correlations avec les particularites de son appareil digestif. *Comptes Rendus Hebdomadaires des Seances de L'Academie des Sciences, Serie D-Sciences Naturelles,* 272, 3191–3194.

Hume I D (1982). *The Digestive Physiology and Nutrition of Marsupials.* Cambridge University Press, Cambridge.

Hume I D and Esson C (1993). Nutrients, antinutrients and leaf selection by captive koalas (*Phascolarctos cinereus*). *Australian Journal of Zoology,* 41, 379–392.

Janis C (1976). The evolutionary strategy of the Equidae and the origins of rumen and cecal digestion. *Evolution,* 30, 757–774.

Jarman P J (1974). The social organization of antelope in relation to their ecology. *Behaviour,* 48, 215–266.

Jeffries R L, Klein D R and Shaver G R (1994). Vertebrate herbivores and northern plant communities: reciprocal influences and responses. *Oikos,* 71, 193–206.

Kerle J A (1985). Habitat preferences and diet of the northern brushtail possum *Trichosurus arnhemensis* in the Alligator Rivers region, Northern Territory. *Proceedings of the Ecological Society of Australia,* 13, 161–173.

Krockenberger A K (1988). *Effects of Ingestion of Cineole Rich Diets on the Common Brushtail Possum (Trichosurus vulpecula) and the Common Ringtail Possum (Pseudocheirus peregrinus)*. Hons thesis, University of Sydney, Sydney.

Landsberg J and Cork S J (1995). Herbivory. In *Eucalypt Ecology: Individuals to Ecosystems.* Eds J Williams and J Woinarski, Cambridge University Press, Cambridge.

Langer P (1988). *The Mammalian Herbivore Stomach*. Gustav Fischer, Stuttgart.

Lawler I R (1993). *Resource Availability and Eucalyptus: The effects of variation in CO_2 soil nutrients and light on nutritional quality of foliage*. Unpublished Honours Thesis, James Cook University of North Queensland, Townsville.

Martindale (1989). *The Extra Pharmacopeia 29th edn*. Pharmaceutical Society. London.

McArthur C (1988). Variation in neutral detergent fiber analysis of tannin-rich foliage. *Journal of Wildlife Management,* 52, 374–378.

Mcarthur C and Sanson G D (1991). Effects of tannins on digestion in the common ringtail possum (*Pseudocheirus peregrinus*), a specialized marsupial folivore. *Journal of Zoology,* 225, 233–251.

Mcarthur C and Sanson G D (December 1993). Nutritional effects and costs of a tannin in two marsupial arboreal folivores. *Functional Ecology,* 7, 697–703.

McArthur C, Sanson G D and Beal A M (1996) Salivary proline-rich proteins inmammals: roles in oral homeostasis and counteracting dietary tannins. *Journal of Chemical Ecology* (in press).

Mclean S, Foley W J, Davies N W, Brandon S, Li D and Blackman A J (1993). Metabolic fate of dietary terpenes from eucalyptus radiata in common ringtail possum (*Pseudocheirus peregrinus*). *Journal of Chemical Ecology,* 19, 1625–1643.

McNab B K (1978). Energetics of arboreal folivores: Physiological problems and ecological consequences of feeding on an ubiquitous food supply. In *The Ecology of Arboreal Folivores.* Ed. G G Montgomery, Smithsonian Institution Press, Washington, D C. pp 153–162.

McNab B K (1986). The influence of food habits on the energetics of eutherian mammals. *Ecological Monographs,* 56, 1–19.

Mehansho H, Butler L G and Carlson D M (1987). Dietary tannins and salivary proline-rich proteins: Interactions, induction and defense mechanisms. *Annual Review of Nutrition,* 7, 423–440.

Milton K (1979). Factors influencing leaf choice by howler monkeys: A test of some hypotheses of food selection by generalist herbivores. *American Naturalist,* 114, 362–378.

Milton K, Casey T M and Casey K K (1979). The basal metabolic rate of the mantled howler monkey (*Alouatta palliata*). *Journal of Mammalogy,* 60, 373–376.

Mole S, Butler L G and Iason G (1990). Defense against dietary tannin in herbivores: a survey for proline rich salivary proteins in mammals. *Biochemical Systematics and Ecology,* 18, 287–293.

Mole S and Waterman P G (1987). Tannins as antifeedants to mammalian herbivores: still an open question? In *Tannins as Allelochemicals in Agriculture and Forestry.* Ed. G R Waller, American Chemical Society, Washington, D C. pp 572–587.

Morrow P A and Fox L R (1980). Effects of variation in *Eucalyptus* essential oil yield on insect growth and grazing. *Oecologia,* 45, 209–219.

Moller E F (1985). Basal metabolic rates in primates — the possible role of phylogenetic and ecological factors. *Comparative Biochemistry and Physiology,* 81A, 707–711.

Nagy K A (1987). Field metabolic rate and food requirement scaling in mammals and birds. *Ecological Monographs,* 57, 111–128.

Nagy K A and Montgomery G G (1980). Field metabolic rate, water flux, and food consumption in three-toed sloths (*Bradypus variegatus*). *Journal of Mammalogy,* 61, 465–472.

O'Brien T P, Lomdahl A and Sanson G (1986). Preliminary microscopic investigations of the digesta derived from foliage of *Eucalyptus ovava* (Labill.) in the digestive tract of the common ringtail possum, *Pseudocheirus peregrinus* (Marsupialia). *Australian Journal of Zoology,* 34, 157–176.

Osawa R (1992). Tannin-protein complex-degrading enterobacteria isolated from the alimentary tracts of koalas and a selective medium for their enumeration. *Applied and Environmental Microbiology,* 58, 1754–1759.

Osawa R and Mitsuoka T (1990). Selective medium for enumeration of tannin-protein complex degrading *Streptococcus* spp in feces of koalas. *Applied and Environmental Microbiology,* 56, 3609–3611.

Pahl L I (1984). Diet preference, diet composition and population density of the ringtail possum (*Pseudocheirus peregrinus cooki*) in several plant communities in southern Victoria. In *Possums and Gliders.* Eds A P Smith and I D Hume, Australian Mammal Society, Sydney. pp 252–260.

Pahl L I and Hume I D (1990). Preferences for *Eucalyptus* species of the New England Tablelands and initial development of an artificial diet for koalas. In *The Biology of the Koala.* Eds A K Lee, K A Handasyde and G D Sanson, Surrey Beatty & Sons Pty Ltd, Sydney. pp 123–128.

Parra R (1978). Comparison of foregut and hindgut fermentation in herbivores. In *The Ecology of Arboreal Folivores.* Ed. G G Montgomery, Smithsonian Institution Press, Washington, D C. pp 205–230.

Pass D M (1992). *Feeding Deterrency towards*

the Common Ringtail Possum Pseudocheirus peregrinus within the Eucalyptus species E. ovata and E. polyanthemos. Hons thesis, James Cook University fo North Queensland, Townsville.

Pausas J G, Braithwaite L W and Austin M P (1995). Modelling habitat quality for arboreal marsupials in the south coastal forests of New South Wales, Australia. *Forest Ecology and Management,* 78, 39–49.

Pratt A (1937). *The Call of the Koala.* Mullens, Melbourne.

Reichardt P B, Bryant J P, Clausen T P and Weiland G D (1984). Defense of winter-dormant Alaska paper birch against snowshoe hares. *Oecologia,* 65, 58–69.

Reichardt P B, Clausen T and Bryant J (1985). Plant secondary metabolites as feeding deterrents to vertebrate herbivores. In *Proceedings of Symposium on Plant-Herbivore Interactions, Snowbird, Utah.* Eds F D Provenza, J T Flinders and E D McArthur, USDA, Ogden, Utah. pp 37–42

Rhoades D F and Cates R G (1976). Toward a general theory of plant antiherbivore chemistry. In *Recent Advances in Phytochemistry.* Eds J W Wallace and R W Mansell, Plenum Press, New York. pp 168–213.

Robbins C T, Hanley T A, Hagerman A E, Hjeljord O, Baker D L, Schwartz C C and Mautz W W (1987). Role of tannins in defending plants against ruminants: reduction in protein availability? *Ecology,* 68, 98–107.

Robinson D, Smith J N and Williams R T (1953). Studies in detoxication: 52: The apparent dissociation constants of some glucuronides, mercapturic acids and related compounds. *Biochemical Journal,* 55, 151–155.

Robinson J G and Redford K H (1986). Body size, diet, and population density of neotropical forest mammals. *American Naturalist,* 128, 665–680.

Smith A P and Ganzhorn J U (1996). Convergence and divergence in community structure and dietary adaptation in Australian possums and gliders and Malagasy lemurs. *Australian Journal of Ecology,* 21, 31–46.

Smith R F C. (1969). Studies on the marsupial glider, *Schoinobates volans* (Kerr). I. Reproduction. *Australian Journal of Zoology,* 17, 625–636.

Southwell I A (1978). Essential oil content of koala food trees. In *The Koala.* Ed. T J Bergin, Zoological Parks Board of NSW, Sydney. pp 62–74.

Southwell I A, Flynn T M and Degabriele R (1980). Metabolism of a and b pinene, p-cymene and 1,8 cineole in the brushtail possum *Trichosurus vulpecula. Xenobiotica,* 10, 17–23.

Specht R L and Rundel P W (1990). Sclerophylly and Foliar Nutrient Status of Mediterranean-Climate Plant Communities in Southern Australia. *Australian Journal of Botany,* 38, 459–474.

Terrill T H, Waghorn G C, Woolley D J, McNabb W C and Barry T N (1994). Assay and digestion of 14C-labelled condensed tannins in the gastrointestinal tract of sheep. *British Journal of Nutrition,* 72, 467–477.

Tyndale-Biscoe C H and Smith R F C (1969a). Studies on the marsupial glider *Schoinobates volans* (Kerr). 2. Population structure and regulatory mechanisms. *Journal of Animal Ecology,* 38, 637–650.

Tyndale-Biscoe C H and Smith R F C (1969b). Studies on the marsupial glider *Schoinobates volans* (Kerr). 3. Response to habitat destruction. *Journal of Animal Ecology,* 38, 651–659.

van Dyck S (1979). Destruction of wild tobacco trees (*Solanum mauritianum* Scopoli) by mountain possums (*Trichosurus caninus* Ogilby). *Memoirs of the Queensland Museum,* 19, 367–368.

van Soest P J (1982). *Nutritional Ecology of the Ruminant.* Durham and Downey, Portland.

Zoidis A M and Markowitz H (1992). Findings from a feeding study of the koala (*Phascolarctos cinereus adjustus*) at the San Francisco Zoo. *Zoo Biology,* 11, 417–431.

Zucker W V (1983). Tannins: Does structure determine function? An ecological perspective. *American Naturalist,* 121, 335–365.

CHAPTER 13

REPRODUCTIVE ANOMALIES IN THE RED KANGAROO IN CENTRAL AUSTRALIA EXPLAINED BY ABORIGINAL TRADITIONAL KNOWLEDGE AND ECOLOGY

Alan Newsome

INTRODUCTION

Two reproductive anomalies in red kangaroos (*Macropus rufus*) are discussed here, high levels of impaired spermatogenesis and low fecundity at times when full fertility could be expected (Newsome 1973). Why, when the environment is at its very best following good rains, should more than half of recently oestrous female red kangaroos fail to become pregnant? Such a response would appear to be maladaptive. My hypothesis is that red kangaroos have come to utilise open landscapes recently converted into drought refugia for them by the cattle industry. In this chapter I show that low fecundity is related to levels of azoospermia that are, in turn, related to habitat quality. I also show that poor habitats are predictable from Aboriginal traditional knowledge of the red kangaroo in central Australia (Newsome 1980a). The initial connection between the Aboriginal and scientific knowledge of red kangaroos (Newsome 1980a) came from Strehlow's (1947) excellent book, *Aranda Traditions* and its equally magnificent sequel, *Songs of Central Australia* (Strehlow 1971). That connection is expanded further here.

REPRODUCTIVE ANOMALIES

Lack of pregnancy in cycling females

Reproduction of the female red kangaroo is highly adaptive to the uncertainty of herbage production in central Australia (Newsome 1965c). In essence, the reproductive strategy includes the maintenance of fertility at the expense of fecundity early in drought, with the reverse prevailing later (Newsome 1965c).

Thus, half of the pouch young die after two to three months of drought; but it takes a further one to two months of drought for half of the mothers to become anoestrus. Any young lost are replaced through implantation of the dormant blastocysts formed after earlier postpartum matings (Newsome 1965a). As the drought deepens that cycle of mortality and replacement may be repeated. Eventually, no Graafian follicles develop in the contralateral ovary during pregnancy so that there can be no postpartum oestrus; the mother has entered anoestrus (Newsome 1964a). Despite anoestrus, such mothers lactate and rear young. Indeed, about 30% of all pouch young that survive droughts were suckled by such mothers. There are extra adaptive advantages because, with food supplies improved again after rain, anoestrous lactating females return to oestrus as do their non-lactating counterparts. If mated, they undergo pregnancy and regain the usual state of lactation including a dormant blastocyst *in utero* (Newsome 1964b). However, as indicated above, there is the anomaly that some non-lactating females returning to oestrus do not become pregnant, and may undergo several oestrous cycles (Newsome 1973). The evidence for that diagnosis was the presence of a large corpus luteum in one ovary, but the animal was not pregnant. It was found that lack of pregnancies in adult females was related to the prevalence of azoospermy in adult males in samples (13.1a). Estimates of total numbers of non-pregnant females used for Figure 13.1a could have been conservative because only those females late in the oestrous cycle were included.

Impaired spermatogenesis

The histology of the seminiferous tubules in 106 of 197 adult male red kangaroos shot in drought and shortly afterwards had a simplified structure (Table 13.1). The lumina of seminiferous tubules were empty or else contained cellular detritus, and no stages of spermatogenesis were present, only basal cell spermatogonia and Sertoli cells (Newsome 1973). That condition was related to the

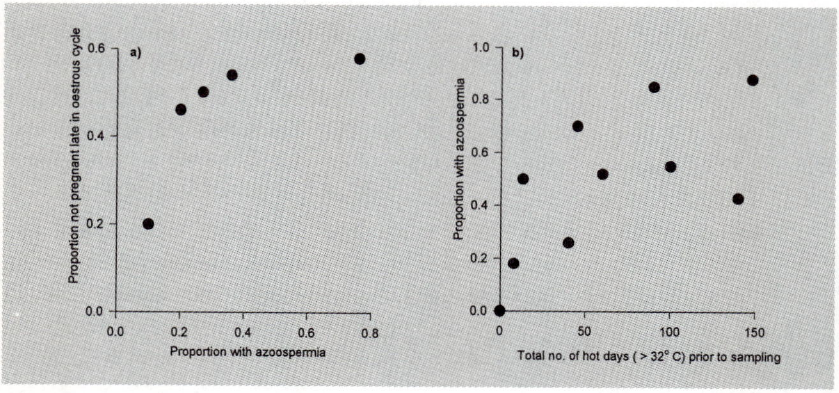

Figure 13.1 Relationships in central Australia between
(a) the proportion of non-pregnant, reproductively cycling adult females and the proportion of adult males with azoospermia; and
(b) the proportion of male red kangaroos with azoospermia and the length of previous periods of hot weather.
Source: Modified from Newsome (1973) using the original data.

number of preceding days with maximum temperature >32°C (Figure 13.1b). Azoospermia was not confined to hot weather, however, but existed in some adult males shot a few weeks after droughts broke. Return to spermatogenesis was not rapid like the return to oestrus, probably due to the time needed to heal what was assumed to be heat damage to the spermatogenetic process. Thus, lack of pregnancy during those circumstances was a consequence of the time taken for males to recover full fertility (Figure 13.1a), although full fecundity might never have been reached, due to the frequency of hot weather experienced at the time. The size of Leydig cells and of their interstitial patches had declined significantly in drought also possibly due to poor nutrition (Newsome 1973). It is now known that effects on tubules are not necessarily independent from effects on the interstitium (de Kretser & Kerr 1994).

THE FREQUENCY OF AZOOSPERMIA BY HABITATS

One test of the hypothesis that male infertility in red kangaroos in central Australia is due to heat exposure could be that azoospermia is most prevalent in open habitats. The causal mechanism implied is that lack of shade trees leads to exposure and to heat infertility. Accordingly, my original data (as for Newsome 1973) were re-examined, each kangaroo being allocated according to where it had been shot (see habitat map, Newsome 1965a). Re-sorting the data in that way provided support for the above hypothesis (Table 13.1). Thus, the incidence of azoospermia was lowest in the well-shaded floodouts and highest in the open treeless floodplains and Mitchell Grass (*Astrebla pectinata*) plains that have the least intrinsic shade. Azoospermia was predominantly present in about 50% of the largest sample (n=124), which had been shot on extensive open savannah (mixed open grassy flats and clumps of mulga *Acacia aneura*) where about 35% of kangaroos were found during drought. Sample sizes were not robust in all cases, and so the results may be influenced by sampling error because the original study did not set out to examine male infertility.

Table 13.1 Incidence of impaired spermatogenesis by habitats*

SHADE LEVELS	AZOOSPERMIA	SPERMATOGENESIS
closed canopy (floodouts)	12	20
partial shade (savannah and gil-gai plains)	69	65
no shade (Mitchell Grass Plains and floodplains)	25	6

Notes: (n = 197 red kangaroos) χ^2 = 14.1; $p < 0.001$
*Re-analysis of basic data used in Newsome (1973).

Habitat preferences

It had been noted that kangaroos flocked to the open landscapes in drought, but left them after good rains. Where did they go, and why did they switch habitats in that way when green herbage was in abundance everywhere? Extensive aerial surveys in drought (October 1961) and after good widespread rains (April 1962) (Newsome 1965a, 1965b) found that there was an inverse relationship between kangaroo numbers on some habitats (Table 13.2). Thus, drainages and the Mitchell Grass plains were the kangaroos' drought refugia, despite their relatively minor areas, but the Mitchell Grass plains were abandoned after rain. The pattern for the adjacent mulga woodlands was the reverse, with few kangaroos there in drought but many after rain when green herbage grew profusely throughout that shady habitat. Kangaroos spread into the surrounding mulga woodlands from the watercourses after rain also. Alternating use of open plains and the mulga has been confirmed by Low *et al.* (1973).

Savannah provided another drought refuge, but had average or better densities all the time. Its mixture of open grassy flats and mulga clumps probably provided good options for the kangaroos at any time. But now there have been changes to such habitats in dry times, because green shoots sprout from cattle grazing, so creating 'marsupial lawns' which favour the kangaroos. The issue of habitat change in historic times is developed further below. I argue that the watercourses were the traditional drought refuges and the open grassy areas are recently created ones.

THE IMPORTANCE OF GREEN HERBAGE AND SHADE

Switching of habitat preferences between drought and good times had the same underlying pattern. In drought the large open plains lacked shade trees, but the green pick in the gil-gai depressions attracted the kangaroos. The kangaroos sheltered in the surrounding mulga woodlands to a distance of 3 to 5 km around the open plains. In fact about 80% of the entire population on the study area were in a band of mulga about 1.5 km deep around the areas of persistent green herbage (Newsome 1965b). That relationship between density of red kangaroos and the availability of green herbage and shade trees in drought was examined in more detail (Table 13.3). Densities were estimated for six categories of habitat, based on three classes of green herbage and two of shade trees (Newsome 1975). The analysis brought out the pre-eminence of floodouts for kangaroos (3.4 km^{-2}) independent of floodplains and the large open plains (1.5 km^{-2}). The terminal floodouts are relatively scarce in the landscape (1 to 2%), but only there does green herbage persist in swards during droughts and it does so under a thick canopy of shady mulga. Savannah by itself, disassociated from other habitats like watercourses (see comment on mixed habitats from Table 13.2), had below average densities, as did middle sections of the watercourses (0.5 and 0.6 km^{-2}). However, the large extent of the savannah landscapes meant that large numbers of kangaroos lived there. It was on such open landscapes that most kangaroos were shot for the reproductive samples (Newsome 1965c, 1973) (Table 13.1).

Table 13.2 Changing habitat preferences between drought and rain in central Australia (study area 1054 km²)

MAJOR HABITATS IN AERIAL SURVEYS	AREA %	DENSITIES OF RED KANGAROOS (NO. KM^{-2})	
		AFTER 0.5 YR DROUGHT (OCT. 1961)	AFTER GOOD RAINS (APRIL 1962)
major drainages	14	2.1	0.6*
large Mitchell Grass Plains	12	2.1	0.2*
savannah (mixed open flats and mulga)	34	0.9	1.9*
mulga scrub	40	0.1	0.8*
mean no. of kangaroos/km^{-2}		0.9	1.1
total estimate of kangaroos		938	1128

*Significantly different ($p < 0.05$).
Source: Newsome (1965a).

ANTHROPOGENIC CHANGES TO OPEN GRASSY HABITATS

The contrast is between occasional and perennial use of habitats during drought and after good rains. All the open grassy areas, floodplains, Mitchell Grass plains and savannah, are grazed extensively by cattle. In drought most kangaroos were found on or near the green herbage of the sub-climax 'marsupial lawns' created by cattle grazing (see above). Dry grass is poor feed for the non-ruminant red kangaroos (McIntosh 1966). Without grazing by cattle, the perennial grasses (*Eragrostis setifolia*, *Astrebla pectinata*) mature, seed and dry off. When cropped short by cattle, those grasses produce green shoots, which red kangaroos feed on. Areas of perennial grass left ungrazed by cattle were hard to find, but a small holding paddock (75 ha) containing tall dry *E. setifolia* was found, with 'marsupial lawn' surrounding it. Kangaroos had easy access to the paddock through the standard three-wire cattle fence. Over 5 nights in 1963, only one kangaroo was counted inside the paddock but 276 were outside on a similarly sized area (Newsome 1971). Exploitation by cattle of those open, but once waterless, landscapes has opened up over ten times more non-traditional drought refugia (the 'marsupial lawns') for red kangaroos than was available naturally. The results in Table 13.2 for the entirely open habitat of the Mitchell Grass plains is particularly instructive. That habitat was as important to kangaroos in drought as the drainages but was abandoned after rain, being no different from the sandplain desert then. There could have been plenty of food but there was no intrinsic shade.

Aboriginal totemic sites and habitats for red kangaroos

To the Aranda, my study area of the Burt Plain is the Ilbalintja Plain, a region rich in food sources and creation legends as related by Strehlow (1947; 1971). Strehlow relates parts of the traditional Aranda legend for the red kangaroo there. In the beginning the totemic ancestors arose from the featureless land and made voyages of creation, leaving behind all the physical and biological features that we know today. Created were not only the red kangaroo and its necessary habitat, but also the human beings to whom they were related, and to whom responsibilities for the red kangaroos were bequeathed. Through traditional commemoration of songs, legends and ceremonial performances at totemic sites, the Aborigines ensured the continuance of the red kangaroos as plentiful and fat.

Thanks to the Hermannsburg Lutheran Mission, I met, among others, an old kangaroo man from a neighbouring linguistic group, the Unmatjera, whose land borders that of the Aranda on the north-east. The old man sang his kangaroo song for me and, with his permission, it was translated by Professor Ken Hale, a world leader on the Aranda language from the Massachusetts Institute of Technology, who worked in central Australia in my time. He pointed out the close links between the Unmatjera and Aranda legends with some stanzas in common, as published by Strehlow (1971). The Unmatjera elders also listed a series of red kangaroo sites associated with the Aranda and Unmatjera myths. The positions of some had been mapped (Strehlow 1947, 1971), and the general localities and descriptions of a few others were provided in discussions

Table 13.3 The importance of food and shade during drought in central Australia (number of kangaroos km^{-2})

SHADE TREES	GREEN GRASS		
	Common	Patchy	Rare
common	3.4 (floodouts)	0.5 (streamlines with sandy channels)	0.2 (mulga)
scarce	1.5 (floodplains and open plains)	0.6 (savannah)	0.01 (desert)

Source: Aerial surveys lumping by quality of habitat, Newsome (1975), Table 6.

with the Aborigines. Knowing the country as I do, it was possible to associate the totemic sites with kangaroo habitats (Table 13.4), as mapped elsewhere (Newsome 1980a). Ten of 14 sites lie along or near the most favoured habitat of red kangaroos, the major watercourses. Half of them are on floodouts, and the most famous totemic site is close to the very best red kangaroo habitat in all of central Australia, an extensive floodout. Three are associated with small patches of savannah in areas remote from major watercourses; and one (shared with another famous totemic ancestor) is in the mulga not far from a streamline (Newsome 1980a). None are on the open plains despite their importance as drought refuges today, rivalling the ancestral floodouts that are so vital to the legends.

I am persuaded that the relationship between habitats and totemic sites is no coincidence. I am further persuaded that the correlation between landscapes and the means of travel utilised by the red kangaroo ancestors is no accident (Newsome 1980a). The Aranda legend follows the line of the best habitat for red kangaroos today, the streamlines. Of equal importance to the argument is the means of travel of the ancestral kangaroos. Although the legends relate basically overland journeys of creation, travelled in the daytime and traversed by natural means, there is a gap of about 120 km (from the second to the third most westerly sites known to me), which is very poor habitat for red kangaroos. The legend describes how a great wind bore one of the heroic kangaroo ancestors across that gap. As well, the Unmatjera legend includes an even longer supernatural means of travel, underground, between two sites separated by desert for a distance of about 350 km. The ancestor finally emerged near today's suitable habitat for the red kangaroo.

CONCLUSIONS

Although it is not possible to be sure, the conclusions are that high levels of azoospermia and its consequence, lowered fecundity, have arisen recently, since the rise of the cattle industry. The kangaroos now occupy extensive habitats that were previously unavailable but that have been improved for them by the cattle industry. Whereas food supplies have been improved greatly, the landscape remains open and exposed. Impaired spermatogenesis due to heat infertility is the penalty, and sub-optimal fecundity is the sequel. The fact that there is now permanent water to drink may have allowed kangaroos to tolerate the exposure to heat in other ways.

These ideas are speculative and require confirmation. It is possible that high levels of azoospermia are natural because central Australia is on the edge of the range for the red kangaroo (see maps of Caughley *et al.* 1987). Nevertheless, the evidence from Aboriginal traditional knowledge (Table 13.4) provides corroboration for the ecological findings in that 10 of the 14 totemic sites known to me follow the line of the perennially most favourable habitat, the major watercourses. (Table 13.3). Habitats there have the best moisture regimes, and are the most fertile of the region. Today, they support high to medium densities of kangaroos regardless of environmental conditions. The incidence of azoospermia was least in the terminal floodouts.

The aerial surveys have shown that red kangaroos do not assort themselves randomly across habitats in central Australia (Table 13.2) and (Table 13.3) (Newsome 1965a, 1965b; Newsome *et al.* 1967). As well, it is known that the older females remain in breeding condition longest into drought, implying habitat selection (Newsome 1980b), and that there are sexual differences in diets (Newsome 1977). Other studies in New South Wales have confirmed that red kangaroos discriminate among habitats by sex, age and reproductive class, depending on pasture conditions (Johnson & Bayliss 1981). The results reported here may indicate habitat preferences that affect fecundity.

Traditionally, drought refuges in central Australia appear to have been fewer and much less substantial than today. They would have been confined to small subsets of drainage lines, perhaps even to the small floodouts. Note that the most famous traditional totemic site is in fact associated with the very best of those floodouts. Although the traditional creation line goes close to the Mitchell Grass plains, which are so important as drought refugia today, *no* totemic site is reported there. Reports from pioneering cattlemen confirm that red kangaroos were not seen often on the open areas in the 1920s, and earlier written descriptions are similar (Newsome *et al.* 1994). As well, the Payne-Fletcher Report on the grazing and agricultural industries of the Northern Territory in 1935 (Commonwealth of Australia 1937) listed dingoes (*Canis familiaris dingo*) and rabbits (*Oryctolagus cuniculus*) as pests, but there is no mention of kangaroos. Yet, within 10 years, they were reported in great numbers on those open landscapes, and were shot in their thousands each year after 1945 until the drought of the late 1950s. That the kangaroos abandon the open areas after good rains (Tables 13.2) strongly indicates that the open grassy plains were not preferred habitats ancestrally. Indeed, after rain, the open plains are as poor a habitat as deserts (Table 13.2); and the absence of

Table 13.4 Location of totemic sites and habitat favourability for red kangaroos in central Australia

HABITATS	FAVOURABILITY	NO. OF TOTEMIC SITES
on or near floodouts	very high (especially in drought)	7*
major water-courses	high	3
savannah	good	3
mulga woodlands	only after rain	1*
open plains	only in drought	0
desert	never	0

*One totemic site shared with other important Aranda totemic ancestors
Source: Modified from Newsome (1980).

kangaroos on open plains at such times indicates that remnants of ancestral sub-optimal characteristics still remain, the shortage of shade trees.

If I understand the Aboriginal traditional knowledge on the red kangaroo in central Australia, it identifies prime habitat for the species, and does so over a much longer timespan than European history can provide. It also aids conclusions about changes to grasslands due to current land management. Traditionally, the totemic sites are sources of commemoration for the creation of red kangaroos. Every speck of dust struck during increase ceremonies gives rise to red kangaroos when next it rains. Traditionally as well, no hunting was permitted for some distance around those sites. So, from the viewpoint of the scientific paradigm, a string of protected sites had been established for the red kangaroo in their finest habitat from the beginning. In the modern tradition that amounts to a network of national parks placed in prime habitat. Thus, I argue, the traditional and scientific knowledge of the red kangaroos on the Ilbalintja/Burt Plain are congruent.

Aboriginal traditional knowledge is in use elsewhere in central Australia for the management of Uluru National Park (Reid *et al.* 1993) and to reveal the basic distributions and habits of marsupials now rare or extinct (Burbidge *et al.* 1988). Environmental management of Kakadu National Park in the monsoonal tropics similarly utilises Aboriginal knowledge and traditional practices for the use of fire in various habitats (Braithwaite 1991; Press 1988). At the other end of Australia, in the Flinders Ranges, Aboriginal language and memory have led to the identification of a suite of meso-marsupials extinct in the late 1800s but extant in colonial times (Tunbridge 1991). Their existence had remained unknown to science until uncovered anthropologically.

Acknowledgments

Above all I wish to acknowledge and thank the Aborigines who helped my understanding so greatly and provided so much information on the eco-mythology of the red kangaroo, and also all those earlier Aborigines who provided Professor Strehlow with the wealth of information. Also deeply acknowledged and thanked are Professor Ken Hale for his skilled translations and knowledge; Mr Dick Kimber for his wisdom and discerning discussions of Aboriginal customs and legends; Professor Geoffrey Sharman for his guidance and fine supervision of my reproductive study of the red kangaroo; Mr Dean Stephens for his immense contributions to our ecological study of red kangaroos in central Australia; Professor Oliver Pearson, University of California, Berkeley, for his incredulity that the reproductive anomalies discussed here conformed with the evolutionary ways of Mother Nature; and to Dr Hugh Tyndale-Biscoe for his enthusiasm for eco-mythology and its inclusion in this volume. I am grateful to, and sincerely thank, Dr Lyn Hinds and Mr Jim Merchant for their help with the manuscript, and Mr Adam McKeown for his computer skills for the Figures.

References

Braithwaite R (1991). Aboriginal fire regimes of monsoonal Australia in the 19th Century. *Search,* 22, 247–249.

Burbidge A, Johnson K, Fuller P and Southgate R (1988). Aboriginal knowledge of the mammals of the central deserts of Australia. *Australian Wildlife Research,* 15, 9–39.

Caughley G, Shepherd N and Short J (1987) (eds). *Kangaroos : Their Ecology and Management in the Sheep Rangelands of Australia.* Cambridge University Press, Cambridge.

Commonwealth of Australia (1937). *Report of the Board of Inquiry appointed to inquire into the Land and Land Industries of the Northern Territory of Australia.* Commonwealth Printer, Canberra.

de Kretser D M and Kerr J B (1994). The cytology of the testis. In *The Physiology of the Testis.* Eds E Knobil and J D Neill. 2nd Ed. Raven Press, New York. Chapter 9.

Johnson C and Bayliss P (1981). Habitat selection by sex, age and reproductive class in the red kangaroo, *Macropus rufus*, in Western New South Wales. *Australian Wildlife Research,* 8, 465–474.

Low W, Birk E, Lendon C and Low B. (1973). Community utilisation by cattle and red kangaroos in mulga near Alice Springs, NT. *Tropical Grasslands,* 7, 149–156.

McIntosh D L (1966). The digestibility of two roughages and the rate of passage of residues by red kangaroos. *CSIRO Wildlife Research,* 11, 145–167.

Newsome A (1964a). Anoestrus in the red kangaroo, *Megaleia rufa* (Desmarest). *Australian Journal of Zoology,* 12, 9–17.

Newsome A (1964b). Oestrus in the lactating red kangaroo, *Megaleia rufa* (Desmarest). *Australian Journal of Zoology,* 12, 315–321.

Newsome A (1965a). The abundance of red kangaroos, *Megaleia rufa* (Desmarest), in central Australia. *Australian Journal of Zoology,* 13, 269–287.

Newsome A (1965b). The distribution of red kangaroos, *Megaleia rufa* (Desmarest), about persistent food and water in central Australia. *Australian Journal of Zoology,* 13, 289–299.

Newsome A (1965c). Reproduction in natural population of the red kangaroo, *Megaleia rufa (Desmarest),* in central Australia. *Australian Journal of Zoology,* 13, 735–759.

Newsome A, Stephens D and Shipway A (1967). Effects of a long drought on the abundance of red kangaroos in central Australia. *CSIRO Wildlife Research,* 12, 1–8.

Newsome A (1971). The ecology of red kangaroos. *Australian Zoologist,* 16, 32–50.

Newsome A (1973). Cellular degeneration in the testis of red kangaroos during hot weather and drought in central Australia. *Journal of Reproduction and Fertility, Supplement* 19,191–201.

Newsome A (1975). An ecological comparison of the two arid-zone kangaroos of Australia, and their anomalous prosperity since the introduction of ruminant stock to their environment. *Quarterly Reviews of Biology,* 50, 389–424.

Newsome A (1977). The ages of non-breeding red kangaroos. *Australian Wildlife Research,* 4, 7–11.

Newsome A (1980a). The eco-mythology of the red kangaroo in central Australia. *Mankind,* 12, 327–333.

Newsome A (1980b). Differences in diets of male and female red kangaroos in central Australia. *African Journal of Ecology,* 18, 27–31.

Newsome A, Nicholls N and Hobbs T (1997). The Uncertain Climate. In *The Centenary of the Horn Expedition to Central Australia.* Ed S Morton. Surrey Beatty & Sons Pty Ltd, Sydney (in press).

Press A (1988). Comparisons of the extent of fire in different land management systems of the Top End of the Northern Territory. *Proceedings of the Ecological Society of Australia,* 15, 167–175.

Reid J, Kerle J and Morton S (1993). Uluru Fauna. The distribution and abundance of vertebrate fauna of Uluru (Ayers Rock-Mount Olga National Park, NT). *Kowari,* 4, 152 pp. Australian National Parks and Wildlife Service.

Strehlow T G H (1947). *Aranda Traditions.* Melbourne University Press, Melbourne.

Strehlow T G H (1971). *Songs of Central Australia.* Angus & Robertson, Sydney.

Tunbridge D (1991). *The Story of the Flinders Ranges Mammals.* Kangaroo Press, Sydney.

SECTION 4

PATHOLOGY AND HOMEOSTASIS

CHAPTER 14

THE LABORATORY OPOSSUM (*MONODELPHIS DOMESTICA*) IN BIOMEDICAL RESEARCH

John L VandeBerg and Edward S Robinson

INTRODUCTION

The South American grey short-tailed opossum, *Monodelphis domestica*, is a small (80 to 120 g), nocturnal marsupial native to Brazil and adjacent countries. It is docile, breeds readily in captivity, and produces large litters (typically 6 to 13). Since the first *M. domestica* were imported into the USA in 1978, efforts to develop this species as a laboratory marsupial have been highly successful (VandeBerg 1983, 1990). The standard laboratory diet is a commercial pelleted fox food provided *ad libitum*, and the animals are maintained in polycarbonate or polypropylene rodent cages under unremarkable conditions (VandeBerg 1990). This is the only marsupial species that has been produced in captivity in very large numbers, i.e., tens of thousands. Breeding colonies currently exist at academic and research institutions in Germany, the United Kingdom, Brazil, Australia, and at several locations in the USA. Because of its many advantageous characteristics as a laboratory animal and its economical production in captivity, *M. domestica* has become a prototype species for basic research on marsupial biology, in much the same way that the laboratory mouse is a prototype species for basic research on the biology of eutherian or placental mammals. Consequently, we now refer to *M. domestica* as the laboratory opossum.

The status of *M. domestica* as a laboratory animal was reviewed by VandeBerg (1990). Much progress has been made since then, both in basic biological research with this species and in its development as an animal model for research on human diseases. The present review focuses on recent progress.

MONODELPHIS DOMESTICA AS A LABORATORY ANIMAL

Current composition of colony

Table 14.1 summarises the origin of the current population of laboratory opossums. There were seven separate introductions of animals into captive breeding between 1978 and 1993; the animals came from four different localities in three states of Brazil, and a single locality in Bolivia. In total, 38 animals were imported, of which 28 have contributed to the current gene pool as founders. Only portions of the genomes of each of the 28 founders are still present, because some genes are lost in each generation by chance. For example, between 1978 and 1984, a computer simulation based on the pedigree structure of the colony estimated that 31% of the genetic material of the original 9 animals imported in 1978 already had been lost (MacCluer *et al.* 1986). Therefore, the genetic variability present in the colony is much less than that which was present in the 28 founders. Van Oorschot *et al.* (1992a) have described in detail the geographic origins of the founders obtained by 1988 and genetic differences between the groups descended from them.

Figure 14.1 illustrates the locations at which the founders were trapped. Some of these populations are separated by major rivers and mountain ranges which may substantially inhibit the transfer of genetic material between populations.

Founding animals from the five populations are designated in our laboratory as from Populations 1, 2, 3, 4 or 5, respectively, as are any animals descended entirely from founders of any one population. Population 1 is designated as the reference population, and interbreeding among populations has been restricted to Population 1 and any one other population. Animals descended from ancestors from Population 1 and Population 2, and having less than 30% of their genetic material derived from Population 1, are designated as members of Admixed Population 2. Admixed Population 3 has up to 37.5% of its genetic material from Population 1, and Admixed Population 4 has 50% of its genetic material from Population 4. The Bolivian animals (Population 5) have not been crossed with the Brazilian stocks.

Production rate

More than 29 000 progeny have been weaned in our colony since 1980. Currently, the colony has approximately 2400 individuals of which many are used in experimental research, rather than in the breeding program. The breeding colony consists of about

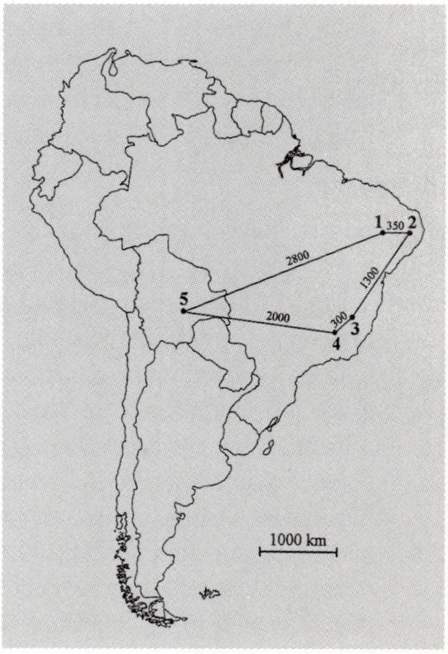

Figure 14.1 Map of South America showing the sites at which each of the 5 sets of founders were trapped. The numbers 1 to 5 refer to the laboratory codes for the population sampled at each of the 5 sites (see Table 14.1). Approximate distances between adjacent sites are indicated in kilometres.

Table 14.1 Founders of the laboratory population of *M. domestica*

MONTH/ YEAR	LOCALITY OF COLLECTION (STATE/TOWN)[1]	LATITUDE, LONGITUDE
05/78	Pernambuco/Exu[2]	7.28°S, 39.45°W
07/84	Paraiba/Piraua[3]	7.29°S, 35.31°W
12/84	Paraiba/Piraua	7.29°S, 35.31°W
10/85	Paraiba/Piraua	7.29°S, 35.31°W
09/88	Paraiba/Piraua	7.29°S, 35.31°W
10/90	Minas Gerais/Joaima	16.39°S, 41.05°W
11/92	Minas Gerais/Conselheiro Mota[4]	18.17°S, 44.00°W
06/93	Santa Cruz/Brecha 3[5]	17.58°S, 63.03°W

[1]All localities are in Brazil, except Santa Cruz, which is in Bolivia.
[2]Exu is 45 km south-west of the city of Crato.
[3]Piraua is 25 km west of the city of Timbaúba.
[4]Conselheiro Mota is 37 km west of the city of Diamantina.
[5]Brecha 3 is 27 km south-east of the city of Santa Cruz.

700 adult females and 450 adult males, and it produces approximately 4600 weanlings per year, in addition to sucklings used experimentally before weaning. Approximately 300 individuals are provided to other institutions each year as experimental subjects or breeding nuclei. The rest are used as research subjects, or as replacement breeders to perpetuate stocks or to continue the development of inbred strains. Detailed records have been kept on each animal in the colony, which is fully pedigreed, and the data base is computerised. Some pedigrees extend back as many as 22 generations.

Inbreeding

Because of the low number of founders and the large number of animals produced over many generations, a limited amount of inbreeding has been unavoidable in the colony at large. In addition, some groups have been purposely inbred over several generations to develop partially inbred strains. Brother-sister matings were used in initial attempts to develop inbred strains. One stock achieved an inbreeding coefficient of 0.911 before it was lost, but attempts to produce fully inbred strains (F=0.986, equivalent to that achieved after 20 consecutive generations of full-sib matings) have not succeeded because of rapid decline in fecundity and infant survival after 5 or 6 generations. We hypothesise that the low recombination rate in females by comparison with eutherian mammals (van Oorschot *et al.* 1992b; 1993) inhibits the formation of new gene combinations sufficiently quickly in brother-sister mating schemes to enable the acquisition of gene combinations compatible with an inbred state. However, we are succeeding in

LABORATORY CODE FOR THE POPULATION	NUMBER SENT TO USA	NUMBER THAT CONTRIBUTED TO EXISTING GENE POOL	PROPORTIONATE CONTRIBUTION TO EXISTING GENE POOL
1	9	9	0.472
2	2	2	0.481
2	11	7	
2	8	3	
2	3	2	
3	2	2	0.013
4	1	1	0.011
5	2	2	0.023
Total	**38**	**28**	**1.000**

gradually increasing the inbreeding coefficients in partially inbred stocks by less severe inbreeding schemes, and are hopeful of achieving fully inbred strains in this manner. At the present time, the colony has 34 partially inbred stocks. Of those that continue to have high fecundity and infant survival rates, the most highly inbred has an average inbreeding coefficient of 0.674. Approximately 600 members of the colony have inbreeding coefficients in excess of 0.6. Since more than 60% of their gene loci have alleles that are identical by descent, these animals are valuable for research designed to detect recessive genes for susceptibility to experimentally induced diseases. They also are especially useful for intercrossing and backcrossing for linkage analysis and gene-mapping purposes.

Increasing use of M. domestica

Animals superfluous to our own research needs have been available for distribution to other laboratories since 1984, and the species has already made important contributions in many areas of basic and biomedical research. The annual number of publications, which were indexed in BIOSIS Previews (all life science subjects), about *M. domestica* rose from 0 in 1978 to 14 in 1986 (VandeBerg 1990). The number has grown steadily since then by approximately 5 per year: 45 were indexed in 1992 and 54 in 1993, making a total of 276 indexed publications since 1978. The possibility that this rate of expansion of the scientific literature on *M. domestica* is simply a reflection of increasing rate of publication on all marsupial species is not substantiated by a count of the numbers of publications about *Macropus eugenii* and *Sminthopsis crassicaudata*, two other commonly used laboratory marsupials; for both of those species the annual number of publications has remained approximately constant with means of 25.2 and 4.2, respectively, during the 5-year-period from 1989 to 1993.

Some areas of investigation have developed because of the marsupial characteristics of this new laboratory animal; for instance, birth after a short gestation enables manipulation of young at an immature stage of development (Tyndale-Biscoe & Janssens 1988; Janssens *et al.* 1995, chapter 4 in this volume). Others arose because of the value of an out-group of mammals for comparative research strategies; for example, marsupials exhibit paternal X-chromosome inactivation rather than random X-chromosome inactivation, as is characteristic of eutherian mammals. Laboratory opossums are used as a model for comparative research on these divergent strategies to accomplish balance between sexes in expression of X-linked genes (reviewed by VandeBerg *et al.* 1987). Still other areas have arisen simply because preliminary studies on physiological systems of *M. domestica* revealed research opportunities that did not exist for the conventional laboratory animal species that were available. These other research opportunities are not related to the marsupial characteristics of laboratory opossums, but arose simply as a consequence of the existence of a new laboratory animal with its own species-specific characteristics.

Ultraviolet radiation-induced skin and eye cancers and dietary induced hyper-cholesterolemia, discussed in this review, are diseases for which *M. domestica* has become an important model species for reasons unrelated to its marsupial characteristics.

M. DOMESTICA AS A MODEL FOR MELANOMA

The incidence of human melanomas is increasing at an alarming rate, indeed at a faster rate than any other type of cancer except lung cancer in women. Ultraviolet radiation (UVR) is believed to be a major causative factor. Unfortunately, research on the role of UVR in inducing melanoma in humans has been hampered by the lack of a mammalian model in which melanoma can be induced by UVR alone. UVR is not a factor in the induction or promotion of melanoma in the Sinclair swine model (Greene *et al.* 1994) and in mice, UVR must be combined with chemical carcinogens to induce melanomas (Kripke 1979). Although a newly developed transgenic mouse appears to be a promising model for research on UVR-induced melanomas (Mintz & Silvers 1993; Klein-Szanto *et al.* 1994), the laboratory opossum has the distinction of being the only naturally existing mammal, other than humans, known to be susceptible to melanoma in response to UVR alone.

Melanocytic naevi

In humans, some types of melanocytic naevi or moles are a risk factor for melanoma; occasionally a cell in these naevi can undergo transformation and become a melanoma cell. Conditions for the production of melanocytic naevi by repeated shaving of adult laboratory opossums and exposure to UVR were developed by Dr R D Ley and his colleagues (see Kusewitt *et al.* 1991 and references therein). The protocol in our laboratory involves exposure to a spectral peak of 302 nm (UVB) at a dose of 125 J/m^2 per exposure. This dose is equivalent to exposure at 40 cm from the lamp for 1 to 1.75 min, depending on the age of the lamp. A few animals exhibited slight erythema after the initial exposures, but none exhibited severe sunburn. In our experience, the incidence of melanocytic naevi after three exposures per week was 7% after 30 weeks and 14% after 45 weeks (VandeBerg *et al.* 1994a). When the same exposure regimen was initiated in weanlings, rather than young adults, the incidence was considerably lower: only 5% after 45 weeks (Robinson *et al.* 1995).

Six cell lines were developed from melanocytic naevi induced in the earlier experiments with adults (Dooley *et al.* 1993). Cytogenetic analyses revealed that 3 of them exhibited aneuploidy involving extra copies of chromosomes 3, 5, 7 and/or 8. This result suggested that aneuploidy of these particular chromosomes might be an early event in the induction of melanocytic naevi.

In an attempt to reduce the time and effort required to produce animals with naevi as potential precursors of melanoma, we devised a technology for irradiating laboratory opossums repeatedly between birth and 19 days of age, when they begin to become furred (Robinson *et al.* 1994). This technology depends on the fact that *M. domestica* are pouchless, so the neonates are fully exposed on the ventral side of their mother. The protocol involves placing the UVR source beneath a wire grid, aimed up at the litter of a mother standing on the grid. With a mean litter size of 7, many individuals can be irradiated in a relatively short time, without the need for anaesthesia or shaving.

Although the optimal exposure protocol remains to be determined, the combined results of all protocols attempted have yielded an incidence of 10% of naevus-bearing animals at, or up to, 1 month after weaning (8 weeks) (Robinson *et al.* 1994). Additional exposure protocols are currently being evaluated, and experiments are in progress to determine if susceptibility to naevus formation is under genetic control.

Melanomas

In some instances, we have further irradiated animals that had been exposed as sucklings. The second irradiation treatment schedule involved shaving and irradiating the weanlings three times per week, as in the earlier experiments. Of the animals that had a naevus at weaning 40% developed malignant melanoma after up to 45 weeks of post-weaning UVR. In each case, the original naevus continued to grow, underwent changes in appearance, and metastasised to one or more lymph nodes.

Cell lines clonally derived from pigmented cells from an affected lymph node provided definitive evidence of malignancy in this model (Robinson *et al.* 1994). Not only were they highly dendritic and pigmented *in vitro* (properties of melanotic metastatic melanoma), but they were moderately tumourigenic and produced pigment after injection into immunoincompetent nude mice.

M. *DOMESTICA* AS A MODEL FOR ANGIOGENESIS AND CORNEAL CANCER

Exposure to environmental or artificial UVR is widely believed to be a contributing factor to loss of transparency (opacification) of the cornea and to angiogenesis (the formation of blood vessels by neovascularisation) (discussed in Applegate & Ley 1991). The laboratory opossum is particularly susceptible to the development of these abnormalities after repeated exposure to low doses of UVR, and provides an excellent model for investigation of the biological mechanisms involved. In many individuals of this species, opacification and neovascularisation are followed by the development of corneal sarcomas. Although UVR-induced corneal cancers are uncommon and rarely metastasise in humans (Erie *et al.* 1986), those that develop in *M. domestica* provide a valuable generalised sarcoma model.

Initiation and progression

We developed a number-coded system for clinical assessment of histopathological changes in corneal phenotype caused by UVR exposure ranging from normal, through opacification and angiogenesis, to an advanced, highly invasive stage (VandeBerg *et al.* 1994b). The development of most types of tumours is accompanied by angiogenesis. Angiogenesis is a clearly identifiable early event in corneas of UV-irradiated *M. domestica* during transformation of normal stromal (fibroblastic) cells into tumor (sarcoma) cells. Blood vessels originating in the limbal region around the periphery of the cornea colonise the stroma and become most pronounced at foci of neoplastic transformation. The resulting sarcoma cells soon spread through the cornea into adjacent ocular structures and at advanced stages they occupy much of the eyeball, eventually invading the orbit and migrating along the optic nerve to the brain. Little is known about the molecular mechanisms by which UVR induces these potentially life-threatening corneal sarcomas. Mutationally activated or amplified *ras* oncogenes (K-*ras*, N-*ras*, and H-*ras*) have been found in experimentally induced animal tumours and naturally occurring human tumours, and activation of K-*ras* may play a role in the development of UVR-induced corneal tumours in *M. domestica* (Sabourin *et al.* 1992). Moreover, it appears that the opossum K-*ras* gene can be mutationally activated in a manner similar to the K-*ras* genes of eutherian mammals (Kusewitt *et al.* 1993).

Genetics

One hundred fifty-one individuals, consisting of 33 sibships belonging to 9 families, completed at least 30 weeks of the UVR protocol, and 137 completed 45 weeks (VandeBerg *et al.* 1994b). Although the sample size is small for genetic analysis, particularly because only a single generation was tested, estimates of heritability (h^2) of corneal sarcoma were high. After 30 weeks of treatment, h^2 was 0.74 with a 95% confidence interval of 0.30 to 1.00; after 45 weeks, h^2 was 1.00 with the 95% confidence interval ranging down to 0.35. It will be important to refine the estimate of heritability with a reduced confidence interval, but already it is clear that susceptibility to UVR-induced corneal sarcoma is highly heritable in this population.

M. DOMESTICA AS A MODEL FOR DIETARY INDUCED HYPERCHOLESTEROLAEMIA

Cardiovascular disease remains the greatest single cause of health problems and mortality in industrialised societies, and characteristics of blood cholesterol are among the most important risk factors. Intensive research on lipid metabolism since the mid-1970's has greatly enhanced our understanding of disorders of human lipid metabolism. Some of that progress is attributable to the use of animal models, including mice, rabbits, swine, and non-human primates. Most of the emphasis and progress in research has related to constitutive lipid disorders, such as familial hypercholesterolaemia, rather than the dietary-induced lipid disorders which are substantially responsible for the most cardiovascular disease. Animal models that exhibit genetic variation in lipemic responsiveness to dietary components are needed to dissect out the genes that are believed to contribute to individual variability in this physiological characteristic.

Research to date with the laboratory opossum indicates that this species provides a unique and valuable model for this purpose. It exhibits extensive individual variation in response to dietary fat and cholesterol, and that variation appears to be largely controlled by a single recessive gene. As in humans and other animal models, hyperlipidaemia can lead to atherosclerotic lesions in *M. domestica* (H C McGill Jr & J L VandeBerg, unpublished observations). To our knowledge, no other animal model exists of a dietary induced hyperlipidaemia that is both as pronounced as in *M. domestica* and as highly regulated by a single gene. Identification of this gene and characterisation of its mechanism of action may lead to a better understanding of dietary-induced changes in human lipoprotein characteristics. Indeed, the omnivorous characteristic of *M. domestica* may make it a better model than conventional laboratory animals for research on physiological effects of dietary fat and cholesterol.

Plasma cholesterol levels

The fasting plasma cholesterol level in laboratory opossums fed a standard laboratory diet (National Complete Fox Food Pellets, Reproduction Diet, Milk Specialties Co., P.O. Box 119, New Holstein, Wisconsin) is 85 ± 22 mg/100 ml (VandeBerg & Cheng 1985a). For experimental dietary challenges, the diet is enriched with lard from the basal level of 8.1 to 17.7% fat (dry weight) and with cholesterol from 0.04 to 0.61% dry weight. This level of dietary fat corresponds to about 40% of calories from fat, equivalent to the mean in many industrialised societies. The level of cholesterol is perhaps six-fold higher than the mean in North America, but not above the high end of the range of human cholesterol consumption. After laboratory opossums have eaten this challenge diet for 8 weeks, the population data for plasma cholesterol levels are distinctly bimodal. In two studies, the means of the low responding group were 109 ± 13 mg/100 ml and 118 ± 17 mg/100 ml (VandeBerg & Cheng 1985a; Rainwater & VandeBerg 1992). The means of the high responding groups in the two studies were 304 ± 213 mg/100 ml and 593 ± 74 mg/100 ml. Many animals attain plasma cholesterol levels in the range of 1000 to 1900 mg/100 ml after eating this diet for 8 weeks (J L VandeBerg, unpublished observations).

For the purpose of comparison, human plasma cholesterol levels in the range of 200 to 239 mg/100 ml are considered to be borderline-high, and levels in excess of 240 mg/100 ml confer high risk of cardiovascular disease. Individuals heterozygous for familial hypercholesterolaemia typically have plasma cholesterol levels of 350 to 400 mg/100 ml.

Distribution of cholesterol among lipoprotein classes

A large compilation of data from animal models and human subjects indicates that the amount of cholesterol carried by very low and low density lipoproteins is a causative factor in cardiovascular disease, whereas that carried by high density lipoproteins is protective (see Gotto *et al.* 1990, and references therein). A strong indicator of risk of cardiovascular disease is the ratio of low density lipoprotein cholesterol to high density lipoprotein cholesterol.

In *M. domestica*, which are low responders, this ratio is relatively constant regardless of diet: 0.40 ± 0.06 mg/100 ml and 0.55 ± 0.11 mg/100 ml for the standard and challenge diets, respectively. However, the ratio is dramatically increased in the high responders fed the challenge diet, from 0.49 ± 0.05 to 6.48 ± 1.28 mg/100 ml (Rainwater & VandeBerg 1992). Thus, responsiveness is primarily a characteristic of low density lipoprotein cholesterol.

The characteristics of *M. domestica* lipoproteins under these two dietary regimens have been investigated in detail (Rainwater & VandeBerg 1992). The close resemblance to human lipoproteins in physical characteristics and specific apolipoprotein contents, indicates that results for the laboratory opossum model can be appropriately interpreted in relation to human lipoprotein metabolism.

Genetics

The initial pedigree data were limited, but were consistent with the hypothesis that plasma cholesterol levels after dietary challenge were largely controlled by a single recessive gene (VandeBerg & Cheng 1985b).

More recently, data from 327 individuals have been analysed by complex segregation analysis (Atwood *et al.* 1994). This analysis focused on low density lipoprotein cholesterol, and concluded that plasma concentration in animals fed the challenge diet is inherited as a major gene. The results further indicated that 68% of the total variance in the population is attributed to variation at this gene locus. Such a high contribution by a single gene to any quantitative character is unusual, and emphasises the potential value of the *M. domestica* model. One or more other genes appear also to influence low density lipoprotein cholesterol concentrations because the residual heritability accounted for 17% of the total variance. Thus, the model may enable an understanding of how a major gene interacts with modifying genes and environmental (dietary) factors in controlling a quantitative phenotype of medical importance.

GENE MAPPING

Considerable effort has been committed to marsupial gene mapping, and the results have been informative in relation to understanding mammalian karyotypic evolution. Most mapping data have been derived by *in situ* hybridisation of tritium-labelled human gene probes to chromosomes of marsupial fibroblasts, and some were obtained by somatic cell genetic techniques. All published data derived for marsupials from these techniques are from macropodids (kangaroos), dasyurids (Australian 'marsupial mice') and phalangerids (Australian possums)(O'Brien *et al.* 1993). The laboratory opossum offers a good opportunity to expand the comparative approach to understanding mammalian genome evolution because American and Australian marsupials are only distantly related, having evolved as separate lineages for an estimated 70 million years (Maxson *et al.* 1975; Lowenstein *et al.* 1981). In contrast, it is estimated that 120 to 156 million years have elapsed since the separation of the marsupial and eutherian lineages (Hope *et al.* 1990; Mark & Marotte 1992). Therefore, in instances where syntenies have been disrupted between orders of eutherian mammals or between Australian marsupials and eutherian mammals, the syntenic relationships in *M. domestica* can be used to determine which synteny represents the ancestral condition.

From the standpoint of biomedical research, the more important aspect of gene mapping is linkage relationships (compared with syntenic relationships). For that reason we have focused our gene mapping strategies on linkage analysis. The intent is to construct a linkage map for localising genes that affect normal physiological characteristics, physiological responses to environmental stimuli, and susceptibility to experimentally induced diseases. If such genes can be localized in the *M. domestica* genome, and if the genome is sufficiently well mapped to relate segments of chromosomes to the human genome, then it may be possible to clone those genes from human libraries and to identify variation in them in relation to human physiological conditions.

Karyotype

M. domestica, like all other marsupials, has a low number of chromosomes (2n=18 in this species), which are large and ideally suited for cytogenetic analysis (Merry *et al.* 1983). Techniques for high resolution G-banding and R-banding have been developed, and 223 bands can be resolved per haploid set of chromosomes (Pathak *et al.* 1993).

Protein polymorphisms

Initial efforts to identify genetic markers focused on blood protein polymorphisms in descendants from Population 1 and Population 2. Fifty-four blood protein markers have been surveyed by electrophoretic- and isoelectric-focusing techniques and polymorphism was detected in 13 of them (ACP1, ACP2, ADH3, AK1, ALDH3, APRT, AT3, C6, C7, DIA1, GAPD, GPT, PI; abbreviations from Human Gene Mapping 11, 1991)(Sevilir *et al.* 1989; van Oorschot & VandeBerg 1989; VandeBerg 1990; Holmes *et al.* 1990, 1991, 1992; Bell *et al.* 1992, van Oorschot *et al.* 1992a,b, 1993). The mean proportion of heterozygous loci for these markers is approximately 5% per individual. This level of polymorphism is consistent with that in most eutherian species and slightly higher than that in macropodids (16.7% of loci heterozygous, 4% mean heterozygosity per individual; Cooper *et al.* 1979).

Numerous additional protein markers and three more populations (3, 4 and 5) remain to be surveyed. The founders of these three populations were trapped long distances from the founders of Populations 1 and 2 (Figure 14.1, page 000), so we anticipate the discovery of considerably more protein polymorphisms in the laboratory population of *M. domestica*.

DNA polymorphisms

Efforts to identify restriction length polymorphisms in genomic DNA of *M. domestica* have been initiated. Preliminary results from several human probes used in Southern blot analyses are promising (P B Samollow, personal communication), and one polymorphic system detected by a probe for the N-*ras* oncogene (ATCC #41030, 1.5 Kb genomic DNA fragment) has been unequivocally established as consistent with Mendelian inheritance. This probe detects two *Hind*III polymorphisms in *M. domestica*. One consists of two alleles represented by 11.5 and 8.5 Kb fragments. Animals of mixed ancestry from Populations 1 and 2 exhibited frequencies of 0.32 and 0.68 for the two alleles (van Oorschot *et al.* 1991). The other is a two-allele system involving 7.0 and 4.5 Kb fragments (Perelygin *et al.* 1994). All animals typed from Populations 1 and 2 were homozygous for the 4.5 Kb fragment whereas the two founders from Joaima (Population 3) were homozygous for the 7.0 Kb fragment. Although the data are too limited to be interpreted as indicative of fixed allelic differences, they do support the notion of substantial differences in gene frequencies between the populations (Perelygin *et al.* 1994).

To increase the number of markers available for linkage analysis more quickly, and to increase the efficiency of typing animals for the purpose of linkage analysis, 10-mer primers are being used to generate random amplified polymorphic DNA markers for typing by the polymerase chain reaction. Initial results have established 9 random amplified polymorphic

DNA markers that segregate as Mendelian markers, indicating that these will be a useful adjunct to protein polymorphisms and restriction length polymorphisms in developing a linkage map (P B Samollow, personal communication). Another strategy that is planned for development is the isolation of *M. domestica* microsatellite sequences for developing polymerase chain reaction primers specific for this species.

Recombination and linkage

Two linkage groups have been identified in *M. domestica*, and both exhibit much lower recombination rates in females than in males (van Oorschot *et al.* 1992b, 1993). AK1 and PI exhibit a recombination rate of 0.087 ± 0.23 (standard error) in females and are unlinked in males, and GPT and the C6-C7 haplotype exhibit a recombination rate of 0.172 ± 0.056 in females and are unlinked in males. C6 and C7 have not yielded recombinants in either sex, and also are very tightly linked in eutherian mammals (Coto *et al.* 1991; Eldridge *et al.* 1983; Orren *et al.* 1985).

Because of the estimated times of divergence of marsupials and eutherian mammals, these 2 linkage groups are informative in regard to mammalian genome evolution. AK1 and PI are not syntenic in mice or humans; C6 and C7 are tightly linked in mice and humans and presumably arose by gene duplication; and GPT is syntenic with C6-C7 in mice (chromosome 15) but not in humans (C6-C7 on chromosome 5, GPT on chromosome 8) (O'Brien *et al.* 1993). Based on these comparisons, it seems likely that the C6-C7-GPT synteny is ancestral to the split between marsupials and eutherian mammals and was broken during evolution of the human genome.

The sex difference in recombination rate is consistent with results of cytogenetic analyses, which revealed that chiasmata are clustered near the telomeres in female meiosis, but occur more uniformly along the length of the chromosomes in male meiosis (Hayman *et al.* 1988). These recombination and cytogenetic characteristics parallel those previously observed in an Australian dasyurid, *Sminthopsis crassicaudata* (Bennett *et al.* 1986).

The sex-specific difference in recombination frequencies will be valuable for detecting syntenies and doing long-range mapping via matings established with multiply heterozygous females, whereas multiply heterozygous males can be used in crosses designed for shorter-range mapping.

FUTURE DIRECTIONS

Colony maintenance and development

We plan to maintain random breeding stocks of animals belonging to Populations 1, 2 and 5, although Population 5 will probably become inbred too quickly to enable its perpetuation without admixing with another population. In addition, random breeding stocks of admixed Populations 2, 3, and 4 will be maintained. These populations will enable the preservation of as much genetic variability as is practical, while simultaneously providing genetically distinct stocks that may serve different purposes.

Although some of the 34 partially inbred stocks will probably be lost, we will pursue the goal of developing fully inbred strains by gradually increasing the mean inbreeding coefficients of the existing stocks.

Melanoma

The capability of screening large numbers of animals for susceptibility to UVR-induced naevi has led us to embark on a genetic analysis of this characteristic. Because the naevus phenotype is scored before sexual maturity in animals irradiated as sucklings, prospective matings based on phenotype can be implemented. Whether the results establish a genetic basis to susceptibility or not, the model makes feasible research on the molecular mechanisms by which UVR can cause melanocytes to become naevi and eventually metastatic melanoma. The whole animal model also may be useful for testing chemopreventative and chemotherapeutic agents for melanoma. An initial investigation of this type supported the feasibility of the model for this purpose (Maenpaa *et al.* 1993).

The malignant melanoma cell lines are attractive as experimental model systems for melanoma research because they are derived from tumours that were initiated and promoted by UVR alone and they retain the ability to produce pigment (unlike most human metastatic melanoma cell lines currently available). These cell lines should be useful reagents for the discovery and development of anti-melanoma therapies, including conventional chemotherapeutic studies of cytotoxic and cytostatic activity *in vitro*. There is also the prospect of successful allogeneic grafting of metastatic cell lines in this species, for use in *in vivo* chemotherapy trials and for testing immuno-modulatory therapies (Robinson *et al.* 1994).

Corneal cancer

UVR-induced sarcomas in the cornea of *M. domestica* provide a valuable generalised sarcoma model, particularly for studies on angiogenesis and oncogenic transformation in this translucent, accessible, and immunologically privileged site. In normal cells, angiogenesis is controlled by a balance between endogenous factors that stimulate new blood vessel growth and factors that inhibit it. In contrast, tumour cells are potently angiogenic as a result of decreased production of inhibitors and increased secretion of inducers (Dameron *et al.* 1994). The anterior chamber of the eye of *M. domestica* is a favourable and readily available site for *in vivo* studies on the effects of reagents that are known or likely to promote or inhibit angiogenesis. For example, it will be possible to examine the effects of known inhibitory factors of angiogenesis (Fidler & Ellis 1994), including the recently identified and characterised antiangiogenesis factor, angiostatin (O'Reilly *et al.* 1994). *M. domestica* corneal sarcoma cells also grow well in culture. It will therefore be possible to combine *in vitro* and *in vivo* strategies to test the efficacy of established or candidate chemotherapy and immunotherapy reagents in this new sarcoma model.

Hypercholesterolaemia

In addition to providing a model for genetic analysis of dietary responsiveness, the laboratory opossum also has clear potential for research on other dietary factors that affect plasma cholesterol levels. For example, the model can be used to examine the lipemic effects of different types of dietary fat alone, in combination with one another, and in combination with dietary cholesterol. As the major gene and modifying gene(s) become better understood, and perhaps even identified, this system could serve as a model not only of biomedical importance, but also of importance to understanding complex genotype-by-genotype and genotype-by-diet interactions. In addition, the model has clear potential for testing the efficacy of cholesterol-lowering drugs.

Gene mapping

Although few linkage groups have been identified in all marsupial species combined, *M. domestica* is well suited for linkage mapping because of its high reproductive capacity. Preliminary data suggest that Population 5 (from Bolivia) may be quite different from the Brazilian populations of *M. domestica* in gene frequencies, and that the Bolivian population may be monomorphic for some alleles not present in the Brazilian population (P B Samollow, personal communication). We plan to produce hybrids and backcross progeny from matings between Populations 1 and 5 and to use this panel of animals as the focus of future linkage mapping efforts. We anticipate that the backcross involving Populations 1 and 5 together with the use of polymerase chain reaction technologies to identify random amplified polymorphic DNA markers and microsatellite polymorphisms will rapidly increase the pace of linkage mapping in the species.

CONCLUSION

Although the laboratory opossum appeared on the research scene less than two decades ago, its value to existing research programs in comparative biology and medicine is evident. Given the wealth of baseline data that now exists for this species, its ready availability, and the substantial number of investigators using it in experimental research, the years ahead are bound to offer exciting new opportunities to exploit this prototype laboratory marsupial.

Acknowledgments

We thank Dr Paul B Samollow for permission to cite unpublished data. Acquisition of data and preparation of the manuscript were supported in part by National Institutes of Health grant PO1 RR09919 and a grant from the Robert J Kleberg, Jr and Helen C Kleberg Foundation.

References

Atwood L D, Kammerer C M, VandeBerg J F, Hackelman S M and VandeBerg J L (1994). A major gene affects V+LDL cholesterol concentrations in *Monodelphis domestica*. *Texas Genetics Society*, 21st Annual Meeting, Houston, Texas. 21(14).

Applegate L A and Ley R D (1991). DNA damage is involved in the induction of opacification and neovascularization of the cornea by ultraviolet radiation. *Experimental Eye Research*, 52, 493–497.

Bell K, Arthur H, van Oorschot R A H and VandeBerg J L (1992). Antithrombin III (AT3) polymorphism in the marsupial *Monodelphis domestica*: identification and genetics. *Biochemical Genetics*, 30, 591–601.

Bennett J H, Hayman D L and Hope R M (1986). Novel sex differences in linkage values and meiotic chromosome behaviour in a marsupial. *Nature*, 323, 59–60.

Cooper D W, Johnston P G, VandeBerg J L, Maynes G M and Chew G W (1979). A comparison of genetic variability of X linked and autosomal loci in kangaroos, man, and Drosophila. *Genetical Research*, 33, 243–252.

Coto E, Martinez-Naves E, Dominguez O, DiScipio R G, Urra J M and Lopez-Larrea C (1991). DNA polymorphisms and linkage relationship of the human complement component *C6*, *C7*, and *C9* genes. *Immunogenetics*, 33, 184–187.

Dameron K M, Volpert O V, Tainsky M A and Bouck N (1994). Control of angiogenesis in fibroblasts by p53 regulation of thrombospondin-1. *Science*, 265, 1582–1584.

Dooley T P, Mattern V L, Moore C M, Porter P A, Robinson E S and VandeBerg J L (1993). Cell lines derived from ultraviolet radiation-induced benign melanocytic nevi in *Monodelphis domestica* exhibit cytogenetic aneuploidy. *Cancer Genetics and Cytogenetics*, 71, 55–66.

Eldridge P R, Hobart M J and Lachmann P J (1983). The genetics of the sixth and seventh components of complement in the dog: polymorphism, linkage, locus duplication, and silent alleles. *Biochemical Genetics*, 21, 81–91.

Erie J C, Campbell R J and Liesegang T J (1986). Conjunctival and corneal intraepithelial and invasive neoplasia. *Ophthalmology*, 93, 176–183.

Fidler I J and Ellis L M (1994). The implications of angiogenesis for the biology and therapy of cancer metastasis. *Cell*, 79, 185–188.

Greene J F Jr, Townsend J S and Amoss M S Jr (1994). Histopathology of regression in Sinclair swine model of melanoma. *Laboratory Investigation*, 71, 17–24.

Gotto A M Jr, LaRosa J C, Hunnunghake D, Grundy S M, Wilson P W, Clarkson T B, Goodman D S and Hay J W. (1990). The cholesterol facts. A summary of the evidence relating dietary fats, serum cholesterol and coronary heart disease. A joint statement by the American Heart Association and the National Heart, Lung and Blood Institute. *Circulation*, 81, 1721–1733.

Hayman D L, Moore H D M and Evans E P (1988). Further evidence of novel sex differences in chiasma distribution in marsupials. *Heredity*, 61, 455–458.

Holmes R S, van Oorschot R A H and VandeBerg J L (1990). Genetics of alcohol dehydrogenase and aldehyde dehydrogenase from *Monodelphis domestica* cornea: further evidence for identity of corneal aldehyde dehydrogenase with a major soluble protein. *Genetical Research*, 56, 259–265.

Holmes R S, van Oorschot R A H and VandeBerg J L (1991). Aldehyde dehydrogenase (ALDH) isozymes in the gray short-tailed opossum (*Monodelphis domestica*): tissue and subcellular distribution and biochemical genetics of ALDH3. *Biochemical Genetics*, 29, 163–175.

Holmes R S, van Oorschot R A H and VandeBerg J L (1992). Biochemical genetics of alcohol dehydrogenase isozymes in the gray short-tailed opossum (*Monodelphis domestica*). *Biochemical Genetics*, 30, 215–231.

Hope R, Cooper S and Wainwright B (1990). Globin macromolecular sequences in marsupials and monotremes. *Australian Journal of Zoology*, 37, 289–313.

Klein-Szanto A J P, Silvers W K and Mintz B (1994). Ultraviolet radiation-induced malignant skin melanoma in melanoma-susceptible transgenic mice. *Cancer Research*, 54, 4569–4572.

Kripke M L (1979). Speculations on the role of ultraviolet radiation in the development of malignant melanoma. *Journal of the National Cancer Institute*, 63, 541–548.

Kusewitt D F. Applegate L A and Ley R D (1991). Ultraviolet radiation-induced skin tumors in a South American opossum (*Monodelphis domestica*). *Veterinary Pathology*, 28, 55–65.

Kusewitt D F, Kelly G, Sabourin C L K and Ley, R D (1993). Characterization of the K-*ras* gene of the marsupial *Monodelphis domestica*. *DNA Sequence*, 4, 37–42.

Lowenstein, J M Sarich, V M and Richardson, B J (1981). Albumin systematics of the extinct mammoth and Tasmanian wolf. *Nature*, 291, 409–411.

MacCluer, J W VandeBerg, J L Read, B and Ryder, O A (1986). Pedigree analysis by computer simulation. *Zoo Biology*, 5, 147–160.

Maenpaa J, Dooley T, Wurz G, VandeBerg J, Robinson E, Emshoff V, Sipila P, Wiebe V, Day C and DeGregorio M (1993). Topical toremifene: a new approach for cutaneous melanoma. *Cancer Chemotherapy and Pharmacology*, 32, 392–395.

Mark R F and Marotte L R (1992). Australian marsupials as models for the developing mammalian visual system. *Trends in Neurosciences*, 15, 51–57.

Maxson L R, Sarich V M and Wilson A C (1975). Continental drift and the use of albumin as an evolutionary clock. *Nature*, 255, 397–400.

Merry D E, Pathak S and VandeBerg J L (1983). Differential NOR activities in somatic and germ cells of *Monodelphis domestica* (Marsupialia, Mammalia). *Cytogenetics and Cell Genetics*, 35, 244–251.

Mintz B and Silvers W K (1993). Transgenic mouse model of malignant skin melanoma. *Proceedings of the National Academy of Sciences USA*, 90, 8817–8821.

O'Brien S J, Peters J, Searle A G, Womack J E, Johnson P A and Graves J A M (1994). Report of the committee on comparative gene mapping. In *Human Gene Mapping* 1993. A compendium. Eds A J Cuticchia and P L Pearson. John Hopkins University Press, Baltimore. pp 846–892.

O'Reilly M S, Holmgren L, Shing Y, Chen C, Rosenthal R A, Moses M, Lane W S, Cao Y, Sage E H and Folkman J (1994). Angiostatin: a novel angiogenesis inhibitor that mediates the suppression of metastases by a Lewis lung carcinoma. *Cell*, 79, 315–328.

Orren A, Hobart M J, Nash H R and Lachmann P J (1985). Close linkage between mouse genes determining the two forms of complement component C6 and component C7, and *Cis* action of a *C6* regulatory gene. *Immunogenetics*, 21, 591–599.

Pathak S, Ronne M, Brown N M, Furlong C L and VandeBerg J L (1993). A high-resolution banding pattern ideogram of *Monodelphis domestica* chromosomes (Marsupialia, Mammalia). *Cytogenetics and Cell Genetics*, 63, 181–184.

Perelygin A A, Samollow P B and VandeBerg J L (1994). A new *Hind*III RFLP of N*RAS* in the laboratory opossum, *Monodelphis domestica*. *Animal Genetics*, 25, 432.

Rainwater, D L and VandeBerg, J L (1992). Dramatic differences in lipoprotein composition among gray short-tailed opossums (*Monodelphis domestica*) fed a high cholesterol/saturated fat diet. *Biochimica et Biophysica Acta*, 1126, 159–166.

Robinson E S, Hubbard G B and VandeBerg J L (1995). Ultraviolet radiation-induced skin lesions in laboratory opossums (*Monodelphis domestica*) exposed from the weanling stage. *Archives of Dermatological Research*, 287, 333–337.

Robinson E S, VandeBerg J L, Hubbard G B and Dooley T P (1994). Malignant melanoma in ultraviolet-irradiated laboratory opossums: initiation in suckling young, metastasis in adults, and xenograft behavior in nude mice. *Cancer Research*, 54, 5986–5991.

Sabourin C L K, Freeman A G, Kusewitt D F and Ley R D (1992). Identification of a transforming *ras* oncogene in an ultraviolet radiation-induced corneal tumor of *Monodelophis domestica*. *Photochemistry and Photobiology*, 55, 417–424.

Sevilir N S, Torgerson E, Guajardo A C, VandeBerg J L and Stone W H (1989). Acid phosphatase polymorphism in a laboratory marsupial, *Monodelphis domestica*. *Biochemical Genetics*, 27, 755–759.

Tyndale-Biscoe C H, and Janssens P A (1988). *The Developing Marsupial. Models for Biomedical Research*. Springer-Verlag, Berlin.

VandeBerg J L (1983). The gray short-tailed opossum: a new laboratory animal. *Institute of Laboratory Animal Resources News*, 26, 9–12.

VandeBerg J L (1990). The gray short-tailed opossum (*Monodelphis domestica*) as a model didelphid species for genetic research. *Australian Journal of Zoology*, 37, 235–247.

VandeBerg J L and Cheng M-L (1985a). A heritable diet-induced hypercholesterolemia in a laboratory marsupial. 7th International Symposium on Atherosclerosis, Melbourne, Program and Abstracts, np.

VandeBerg J L and Cheng M-L (1985b). Dyslipoproteinemia in a laboratory marsupial, *Monodelphis domestica*. *Isozyme Bulletin*, 18, 66.

VandeBerg J L, Robinson E S, Samollow P B and Johnston P G (1987). X-linked gene expression and X-chromosome inactivation: marsupials, mouse and man compared. *Isozymes: Current Topics in Biological and Medical Research*, 15, 225–253.

VandeBerg J L, Williams-Blangero S, Hubbard G B, Ley R D and Robinson E S (1994a). Genetic analysis of ultraviolet radiation-induced skin hyperplasia and neoplasia in a laboratory marsupial model (*Monodelphis domestica*). *Archives of Dermatological Research*, 268, 12–17.

VandeBerg J L, Williams-Blangero S, Hubbard G B and Robinson E S (1994b). Susceptibility to ultraviolet-induced corneal sarcomas is highly heritable in a laboratory opossum model. *International Journal of Cancer*, 56, 119–123.

van Oorschot R A H and VandeBerg J L (1989). Genetic variation in adenylate kinase 1, glyceraldehyde-3-phosphate dehydrogenase, and glutamic-pyruvate transaminase in the marsupial *Monodelphis domestica*. *Biochemical Genetics*, 27, 761–765.

van Oorschot R A H, Birmingham V and VandeBerg J L (1991). HINDIII RFLP for NRAS in *Monodelphis domestica*. *Nucleic Acids Research*, 19, 6976.

van Oorschot R A H, Williams-Blangero S and VandeBerg J L (1992a). Genetic diversity of laboratory gray short-tailed opossums (*Monodelphis domestica*): effect of newly introduced wild-caught animals. *Laboratory Animal Science*, 42, 255–260.

van Oorschot R A H, Porter P A, Kammerer C M and VandeBerg J L (1992b). Severely reduced recombination in females of the South American marsupial *Monodelphis domestica*. *Cytogenetics and Cell Genetics*, 60, 64–67.

van Oorschot R A H, Birmingham V Porter P A, Kammerer C M and VandeBerg J L (1993). Linkage between complement components 6 and 7 and glutamic pyruvate transaminase in the marsupial *Monodelphis domestica*. *Biochemical Genetics*, 31, 215–222.

CHAPTER 15

ONTOLOGICAL WOUND-HEALING STUDIES IN *MONODELPHIS DOMESTICA* FROM BIRTH TO ADULTHOOD

J R Armstrong and M W J Ferguson

INTRODUCTION

Wounds made in fetal tissues heal via different mechanisms and result in much less scarring than equivalent wounds in adult tissues (Somasundaram & Prathap 1969; Krummel *et al.* 1987; Whitby & Ferguson 1991a; Mast & Krummel 1992). The factors that influence this fetal healing phenotype are still incompletely understood. Fetal tissues must undergo a transition at some time during gestation, leading to the characteristic features of repair with collagenous material seen in adult wound healing. The timing may be different in different species. Identification and study of this period of fetal life in which the transition from fetal to adult-healing phenotype occurs may clarify some of the mechanisms important to scar-free fetal-healing. Although the concept of a gradual transition and even of the nonscarring phenomenon itself is not universally accepted (Horne *et al.* 1992; Kumta *et al.* 1994), there is general support for the idea of a gradual prenatal change in the pattern of healing (Ihara *et al.* 1990; Longaker *et al.* 1990; Armstrong & Ferguson 1995).

The advantages of using a marsupial model for wound healing studies have been expounded previously (Ferguson & Howarth 1991) but briefly include accessibility and ease of repeated manipulations. Earlier work from this laboratory has shown that 28 day postnatal *Monodelphis domestica* heal by scarring, that is, an adult phenotype, whereas 2 day young heal without a macroscopic scar. Furthermore, the wounds in 2 day young showed very little inflammatory response (Ferguson & Howarth 1991) in keeping with findings in prenatal mice (Whitby & Ferguson 1991b). For these reasons, early neonatal marsupials are considered equivalent, with respect to wound healing, to fetal stages of eutherian species such as rodents.

BRIEF OVERVIEW

Postnatal *Monodelphis* begin to display healing by scarring from around 9 days (Figure 15.1). Some 'noise' was present in the precise timing of the onset of the ability to form scars, indicating considerable inter-animal variability. There were many elements of the skin which showed maturation at around the same time, such as dermal collagen structure, development of an adipose layer, a diminution of epidermal thickness, and the appearance of hair follicles. However, of most interest was the finding that the inflammatory reaction in response to wounding became rapidly more pronounced and more prolonged in young wounded after 9 days (Armstrong & Ferguson 1995). It has been argued that the lack of inflammation in fetal wounds may be related to their sterile environment (Sopher 1975; Frantz *et al.* 1993). This, of course, cannot be true in a metatherian model of fetal healing, where the wounds exist in an open, contaminated environment. Results of an experiment in which zymosan was injected intra-dermally into young *Monodelphis* suggested that the poor inflammatory reaction at less than 9 days was absolute and was not due to a potentially weak stimulus such as an incisional wound. Immunological immaturity has also been implicated in connection with the fetal inflammatory response (Adolph *et al.* 1993) and an abnormal immune system certainly has implications in terms of the inflammatory response and the fibrotic reaction (Fishel *et al.* 1983; Peterson *et al.* 1987; Wahl & Allen 1988). Inflammatory cells are a potent source of cytokines, which in other animals are critical in determining whether the wounds scar or not (Adzick *et al.* 1985; Shah *et al.* 1991, 1992, 1994; Houghton *et al.* 1994). We therefore investigated the leucocyte profile in *Monodelphis* wound sites and peripheral blood. Considerable difficulties were encountered in labelling and identifying leucocyte subpopulations in *Monodelphis* (outlined below). Some of our data suggested that leucocytes from early neonates were functionally immature.

Previous work in a mouse model documented clear differences between fetal, neonatal and adult, wounds in terms of their extracellular matrix composition and growth factor localisation (Whitby & Ferguson 1991a, b). Similar techniques of immunocytochemistry on cryosections were applied to the *Monodelphis* wound model. Results (see below) supported these earlier findings and further demonstrated a fundamental divide between younger (less than 9 days) and older (more than 9 days) animals.

During the study of the ontogeny of the scarring phenotype it became increasingly obvious and important to try to define and measure scars at a microscopical level. Biochemical measurement of collagen gives no information about the morphological structure of a scar, which is so important to its appearance and functional behaviour. The distinction at either end of the healing phenotype spectrum (scar or no scar) was relatively clear and easy to make, but within the transitional phase when a

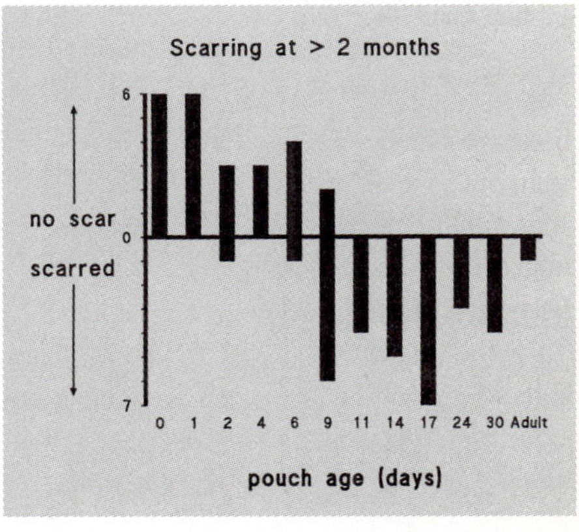

Figure 15.1 The macroscopic scarring phenotype at 2 months post-wounding according to age at time of wounding. Note: The transition into a scarring phenotype occurs at about 9 days.

non-scarring response was changing to healing with a scar, distinguishing the two was extremely difficult. A quantitative measure is also pertinent for other studies in which wounds are manipulated to down-regulate the scarring response. Within this context, further work was undertaken to define and examine the transitional phase between immature and adult-healing phenotypes. Emphasis was placed on the inflammatory reaction, including different types of leucocytes, the quality of healing and immunocytochemical data within the context of both the developing integument and the young as a whole. Some of these data relating to staining of leucocytes and the immunocytochemistry are presented here.

EXPERIMENTAL EVIDENCE

Staining techniques for leucocytes

Considerable difficulties were encountered to stain macrophages, their functional precursors, monocytes and polymorphonuclear leucocytes (PMN) in *Monodelphis*. This was probably due to subtle, but important, differences in the plasma membrane glycoproteins and lysosomal enzyme systems of the pertinent cells in metatherians compared to eutherians.

Initially, attempts were made to identify and quantify white cells from peripheral blood using anti-mouse and anti-rat monoclonal antibodies (mAb) for fluorescence activated cell sorting (FACS). Then a range of anti-macrophage and leucocyte antibodies were assayed using immunocytochemical techniques on cryosections. Reported evidence has suggested that a particular lectin may specifically label macrophages, including immature macrophages, in fetal tissues (Honda *et al.* 1989; Sorokin & Hoyt 1992; Sorokin *et al.* 1992). This approach was also tried. Finally, various protocols for histochemically staining inflammatory cells, by means of localising relatively cell-specific enzymes were performed. Non-specific esterase, chloroacetate esterase and myeloperoxidase were assayed on both paraffin sections and cryosections. The method which was finally used in this study is not ideal and further research into this topic is warranted.

Immunostaining

A panel of various anti-leucocyte antibodies was assayed on 7 μm cryosections of wounds of both adults and young at different times post-wounding (1 to 15 days): ED1, anti-monocyte, anti-macrophage; ED2, anti-macrophage, from Serotec; anti-macrophage, MAS 035 (Sera Lab); KP1 clone, anti-CD68 (Dako); CD11b, anti-3b receptor; OX43; F4/8D; LP9, LCA, anti-leucocyte common antigen. None of these gave positive results.

Fluorescence activated cell sorting (FACS)

FACS was used to try and identify and quantify white blood cell subsets. Leucocytes were prepared from peripheral blood samples from *Monodelphis* adults after red cell lysis. The antibodies used were: OX21, OX7, OX8, OX19, OX34, W3/25, F4/80, H81. None of the antibodies gave a positive staining reaction on repeated testing.

Staining with lectin

The methods utilised were all derived from those described by Sorokin (Sorokin & Hoyt 1992) and Honda (Honda *et al.* 1989) in the rat. The lectin employed was from *Bandeiraea simplificolia* BS-I, isolectin B4 (BSA-IB4) (Sigma Chemical Co.) sometimes also referred to as *Griffonia simplificolia*. Many attempts were made to label macrophages and their precursors by altering various parameters in these protocols. The essential difference between the methods of Sorokin and Honda is that Honda incubated the specimens with the lectin after sectioning of the tissue, whereas Sorokin incubated before sectioning.

A variety of fixation regimens was tested to exclude the possibility of binding sites being masked or destroyed by a particular fixative: glutaraldehyde, 0.125 to 1.0%, 4% fomol saline, 6% $HgCl_2$, acetone, methanol. Low temperature (45 ^{0}C) wax (TAAB Laboratory Equipment, UK) and cryosections were used to circumvent possible destruction by heat during normal processing into wax blocks. A range of lectin concentrations from 1 to 200 µg/ml was tried on both wax and cryosections. Most of these tests were assayed using a peroxidase-conjugated lectin. However, a biotinylated lectin was also tested on cryosections over a range of concentrations from 1 to 20 µg/ml. Controls consisted of incubation without the lectin and also incubation in a 50/50 mix of lectin and D-galactose. D-galactose competitively binds to the same receptor as BSA-IB4 and should therefore reversibly inhibit the lectin binding. None of the assays with BSA-IB4 stained any cell in any of the section types or blood smears. Because of the previous positive findings of binding of this lectin to macrophages in a variety of tissues in the rat (Honda *et al.* 1989; Sorokin & Hoyt 1992) we compared serial paraffin and cryosections of *Monodelphis* and rat skin stained with ED1 antibody (specific for macrophages), von Willebrand's factor antibody (VWF, stains blood vessels) and the BSA I-B4 lectin. The lectin did not stain the vessels or any cells in either *Monodelphis* or rat tissue. ED1 stained macrophages well in rat, but not in *Monodelphis*. VWF antibody stained different structures from ED1 in rat and also stained blood vessels well in cryosections of *Monodelphis* tissue (see below).

Histochemical staining for inflammatory cells

Methods for localising non-specific esterase, chloroacetate esterase and myeloperoxidase were tested on conventional paraffin sections, cryosections and blood smears. Again, several different protocols and substrates were assessed for each enzyme. In eutherians, non-specific esterase is considered the most specific enzyme for the macrophage series including the tissue based macrophage or histiocyte (Schmalzl & Braunsteiner 1970). There is little non-specific esterase activity in any of the polymorphonuclear cell types (Yam *et al.* 1971; Kaplow 1981). Chloroacetate esterase is localised to neutophils (including precursors) and tissue mast cells. It is largely absent from eosinophils and basophils as well as the monocytic lineage (Moloney *et al.* 1960; Li *et al.* 1973). Myeloperoxidase is less specific and is found predominantly in eosinophils, to a lesser extent in monocytes and rarely in neutrophils. It is not found in basophils or other leucocytes (Kaplow 1965).

Three different protocols for non-specific esterase were assayed: a) following Kaplow (1981), b) that of Freemont & Denton (1991) and c) one using hexatised parasanalin. Each method and also the one for chloroacetate esterase was assayed with rat tissue or peritoneal aspirate cytospin preparations as positive controls. Results of these controls were always positive and predictable.

Chloroacetate esterase

The protocol for chloroacetate esterase was based on a description by (Yam *et al.* 1971). This method was sometimes combined as the second stage of a dual staining procedure for both non-specific esterase and chloroacetate esterase. A blue reaction product was obtained for the latter.

Myeloperoxidase

The method for myeloperoxidase relied on the oxidation of diaminobenzidine (DAB), originally described by (Graham & Karnovsky 1965) and was modified from the method given by (Dacie and Lewis 1991).

RESULTS

Non-specific esterase

There were consistent differences between paraffin and cryosections for the non-specific esterase staining and minor differences between the 3 protocols for non-specific esterase. In normal tissue, there was a positive reaction in some cells of the liver, scattered cells in connective tissues and dermis, some muscle tissue (irregular pattern and distribution, usually around the periphery of a muscle) and sebaceous glands next to the hair follicles. Much of this was only seen in cryosections and was lost in the processing to paraffin wax, although the scattered dermal cells were still obvious. These cells were identified to be tissue histiocytes by virtue of their distribution, histological appearance and staining characteristic. There were changes throughout development: greater numbers of positive cells in the liver were present with maturation, but there was diminished staining in muscle. Sebaceous glands were negative at 4 days but became weakly positive by 9 days. With respect to wound and zymosan injection sites, there were no increased numbers of non-specific esterase positive cells at such sites. These were counted using a squared graticule. There was no consistent trend in the number of histiocytes during the 10 days post-wounding in any of the 4 different age groups analysed, nor was there any difference between the numbers of histiocytes at zymosan injection sites and normal tissue at any of the 4 ages assayed. Varying the fixative (formol calcium, formol saline, acetone and an acetone/formaldehyde mix) used for cryosections had little significant effect on the overall staining pattern. None of the assayed protocols or variations for non-specific esterase produced labelling of inflammatory cells.

Chloroacetate esterase

Staining for chloroacetate esterase consistently localised normal tissue mast cells in paraffin and cryosections. These were also identified by virtue of their size, distribution and staining pattern. There was no apparent increase in numbers of these mast cells at inflammatory sites, although numbers were not actually quantified. The method did not consistently stain inflammatory neutrophils in wounds or at sites of zymosan injection. Sections which underwent dual staining for non-specific esterase and chloroacetate esterase showed two mutually distinct cell types demonstrating that although having similar tissue distributions, the non-specific esterase and chloroacetate esterase positive cells were, in fact, different cells.

Myeloperoxidase

This method gave consistent results and reliably stained some inflammatory cells in the incisional wounds. However, it was obvious when viewing the wound sites at high power that many of the PMN leucocytes were not positively stained. Because the method proved reliable and consistent it was chosen as the method of choice for assessing the inflammatory reaction in this study. Full accounts of the distribution and dynamics of myeloperoxidase positive cells in wounds and their numbers in peripheral blood are due to be described elsewhere. Briefly however, there were few myeloperoxidase positive cells in wounds made up to 9 days, which accumulated very close to the wound margins and rapidly disappeared again. Wounds made later than 9 days had far greater numbers of positive cells which were distributed much farther from the wound margins and remained present far longer. The numbers of myeloperoxidase positive cells in peripheral blood displayed little variation throughout postnatal life.

DISCUSSION

Immunostaining, FACS and lectin staining

Results for all the antibodies assayed in either of these methods were negative. It is not clear what this lack of interspecies cross-reactivity is due to; however, it is presumably not a developmental phenomenon since both adult and immature tissues were negative. The cause of the negative results obtained with the lectin GSA I-B4 is unclear. It may have been due to a technical problem, as staining was also not obtained when rat tissues, which others have shown to bind this lectin, were used. A second possibility, which was not explored, is that macrophages at wound sites differ in their surface galactose residues compared to their visceral counterparts.

Histochemical staining

The results from the histochemical studies in *Monodelphis* were not entirely as predicted from the well-established eutherian data. Only one previous report was found which contained very limited information on the histochemical makeup of white cells in a metatherian, *Setonix brachyurus* (Cockson & McNeice 1980). In *Monodelphis*, the non-specific esterase probably stained tissue histiocytes and hair follicle-associated sebaceous glands in the dermis. In eutherians this enzyme is found predominantly in monocytes and macrophages (Braunstein 1959; Moloney *et al.* 1960; Schmalzl & Braunsteiner 1970; Yam *et al.* 1971; Kaplow 1981) but also in histiocytes and megakaryocytes (Yam *et al.* 1971) which partly supports the present observation of positive histiocytes, but begs the question of why the monocytes and macrophages attracted to a wound site did not stain for non-specific esterase. The most likely reason is one of substrate specificity. Small changes in the length of the aliphatic chain of the substrate or its side groups have been shown to significantly alter the strength of the reaction (Li *et al.* 1973; Freemont & Davies 1982). Significant species differences (within eutherians) have been noted previously, although the authors do not discuss possible reasons for this (Moloney *et al.* 1960). The negative results are unlikely to have been due to weak or low levels of the enzyme or due to destruction during processing into wax (Higuchi *et al.* 1979) since long incubation times were performed (up to 5 h) and the cryosections were similarly negative. Also the rat controls were positive. The pH of the incubation mixture is also important (Li *et al.* 1973) and *Monodelphis* may have different requirements for particular enzyme reactions.

Chloroacetate esterase: the PMNs which were expected (Li *et al.* 1973) to stain for chloroacetate esterase were negative. Mast cells are well recognised to contain large amounts of chloroacetate esterase (Moloney *et al.* 1960; Yam *et al.* 1971) and this was confirmed in *Monodelphis* by virtue of their positive reaction, their size, cytoplasmic granules and tissue distribution. However, the inflammatory cells which accumulated at wounds or zymosan injection sites were negative. Reasons for the lack of success with chloroacetate esterase are similar to those outlined for non-specific esterase above.

Myeloperoxidase: the observations with myeloperoxidase were consistent with those in the eutherian literature (Kaplow 1965; Dacie & Lewis 1991). A fair percentage of the inflammatory cells at wound sites were positive, but there were many polymorphonuclear cells which did not stain. Results using this stain on smears of peripheral blood, revealed that there was a correlation between the numbers of positive cells and numbers of eosinophils within individual animals. This suggests that eosinophils were the cells being labelled in the *Monodelphis* wounds (Armstrong & Ferguson, in preparation).

IMMUNOCYTOCHEMISTRY OF *MONODELPHIS* WOUNDS

Immunocytochemistry on cryosections of the *Monodelphis* wound sites was performed for selected extracellular matrix (ECM) components and transforming growth factors $ß_1$ and $ß_3$ (TGF-ß) to obtain qualitative information on the temporal changes through postnatal period and differences between the young and adult tissues. The anti-ECM component antibodies used were against tenascin, fibronectin, laminin and VWF. Earlier evidence from this laboratory and elsewhere has shown differences in the ECM between fetal and adult wounds (Longaker *et al.* 1989; Whitby & Ferguson 1991b). Growth factors are also altered in these two types of wounds (Whitby & Ferguson 1991a). However, further data were sought for the temporal changes of these molecules in wounds made at different pouch ages, in addition to the absolute differences between fetal and adult wounds previously noted. The contrast in wound inflammation from fetal to adult tissues strongly suggests that fetal wounds should have a very different growth factor profile, as inflammatory cells are such potent sources of these factors (Rappolee *et al.* 1988; McGrath 1990; Martin *et al.* 1992; Bennett & Schultz 1993). Further, results from this lab have suggested that injected TGF-$ß_3$ and neutralising antibody to TGF-$ß_1$ and TFG-$ß_2$ modulate the scarring response (Shah *et al.* 1994, 1995), so a similar experiment was undertaken using the pouch young *Monodelphis* wound model system. The aim of this experiment was to manipulate the healing phenotype on either side of the transitional pouch age: by down or up regulating the scarring, using injections of either a) TGF-$ß_3$ or neutralising Ab to TGF-$ß_1$ and TGF-$ß_2$, or b) TGF-$ß_1$ and TGF-$ß_2$ peptides into the wounds. Since the inception of this experiment, a similar study with supportive results has been published (Houghton *et al.* 1994). Details of the antibodies used for the immunocytochemistry are given in Table 15.1.

RESULTS

Fibronectin staining

Adults: there was little staining by 1 day (post-wounding) at the wound site but this increased rapidly over the next 72 hours (Figures 15.2a and b). Staining occurred throughout the wound but was maximal in the superficial regions. Some positive staining was also seen extending laterally away from the wound site underneath the panniculus carnosus muscle. This was maintained until 7 days post-wounding, but was less marked by 10 days. At this time the brightest signal was located just beneath the new epithelium. There was little in the deeper regions of the wound site. The pattern at 15 days remained very similar.

17 days postnatal: the staining pattern and time course of changes in wounds made at this age were similar to those of adult wounds.

9 days postnatal: there was widespread staining in and around the wound by 1 day

Table 15.1 Antibodies used for extracellular matrix and growth factor staining

ANTIGEN	COMPANY	SOURCE OF 1°	1°AB AB DILUTION	2°AB
VWF (human)	Dako	rabbit	1:5	donkey anti-rabbit biotinylated 1:200
tenascin (human)	Chemicon	rabbit	1:50	swine anti-rabbit biotinylated 1:20
fibronectin (rat)	Chemicon	rabbit	1:100	swine anti-rabbit biotinylated 1:40
laminin (human)	Chemicon	rabbit	1:50	swine anti-rabbit biotinylated 1:40
TGF-β_1 (human)	R and D Systems	chicken	1:50	goat anti-chicken biotinylated 1:200
TGF-β_3 (human)	Santa Cruz Biotechnology Inc.	rabbit	1:20	donkey anti-rabbit biotinylated 1:200

post-wounding (Figure 15.3). The high intensity subepithelial staining was apparent at 2 days post-wounding and this intense staining had spread through the full depth of the wound by 4 days, contrasting with the more limited staining noted in the wounds of 17-day postnatal and adult animals (Figure 15.4). By 15 days post-wounding the staining was less extensive again, being restricted to the area of the new dermis. However, it was still more widespread than the comparable adult wound sites.

4 days postnatal: the results of fibronectin staining in wounds at this age were inconclusive due to poor staining of the tissue.

Laminin staining

Adults: laminin stained the basement membrane of the epidermis, the blood vessels and the perimysium. There was decreased staining in the wound site at 1 day post-wounding, which persisted at least until 4 days (Figure 15.5). The epidermal basement membrane was renewed by 7 days. New capillaries were apparent in the central granulation tissue by the same time and these increased dramatically in number up to 15 days (Figure 15.6).

17 days postnatal: there was no staining in the wound site at 1 day post-wounding. A few new vessels were apparent by 4 days and became more numerous by 7 days. The density of capillaries continued to increase up to 15 days, predominantly in the deep dermal region, at which time point they were more numerous than in the adjacent normal dermis (Figure 15.7). The epidermal basement membrane had re-formed by 7 days, as in the adults.

9 days postnatal: there was little staining in the wound sites of 9-day postnatal animals until 4 days post-wounding, although signs of early epidermal basement membrane were visible by 2 days and complete by 4 days. New vessels were also present by 4 days

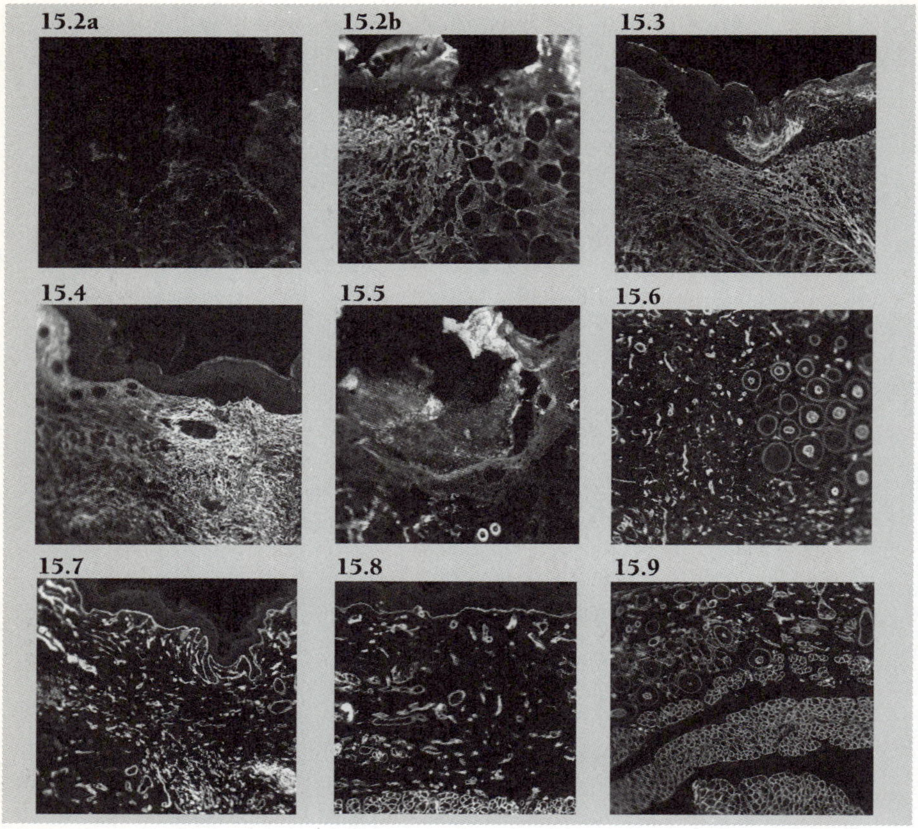

Figure 15.2 (a)Adult tissue stained for fibronectin on 1 day post-wounding (x70). (b) Adult tissue 4 days post-wounding, stained for fibronectin. More signal is present at the wound margins (x75).

Figure 15.3 Fibronectin stain, 9 days postnatal and 1 day post-wounding with bright signal around the wound (x75).

Figure 15.4 Fibronectin stain, 9 days postnatal and 4 days post-wounding illustrating extensive, but localised signal throughout the wound (x70).

Figure 15.5 Laminin stain, adult wound and 4 days post-wounding. There is little staining directly under the wound site at this time (x75).

Figure 15.6 Laminin stain, adult wound site on 15th day post-wounding. A marked increase in new vessels is visible throughout the depth of the wound (x70).

Figure 15.7 Laminin stain, 17 days postnatal on 15th day post-wounding. Many new vessels are apparent in the wound, particularly at a deep dermal level (x85).

Figure 15.8 Laminin stain, 9 days postnatal on 4th day post-wounding. The wound has healed and there are a few stained vessels in the wound site, but in no greater numbers than in the uninjured adjacent tissue (x75).

Figure 15.9 Laminin stain, 9 days postnatal on 15th day post-wounding. Apart from disruption in the panniculus carnosus the staining pattern is the same as normal dermis (x75).

but in numbers comparable to the normal dermis (see Figure 15.8). Unlike wounds in the adult and 17-day animals, the density of new vessels never exceeded that of adjacent tissue in wounds in 9-day animals. The pattern of staining at the wound site was indistinguishable from normal by 15 days post-wounding (Figure 15.9).

4 days postnatal: the results for these animals were very similar to those of the 9-day animals.

Von Willebrand's factor staining

Adults: antibodies to VWF localise to the endothelium of blood vessels. The pattern of blood vessel staining with this antibody was essentially the same as that for laminin. There was also staining of the wound margins in the first 2 days post wounding. This had disappeared by 4 days and new vessels appeared by 7 days. The numbers of vessels continued to increase up to 15 days.

17 days postnatal: the staining pattern in 17-day animals was the same as for adults.

9 days postnatal: very bright staining was seen throughout the re-epithelialised wound by 2 days post-wounding but had returned to normal by 4 days. Vessels in the granulation tissue were apparent by 4 days but in a density similar to the normal dermis. There was no difference in the staining pattern between wound and normal sites by 7 days.

4 days postnatal: staining for VWF in wounds of 4-day animals was very similar to that in 9-day animals.

Tenascin staining

Adults: there was faint staining in the wounds by the 4th day post-wounding in adults. At 7 days there was some staining of the epidermal keratin layers at the wound site and also of the nearby hair follicle-associated sebaceous glands (Figure 15.10). By 15 days the staining was present in similar tissues but further laterally away from the wound site. There was a little faint staining deep in the granulation tissue at this time.

17 days postnatal: positive staining was apparent at 24 hours in these wound sites. The new granulation tissue, sebaceous glands and epidermal keratin were all stained by 4 days, but staining had virtually disappeared by 7 days.

9 days postnatal: similar to the 17-day animals, there was good staining of the new keratin, granulation tissue and in particular, the new basement membrane region by day 4 post-wounding in the wounds of 9-day animals (Figure 15.11). Staining also disappeared by the 7th day. Although hair follicles at the wound site were negative at this time, they were positive elsewhere. The staining of these wound site follicles reverted to normal by 15 days (Figure 15.12).

4 days postnatal: the pattern and time course of changes in wounds at this age were very similar to those in 9-day animals.

TGF-$ß_1$ staining

Adults: normal adult skin showed positive cells scattered sparsely throughout the dermis, adipose layer and occasionally in the epidermis. At 1 day post-wounding there were a few positive cells at the wound margins and the cut epithelial edges also stained positively. Huge numbers of large round, inflammatory-type positive cells had accumulated under the eschar by 2 days (Figure 15.13). The 4th day post-wounding showed a wider distribution of the same cell types around the wound area and many similar positive cells streaming in from under the panniculus carnosus muscle. There was no change in distribution at 7 days but the numbers of positively staining cells were greater. Positive,

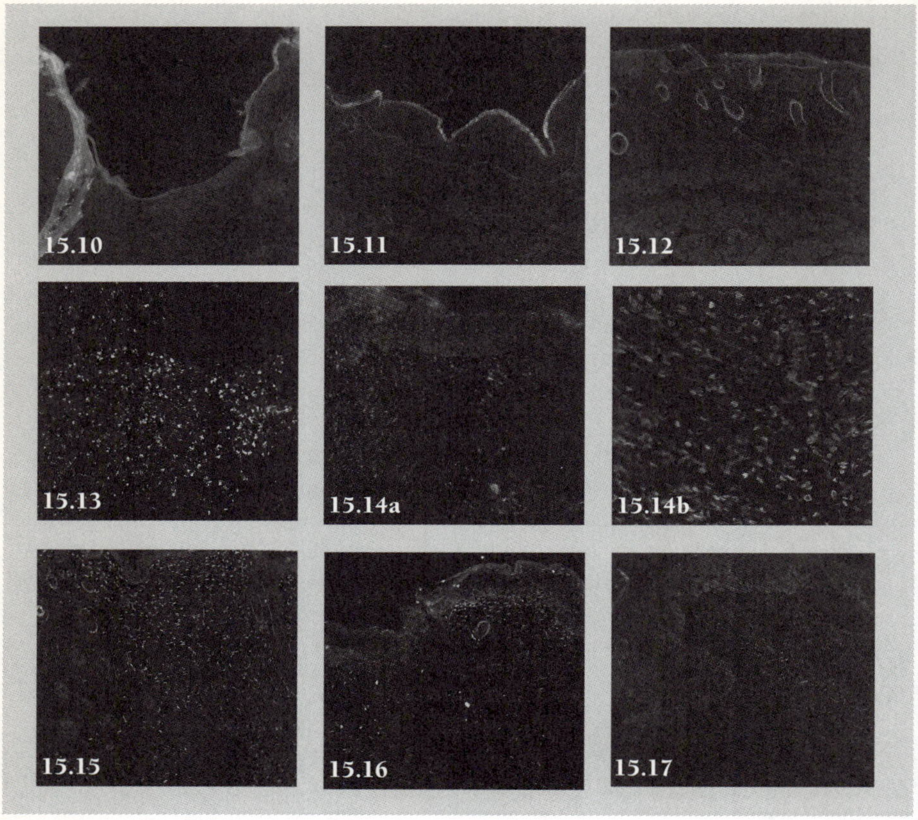

Figure 15.10 Tenascin stain, adult wound on the 7th day post-wounding. There is a very faint stain in the epidermis and nearby sebaceous glands (x70).

Figure 15.11 Tenascin stain, 9 days postnatal on 4th post-wounding day. The staining is very similar to the same time in 17 days postnatal with a faint signal from the new granulation tissue (x80).

Figure 15.12 Tenascin stain, 9 days postnatal on 15th day post-wounding. The hair follicles at the wound site are stained similarly to those away from the wound (x80).

Figure 15.13 TGF-$ß_1$ stain, adult on 2nd day post-wounding. Large numbers of round cells present in the wound site directly under the eschar (x75).

Figure 15.14 (a) TGF-$ß_1$ stain, adult wound on 10th day post-wounding. Many positive cells visible streaming in from the deep layers under the panniculus carnosus (x70).
(b) High power view of Figure 15.14a illustrating that many of the cells were spindle-shaped (x140).

Figure 15.15 TGF-$ß_1$ stain in adult wound on 15th day post-wounding depicting positive cells predominating in the superficial layer of the wound and being mostly spindle-shaped (x70).

Figure 15.16 TGF-$ß_1$ stain, 9 days postnatal wound on 2nd day post-wounding. Relatively few positive cells were present under the nearly re-epthelialised wound (x75).

Figure 15.17 TGF-$ß_1$ stain, 9 days postnatal wound on 4th day post-wounding. The cells are more weakly stained than in adult tissue and are seen entering the wound site from the lateral margins rather than the deep aspect (x85).

spindle-shaped cells were predominant in the granulation tissue by 10 days with fewer of the larger, rounder cells than previously. These spindle-shaped cells displayed a degree of alignment with each other and were presumed to be fibroblasts, being quite distinctive from the early, rounded-type inflammatory cells (Figures 15.14a and b). By 15 days the number of positive spindle-shaped cells in the granulation tissue had diminished and those present were concentrated in the more superficial region of the wound site. There were very few round positive cells visible in the 15-day wounds (Figure 15.15).

17 days postnatal: wounds made at this age stained in a similar fashion to adult wounds but with the clear difference that far fewer cells stained positively in the wounds of 17-day animals.

9 days postnatal: a few positive cells had collected at the wound edges by 24 hours in these wounds but there was no increase during the next 24 hours (Figure 15.16). By 4 days post-wounding there were positive cells coming into the wound site from the lateral margins, in contradistinction to the cells recruited into the adult wounds many of which come from the deeper layers (Figure 15.17). The other marked difference was that the cells stained much less strongly compared to adult cells. Seven days post wounding the wounds had healed and there were fewer positive cells in the region of the new dermis compared to normal dermis. The positive cells which were present lay mostly in the superficial dermis. The appearance at 10 days was similar to that of 7 days post-wounding. A few positive spindle cells were still visible at the lateral wound margins.

4 days postnatal: wounds at this age had healed by 2 days, at which time the wound site contained only a few weakly positive cells with a few more present in the lateral margins. The staining pattern from 4 days onwards was similar to normal dermis (Figures 15.18a and b).

TGF-ß$_3$ staining

Adults: large round positive cells were present in adult wounds from 1 day and increased in number until 10 days, but had dissipated again by 15 days (Figures

Figure 15.18 (a) and (b) TGF-ß$_1$ stain, 4 days postnatal wound on 4th day post-wounding and normal 4-day postnatal tissue respectively. (Note the similar appearance of the wound site compared with normal (a) x80 (b) x75.

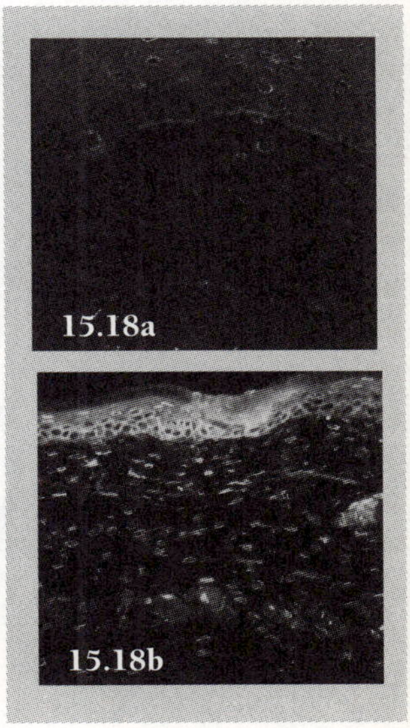

15.19a and b). The normal epithelium was positive and this did not return to normal at the wound site until about the 7th to 10th day post-wounding. By the 15th day some of the hair follicles at the wound site were showing positive staining.

17 days postnatal: the overall pattern of staining in this age group was similar to the adult wounds although the intensity of the staining was generally weaker.

9 days postnatal: staining was only moderate in the wounds of 9-day animals. There was a positive reaction by 1 day, concentrated in the superficial regions of the wound. By 7 days the new dermis still showed more staining than adjacent (original) dermis but the wound site looked normal by 10 days.

4 days postnatal: on 1 day and 2 days post-wounding there was increased staining but by 4 days there were only a few positive cells present in the wound site dermis (Figure 15.20). Their numbers were not greatly different from the normal dermis.

DISCUSSION

Immunostaining of ECM components

The results of the immunostaining for all the ECM components revealed a clear difference between the early pouch young (9 days and 4 days) and older animals. Tenascin and fibronectin were both more prominent in the wounds of 4 day and 9 day animals

15.19a

15.19b

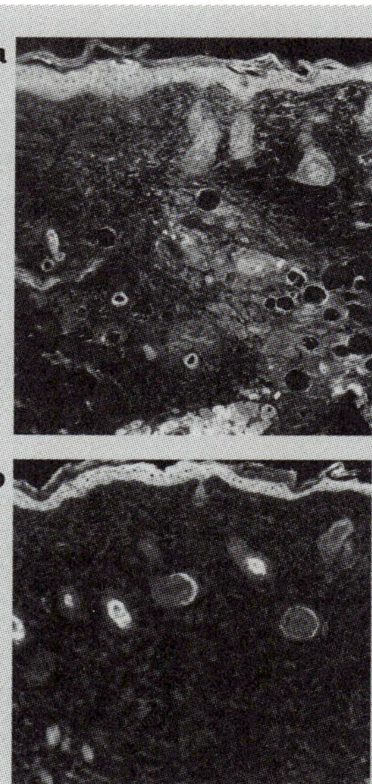

Figure 15.19 (a) TGF-ß$_3$ stain, adult wound on 10th day post-wounding. Positive cells in maximal numbers were present at this time (x40).
(b) TGF-ß$_3$ stain, 15 day-old adult wound. (Note the reduced numbers of positive cells compared with Figure 15.19a (x65).)

15.20

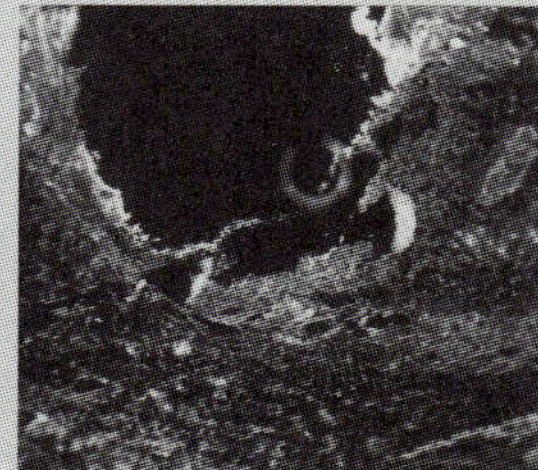

Figure 15.20 TGF-ß$_3$ stain, 4 days postnatal wound on 1st day post-wounding. Positive staining was present by this time throughout the wound site but had dissipated by the 4th day post-wounding (x50).

compared with adult wounds. These findings generally consistent with similar experiments in a mouse model (Whitby & Ferguson 1991b). Fibronectin is abundant in embryonic morphogenesis (Wartiovaara *et al.* 1979; Gibson *et al.* 1983), inflammation (Clark *et al.* 1981) and wound healing (Grinnell *et al.* 1981). These are all examples of tissues which are undergoing structural change. Clark postulates that fibroblasts and epidermal cells use fibronectin as a contact guidance system (Clark *et al.* 1982a; Clark 1983). The present result of more widespread fibronectin staining in neonatal wounds, although not quantitative, suggests that the fibronectin plays a crucial role in the rapid re-epithelialisation and repair so characteristic of immature ('fetal') wounds. Not only is the relative concentration of fibronectin greater in 'fetal' wounds but it's appearance at the wound site is earlier in the post wounding period (Longaker *et al.* 1989; Whitby & Ferguson 1991b). Interestingly, the initial fibronectin at an inflammatory site is derived from a plasma source (Clark *et al.* 1981) whereas later fibronectin is produced by endothelial and fibroblast cells (Clark *et al.* 1982b). The implication for 'fetal' wounds is either, that there is higher endogenous plasma levels of fibronectin and/or, that fibronectin synthesis is ongoing within the developing dermis. It is also possible that the 'fetal' cells react to the fibronectin differently in order to effect the quicker wound repair.

The findings for laminin staining correlate very closely with those for VWF with respect to new blood vessel formation in the wounds. The striking difference between wounds made earlier than 9 days and those made after 9 days was the reduction of new vessels in the former wounds. Sparse neovascularisation in fetal mouse wounds has also been reported (Whitby & Ferguson 1991b). It is tempting to link the dynamics of fibronectin in fetal wounds, the late endothelial production of fibronectin and the lack of neovascularisation: the paucity of endothelial cells in the fetal wound site may limit the duration of fibronectin presence at the wound site. As fibronectin is chemotactic for fibroblasts (Postlethwaite *et al.* 1981) as well as providing contact guidance, the early but time-limited presence of fibronectin in fetal wounds may have a role in controlling early collagen deposition: this in turn could minimise the later scarring.

Immunostaining of TGF-ß

As with the staining of the ECM components, the pattern obtained for TGF-ß staining displayed a clear watershed between the pouch young less than 9 days and those aged greater than 9 days. This was particularly marked for TGF-$ß_1$. The main differences were a) far fewer positive cells at 9 days or younger; b) much shorter duration of presence of positve cells in those aged 9 days or younger and c) a difference in source of positive cells; laterally from the dermis in the more immature young versus a deeper subcutaneous source in the older young and adults. A similar difference in TGF-ß staining exists between fetal and neonatal mouse wounds (Whitby & Ferguson 1991a). TGF-ß is constitutively expressed by monocytes, but requires activation of the cell for active peptide secretion (Assoian *et al.* 1987). The natural deduction therefore, is that fetal tissues contain fewer cells of the macrophage lineage or that the cells are immature and do not yet express TGF-$ß_1$. Our unpublished data have shown that monocytes were present in the peripheral blood in similar numbers throughout postnatal life in *Monodelphis*. It therefore seems likely that the monocytes of early postnatal *Monodelphis* are functionally immature. Since monocytes and macrophages are a potent source of TGF-ß in wounds, (Assoian *et al.* 1987), diminished levels of TGF-ß in 'fetal' wounds are possible, which, in turn, would profoundly affect collagen synthesis and deposition. As corroborative evidence, exogenous addition of platelet-derived human TGF-ß into subcutaneous implants in fetal rabbits leads to increased fibrosis measured at 7 days (Krummel *et al.* 1988) and similarly TGF-ß added to fetal limb organ cultures had an equivalent effect (Houghton *et al.* 1994). The increased collagen deposition subsequent upon the addition of TGF-ß to a wound, translates into enhanced wound breaking strength during the first 21 days post wounding (Mustoe *et al.* 1987). However, TGF-ß also stimulates a faster healing rate in TGF-ß treated wounds (Sporn *et al.* 1983; Mustoe *et al.* 1987). Fetal wound healing is therefore paradoxical: fetal wounds probably contain less TGF-ß but still heal very rapidly. Clearly, there must be other factors which allow the rapid healing typical of fetal wounds.

FUTURE DIRECTIONS

Marsupials and in particular, *Monodelphis*, have great future potential for wound- healing studies. They also offer excellent models for other ontological studies especially the fascinating interface between the immune system and wound healing. Future research should include not only fundamental immunological research comparing metatherians and eutherians, in terms of the major histocompatability complex and the metatherian immune response, but also the development of monoclonal antibodies against the equivalent of CD determinants identified in eutherians, for use as research tools in marsupials.

Further research on leucocyte ontogeny, including functional ontogeny, would be of great interest and value in the wound healing field. Additionally, attention needs directing at quantifying the microscopic appearance of scars. The present state of subjectively assessing scar quality or, at best, applying a crude scoring scheme leaves much to be desired, especially during the study of the fetal-to-adult-healing transition. Development of accurate measuring techniques to assay specific cytokine levels in wound tissue would allow a clearer picture of the wound healing cytokine profile. This is a particular challenge in wound-healing studies in immature animals, as the quantity of tissue is generally much smaller than in comparable adult studies.

Finally, lessons learned from scar-free eutherian fetal wound healing or metatherian postnatal wound healing may allow the development of novel therapeutic strategies to minimise scarring during adult wound healing (Shah *et al.* 1994, 1995). This would have enormous potential in humans to minimise the adverse functional, growth-impairing and cosmetic sequelae of scarring.

Acknowledgments

This work was funded through a Wellcome Clinical Research Fellowship. Thanks are also given to the staff in the Animal Unit at Manchester University and the staff in the Photographic Department of the School of Biological Sciences.

References

Adolph V R, DiSanto S K, Bleacher J C, Dillon P W and Krummel T M (1993). The potential role of lymphocytes in fetal wound healing. *Journal of Pediatric Surgery,* 28, 1316–1320.

Adzick N S, Harrison M R, Glick P L, Beckstead J H, Villa R L, Scheuenstuhl H and Goodson W H III (1985). Comparison of fetal, newborn and adult wound healing by histologic, enzyme-histochemical and hydroxyproline determinations. *Journal of Pediatric Surgery*, 20, 315–319.

Armstrong J R and Ferguson M W J (1995). Ontogeny of the skin and transition from a scar-free to a scarring phenotype during wound healing in the pouch young of a marsupial, *Monodelphis domestica. Developmental Biology,* 169, 242–260.

Assoian R K, Fleurdelys B E, Stevenson H C, Miller P J, Madtes D T, Raines E W, Ross R and Sporn M B (1987). Expression and secretion of type β transforming growth factor by activated human macrophages. *Proceedings of the National Academy of Sciences USA,* 84, 6020–6024.

Bennett N T and Schultz G S (1993). Growth factors and wound healing: Biochemical properties of growth factors and their receptors. *American Journal of Surgery*, 165, 728–737.

Border W A, Noble N A, Yamamoto T, Harper J R, Yamaguchi Y, Pierschbacher M D and Ruoslahti E (1992). Natural inhibitor of transforming growth factor-β protects against scarring in experimental kidney disease. *Nature*, 360, 361–364.

Braunstein H (1959). Esterase activity in leucocytes. *Journal of Histochemistry and Cytochemistry,* 7, 202.

Clark R A F (1983). Fibronectin in the skin. *Journal of Investigative Dermatology*, 81, 475–479.

Clark R A F, Dvorak H F and Colvin R B (1981). Fibronectin in delayed type hypersensitivity skin reactions: association with vessel permeability and endothelial cell activation. *Journal of Immunology*, 126, 787–793.

Clark R A F, Lanigan J M, DellaPelle P, Manseau E, Dvorak H F and Colvin R B (1982a). Fibronectin and fibrin provide a provisional matrix for epidermal cell migration during wound healing. *Journal of Investigative Dermatology*, 79, 264–269.

Clark R A F, Quinn J H, Winn H J, Lanigan J M, DellaPelle P and Colvin R B (1982b). Fibronectin is produced by endothelial cells in response to injury. *Journal of Experimental Medicine*, 156, 646–651.

Cockson A and McNeice R (1980). Survival in the pouch: the role of macrophages and maternal milk cells. *Comparative Biochemistry and Physiology*, 66A, 221–225.

Dacie J V and Lewis S M (1991). In *Practical Haematology*. Eds JV Dacie and SM Lewis. 7th edn. Churchill Livingstone, Edinburgh. p126.

Ferguson M W J and Howarth G F. (1991). Marsupial models of scarless fetal wound healing. In *Fetal Wound Healing*. Eds N S Adzick and M T Longaker. 1st edn. Elsevier, Amsterdam. pp 95–124.

Fishel R, Barbul A, Wasserkrul H L, Penberthy L T, Rettura G and Efron G (1983). Cyclosporine A impairs wound healing in rats. *Journal of Surgical Research,* 34, 572–575.

Frantz F W, Bettinger D A, Haynes J H, Johnson D E, Harvey K M, Dalton H P, Yager D R, Diegelmann D F and Cohen K I (1993). Biology of fetal repair: The presence of bacteria in fetal wounds induces an adult-like healing response. *Journal of Pediatric Surgery*, 28, 428–434.

Freemont A J and Davies S J (1982). Acid phosphatase activity in lymphocytes and other cells: a comparison of six a naphthyl based substrates. *Medical Laboratory Science,* 39, 405–407.

Freemont A J and Denton J (1991). In *Atlas of Synovial Fluid Cytology.* Vol. 18 of Current Histopathology Series. Khuwer Academic Publishers, Dordrecht. pp 25–26.

Gibson W T, Couchman J R and Weaver A C (1983). Fibronectin distribution during the development of fetal rat skin. *Journal of Investigative Dermatology*, 81, 480 485.

Graham R C and Karnovsky M J (1965). The early stages of absorption of injected horseradish peroxidase in the proximal tubules of mouse kidney: ultrastructural cytochemistry by a new technique. *Journal of Histochemistry and Cytochemistry*, 14, 291–302

Grinnell F, Billingham R E and Burgess L (1981). Distribution of fibronectin during wound healing in vivo. *Journal of Investigative Dermatology*, 76, 181–189.

Higuchi S, Suga M, Dannenburg A M and

Schofield B H (1979). Histochemical demonstration of enzyme activities in plastic and paraffin embedded tissue sections. *Stain Technology*, 54, 5–12.

Honda T, Schulte B A and Spicer S S (1989). Glycoconjugate with terminal galactose. A selective property of macrophages in developing rat lung. *Histochemistry*, 91, 61–67.

Horne R S C, Hurley J V, Crowe D M, Ritz M, O'Brien B McC and Arnold L I (1992). Wound healing in foetal sheep: a histological and electron microscope study. *British Journal of Plastic Surgery*, 45, 333–344.

Houghton P E, Keefer K A and Krummel T M (1995). The role of transforming growth factor beta (TGFß) in the conversion from scarless healing to healing with scar formation.*Wound Repair and Regeneration,* 3, 229–236.

Ihara S, Motobayashi Y, Nagao E and Kistler A (1990). Ontogenetic transition of wound healing pattern in rat skin occurring at the fetal stage. *Development*, 110, 671–680.

Kaplow L S (1965). Simpified myeloperoxidase stain using benzidine dihydrochloride. *Blood*, 26, 215–219.

Kaplow L S. (1981). Cytochemical identification of mononuclear macrophages. In *Manual of Macrophage Methodolgy*. Eds H B Herscowitz *et al.* 1st edn. Marcel Dekker Inc, New York. pp 199–207.

Krummel T M Michna B A, Thomas B L, Sporn M B, Nelson J M, Salzberg A M, Cohen I K and Diegelmann R F (1988). Transforming growth factor beta (TGF-ßeta) induces fibrosis in a fetal wound model. *Journal of Pediatric Surgery*, 23, 647–652.

Krummel T M, Nelson J M, Diegelmann R F, Lindblad W J, Salzberg A M, Greenfield L J and Cohen I K (1987). Fetal response to injury in the rabbit. *Journal of Pediatric Surgery*, 22, 640–644.

Kumta S, Ritz M, Hurley J V, Crowe D, Romeo R and O'Brien BMcC (1994). Acute inflammation in foetal and adult sheep: the response to subcutaneous injection of turpentine and carrageenan. *British Journal of Plastic Surgery*, 47, 360–368.

Li C Y, Lam K W and Yam L T, (1973). Esterases in human leukocytes. *Journal of Histochemistry and Cytochemistry*, 21, 1–12.

Longaker M T Whitby D J, Adzick N S, Crombleholme T M, Langer J C, Duncan B W, Bradley S M, Stern R, Ferguson M W J and Harrison M R (1990). Studies in fetal wound healing: VI. Second and early third trimester fetal wounds demonstrate rapid collagen deposition without scar formation. *Journal of Pediatric Surgery*, 25, 63–69.

Longaker M T, Whitby D J, Ferguson M W, Harrison M R, Crombleholme T M, Langer J C, Cochrum K C, Verrier E D and Stern R(1989). Studies in fetal wound healing: III. Early deposition of fibronectin distinguishes fetal from adult wound healing. *Journal of Pediatric Surgery*., 24, 799–805.

Martin P, Hopkinson-Wooley J and McLuskey J (1992). Growth factors in cutaneous wound repair. *Progress in Growth Factor Research*, 4, 1–20.

Mast B A and Krummel T M (1992). Acute inflammation in fetal wound healing. In *Fetal Wound Healing*. Eds N S Adzick and M T Longaker. 1st edn. Elsevier, Amsterdam. pp 227–240.

McGrath M H (1990). Peptide growth factors and wound healing. *Clinics in Plastic Surgery*, 17, 421–430.

Moloney W C, McPherson K and Fliegelman L (1960). Esterase activity in leukocytes demonstrated by the use of naphthol AS-D chloroacetate substrate. *Journal of Histochemistry and Cytochemistry,* 8, 200–207.

Mustoe T A, Peirce G A, Thomason A, Gramates P, Sporn M B and Deuel T F (1987). Accelerated healing of incisional wounds in rats induced by transforming growth factor-ß. *Science*, 237, 1333–1336.

Peterson J M, Barbul A, Breslin R J, Wasserkrug H L and Efron G (1987). Significance of T lymphocytes in wound healing. *Surgery*, 102, 300–305.

Postlethwaite A E, Keski-Oja J, Balian G and Kang A H (1981). Induction of fibroblast chemotaxis by fibronectin. *Journal of Experimental Medicine*, 153, 494–499.

Rappolee D A, Mark D, Banda M J and Werb Z (1988). Wound macrophages express TGFα and other growth factors in vivo: analysis by mRNA phenotyping. *Science*, 241, 708–712.

Schmalzl F and Braunsteiner H (1970). The cytochemistry of monocytes and macrophages. *Series Haematologica*, 111, 93–131.

Shah M, Foreman D M and Ferguson M W J (1992). Control of scarring in adult wounds by neutralising antibody to transforming growth factor ß. *Lancet*, 339, 213–214.

Shah M, Foreman D M and Ferguson M W J (1994). Neutralising antibody to TGF-$ß_{1,2}$ reduces cutaneous scarring in adult rodents. *Journal of Cell Science*, 107, 1137–1157.

Shah M, Foreman D M and Ferguson M W J (1995). Neutralisation of TGF-$ß_1$ and TGF-$ß_2$ or exogenous addition of TGF-$ß_3$ to cutaneous rat wounds reduces scarring. *Journal of Cell Science*, 108, 985–1002.

Shah M,Foreman D and Ferguson M W J (1991). Reduction of scar tissue formation in adult rodent wound healing by manipulation of the growth factor profile. *Journal of Cellular Biochemistry*, Suppl.15F, 198.

Somasundaram K and Prathap K (1969). Intra-uterine healing of skin wounds in rabbit foetuses. *Journal of Pathology*, 100, 81–86.

Sopher D (1975). A study of wound healing in the foetal tissues of the Cynomolgus monkey. *Laboratory Animals Handbook*, 6, 327–335.

Sorokin S P and Hoyt R F (1992). Macrophage development: I. Rationale for using *Griffonia simplificolia* isolectin B4 as a marker for the line. *Anatomical Record*, 232, 520–526.

Sorokin S P, Hoyt R F, Blunt D G and McNelly N A (1992). Macrophage development: II. Early ontogeny of macrophage populations in brain, liver and lungs of rat embryos as revealed by a lectin marker. *Anatomical Record*, 232, 527–550.

Sporn M B, Roberts A B, Shull J H, Smith J M and Ward J M (1983). Polypeptide transforming growth factors isolated from bovine sources and used for wound healing in vivo. *Science*, 219, 1329–1330.

Stone W H, Samples N K, Kraig E and VandeBerg J L (1987). Definition of class I antigens of the MHC of a marsupial (Monodelphis domestica). *Archivos de Zootecnia*, 36, 219–235.

Wahl S M and Allen J B (1988). T Lymphocyte-dependent mechanisms of fibrosis. In:*Growth factors and other aspects of wound healing: Biological and clinical implications*. Eds A Barbul *et al.*, Alan R Liss, New York. pp 147–60.

Wartiovaara J, Leivo I and Vaheri I (1979). Expression of cell surface associated glycoprotein fibronectin in the early mouse embryo. *Developmental Biology*, 69, 247–257.

Whitby D J and Ferguson M W J (1991a). Immunohistochemical localization of growth factors in fetal wound healing. *Developmental Biology*, 147, 207–215.

Whitby D J and Ferguson M W J (1991b). The extracellular matrix of lip wounds in fetal, neonatal and adult mice. *Development*, 112, 651–668.

Yam L T, Li C Y and Crosby W H (1971). Cytochemical identification of monocytes and granulocytes. *American Journal of Clinical Pathology*, 55, 283–290.

CHAPTER 16

ANTIBODY TRANSPORT ACROSS THE GUT OF THE SUCKLING OPOSSUM, *MONODELPHIS DOMESTICA*

Arthur E Wild

INTRODUCTION

As part of the process of combating infections, mammals, like other vertebrates, display humoral immune responses in which antibodies are produced. This involves a complex learning process requiring stimulation of appropriate B cells by antigen. Since this process takes time, newborn mammals in their immunologically immature state would be vulnerable to infection at birth were it not for the fact that their own immune responses are learnt under cover of a protective panoply of antibodies acquired from the mother. Such acquisition of passive immunity has been well studied in eutherian mammals (reviewed by Brambell 1970; Baintner 1986; Wild 1993). Here, antibodies are transcytosed across epithelial barriers comprised of visceral yolk sac endoderm, syncytiotrophoblast and gut enterocytes depending both upon the species concerned and the time they are transported relative to birth. Our understanding of the cellular mechanisms involved is still limited and controversial (Wild 1993) although some common findings have emerged. It is against this background knowledge derived from eutherian mammals that more recent findings for the South American grey short-tailed opossum (*Monodelphis domestica* - also known as the laboratory opossum because it can be bred easily in captivity) together with those for other marsupials where relevant, are described. Unless otherwise stated, where 'opossum' is referred to, it is the laboratory opossum that is considered here.

TRANSPORT OF ANTIBODIES OCCURS POST-NATALLY

Like the related North American Virginia opossum (*Didelphis virginiana*), the laboratory opossum has a very short gestation period (14 to 15 days for *Monodelphis*, 13 days for *Didelphis*, Harder *et al.* 1993). Consequently it is reasonable to assume that the gut would be differentiated rapidly in these species to meet physiological needs. Pre-suckling

Monodelphis pups born of mothers hyperimmunised against sheep red blood cells (SRBC) prior to mating and collected within minutes of birth, were shown by Samples *et al.* (1986) not to have agglutinating antibodies present in homogenised tissue extracts when compared to extracts of litter mates that had been allowed to suckle. This finding points to there being no transmission of antibodies before birth in *Monodelphis*, but that, consequent upon suckling, antibodies are transported from the milk across the gut to the blood. Some caution is needed in interpreting these results, however, since the class of agglutinating antibody (IgM or IgG) was not described and non-agglutinating sensitising antibodies (often IgG) were not tested for. This is important since in humans, rats and mice, only maternal IgG antibody is transported prenatally or postnatally, although in the rabbit both IgG and IgM are transported to the fetus (Merad & Wild 1992). Furthermore, evidence for transport of antibodies across the gut, from the milk, was deduced by Samples *et al.* (1986) from detection of agglutinins in homogenised tissue. This would have included stomach and gut contents likely to have contained milk (shown to have anti-SRBC antibodies present) and which most likely would have contaminated the tissue extracts. However, more recently we have fed suckling *Monodelphis* with homologous anti-SRBC serum orally and confirmed transport of sensitising antibody (IgG) to the blood prior to 55 days of age (Wild *et al.* 1994).

Immunoelectrophoretic (IE) analysis of the serum of newborn *Didelphis* prior to suckling has also revealed an absence of immunoglobulins (IgG,IgA and IgM) but the presence of IgG and IgA precipitin lines at 12 to 14 hours post-suckling similar to those seen in milk (Hindes & Mizell 1976). Lack of circulating lymphocytes and organised lymphoid tissue in *Didelphis* at birth (Rowlands *et al.* 1972). makes it likely these immunoglobulins are maternally derived and not synthesised endogenously. Absence of IgG from prenatal and newborn blood prior to suckling has also been reported by Yadav *et al.* (1971) for the quokka (*Setonix brachyurus*) and although immunoglobulins have been detected in uterine fluid in this species, they have not been detected in fluids of the yolk sac, amnion and allantois (Yadav & Eadie 1973). Transport of antibody (IgG) across the gut to the blood has been well demonstrated in this species by feeding pouch young antibody fractions via stomach tube (Yadav 1971); similarly, transport was also demonstrated in the tammar wallaby (*Macropus eugenii*) and the brushtail possum (*Trichosurus vulpecula*). In contrast to these findings, Renfree (1973) found IgG present in fetal tammar serum obtained two days prior to birth as revealed by IE. Quantitative estimates made by Deane & Cooper (1988) indicate the amount of IgG present is relatively low (0.06mg/ml) and placental transport of this fetal IgG has not yet been proven in the tammar. Seen in conjunction with findings for *Didelphis*, what limited evidence there is available suggests that in *Monodelphis* there is no significant prenatal transport of antibody to the embryo across the trilaminar yolk sac placenta. In the likely event that marsupials gain most, if not all of, their maternal passive immunity after birth, they are similar to pigs, horses and ruminants where the young are also normally born with no maternally derived immunoglobulin (Baintner 1986).

WEANING IS ASSOCIATED WITH CESSATION OF ANTIBODY TRANSPORT

In studies aimed in part at investigating the period for which antibody is transported across the gut of the suckling opossum, we fed pups orally at ages ranging from 16 to

62 days with homologous anti-SRBC serum (Wild *et al.* 1994). Whilst IgG sensitising antibody was found to be present in the serum of blood collected up to 3 hours after feeding from pups at 52 days and earlier, antibody could not be detected at 55 days and later. In *Monodelphis* weaning has been reported to occur between 7 to 9 weeks (Adam *et al.* 1988) and at 56 days (Harder *et al.* 1993). Although our findings were for single animals at each age tested, they correlate well with the time weaning commences and also with the time when putative Fcγ receptors (considered later) disappear from gut enterocytes. Yadav (1971) also found that weaning was associated with cessation of antibody (IgG) transport in the quokka and brushtail possum and in which the transport process endures for remarkably long periods (170 to 200 and 98 to 145 days respectively). In the rat and mouse antibody transport occurs across the neonatal gut for up to 21 and 16 days respectively, the time when weaning also commences (Brambell 1970). Here, too, it is correlated with disappearance of Fcγ receptors (Wild & Richardson 1979; Griffin & Wild 1987) present on enterocytes in the proximal small intestine. Cessation of antibody transport in the rat is effected prematurely by injection of cortisone acetate (Halliday 1959; Morris & Morris 1976), as is loss of Fcγ receptors (Wild 1981). Induced cell replacement and/or repression of receptor expression have been suggested as possible explanations (Morris & Morris 1976) and increasing levels of corticosteroids as the young mammal ages might normally trigger such events (Patt 1977). Changes which take place in the ultrastructure of opossum gut enterocytes at about the time of weaning and which are clearly related to cessation of endocytic uptake of immunoglobulin, are described later.

EVIDENCE THAT TRANSPORT OF ANTIBODY IS SELECTIVE IN *MONODELPHIS* AND RELATED TO PRESENCE OF FCγ RECEPTORS

By what mechanism are antibodies transported across the gut of suckling laboratory opossums? The answer to this question can first usefully be considered in the light of what we know about similar processes in eutherian mammals. Transport of antibodies across the gut of pigs, horses and ruminants, takes place from the colostrum only within the first 36 hoursof birth and is non-selective. A variety of experimentally fed macromolecules, including IgM and the inert polymer polyvinylpyrrolidone, can be transcytosed across gut enterocytes during this period (Jeffcott 1972; Baintner 1986), implying that specific receptors are not involved in these species of mammal. Specific Fcγ receptors are involved however, in selectively transcytosing IgG isotypes across mammary gland epithelium as the colostrum is formed (Kemmler *et al.* 1975; Sasaki *et al.* 1977). In contrast, in rats and mice it is well established that IgG is the only immunoglobulin to be transported across enterocytes following Fcγ receptor binding and subsequent endocytosis; heterologous IgG is also transported, but to an extent dependent upon its species of origin (Mackenzie 1984; Wild 1993). *Monodelphis* pups appear more to resemble their eutherian rodent counterparts in that they too have Fcγ receptors present on gut enterocytes and show selective binding and transport of IgG according to its species of origin (Wild *et al.* 1994). To detect these receptors in suckling opossum gut we used an erythrocyte-antibody (E-A) rosette assay in which SRBC were sensitised with sub-agglutinating amounts of opossum or rabbit IgG and reacted with enterocytes isolated from proximal and distal small intestine of opossums aged 12 to 73 days. Indicator SRBC binding was studied at pH 6.0 and 7.2 since in rodents binding of IgG to the Fcγ receptor is highly acid pH-dependent and restricted to enterocytes of the proximal small intestine (Rodewald 1976). A similar E-A rosette assay was previously used to demonstrate these acid pH-dependent Fcg receptors in suckling rats (Wild 1981; Griffin & Wild 1987) using indicator SRBC sensitised with whole rabbit IgG or with F(ab')2 fragment of IgG as control. It has not yet been possible to prepare F(ab')2 fragment from opossum anti-SRBC IgG and we have relied on using normal SRBC as a control. Findings of selective binding of indicator SRBC and inhibition of rosette formation by IgG which is correlated with IgG transport, do, however, support the contention that Fcγ receptors are the entity being detected in the opossum.

Indicator SRBC binding resulting in rosette formation (see Figure 16.1) occurs only when SRBC are sensitised with opossum IgG (detected by means of a class specific antiserum), not with rabbit IgG. Furthermore, when the same rabbit IgG anti-SRBC reagent used to sensitise SRBC was fed to suckling opossums, but in undiluted form, no subsequent transport to the blood could be found (Wild *et al.* 1994). The percentage of enterocytes that form rosettes at any particular age is extremely variable,

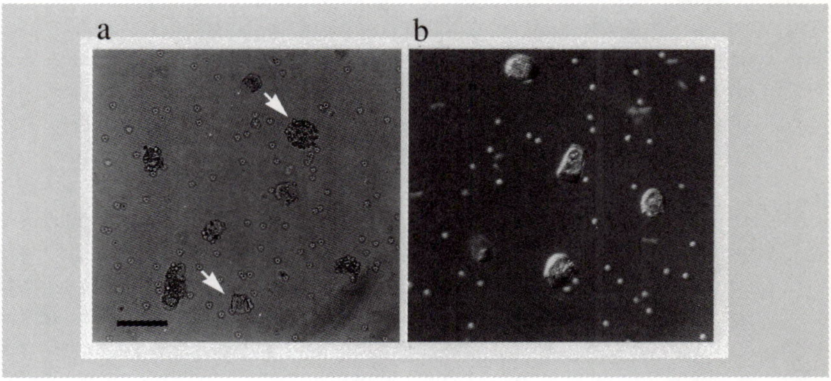

Figure 16.1 (a) Examples of E-A rosettes formed at pH 7.2 with enterocytes isolated from 35-day opossum proximal small intestine (SI) as described in note 1 and following reaction with opossum IgG sensitised sheep red blood cells (SRBC). Indicator SRBC binding occurs to lateral and basal as well as apical plasma membranes. Two cells (see arrows) show no or little binding of indicator SRBC.
(b) Absence of rosette formation when enterocytes are reacted with normal SRBC or SRBC sensitised with rabbit IgG.
Note: (a) photographed using phase and (b) using Nomarski optics.
Scale bar=40 µm.

Note 1
In the examples of rosettes shown in Figure 16.1a, and for the data compiled in Table 16.1, enterocytes from 35 day suckling opossums were isolated by a modification of the technique previously described (Wild 1981; Wild *et al.* 1994). Slit pieces of proximal and distal small intestine were rinsed with PBS, pH 7.2, immersed in PBS/hyaluronidase (0.5 mg/ml) for 1 h, washed x2 in PBS by centrifugation and resuspended in PBS containing 5 mM EDTA. This suspension was repeatedly syringed with a 23-gauge needle until free enterocytes with intact brush borders were seen and the suspension passed through a nylon tea-strainer to remove large clumps. Enterocytes obtained in this way can conveniently be stored at -20°C by resuspending them in DMEM containing 10% FCS and 10% DMSO, and aliquoting into 1.5 ml Eppendorf centrifuge tubes. Although loss of apical microvilli may occur on thawing and resuspension in PBS following washing, as can be seen such enterocytes still retain their capacity to form rosettes.

perhaps relating to receptor occupancy by ligand and vagaries of the technique of cell isolation. However, unlike the situation in rodents, indicator SRBC binding does not appear to be particularly acid pH-dependent (see for example Figure 16.1 and also Table 16.1 where results were obtained at pH 7.2, as well as data in Wild *et al.* 1994). We have recently used the relatively crude, but effective, technique of slitting open the small intestine of five sibling 44 day old opossum pups immediately following dissection and applying small pieces of Whatman pH paper (narrow range 6.0 to 8.0) to the wet mucosal surface, to determine pH. The most proximal part (duodenum) of the proximal small intestine was consistently found to be pH 6.0, as was the most distal part of the proximal small intestine. The most proximal part of the distal small intestine ranged from pH 6.0 to 6.5 and the most distal, pH 6.5 to 7.0. Therefore, despite a capacity to bind at pH 7.2 in the rosette assay *in vitro,* binding of IgG to Fcγ receptors *in vivo* within the gut lumen would normally be expected to occur at an acid pH if the values we have found at 44 days reflect those occurring in the gut lumen at earlier and later stages of development.

An advantage of the E-A rosette assay is that binding of immunoglobulins and their fragments to Fcγ receptors on target cells can easily be determined from their capacity to inhibit binding of IgG sensitised SRBC. As can be seen from Table 16.1 and as expected from the failure of rabbit IgG sensitised SRBC to bind to isolated enterocytes, rabbit IgG showed little inhibition. This is less true for human and bovine IgG but surprisingly, mouse IgG appears to bind very effectively to the receptor, causing complete inhibition of rosette formation at the concentrations used. Opossum serum, previously heated to destroy complement and prevent red cell lysis, also inhibited rosette formation presumably by virtue of its contained IgG. Bennett's wallaby serum was also very effective suggesting the likelihood of much closer similarity of marsupial IgG in its binding sequences. It will be interesting in future to see if any of the four mouse IgG isotypes preferentially inhibit rosette formation. Differences have been reported for Fcγ receptor mediated binding of mouse IgG isotypes to rat jejunal enterocytes (Borthistle *et al.* 1977) and rat IgG isotypes to rat proximal small intestine enterocytes (Peppard *et al.* 1985). In separate experiments to those shown in Table 16.1, mouse IgG Fab fragments were found not to inhibit rosette formation, but we have not yet been able to obtain and study mouse IgG Fc. However, in the light of mouse IgG proving to be inhibitory and rabbit IgG not so, experiments have been conducted in which approximately 200 µl of a mixture of mouse and rabbit IgG (each at a final concentration of 1 mg/ml PBS pH 7.2) was fed to four suckling sibling 31 day old opossums following removal from the teat. These were decapitated after 30, 45, 60 and 75 min, exuded blood was obtained from the cut surfaces, and the serum subjected to immunodiffusion analysis using rabbit anti-mouse IgG and mouse anti-rabbit IgG. As can be seen from Figure16.2, a precipitin line indicative only of mouse IgG was detected in the serum from the opossum exposed to the IgG mixture for 75 min. This experiment was repeated with three 33 day old siblings decapitated after 30, 60 and 90 mins. Whilst again no rabbit IgG could be detected in the serum, mouse IgG was detected only in the case of the 60 mins exposure and not that of 90 min. This may relate to the fact that standardisation of the dose fed is difficult to achieve in suckling opossums by oral feeding since some regurgitation of the sample sometimes occurs. Also, when previous feeding experiments were conducted (Wild *et al.* 1994), we took the precaution of leaving the opossums isolated from the mother for 1 hours in the hope that competing opossum IgG in the milk would be cleared from the gut in this

Table 16.1 Inhibition of 35 day opossum proximal small intestine enterocyte- E-A rosette[1] formation by IgG and serum of different mammalian species

SPECIES OF ORIGIN OF IgG [2] OR SERUM(S)	% ROSETTES 1	2	3	MEAN	% INHIBITION[3]
PBS control	60	75	46	60	—
human	50	45	50	48	20
bovine	42	41	49	47	22
rabbit	53	60	60	58	3
mouse	0	0	0	0	100
opossum(s)	1	3	0	1	98
wallaby(s)	0	0	0	0	100

[1] E-A rosettes formation and inhibition carried out at pH 7.2 in experiments set up in triplicate using opossum-sensitised SRBC as described in Wild *et al.* (1994), but using enterocytes isolated and stored as described in note 1 of the present chapter. In separate experiments carried out using PBS pH 6.0, the mean percentage rosette formation was 50%. No rosettes were observed using normal SRBC or SRBC sensitised with rabbit IgG. (s)=serum

[2] IgG (Miles Laboratories) used at 2mg/ml in PBS pH 7.2.

[3] Percentage inhibition calculated from the formula: 100 (percentage rosettes formed in presence of competing protein /percentage rosettes formed in absence of competing protein x 100).

time. This was not possible in the experiments described here and could have affected the findings. Nevertheless, the experiments provide evidence that selective transport of mouse IgG does occur and confirm that rabbit IgG does not, and as a consequence support the premise that IgG transport is Fcγ receptor mediated.

There are also indications that transport is selective in the quokka, and therefore likely to be Fcγ receptor mediated. This can be construed from the finding of Yadav (1971) that following feeding by stomach tube, IgM antibodies to SRBC, *Salmonella adelaide*, and I x 174 phage are not transported, at least in pouch young older than 75 days. Transport of heterologous possum anti-*S. adelaide* and anti-phage antibodies (both IgG) to 60-day-old tammars was also demonstrated in these studies, but there has as yet been no systematic quantitative analysis carried out in a way comparable to eutherians, to determine whether selectivity of marsupial IgG transport occurs within marsupials.

SITE OF IgG TRANSPORT WITHIN THE GUT

As might be expected from its much shorter gestation period, the small intestine of the suckling opossum is much reduced compared to that of equivalent aged rats (approximately 3.5 and 25 cm respectively at 12 days) and a clear demarcation, based on colour differences between jejunum (white) and ileum (brownish-yellow), at least prior to 20 days, is not easy to achieve. In the suckling rat, where this colour change is very clear, enabling confident separation to be made, Fcγ receptors detectable by E-A rosettes are found to be much restricted to the duodenum and jejunum (Wild 1981;

Figure 16.2 Immunodiffusion analysis of 31-day-old opossum serum obtained from blood collected at the times indicated (in min.) and following oral feeding of a mixture of rabbit and mouse IgG at a final concentration of 1 mg/ml in PBS pH 7.2.
Key:
aM=rabbit anti-mouse IgG
aR=mouse anti-rabbit IgG
R=rabbit IgG (1 mg/ml)
M=mouse IgG (1 mg/ml

Griffin & Wild 1987). Injection *in situ* of labelled IgG into proximal and distal small intestine separated by ligature has also shown transport of intact IgG only from the proximal (duodenum and jejunum) region (Rodewald 1973; Morris & Morris 1974). We resorted to bisecting the small intestine of suckling opossums as a means of obtaining proximal and distal segments and determined the percentage of rosettes formed with enterocytes obtained from each half. However, it is not certain how much contamination by jejunal enterocytes in the distal half would have occurred. Our findings (Wild *et al.* 1994), whilst again very variable, overall indicated that there was a statistically greater chance of finding Fcγ receptor bearing cells in the proximal half. Such findings must, however, also be considered in the light of localisation of homologous and heterologous IgG within gut enterocytes (see later) and which point more strongly to the proximal region being most involved as a site of transport.

Albumin has also been reported to be transported to the suckling opossum, as inferred from a peak in homologous albumin concentration at 8 to 10 days and detection in the plasma of selectively transported human albumin previously injected into the *Monodelphis* mother (Knott *et al.* 1993); there is evidence such transport takes place in the stomach (N R Saunders, personal communication). However, in observations so far limited to 13 day opossum stomach we have seen little indication of uptake and transport of endogenous opossum IgG as revealed by indirect immunofluorescence microscopy (see Figure 16.3a). In view of our finding that mouse IgG is transported to the blood of the suckling opossum, it will be interesting in future studies to co-localise human albumin and mouse IgG in the stomach, following feeding, to see if and where the selective barrier operates. Yadav (1971) isolated the rumen of a 95 day old quokka from the intestine by means of ligatures before feeding antibodies (IgG) by stomach tube; no antibodies could be detected in serum when tested at 4 hoursand 24 hoursfollowing ingestion, suggesting that the rumen wall at least is a barrier to transport. It is not known in which part of the small intestine antibody transport takes place in the quokka.

LOCALISATION OF IgG WITHIN GUT ENTEROCYTES

When our investigations first commenced on *Monodelphis* we did not have available antiserum specific to opossum IgG or know of any immunoglobulin other than opossum IgG that might be transported. In an attempt to gain insight into where and how it was selectively transported, opossum IgG was first isolated from normal serum by DEAE ion exchange chromatography following the method described by Hindes and Mizell (1976) for *Didelphis*. We then directly conjugated this to fluorescein isothiocynate

(FITC) and compared its localisation to a commercial preparation of FITC-labelled rabbit IgG, in proximal and distal small intestine, following feeding separately to 14 and 20 day old opossums and fixing and processing tissue 1 hour later according to the method of Sainte Marie (1962). Specific fluorescence taken to be indicative of both opossum and rabbit IgG was observed in what appeared to be apical vesicles in enterocytes of proximal small intestine but there was much less evidence of uptake by enterocytes in distal small intestine (Wild *et al.* 1994). Supranuclear vesicles characterise many proximal small intestine and all distal small intestine enterocytes in suckling opossums prior to weaning (Figures 16.4a and b) and it was our impression that these may have been labelled in the case of rabbit IgG, but not opossum IgG. Presence of labelled opossum IgG, but not rabbit IgG, was also indicated in intercellular spaces, but in neither case was labelled IgG evident in the villus core suggesting no transport to the blood. These findings may have been influenced by the presence of the fluorochrome affecting binding to Fcγ receptors (although in the case of FITC-opossum IgG it inhibited rosette formation), fluorochrome dissociating from IgG following proteolysis, and the fact that enterocytes of the proximal small intestine would have probaly been exposed to the conjugate for a longer time interval than those of distal small intestine. Studies of selective transport of FITC-labelled homologous and heterologous IgG across rabbit yolk sac endoderm (Wild 1970) have, however, given similar findings compared to their localisation by indirect immunofluorescence.

When a rabbit antiserum to opossum IgG had been prepared, we investigated the distribution of endogenous opossum IgG in 13-day-old opossum small intestine using immunofluorescence (Wild *et al.* 1994). A similar labelling pattern to that of FITC-labelled opossum IgG was observed in enterocytes of the proximal small intestine (apical vesicles and intercellular space) but labelling indicative of IgG was now also observed in the villus core. In the distal small intestine only the villus core and basal intercellular spaces appeared labelled. More recently we have compared the distribution of endogenous opossum IgG in 31 day suckling opossum small intestine with that in 58 day weaned opossum using confocal as well as conventional fluorescence microscopy to aid interpretation of events that might take place during uptake and transport of IgG. A strong reaction indicative of opossum IgG was evident on the microvillous apical border of enterocytes of some, but not all, villi of 31 day proximal small intestine, in intracellular vesicles, in intercellular spaces and in the villus core (see Figures 16.3a–e). No reaction could be discerned in supranuclear vesicles when present. Enterocytes of distal 31 day small intestine exhibited occasional reaction in apical vesicles and a reaction in the villus core but there was little reaction in intercellular spaces. Proximal and distal 58 day small intestine both gave similar findings. Specific fluorescence indicative of IgG was now seen only intraluminally, in the villus core, and occasionally on the apical brush border surface; none was seen intracellularly or intercellularly (Figure 16.3f).

Better indirect immunofluorescence localisation of endogenous IgG in enterocytes of opossum small intestine has been obtained using 4% paraformaldehyde fixation followed by equilibration in 20% buffered sucrose, mounting in optimum cutting tissue compound, freezing in liquid N_2, and sectioning in a cryostat as described in Heath *et al.* (1992). Certain cells in the villus core do, however, autofluoresce compared to fixation with ethanol as used in the Sainte Marie (1962) technique and it is not easy to get perfectly flat, thin sections, so that they are best observed using confocal microscopy. In sections of 31 day suckling opossum proximal small intestine processed in this way and

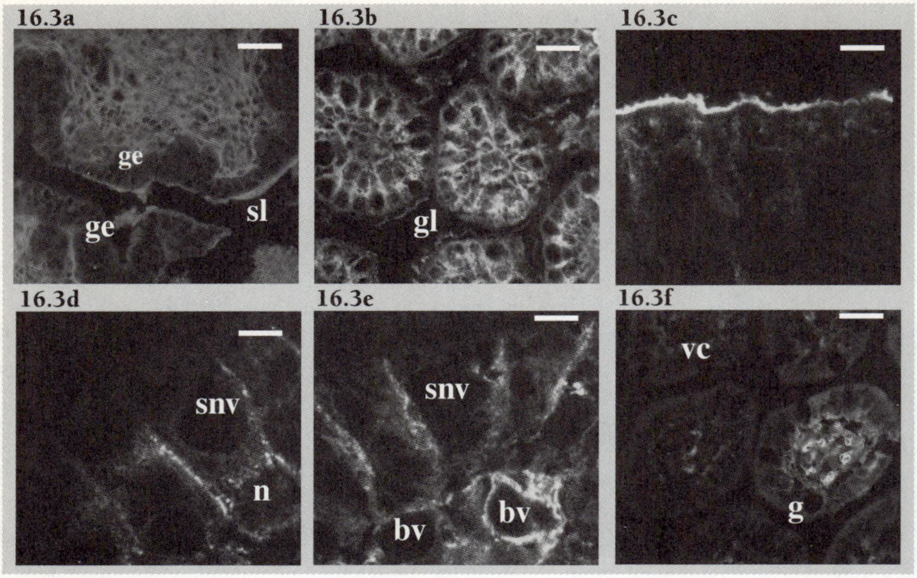

Figure 16.3 (a) Section of 13-day opossum stomach reacted with rabbit anti-opossum IgG and FITC goat anti-rabbit IgG as described in Wild et al. (1994) and viewed by conventional fluorescence microscopy. Specific fluorescence indicative of opossum IgG is present in milk in the stomach lumen, on the surface of gastric epithelial cells and in the underlying mesenchymal tissue. Epithelial cells contain no intracellular specific fluorescence. Scale bar=60 μm.

(b) Transverse sections through villi of 31-day opossum proximal SI reacted for opossum IgG localisation as described in (a) and viewed by conventional fluorescence microscopy. Opossum IgG is present in the gut lumen, in the villus core, intracellularly within enterocytes and intercellularly between enterocytes. Scale bar=60 μm.

[(c), (d) and (e)] High power views of enterocytes of villi similar to those shown in (b) but viewed by confocal fluorescence microscopy. In all cases large supranuclear vesicles appear devoid of IgG.

(c) Opossum IgG is mainly confined to the apical plasma membrane. Scale bar=12 μm.

(d) it is mainly confined to the intercellular space and small vesicles above the nucleus. Scale bar=12 μm.

(e) IgG can also be seen present in a blood vessel and mesenchymal tissue in the villous core.

Scale bar=20 μm.

(f) Transverse sections through villi of 58-day opossum proximal SI reacted for IgG localisation as described in (a) and viewed by conventional fluorescence microscopy. Specific fluorescence indicative of IgG is now confined to mesenchymal tissue and blood vessels within the villus core. Scale bar=60 μm.

Key:
ge=gastric epithelium
mt=mesenchymal tissue
sl=stomach lumen
gl=gut lumen
vc=villus cores
nv=supranuclear vesicle
n=nucleus
bv=blood vessel
gc=goblet cell

Note: In all the figures shown tissue was fixed in 95% ethanol and processed according to the method of Sainte-Marie (1962).

subjected to indirect immunofluorescence, opossum IgG was again observed in enterocyte apical endocytic vesicles, in the intercellular space and villus core (Figures 16.5a and b). This suggests that the localisation of IgG previously seen, particularly in the intercellular space, is not an artefact of ethanol fixation and subsequent paraffin embedding procedures. Small intestine from experiments referred to earlier in which opossums were fed a mixture of mouse and rabbit IgG have been processed in this way and by ethanol fixation. Although both rabbit and mouse IgG have been detected co-localised in apical vesicles of enterocytes of proximal small intestine (Figure 16.5c and d), we have not yet obtained cytological evidence of mouse IgG transport and investigations are still in progress to try and locate transported mouse IgG. These preliminary findings do, however, suggest that much of the uptake of IgG by enterocytes may be non-selective.

POSSIBLE MECHANISM OF SELECTIVE IgG TRANSPORT

Since opossum IgG (detected as anti-SRBC antibody or as endogenous IgG by indirect immunofluorescence) appears not to be transcytosed across enterocytes after weaning, examination of the ultrastructure of enterocytes before and after this time may illuminate mechanisms involved. With this aim in mind, so far we have examined proximal and distal small intestine of 46 and 58 day old opossums. As can be seen from stained 0.7 µm plastic sections of villi illustrated in Figure 16.4, enterocyte supranuclear vesicles are fewer and smaller in 46 day proximal small intestine compared to those in distal small intestine, although apical endocytic vesicles are apparent in both regions. In enterocytes of 58 day old opossum distal and proximal small intestine such large supranuclear vesicles are almost entirely absent, as are apical endocytic vesicles. At the ultrastructural level an apical tubular canalicular system (ATCS) together with endocytic vesicles can be seen beneath the terminal web of enterocytes of both proximal and distal 46 day small intestine (Figures 16.6a–d). Endocytic vesicles are often multivesiculate in enterocytes of distal small intestine (Figure 16.6b) and occasionally tubules can be seen connecting to the apical plasma membrane providing an entry for proteins. A similar endocytic complex of vesicles and ATCS is present in both proximal and distal enterocytes of pre-weaned rat small intestine. In proximal small intestine this complex has been shown to be involved in the endocytosis (Rodewald 1973) and sorting (Abrahamson & Rodewald 1981) of IgG from other non-selectively endocytosed proteins by injection of ferritin labelled rat IgG and rat anti-HRP antibodies directly into the gut lumen. Sorting from non-selectively endocytosed proteins is effected by clathrin coated vesicles (CVs) containing IgG (initially bound to Fcγ receptors present on microvilli, Berryman & Rodewald 1990) budding off from the ATCS (Abrahamson & Rodewald 1981). Such CVs have been observed close to, and apparently fusing with, the lateral plasma membrane, and discharging IgG into the intercellular space where the predicted more neutral pH has been suggested to favour Fcγ receptor-ligand dissociation (Rodewald 1973; Abrahamson & Rodewald 1981). Although enterocytes of rat distal small intestine are also able to endocytose IgG, no binding takes place to microvillous, tubule, or vesicle membranes; IgG does not leave the cell segregated into CVs, but instead is degraded along with other non-selectively endocytyosed proteins in the large phagolysosomes that are prominent here (Rodewald 1973).

So far we have not seen convincing evidence of CVs forming from the ATCS of opossum enterocytes but what appear to be clathrin coated pits, containing electron-dense material, have been observed at the apical plasma membrane, particularly so in the case of enterocytes from 46 day proximal small intestine (Figure 16.6c). CVs are also present at this age

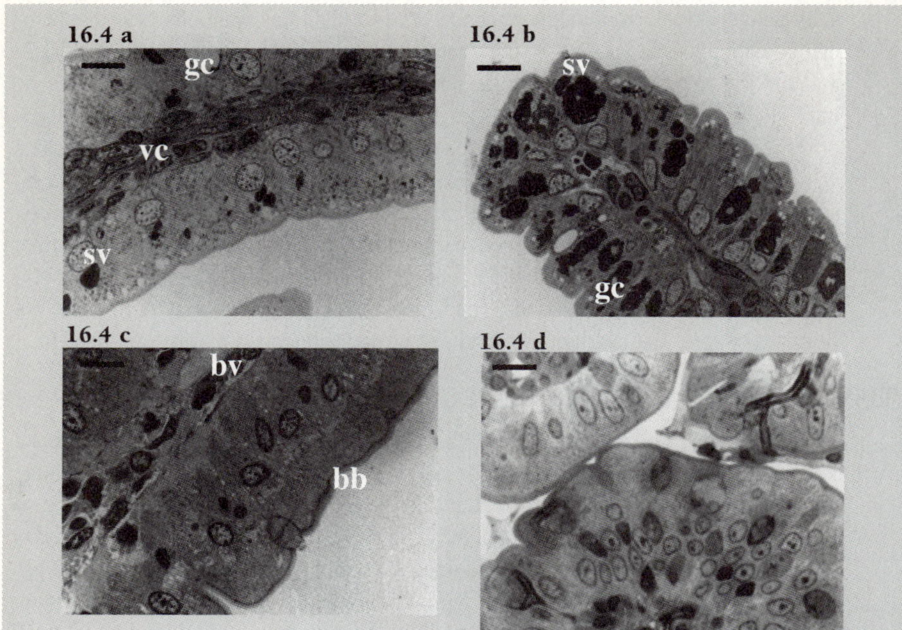

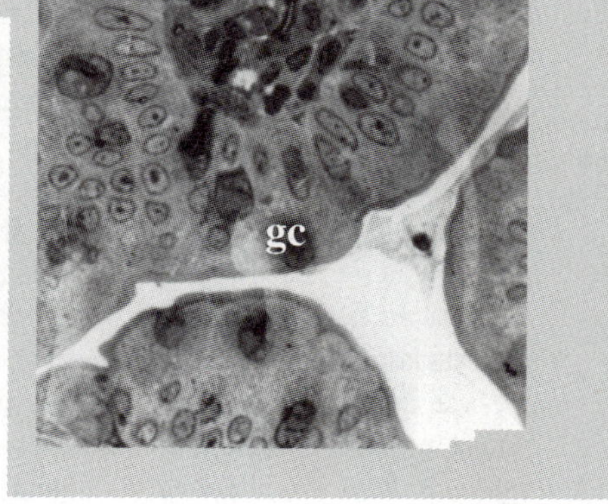

Figure 16.4 (a) and (b) Plastic (0.7 µm) sections of villi of 46-day proximal (a) and distal (b) SI and 58-day proximal (c) and distal (d) SI stained with toluidine blue. Enterocytes of 46-day proximal SI can be seen to have fewer, and smaller, densely staining supranuclear vesicles compared to those of distal SI. Translucent apical endocytic vesicles are evident in enterocytes of both regions and mesenchymal cells and blood vessels can be seen in the villus core. Scale bars=25 µm.
(c) and (d) Supranuclear dense vesicles and endocytic vesicles are absent from enterocytes of 58-day proximal and distal SI. Villi of 58-day distal SI seen in transverse section in (d) have an increased number of goblet cells compared to pre-weaned stages. Scale bar=17 µm.
Labels: gc=goblet cell; sv=supranuclear vesicle; vc=villus core; bv=blood vessel; bb=microvillous brush border.

and site, free within apical cytoplasm (Figure 16.6c and Figure 16.6d) and have been observed close to, and apparently fusing with, the lateral plasma membrane, particularly at the level of the basally situated nucleus (Figures 16.6e and f). Apical-coated pit formation is less apparent in enterocytes of 46 day distal small intestine but again CVs can sometimes be seen close to, and apparently fusing with, the basal lateral plasma membrane. In view of the binding of indicator SRBC to the lateral plasma membrane of isolated enterocytes consequent on rosette formation, the localisation of opossum IgG within the intercellular space as detected by indirect immunofluorescence, and the comparative evidence derived from rat studies described above, it seems reasonable to suppose that such CVs as we have observed also contain opossum IgG and transcytose it in association with Fcγ receptors to

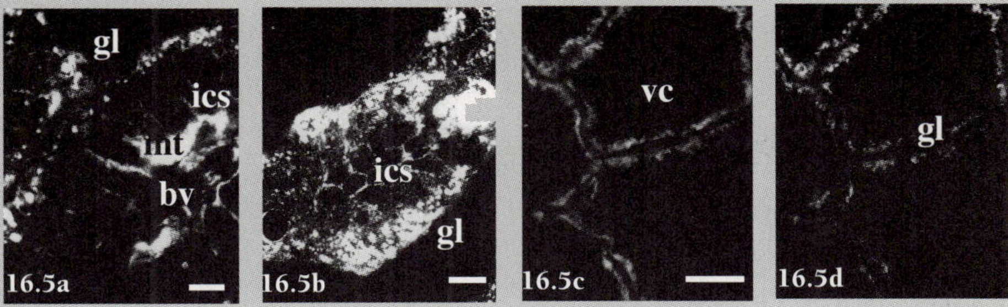

Figure 16.5 [(a) and b)] Sections of villi of 31-day opossum proximal SI reacted with rabbit anti-opossum IgG and TRITC-labelled goat anti-rabbit IgG and viewed by confocal fluorescence microscopy. (a) Parts of villi cut longitudinally. Opossum IgG is localised within enterocytes in apical endocytic vesicles, in intercellular spaces and within mesenchymal tissue and a blood vessel in the villus core. Scale bar=25 μm.
(b) The section passes obliquely through a villus. At the periphery, where enterocytes are cut through the apical region, IgG can be seen present in endocytic vesicles varying greatly in size. In the centre, where enterocytes are cut transversely in what is presumed to be the mid-region, IgG appears confined to intercellular spaces. Scale bar=25 μm.
(c) and (d) Confocal images of sections of villi in proximal SI of a 31-day opossum fed a mixture of mouse and rabbit IgG and the SI removed and fixed 60 min later. The section was reacted with FITC sheep anti-mouse IgG followed by TRITC goat anti-rabbit IgG and viewed for specific fluorescence indicative of mouse IgG in (a) and rabbit IgG in (b). Mouse and rabbit IgG appear to be co-localised in apical endocytic vesicles but there is no evidence of transport to the intercellular space or villus core. Scale bar=45 μm, for (c) and (d).
Key: gl=gut lumen mt=mesenchymal tissue bv=blood vessel ics=intercellular space vc=villus core
Note: In both figures, tissue was fixed in 4% paraformaldehyde and processed as described in Heath *et al.* (1992).

the basolateral plasma membrane. However, in view of the apparent ability of opossum IgG to bind at pH 7.2, some mechanism other than a change in pH may be needed to effect ligand-receptor release. The same considerations also apply to transcytosis of IgG across rabbit yolk sac endoderm, where binding to Fcγ receptors has been shown to be equally effective at pH 6.0 and pH 7.2 irrespective of whether they are located at the apical or lateral plasma membrane (Meads & Wild 1994). The possibility that intercellular macromolecular transport takes place from the gut lumen seems unlikely, since enterocytes are always sealed by tight cell junctions (see Figure 16.6d and Figure 16.7c). However, IgG entering the villus core from a systemic source could conceiveably diffuse basally from here into the intercellular space. This might account for some or all of the localisation of IgG sometimes observed in this site in enterocytes of distal small intestine. It is likely that milk proteins other than IgG enter the large supranuclear vesicle present in enterocytes and that this is where most intracellular digestion takes place. To what extent opossum IgG is also degraded here, remains to be discovered.

Coated pits and apical invaginations are very infrequently observed in enterocytes of 58 day proximal and distal small intestine and the endocytic complex of ATCS and vesicles has virtually disappeared at this age (Figures 16.7a–d). Numerous small, translucent apical and larger mid-region vesicles, probably representing lipid droplets extracted during fixation, as well as electron dense residual bodies, are now present in enterocytes of proximal small intestine. Larger residual bodies are also present in enterocytes of distal small intestine (Figure 16.7d) and may be remnants of the supranuclear vesicles. In a detailed ultrastructural study of the small intestine from newborn *Didelphis* and later stages up to weaning, Krause *et al.* (1977) also reported a reduction in the endocytic complex starting at about 75 days and continuing until about 95 days when weaning commences. They could only suppose this to be related to cessation of absorptive capacity. The combined indirect immunofluorescence and ultrastructural observations obtained here for *Monodelphis* places this supposition on a more certain footing.

FUTURE DIRECTIONS

The Fcγ receptor (now designated FcRn, 'n' denoting neonatal) has been isolated from enterocytes of suckling rat proximal small intestine, cloned, sequenced, shown to be associated with $ß_2$- microglobulin, and the a-chain shown to have considerable homology with Class 1 major histocompatibility (MHC) molecules in the extracellular domains that form the peptide binding groove (Simister & Mostov 1989). A very similar receptor has been shown to be present in enterocytes of mouse proximal small intestine and yolk sac endoderm, but the gene lies outside the MHC, on chromosome 7, indicating divergence from MHC early in the mammalian lineage (Ahouse *et al.* 1993). Crystallisation of the receptor has also been effected (Gastinel *et al.* 1992; Huber *et al.* 1993) and study of the binding site now appears to be in progress. The extracellular domains in association with $ß_2$- microglobulin have been expressed in Chinese hamster ovary cells and stoichiometric analysis of the secreted complex bound with Fc piece of IgG indicates that two FcRn heterodimers bind to a single Fc (Huber *et al.* 1993); whether or not binding occurs to the peptide groove or elsewhere in the extracellular domains, remains to be determined. In view of the knowledge that has now been gained from studies on suckling rats, it should not be too difficult to apply this to marsupials and in particular to *Monodelphis*. Class 1 MHC antigens have been defined in *Monodelphis* (Stone *et al.* 1987 quoted in Jurd 1994) and it will be interesting to see if homology with Class 1 MHC is also retained in the suckling opossum gut Fcγ receptor and to what extent it resembles FcRn. If an absence of acid pH dependency for ligand binding can be confirmed using other binding assays, sequencing of the opossum Fcγ receptor might, on comparison, shed light on how an acid pH is required for it to occur in the case of FcRn. Such investigations will require opossum receptor isolation which, in view of the findings reported here, could now conveniently be carried out using immobilised mouse or tammar IgG, rather than opossum IgG which is more difficult to obtain. Transcytosis of CVs containing IgG bound to receptor and selected through formation of coated pits at the cell surface, has been suggested to be the mechanism operating for transport of maternal IgG in rabbit yolk sac endoderm (Wild 1975; Moxon *et al.* 1976). Does a similar mechanism operate in suckling opossum gut enterocytes or does sorting take place intracellularly as deduced in the suckling rat? Dual EM immunogold detection of mouse and rabbit IgG fed to suckling opossums, or injected into proximal and distal small intestine should provide an answer to this question. Since rabbit IgG shows no Fc binding, antibodies could conveniently be raised in rabbit to the isolated opossum Fcγ receptor, thus avoiding the necessity for F(ab')2 production. Such antibodies, particularly if directed against the ligand binding site, could be expected to inhibit transport of IgG when fed to suckling opossums. This would provide important physiological proof that the isolated

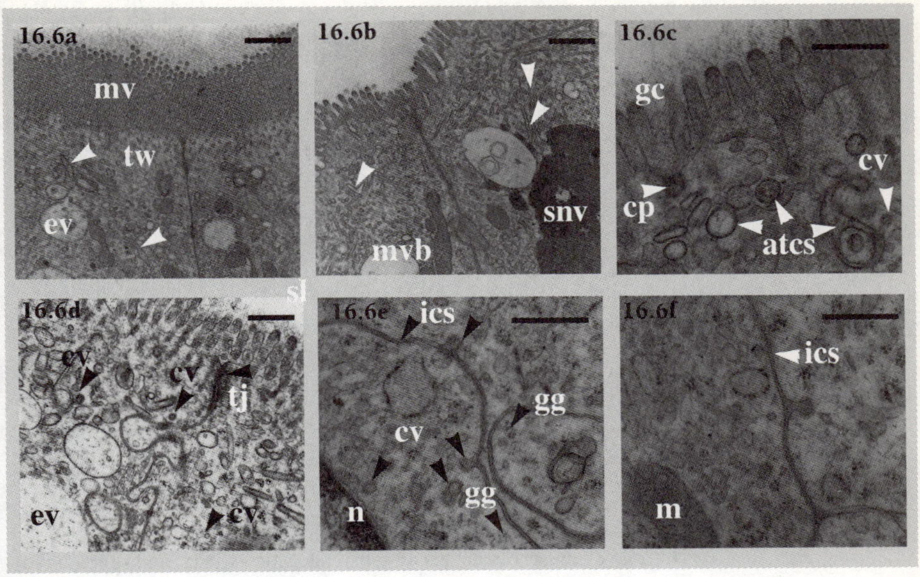

Figure 16.6 Electronmicrographs of enterocytes of 46-day opossum proximal (a, c, d, e and f) and distal (b), SI.
(a) Numerous microvilli can be seen projecting from the terminal web region; elements of the apical tubular canalicular system (see arrows) and endocytic vesicles interspersed with mitochondria are also evident.
Scale bar=1 µm.
(b) Distal SI enterocytes shown here have a high concentration of apical tubules (see arrows) and part of a densely staining supranuclear vesicle is present in one cell; apical vesicles often appear as multivesiculate bodies.
Scale bar=1 µm.
(c) Seen at higher power, microvilli appear coated with a fibrillar glycocalyx and coated pits containing electron-dense material arise as invaginations from the apical plasma membrane; vesicles representative of cross-sections of the apical tubules, and coated vesicles, are seen in the apical cytoplasm.
Scale bar=0.5 µm.
(d) Further examples of coated vesicles in apical cytoplasm are also highlighted here, together with a tight junction and endocytic vesicles.
Scale bar=0.5 µm.
(e) and (f) Examples of the lateral surface of enterocytes at the level of the nucleus; coated vesicles are found both free in the cytoplasm and opening (arrowheads) into the intercellular space; clusters of glycogen granules are dispersed throughout the cytoplasm.
Scale bar=0.5 µm.
Key: mv=microvilli tw=terminal web ev=endocytic vesicle mvb=multivesiculate body snv=supranuclear vesicle gc=glycocalyx cp=coated pit atcs=apical tubular canalicular system cv=coated vesicle tj=tight junction ics=intercellular space m=mitochondrion gg=glycogen granules

receptor functions *in vivo* in transporting maternal IgG. Such antibodies would also enable ultrastructural localisation of the receptor to be effected in a manner comparable to that which Berryman & Rodewald (1990) have already achieved in their studies on the suckling rat.

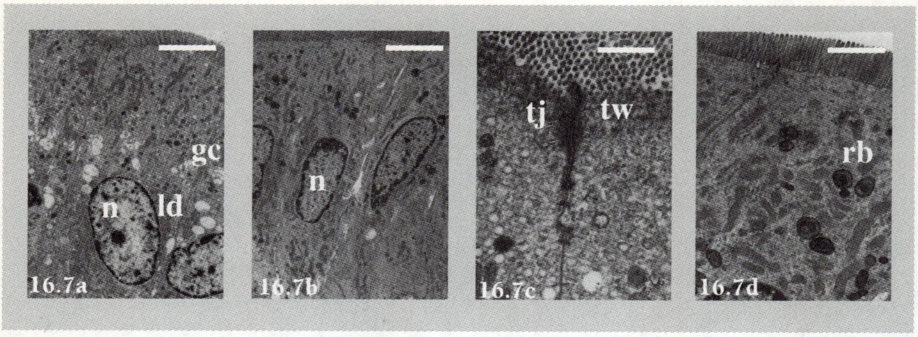

Figure 16.7 Electronmicrographs of enterocytes of 58-day opossum proximal (a) and (c) and distal (b) and (d), SI.
(a) large supranuclear vesicles and endocytic vesicles are now largely absent from enterocytes, as is the apical tubular canalicular system; what appear to be large, extracted lipid droplets and a Golgi complex are evident immediately above the nucleus.
Scale bar=5 μm.
(b) The apical tubular canalicular system, endocytic vesicles, and supranuclear vesicles are also absent from distal SI enterocytes shown here.
Scale bar=5 μm.
(c) The apical cytoplasm of proximal enterocytes shown in higher power here is full of many small (200 nm diameter) translucent vesicles (small arrowheads) also probably representative of extracted lipid droplets; small, electron-dense bodies (large arrowhead) probably representing residual bodies can also be seen; the terminal web region is now more electron-dense (compare to Figures 16.6a and c) presumably due to an increase in actin filaments in this region.
Scale bar=1 μm.
(d) When viewed at higher power, the increase in electrondensity of the terminal web can be seen not to extend to distal SI enterocytes and residual bodies in these cells are larger and more complex; lipid droplets are also absent from these cells.
Scale bar=2 μm.
Key: n=nucleus gc=Golgi complex ld=lipid droplets tj=tight junction tw=terminal web rb=residual body

Acknowledgments

I am grateful to Professor Norman Saunders for recently introducing me to the laboratory opossum as an experimental animal; to Vikki Stauber for help with the electronmicroscopy; to my colleague Dr Tom Fleming for help with confocal microscopy; to Jane Booth for some technical assistance; and to the Royal Society for providing a study visit grant enabling me to attend the Hugh Tyndale-Biscoe 65th birthday meeting, which also resulted in the opportunity to contribute to this book.

References

Abrahamson D R and Rodewald R (1981). Evidence for sorting of endocytic vesicle contents during receptor-mediated transport of IgG across the newborn rat intestine. *Journal of Cell Biology*, 91, 270–280.

Adam E, Bunce C.E, Saunders N R and Snart C (1988). *Monodelphis domestica* (Grey short-tailed opossum): a model for studies of very early stages of development. *Journal of Physiology*, 400, 13P.

Ahouse J J, Hagerman C L, Mittal P, Gilbert D J, Copeland N G, Jenkins N A and Simister N E (1993). Mouse MHC Class 1-like Fc receptor encoded outside the MHC. *Journal of Immunology*, 151, 6076–6088.

Baintner K (1986) *Intestinal Absorption of Macromolecules and Immune transmission from Mother to Young*. : CRC Press, Boca Raton, Florida.

Berryman M A and Rodewald R D (1990). An enhanced method for post-embedding immunocytochemical staining which preserves cell membranes. *Journal of Histochemistry and Cytochemistry*, 38, 159–170.

Borthistle B K, Kubo R T, Brown W R and Grey H M (1977). Studies on receptors for IgG on epithelial cells of rat intestine. *Journal of Immunology*, 119, 471–476.

Brambell F W R (1970). *The transmission of passive immunity from mother to young*. Elsevier, Amsterdam.

Deane E M and Cooper D W (1988). Immunological development of pouch young marsupials. In *The Developing Marsupial. Models for Biomedical Research*. Eds C H Tyndale-Biscoe and P A Janssens. Springer-Verlag, Berlin. pp 190–199.

Gastinel L N, Simister N E and Bjorkman P J (1992). Expression and crystallization of a soluble and functional form of an Fc receptor related to class 1 histocompatibility molecules. *Proceedings of the National Academy of Sciences, USA*, 89, 638–642.

Griffin P and Wild A E (1987). Ontogeny of the Fcγ receptor in rat small intestine. *Journal of Reproductive Immunology*, 11, 287–306.

Halliday R (1959). The effect of steroid hormones on the absorption of antibody by the gut of the young rat. *Journal of Endocrinology*, 18, 56–66.

Harder J D, Stonerook M J and Pondy J (1993). Gestation and placentation in two new world opossums: *Didelphis virginiana* and *Monodelphis domestica*. *The Journal of Experimental Zoology*, 266, 463–479.

Heath L, Wild A E and Thorogood P (1992). Monoclonal antibodies raised against premigratory neural crest reveal population heterogeneity during crest development. *Differentiation*. 49, 151–165.

Hindes R D and Mizell M (1976). The origin of immunoglobulins in opossum "embryos". *Developmental Biology*, 53, 49–61.

Huber A H, Kelley R F, Gastinel L N and Bjorkman P J (1993). Crystallization and stoichiometry of binding of a complex between a rat intestinal Fc receptor and Fc. *Journal of Molecular Biology*, 230, 1077–1083.

Jeffcott L B (1972). Passive immunity and its transfer with special reference to the horse. *Biological Reviews*, 47, 439–464.

Jurd R D (1994). "Not proper mammals": immunity in monotremes and marsupials. *Comparative Immunology and Microbiology of Infectious Diseases*, 17, 41–52.

Kemmler R, Mossman H, Strohmaier U, Kickhofen B and Hammer D K (1975). *In vitro* studies on the selective binding of IgG from different species to tissue sections of the bovine mammary gland. *European Journal of Immunology*, 5, 603–608.

Knott G, Dziegielewska K M, Habgood M D and Saunders N R (1993). Transfer of albumin from milk across the immature gut of postnatal *Monodelphis domestica*. *Proceedings of the Australian Physiological and Pharmacological Society*, 24, 43P.

Krause W J, Cutts H J and Leeson C R (1977). The postnatal development of the alimentary canal in the opossum. *Journal of Anatomy*, 123, 21–45.

Mackenzie N (1984). Fc receptor-mediated transport of immunoglobulin across the intestinal epithelium of the neonatal rodent. *Immunology Today*, 5, 364–366.

Meads T J and Wild A E (1994) Evidence that Fcγ receptors in rabbit yolk sac endoderm do not depend upon an acid pH to effect IgG binding and transcytosis in vitro. *Placenta*, 15, 525–539.

Merad Z and Wild A E (1992) The route of maternal IgM transport to the rabbit fetus. *Placenta*, 13, 291–304.

Morris B and Morris R (1974). The absorption of

^{125}I labelled immunoglobulin G by different regions of the gut in young rats. *Journal of Physiology*, 241, 761–770.

Morris B and Morris R (1976). The effect of corticosterone and cortisone on the uptake of polyvinyl pyrrolidone and the transmission of immunoglobulin G by the small intestine in young rats. *Journal of Physiology*, 255, 619–634.

Moxon L A, Wild A E and Slade B S (1976). Localisation of proteins in coated micropinocytotic vesicles during transport across rabbit yolk sac endoderm. *Cell and Tissue Research*, 171, 175–193.

Patt J A (1977). Factors affecting the duration of intestinal permeability to macromolecules in newborn animals. *Biological Reviews*, 52, 411–429.

Peppard J V, Jackson L E and Mackenzie N (1985). Comparison of *in vitro* binding to enterocyte brush borders of rat IgG subclasses and their transmission *in vivo* in the rat. *Immunology*, 53, 385–393.

Renfree M B (1973). The composition of fetal fluids of the marsupial *Macropus eugenii*. *Developmental Biology*, 33, 62–79.

Rodewald R (1973) Intestinal transport of antibodies in the newborn rat. *Journal of Cell Biology*, 58, 189–211.

Rodewald R (1976). pH-Dependent binding of immunoglobulins to intestinal cells of the neonatal rat. *Journal of Cell Biology*, 71, 666–670.

Rowlands D T, Blakeslee D and Liu Hshin-Hshing (1972). The early immune response and immunoglobulins of opossum embryos. *Journal of Immunology*, 108, 941–946.

Sainte-Marie G (1962). A paraffin embedding technique for studies employing immunofluorescence. *Journal of Histochemistry and Cytochemistry*, 10, 250–256.

Samples K, Vandeburg J L and Stone W H (1986). Passively acquired immunity in the newborn of a marsupial (*Monodelphis domestica*). *American Journal of Reproductive Immunology and Microbiology*, 11, 94–97.

Sasaki M, Lawson B L and Nelson D R (1977). Kinetic analysis of the binding of immunoglobulin IgG1 and IgG2 to bovine mammary cells. *Biochimica Biophysica Acta*, 497, 160–170.

Simister N E and Mostov K E (1989). An Fc receptor structurally related to MHC class 1 antigens. *Nature*, 337, 184–187.

Stone W H, Samples N K, Kraid E and Vandeburg J L (1987). Definition of class 1 antigens of the MHC of a marsupial (*Monodelphis domestica*). *Archives Zootecnia*, 36, 219–235.

Wild A E (1970) Protein transmission across the rabbit foetal membranes. *Journal of Embryology and Experimental Morphology*, 24, 313–330.

Wild A E (1975) Role of the cell surface in selection during transport of proteins from mother to foetus and newly born. *Philosophical Transactions of the Royal Society of London Series B*, 271, 395–410.

Wild A E (1981). Distribution of Fcγ receptors on isolated gut enterocytes of the suckling rat. *Journal of Reproductive Immunology*, 3, 283–296.

Wild A E (1993) The role of yolk sac and gut epithelial cells in maternal immunoglobulin transport. In *Epithelial Organization and Development*. Ed T P Fleming, Chapman & Hall, London. pp 273–298.

Wild A E, Burrows T D and Brand J (1994) IgG transport across the gut of the suckling opossum (*Monodelphis domestica*). *Developmental and Comparative Immunology*, 18, 75–88.

Yadav M (1971) The transmissions of antibodies across the gut of pouch-young marsupials. *Immunology*, 21, 839–850.

Yadav M, Eadie M and Stanley N F (1971). The thymus glands of a marsupial, *Setonix brachyurus* (quokka) and their role in immune responses. V1 Maternal transfer of immunoglobulins. *Australian Journal of Experimental Biology and Medical Sciences*. 49, 322–332.

Yadav M and Eadie M (1973). Passage of maternal immunoglobulins to the pouch young of a marsupial, *Setonix brachyurus*. *Australian Journal of Zoology*, 21, 171–181.

CHAPTER 17

THE NORTH AMERICAN OPOSSUM *DIDELPHIS VIRGINIANA*: AN IMPORTANT MODEL FOR MECHANISTIC STUDIES OF SECRETORY DIARRHOEA AND FLUID-VOLUME HOMEOSTASIS

William J Krause, Leonard R Forte, F Kent Hamra and Ronald H Freeman

INTRODUCTION

The North American opossum, *Didelphis virginiana*, has been used extensively for a number of years by gastroenterologists as an experimental animal model in studies of the upper and lower oesophageal sphincters as well as other sphincteric areas along the gastrointestinal tract (Krause & Cutts 1992). The developing opossum also has been shown to be a valuable experimental model in studies of neuroplasticity (Martin & Wang see chapter 21, this volume). Most recently *Didelphis* has proven to be instrumental in establishing an understanding of the regulation of diverse epithelia by the cell-surface receptor-enzyme, guanylate cyclase, and the newly discovered endogenous hormones that regulate the activity of this receptor. Bacterial peptides mimic the endogenous peptides subverting normal intestinal epithelial cell function resulting in secretory diarrhoea. The first clue that endogenous peptide hormones existed came from the discovery that the kidney of *Didelphis* has receptors in cells forming the proximal tubules for bacterial, diarrhoea-inducing peptide toxins. It was clear that the kidney receptor-guanylate cyclase was not exposed to bacterial toxins *in vivo*. Use of *Didelphis* has provided information which has led to the discovery of two new intestinal peptide hormones, guanylin and uroguanylin, the endogenous regulators of guanylate cyclase. Uroguanylin was first isolated and purified from opossum urine. The discovery of this peptide in opossum urine and intestinal mucosa has led to the development of a postulate suggesting that uroguanylin links the intestinal mucosa to the kidney in the regulation of fluid and electrolyte homeostasis. Moreover, the widespread nature of the uroguanylin/guanylin receptor was first reported in *Didelphis* which has broadened this field of inquiry to include the reproductive tract, liver and respiratory epithelia

Pathogenic strains of bacteria such as *Yersinia enterocolitica, Vibrio cholerae* and *Escherichia coli* synthesise and secrete enterotoxins that bind to apical cell membranes and cause similar forms of secretory diarrhoea by activating signalling mechanisms within intestinal epithelial cells lining the small intestine and colon (Giannella 1981; Greenburg & Guerrant 1981). It has been established that specific enterotoxins produced by specific strains of bacteria utilise different intracellular signalling pathways. Heat-labile enterotoxins, such as cholera toxin, activate an intestinal adenylate cyclase pathway, whereas heat-stable enterotoxins stimulate a guanylate cyclase mechanism (Cassel & Pfeuffer 1978; Field *et al.* 1978; Hughes *et al.* 1978). Both signalling pathways appear to converge and activate a cellular cyclic adenosine 5-monophospate (cAMP)-dependent kinase which amplifies cellular processes leading to enhanced intestinal secretion. *Escherichia coli (E. Coli)* heat-stable enterotoxins are 18- to 19- amino acid peptides that contain six cysteines and three intramolecular disulphide bonds (Gari'epy *et al.* 1987; Ozaki *et al.* 1991). These bacterial peptides affect water and solute transport across the epithelial lining of the small intestine leading to a net secretion of fluid and electrolytes into the lumen resulting in secretory diarrhoea (Rao *et al.* 1981; Huott *et al.* 1988; Field *et al.* 1989). Heat-stable enterotoxins are a major cause of secretory diarrhoea in children (Sack *et al.* 1975; Donta *et al.* 1977; Black *et al.* 1982a,b) as well as in domestic and laboratory animals (Burgess *et al.* 1978). Receptors for this class of enterotoxin have been localised in the plasmalemma forming the microvillus border of intestinal epithelial cells (Giannella *et al.* 1983; Frantz *et al.* 1984; Kuno *et al.* 1986), and these receptors are linked to the stimulation of the enzyme, guanylate cyclase (Guerrant *et al.* 1980; Rao *et al.* 1981; Eldieb *et al.* 1986). These cell-surface receptors were termed guanylate cyclase-C (GC-C) by Garbers (1991) and Schulz *et al.* (1989,1990) because previously identified guanylate cyclase-linked receptors for atrial natriuretic peptides had been named GC-A and GC-B. Cyclic guanosine 3', 5'-monophosphate (cGMP) and its analogues have similar effects on intestinal epithelial cell transport as observed with *E. coli* heat-stable enterotoxin (Hughes *et al.* 1978; Guandalini *et al.* 1982; Inoue *et al.* 1984) suggesting that increased cGMP levels in enterocytes caused by heat-stable enterotoxin play a major role in increased water and solute secretion that occurs during secretory diarrhoea. Although receptors for *E. coli* heat-stable enterotoxin and the cGMP signal transduction mechanism were considered specific features of polarized intestinal epithelial cells, transformed cell culture lines also were found to express receptors for heat-stable enterotoxins (Inoue *et al.* 1983,1984; Thomas & Knoop 1983) suggesting that these receptors might be expressed in other tissues as well. While studying the regulation of cGMP synthesis in cultured kidney cells, it was observed that *E. coli* heat-stable enterotoxin caused a major increase in cGMP accumulation in the opossum kidney (OK) cell line. Cells of the cultured opossum kidney cell line have renal properties such as the expression of apical microvilli, the polarised distribution of plasma membrane proteins, and the presence of renal transport systems that are characteristic of cells forming the renal proximal tubule. The use of this proximal tubular-like cell line derived from the kidney of the *Didelphis* (Koyama *et al.* 1978), as well as another kidney cell line PTK-2 derived from an Australian marsupial species, *Potorous tridactylus*, clearly demonstrate that both OK and PTK-2 cell lines have high affinity receptors for *E. coli* heat-stable enterotoxin that are coupled to the stimulation of cGMP production (Forte *et al.* 1988). In addition, it was observed that heat-stable enterotoxin markedly increased cGMP production in slices of opossum kidney cortex as well as suspensions of isolated intestinal mucosa from this species. *E. coli* heat-stable enterotoxin also caused a large increase in urine cGMP excretion (10 to 50-fold over control) when this peptide was administered intravenously to *Didelphis* (Forte *et al.* 1988,1989). The increased renal excretion of cGMP *in vivo* was associated with a two to three-fold increase in urinary sodium excretion (Freeman *et al.* 1994).

The occurrence of receptors for *E. coli* heat-stable enterotoxin linked to the production of cGMP in epithelial cells lining the intestinal tract, as well as those forming the proximal tubules of the opossum kidney strongly suggested the presence of endogenous enterotoxin-like molecules that would function as the naturally occurring agonists for these receptors (Forte *et al.* 1988). The use of the opossum kidney (OK) cell line and the T_{84} cell line (a human crypt intestinal epithelial cell) have become important *in vitro* models in evaluating the potency and efficacy of heat-stable enterotoxins and their closely related endogenous peptides and/or precursors by measuring chloride secretion and cGMP accumulation in component cells (Forte *et al.* 1992 a, b; 1993 a, b.)

CELL SURFACE GUANYLATE CYCLASE–C RECEPTORS

It is well established that bacterial heat-stable enterotoxins act as molecular mimics at the guanylate cyclase-C (GC-C) receptor targeted by at least two naturally occurring endogenous peptides, guanylin and uroguanylin. Previous studies utilizing *in vivo* and *in vitro* autoradiography of ^{125}I-enterotoxin binding in opossum kidney demonstrated the greatest concentration of these receptors to be confined to epithelial cells of the proximal tubule (Forte *et al.* 1989; White *et al.* 1989; Krause *et al.* 1990). As in the enterocytes lining the small intestine and colon of *Didelphis*, greater receptor density appeared to occur in the region of the microvillus border of proximal tubular epithelial cells.

To determine if the enterotoxin receptor protein was indeed more broadly expressed in *Didelphis*, a variety of tissues and organs were examined utilising the *in vitro* autoradiographic technique (Table 17.1). In this regard the opossum proved to be an invaluable model in demonstrating widespread and highly specific ^{125}I-enterotoxin binding because a variety of tissues from this species express large amounts of the receptor. Heat-stable enterotoxin/guanylin/uroguanylin binding sites were observed in epithelial cells lining the small intestine, colon, gall bladder, cystic duct, common bile duct

Figure 17.1 An autoradiograph of a non-fixed, unstained cryostat section taken from the proximal small intestine of *Didelphis* following incubation with $^{125}I_{1\text{-}19}$ heat-stable enterotoxin illustrates that specific bindings sites are confined primarily to the intestinal epithelium. Intestinal epithelial cells forming the intestinal glands (I) (crypts of Lieberkühn) and covering villi (V) exhibit intense labelling. Smooth muscle cells of the muscularis externa (ME) and muscularis mucosae (M) as well as connective tissue elements forming the submucosa (S) and lamina propria show little, if any, receptor binding. The lumen (L) of the small intestine is to the right. Bright field. (x40)
Source: Forte *et al.* (1989). *American Journal of Physiology*, 257, F874–F881.

Figure 17.2 A region of intestinal mucosa from the preparation shown in Figure 17.1 but viewed under dark field microscopy illustrates that the most intense labelling occurs in the region of the microvillus border (arrows) of intestinal epithelial cells. $^{125}I_{1\text{-}19}$ heat-stable enterotoxin. (x190.)

Figure 17.3 A portion of the trachea from *Didelphis* labelled with $^{125}I_{1\text{-}19}$ heat-stable enterotoxin illustrates strong radioligand binding in the tracheal epithelium, but not in connective tissue elements of the lamina propria (LP) or in underlying hyaline cartilage (HC). The lumen (L) is shown at the top of the photomicrograph. Bright field. (x260.)

Figure 17.4 An autoradiograph of a non-fixed, unstained, frozen section from the adult kidney (metanephros) of *Didelphis* illustrates the intense $^{125}I_{1\text{-}19}$ heat-stable enterotoxin binding in proximal tubules of the cortex (CO) with little demonstrable binding of the radioligand occurring in the medulla (ME). Bright field. (x40.)
Source: Forte *et al.* (1989). *American Journal of Physiology*, 257, F874–F881.

17.1

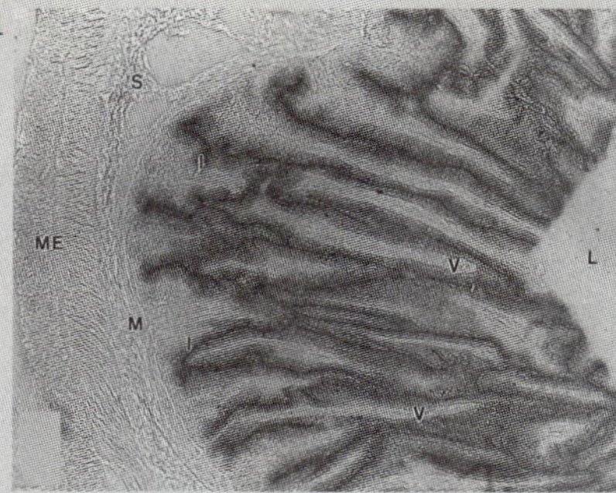

17.2

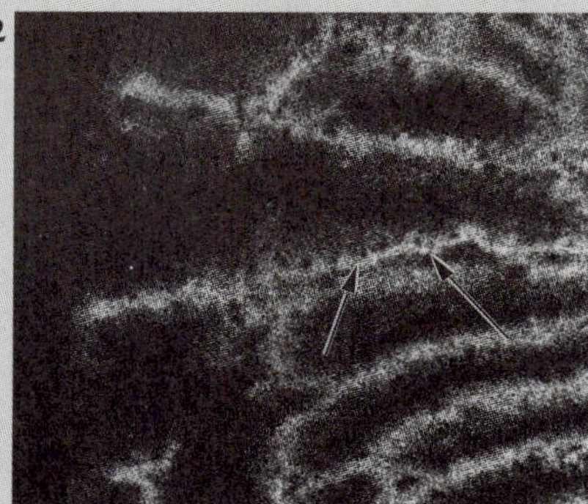

17.3

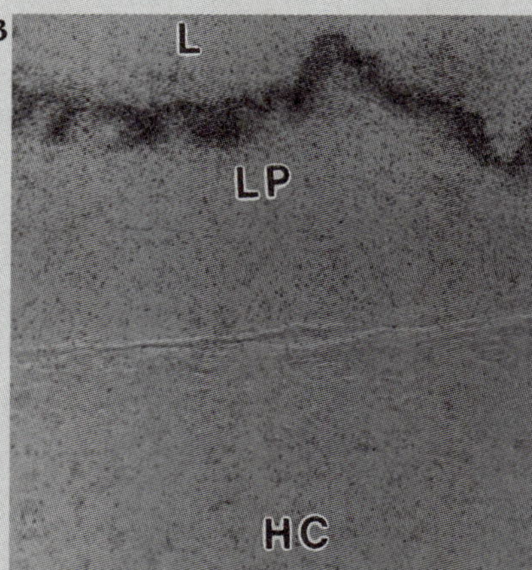

17.4

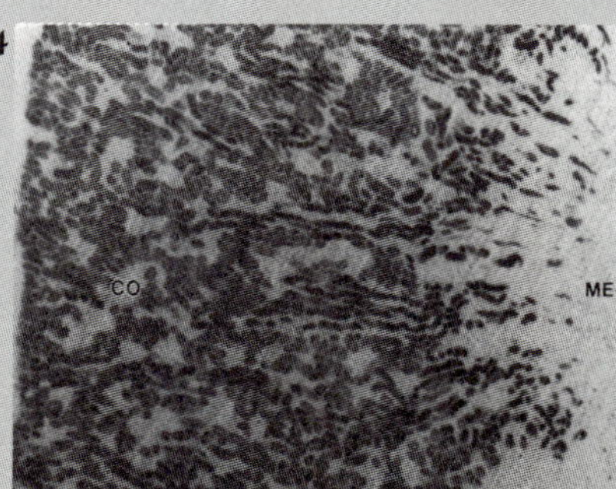

Table 17.1. Distribution of E. coli enterotoxin-specific binding sites in opossum tissues.

submandibular gland	—	metanephros	++++	nasal cavity	—
oesophagus	—	mesonephros	++++	trachea	+++
				bronchi	—
fundic stomach	—	ureter	—	lungs	—
small intestine	++++	urinary bladder	—	cerebral cortex	—
duodenal glands	+++	testis	+	cerebellar cortex	—
colon	++	epididymis	—	midbrain	—
choroid plexus	—	liver	+	prostate	—
pituitary	—	cystic duct	+++	ovary	—
thymus	—	gall bladder	+++	oviduct	—
spleen	—	common bile duct	+++	uterus	—
lymph node	—	pancreas	—	mammary gland	—
cardiac muscle	—	skin	—	eye	—
skeletal muscle	—	anal gland	—	Harderian gland	—
smooth muscle	—	adrenal gland	—	thyroid gland	—

Key: —= absent (not detected) + = light ++= moderate +++= strong ++++= intense

Source: Modified from Krause *et al.* (1990), *Cell Tissue Research.* 260, 387–394.

and trachea. and also in epithelial cells forming the duodenal (Brunner's) glands, seminiferous tubules of the testis, proximal tubules of the kidney and hepatocytes of the liver (Figures 17.1 to 17.4) (Krause *et al.* 1990). In addition to intestinal epithelial cells and cells forming the proximal tubules of the kidney (Forte *et al.* 1988,1989), cells of the liver and testis of *Didelphis* also expressed receptor-guanylate cyclase responsiveness to *E. coli* heat-stable enterotoxin as demonstrated by increased intracellular levels of cGMP (Krause *et al.* 1990). Examination of sections taken from a newborn opossum (~13 days post-conception) revealed that this receptor is expressed quite early in development and is abundant in intestinal epithelial cells lining both the small intestine and colon, as well as in epithelial cells forming the proximal tubules of the mesonephros (Figures 17.5 and 17.6). Comparative studies utilising *in vitro* autoradiography and cGMP stimulation bioassays, have demonstrated the enterotoxin receptors not only in intestinal epithelial cells lining the intestinal tract of all mammals and birds examined to date (Krause *et al.* 1994, 1995), but also in intestinal epithelial cells of some reptiles (Krause *et al.* 1993). Recently, we have demonstrated that the intestinal epithelium of some fish (yellow perch, *Perca flavescens*, and pallid sturgeon, *Scaphirhynchus albus*) have enterotoxin receptors and that extracts from the intestinal mucosa of the yellow perch contains peptides that activated the guanylate cyclase found in human T^{84} cells. In most vertebrates examined there appears to be a decrease in receptor density distally along the length (longitudinal axis) of the small intestine although labelling of receptors by ^{125}I-enterotoxin is observed in enterocytes throughout its length as well as in the colon. In all species examined

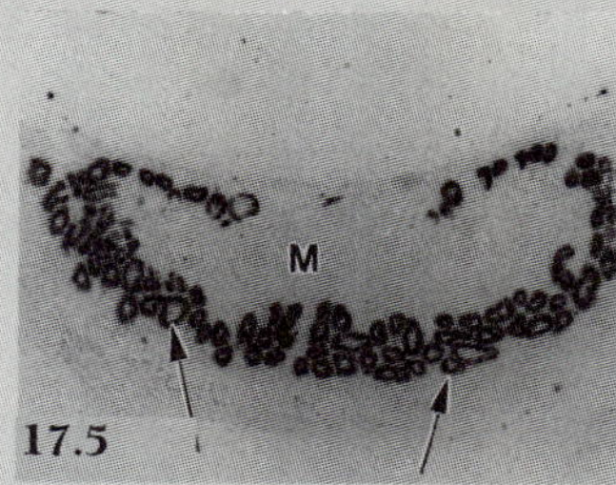

Figure 17.5 A non-fixed, frozen section taken through the longitudinal axis of the mesonephros (M) of a newborn opossum following incubation with $^{125}I_{1\text{-}19}$ heat-stable enterotoxin shows that specific binding sites occur primarily in the proximal tubules (arrows). Bright field. (x70.)

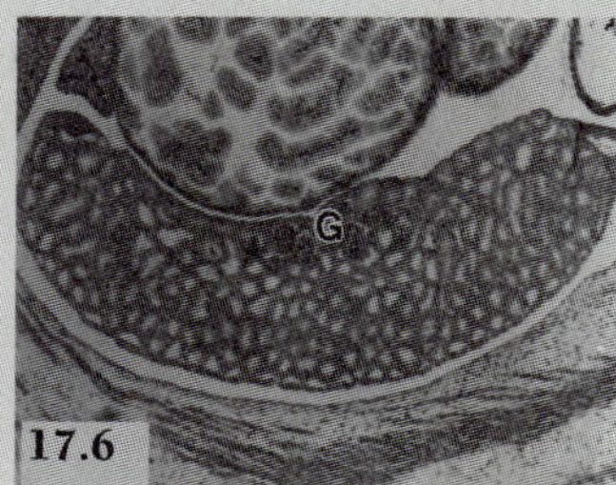

Figure 17.6 A section through the longitudinal axis of a Bouin-fixed, paraffin-embedded mesonephros demonstrates its internal architecture in the newborn opossum. Glomeruli (G) are restricted to the dorsomedial region (cf. Figure17.5, hematoxylin and eosin). (x50.)

to date, the most intense labelling of receptors by ^{125}I -enterotoxin occurs in the region of the microvillus border of individual enterocytes.

To test further whether the enterotoxin receptors were more abundant in the proximal small intestine, isolated samples of mucosa from different segments of the intestinal tract of *Didelphis* as well as other species of mammals, birds, and reptiles were treated with either 1 µM *E. coli* heat-stable enterotoxin or 10 µM of rat guanylin. Cellular cGMP accumulation responses of the mucosal samples were then measured by radioimmunoassay. Treatment of intestinal mucosa with 1 µM heat-stable enterotoxin elicited a large increase in cGMP levels in the proximal small intestine of *Didelphis* as well as the other vertebrate species examined with a much-reduced cGMP accumulation response occurring in the distal small intestine and colon (Krause *et al.* 1994,1995). In comparison to heat-stable enterotoxin, 10 µM of rat guanylin was less effective in the stimulation of cGMP production in identical regions of the small intestine and colon of all species examined (Figures17.7 and 17.8). The reason for the apparent decrease in receptor density distally along the longitudinal axis of the intestinal tract is unclear and may reflect either a true decrease in receptor density distally per enterocyte or alternatively, the distal decrease in measurable receptor density may reflect a decrease in number of enterocytes, as the goblet cell population increases distally throughout the length of the small intestine of *Didelphis* (Krause *et al.* 1977) and other species. In this case the receptor density per enterocyte may be equivalent in different regions of the small intestine, but the total number of enterocytes less.

Along the vertical axis of the intestinal tract (the villus/crypt unit) receptor density is usually greatest in enterocytes located near the basal half of villi and in those enterocytes forming the proximal half of the intestinal glands of humans and other species such as the Western grey kangaroo (*Macropus fuliginosus*) and laboratory rat (Cohen *et al.* 1992; Krause *et al.* 1994). In species

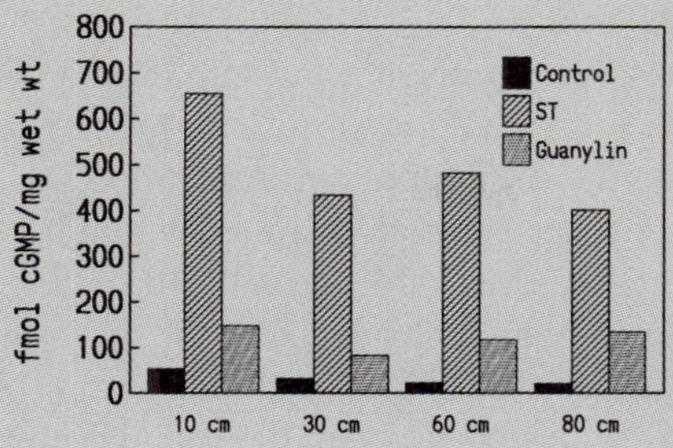

17.7

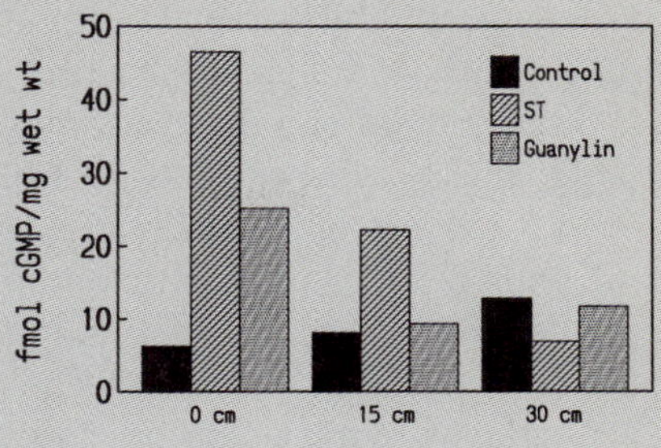

17.8

Figure 17.7 Guanylin and *E. coli* heat-stable enterotoxin stimulation of cGMP production in intestinal mucosa isolated from equidistant segments along the length of the small intestine of *Didelphis*. 1 μM of enterotoxin and 10 μM of guanylin were used in this representative experiment.

Figure 17.8 *E. coli* heat-stable enterotoxin (1 μM) and guanylin (10 μM) stimulation of cGMP production in the intestinal mucosa isolated from 3 equidistant segments along the length of the opossum colon. Because the magnitude of cGMP production in the colon is considerably less than that of the small intestine, an expanded scale has been used for its demonstration in this representative experiment (cf. Figure 17.7).

such as *Didelphis*, with very shallow intestinal glands (Krause *et al*. 1977), this regional variation is not as apparent, although there is a definite tendency for receptor density to increase in enterocytes near the base of villi and in those forming the intestinal glands. The distribution of enterotoxin receptors, as indicated by *in vitro* autoradiography, includes enterocytes covering the tips of villi. The density of receptors in enterocytes at this location is less than those associated with enterocytes found nearer the base of villi or in the proximal regions of the intestinal glands. Such an observation is consistent with the view that enterocytes forming the intestinal glands are involved primarily with chloride secretion (Welsh *et al*. 1982). However, recent evidence does suggest that enterocytes covering the villi also may participate in chloride secretion to some extent (Kockerling & Fromm 1993). The fewer number of receptors present in enterocytes at this location suggests that the cGMP mediated control of transport in this region may be of less magnitude as compared to enterocytes forming the intestinal glands.

Like enterocytes on the tips of villi, those lining the lumen of the colon and rectum appear to have fewer GC-C receptors. Although the receptors are readily demonstrable throughout the length of the colon by *in vitro* autoradiography using ^{125}I-enterotoxin as the probe, guanylate cyclase activation experiments using both heat-stable enterotoxin and guanylin peptides show much less accumulation levels of cGMP in *Didelphis* and other vertebrates when compared to stimulated levels of cGMP in the small intestine (Figures 17.7 and 17.8) (Forte *et al*. 1988; Krause *et al*. 1994,1995). Why receptors in the colon appear less responsive is unclear, particularly in light of the fact that local levels of guanylin in the colon of *Didelphis* are relatively high (Hamra *et al.* 1994) and guanylin mRNA levels of the rat as well as humans are most abundant in the colon (Weigand *et al.* 1992a, b).

It is now well established that an intestinal binding site for both heat-stable enterotoxins and the guanylin/uroguanylin peptides is the surface membrane guanylate cyclase-C (GC-C), which belongs to an important class of receptors for a variety of peptide hormones. These receptor proteins are known to have an N-terminal domain that projects from the exterior of the cell and binds to specific peptide ligands. A single transmembrane domain consisting of a span of hydrophobic amino acids links the extracellular N-terminal domain to the carboxyl-terminal portion of the protein which occupies an intracellular position. The latter is the highly conserved guanylate cyclase catalytic domain that catalyses cGMP production. The binding of peptides to the extracellular domain initiates changes that markedly increase cGMP production resulting in a profound elevation of intracellular cGMP in target cells. To date six different membrane guanylate cyclases and a soluble guanylate cyclase have been identified in vertebrates utilising molecular cloning. The GC-C receptor, is an intestinal cytoskeletal-associated form of membrane guanylate cyclase belonging to the guanylate cyclase family of receptor proteins that differ considerably in their selectivity and activation by ligands. In addition to GC-C, this family includes two other particulate forms of the enzyme (GC-A and GC-B) that selectively bind atrial natriuretic peptides and a soluble form of guanylate cyclase (GC-S), that binds nitric oxide (Drewett & Garbers 1994). The other membrane guanylate cyclase proteins are found in the retina and olfactory mucosa (Ruey-Bing *et al.* 1995). All these enzymes catalyse the production of cGMP and through this mechanism influence cellular function.

E. coli heat-stable enterotoxins as well as the two enterotoxin-like endogenous peptide molecules, guanylin and uroguanylin, bind to the extracellular ligand binding domain of the intestinal isoform of guanylate cyclase-C and activate the intracellular catalytic domain causing increased cellular levels of cGMP. The increased intracellular levels of this second messenger molecule activate a cAMP-dependent protein kinase (Forte *et al.* 1992b; Tien *et al.* 1992). The cAMP-dependent protein kinase in turn controls the state of phosphorylation of a functional cystic fibrosis transmembrane conductance regulator (CFTR) protein (Picciotto *et al.* 1992). The latter protein belongs to a family of transporter proteins and membrane pumps (Welsh 1987; Riordan *et al.* 1989) and functions as a chloride channel involved with the transport of the chloride ion through the apical cell membrane (Bear *et al.* 1992; Tilly *et al.* 1992). The CFTR protein has 12 transmembrane domains and a large intracellular portion that contains several phosphorylation sites (Riordon *et al.* 1989). Phosphorylation of the CFTR protein following activation of cAMP-dependent protein kinase is thought to open a transmembrane channel for chloride efflux from the cytoplasm of the intestinal epithelial cells and into the lumen of the intestine. Both cAMP and cGMP regulate the state of CFTR phosphorylation via a cAMP-dependent protein kinase. The net effect of either bacterial heat-stable enterotoxin, guanylin or uroguanylin in the intestinal tract of *Didelphis* and in other species, is to promote chloride secretion and to inhibit sodium and water absorption resulting in secretory diarrhoea (Field *et al.*

1978,1989; Guandalini *et al.* 1982; Rao *et al.* 1981; Huott *et al.* 1988; Currie *et al.* 1992; Hamra *et al.* 1993b).

It is thought that the secretion of salt and fluid is an important function of intestinal epithelial cells forming the intestinal glands (crypts of Lieberkühn) and to a lesser extent in enterocytes lining the intestinal lumen (Welsh *et al.* 1982). The secretory process is driven by the active transport of chloride across the basolateral cell membranes of intestinal and airway epithelium from the blood and into the intracellular compartment. Chloride ion then passes through chloride channel proteins localised in the apical plasmalemma of the epithelial cells (Welsh 1987; Anderson *et al.* 1992). Sodium/potassium ATPase pumps and potassium channel activities in the basolateral domain of the plasmalemma are required for this secretory process to occur.

Intestinal chloride ion secretion also appears to be regulated by multiple endocrine and neural pathways suggesting the fundamental importance of chloride secretion in the physiological function of the intestinal mucosa. The magnitude of intestinal chloride secretion can be observed when bacterial enterotoxins subvert the normal physiological regulation for this process in diseases such as cholera and traveller's diarrhoea. Both cyclic AMP and cyclic GMP are important regulators of chloride secretion in the intestine. Hormones, such as vasoactive intestinal peptide and prostaglandin E_2, normally activate adenylate cyclase by interacting with receptors located in the basolateral membranes of target cells (Dharmsathaphorn *et al.* 1985; Weymer *et al.* 1985). The adenyl cyclase mechanism of action is utilised by the diarrhoeal producing heat-labile toxins such as the cholera toxin in the disease cholera (Cassel & Pfeuffer 1978). Heat-stable enterotoxins, on the other hand, are small peptides secreted by pathogenic strains of *E. coli* and other bacteria that activate the intestinal guanylate cyclase mechanism to produce secretory diarrhoea. Although the precise mechanism for the phosphorylation of the CFTR protein chloride channel that is regulated by heat-stable enterotoxin/guanylin/uroguanylin needs to be clarified, it seems clear that either cAMP or cGMP-dependent mechanisms for controlling intestinal chloride secretion require functional CFTR proteins located in the apical plasmalemma. The CFTR protein appears to be the rate-limiting step for chloride secretion. Intestinal epithelial cells of patients with cystic fibrosis are unresponsive to either cholera toxin (the cAMP pathway) or *E. coli* enterotoxin (the cGMP pathway) whereas normal tissue shows substantial changes in transepithelial potential difference *in vitro* (Baxter *et al.* 1988; Taylor *et al.* 1988). Intestinal epithelial cells of human colon secrete chloride in response to heat-stable enterotoxin or cGMP in normal subjects, but this response is absent in patients with cystic fibrosis (Goldstein *et al.* 1994). Thus, at least three human conditions: cholera, traveller's-secretory diarrhoea, and cystic fibrosis, appear linked to functional CFTR proteins in regulating chloride ion secretion. Cholera and *E. coli* heat-stable enterotoxins appear to activate the chloride channel-CFTR protein via different intracellular routes to elicit their effects. Cystic fibrosis, on the other hand, results from a lack of CFTR protein, itself due to mutations of the cystic fibrosis gene (Gabriel *et al.* 1994).

THE ENDOGENOUS PEPTIDE LIGANDS: GUANYLIN AND UROGUANYLIN

Guanylin and uroguanylin are members of a recently discovered family of peptides that function as endogenous ligands for the intestinal receptor GC-C (Currie *et al.* 1992; Hamra *et al.* 1993b; Khare *et al.*1994). Both peptides stimulate increases in intracellular levels of cGMP by activating the GC-C receptor and are similar in structure and function to a number of heat-stable enterotoxins that are molecular mimics.

Guanylin, initially isolated as a peptide 15 amino acids in length from rat jejunum, was the first endogenous activator of the GC-C receptor isolated (Currie *et al.* 1992). Subsequently, uroguanylin, a closely related 15 amino-acid peptide that also activates the GC-C receptor was described in *Didelphis* (Hamra *et al.* 1993b). Although uroguanylin was initially isolated and purified from opossum urine, hence its name, it also has been isolated from the intestinal mucosa of the opossum. In addition, a second 15 amino acid peptide was isolated and purified from both opossum intestinal mucosa and urine which is closely related to rat and human guanylin and is thought to be the opossum homologue of guanylin (Hamra *et al.* 1993a). Since the initial isolation and purification of uroguanylin from the opossum, it now has been isolated from both rat and human urine and the amino acid sequence of these peptides determined (Hamra *et al.* 1993b; Kita *et al.* 1994). The primary structures of the known active guanylin and uroguanylin molecules, as well as the molecular structure of some of the enterotoxins, are shown in Figure 17.9.

Uroguanylins are highly acidic molecules that lack the aromatic amino acids, phenylanine (F) and tyrosine (Y), that occur at position 9 in guanylin peptides. The 4 cysteine (C) positions together with glutamic acid (E) following the cysteine at position 4 are highly conserved and characterise both guanylin and uroguanylin peptides. Opossum guanylin is the more basic when compared to guanylin peptides of other species due to the presence of histidine (H) at position 2.

Opossum uroguanylin stimulates cGMP accumulation in T_{84} cells (Figure 17.10). Both opossum uroguanylin and guanylin synthetic peptides have been shown to be about ten-fold more potent than rat or human guanylin when assessed by the T_{84} intestinal cell cGMP accumulation bioassay. Synthetic heat-stable enterotoxin 5–17 is ten-fold more potent than opossum uroguanylin and 100-fold more potent than guanylin when measured by the T_{84} intestinal cell bioassay suggesting that the T_{84} cell GC-C receptor is a uroguanylin-selective subtype receptor and infers that guanylin-selective receptor subtypes may exist in the cell membranes of other target cells.

Receptor-mediated cGMP accumulation responses to *E. coli* heat-stable enterotoxin in the OK cell line suggested that this cell line would be a useful experimental model for characterising the biological activity of the guanylin and uroguanylin peptides. Opossum guanylin was found to be about three-fold more potent than uroguanylin when these peptides were evaluated using the OK cell line cGMP accumulation bioassay system. *E.coli* heat-stable enterotoxin was more potent than either endogenous peptide. Interestingly, uroguanylin is about ten-fold more potent than opossum or rat guanylin when slices of opossum kidney cortex are incubated with these peptides *in vitro*. These cGMP bioassay data suggest that the renal cortex has a predominance of uroguanylin-selective receptors although guanylin-selective receptors also could be present in this tissue. One interpretation of these data is that cultured OK cells primarily express a guanylin-selective receptor whereas the actual opossum kidney cortex has two classes of receptors, one of which is uroguanylin-selective. Thus, there appears to be evidence for the presence of two receptor subtypes that have greater affinities for either uroguanylin or guanylin in OK cells and in cells of the opossum kidney cortex.

Isolated intestinal mucosa from the opossum duodenum when treated with heat-stable enterotoxin, guanylin or opossum uroguanylin, give cGMP accumulation responses with the rank-order potency of heat-stable enterotoxins>uroguanylin> guanylin. If the efficacy of synthetic rat guanylin is compared to *E. coli* heat-stable enterotoxin using the anaesthetised opossum *in vivo*, intravascular injection of rat guanylin caused a modest increase in urine cGMP excretion, whereas a similar amount of enterotoxin caused a substantial increase (20-fold) in urine cGMP content and a sustained (about a two-fold) increase in sodium excretion. No change in the glomerular filtration rate (clearance of creatinine) was observed. Precisely how these peptides influence nephron function is unknown at present, but they appear to target epithelial cells of the tubular component.

Prohormones (115 or 116 amino acid polypeptide precursor molecules) for rat, mouse (Schulz *et al.* 1992; Wiegand *et al.* 1992a), and human guanylins (DeSauvage *et al.* 1992; Wiegand *et al.* 1992b; Garcia *et al.* 1993; Kuhn *et al.* 1993) have been isolated and studied, but these are largely inactive forms until cleaved to the shorter 15 amino acid carboxyl-terminal peptides. Thus, the active form of rat guanylin (guanylin 101 to 115) represents the carboxyl-terminal 15 amino acids of an inactive proguanylin molecule. Synthetic preparations of guanylin 93–115 and 101–115 peptides, are equipotent in T_{84} cell cGMP and chloride ion secretion bioassays.

Prohormone forms of both uroguanylin and guanylin have been isolated and purified from tissue extracts of the small intestine and colon of the opossum (Hamra *et al.* 1994). The prohormone forms appear to be most abundant in the colon, with lesser amounts occurring in the proximal small intestine (Figure 17.11). In addition, guanylin and uroguanylin mRNAs have been shown to be expressed in the mucosa lining the small intestine and colon of *Didelphis*. It now appears that these inactive prohormones, as well as the active forms of uroguanylin and guanylin, are produced primarily by the mucosa lining the small intestine and colon. Circulating forms of prouroguanylin and proguanylin occur in opossum plasma; however, opossum urine contains the active form of uroguanylin and smaller amounts of guanylin. Renal guanylate cyclase receptors show a greater selectivity for uroguanylin over guanylin in opossum kidney and, when activated *in vivo,* can elicit a natriuresis that is accompanied by increased urine cGMP and sodium excretion (Freeman *et al.* 1994; Forte *et al.* 1994a). These observations suggest that uroguanylin is synthesised by cells within the intestinal mucosa and is secreted into the blood as an inactive prohormone form. Following filtration by renal glomeruli,

V. cholerae 01-Hakata enterotoxin	• I D C C E I C C N P A C F G C L N
Y. enterocolitica enterotoxin	– • • • • W D C C D Y C C N P A C A G C
E. coli enterotoxin$_p$	• • F Y C C E L C C N P A C A G C Y
E. coli enterotoxin$_h$	• • • N Y C C E L C C N P A C T G C Y
	1 5 10 15
Uroguanylin (opossum)	Q E D C E L C I N V A C T G C
Uroguanylin (human)	N D D C E L C V N V A C T G C L
	1 5 10 15
Guanylin (opossum)	S H T C E I C A F A A C A G C
Guanylin (human)	P G T C E I C A Y A A C T G C
Guanylin (rat, mouse)	P N T C E I C A Y A A C T G C

Figure 17.9 Comparison of the primary structures (active C-terminal portions, single letter abbreviations) of several heat-stable enterotoxins as well as guanylin and uroguanylin molecules.

prouroguanylin is thought to enter the lumina of the proximal tubules and is cleaved to an active form by proteolysis at this location. The active form of uroguanylin then binds to and activates receptor guanylate cyclases located in the apical plasmalemma of the proximal tubular cells. Following activation of this receptor, the second messenger molecule, cGMP, is produced, which through a cascade of intracellular events influences the function of the tubular component of the nephron. These observations are consistent with the hypothesis that uroguanylin may be an intestinal hormone that influences the kidney in fluid/volume homeostasis via an endocrine mechanism.

FUTURE DIRECTIONS

Guanylin and uroguanylin stimulate GC-C receptor proteins and through cGMP mediated processes act as paracrine and/or endocrine regulators of electrolyte transport in several types of epithelia. Both peptides, and perhaps other yet undiscovered closely related peptides of this family of proteins, may play an important role in the regulation of cell function not only by acting through the GC-C receptor, but also through other homologous receptor proteins. A precedent for such speculation is warranted as it is well established that the natriuretic peptide family of receptors with guanylate cyclase activity have been shown to consist of two subtypes with different tissue distributions and each exhibiting selective affinities for 3 different members of the natriuretic peptide family (Garbers 1991; Schulz *et al.* 1989; Waldman & Murad 1987). A similar situation may exist for peptides of the guanylin/uroguanylin family to act on membrane guanylate cyclases and through cGMP modulate fluid and electrolyte homeostasis. Because of the similarities of the natriuretic peptide, guanylate cyclase system to the guanylin/uroguanylin system, potential additional members of the latter peptide family may exist. Future investigation will

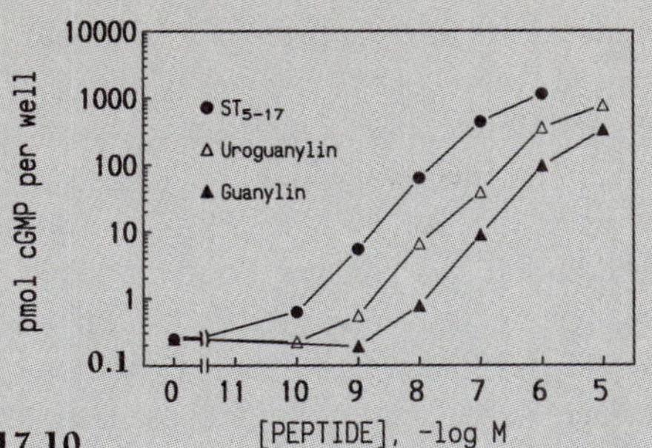

17.10

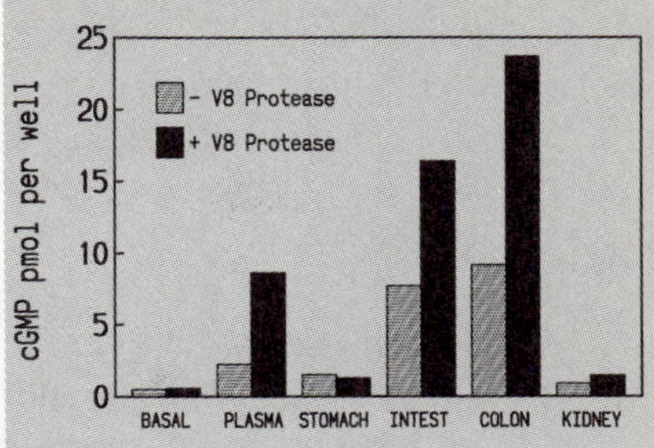

17.11

Figure 17.10 A representative experiment in testing the potency and efficacy of opossum uroguanylin, guanylin and *E. coli* heat-stable enterotoxin (ST) in the activation of cGMP accumulation in T84 intestinal epithelial cells.
Source: Hamra *et al.* (1993). *Proceedings of the National Academy of Sciences USA*, 93, 10464–10468.

Figure 17.11 Bioactivity of uroguanylin/guanylin molecules as demonstrated by the T84 intestinal epithelial cell bioassay in extracts from opossum tissues following activation of the contained prohormones by V8 protease. In addition to the intestinal mucosa, blood plasma exhibits increased bioactivity following protease cleavage.

undoubtedly demonstrate an intricate paracrine and/or endocrine network involved in the regulation of epithelial cell function through the action of these guanylin/uroguanylin-like peptides and their receptors.

Identification of a second receptor regulated by the guanylin/uroguanylin family of peptides will be an important area of future investigation. Uroguanylin has several biological properties in common with guanylin and heat-stable enterotoxin. Uroguanylin stimulates cGMP production and competes for binding sites on T_{84} cells with heat-stable enterotoxin. The more potent effect of uroguanylin on binding displacement and activation of GC-C receptor in the T_{84} cell line may be explained, in part, by the presence of multiple receptors with different binding affinities. A 56kDa protein purified from rat intestinal epithelial cell membranes that binds *E. coli* heat-stable enterotoxin which does not generate guanylate cyclase activity has been reported (Hakki *et al.* 1993). Confirmation of a receptor binding protein that lacks the guanylate cyclase (catalytic) domain would provide further evidence for a second class of uroguanylin/guanylin receptors. Whether similar receptors exist that act as possible clearance receptors for the guanylin/uroguanylin peptides or represent receptors with yet undefined functional activities are important questions that need to be addressed.

The luminal environment to which the intestinal GC-C receptor on the apical surface of enterocytes is exposed may allow for selective activation by either guanylin or uroguanylin dependent on the pH of the environment. In cGMP accumulation bioassays utilising the T_{84} cell line, the potency of opossum guanylin is lowest in an acidic medium (pH 5.0) and highest in a more basic (pH 8.0) medium. In contrast, uroguanylin shows greater potency in stimulating cGMP when the extracellular medium is acid and least, when basic. Recently, segmental differences in the effects of guanylin and *E.coli* heat-stable enterotoxin on chloride secretion in the human gut have been reported (Kuhn *et al.* 1994). Such observations suggest that the abundance of specific subtypes of receptors may change not only as a function of location along the longitudinal axis of the intestinal tract, but also as a function of the pH of the local environment at any given location. In addition, pH changes along the vertical axis (the villus/crypt unit) must be considered in influencing receptor affinity for the guanylin/uroguanylin peptides due to possible conformational changes of the peptides and the receptors themselves. As both guanylin and uroguanylin peptides are found in the intestinal mucosa of *Didelphis* both may act as autocrine/paracrine hormones in the intestinal epithelium, each selectively activating its own receptor. In addition, both hormones and/or their precursors appear to be released into the circulation and may selectively activate receptors of distant target tissues, such as the kidney, via an endocrine mechanism. Studies addressing the mechanisms involved in precisely how these peptides act locally, as well as at a distance, are important questions for future research.

The precise cellular site where either guanylin or uroguanylin is synthesised and released is unknown. Initially it was reported that guanylin mRNA was found in Paneth cells of humans (DeSauvage *et al.* 1992), a cell type that tends to bind non-specifically to nucleic acid probes (Garrett *et al.* 1992). Moreover, classical Paneth cells do not occur in all species (Wheeler & Wheeler 1964; Fawcett 1994) and in those species where present are restricted primarily to the small intestine and are not present in the colon. In contrast to the Paneth cell proposal, guanylin peptides are most abundant in the mucosa lining the ileum and colon of *Didelphis* and most other species examined (Kato *et al.* 1993; Wiegand *et al.* 1992a,b; Hamra *et al.* 1994). Such observations would argue against Paneth cells as being the source of the guanylin peptide. Recently, guanylin-immunoreactivity has been demonstrated in enterochromaffin cells within the intestinal lining epithelium of the guinea pig (Cetin *et al.* 1994). Further work needs to be done as better probes become available utilising the combined techniques of immunohistochemistry and *in situ* hybridisation to clearly establish which cell types in the intestinal mucosa as well as in other epithelia, are involved in the synthesis, storage and release of the guanylin peptide and/or its precursor molecule.

As is the case for guanylin, the cellular site of uroguanylin synthesis and release is unknown. Thus far, evidence does exist that suggests that uroguanylin is produced primarily by epithelial cells within the intestinal lining epithelium of the small intestine and colon as it has been isolated from the mucosa of these organs. Examination of the distribution of guanylin and uroguanylin mRNAs in opossum tissues have shown that mRNAs are most abundant in the mucosa of the colon when compared to the proximal small intestine (Fan *et al.* 1994). Uroguanylin mRNA could not be demonstrated in other tissues examined which included the OK cell line, kidney, heart, and adrenal gland. Inactive prohormone forms of both guanylin and uroguanylin have been purified from extracts of the small intestine and colon of *Didelphis* and both prohormone forms can be activated following pretreatment with V_8

protease to liberate the active carboxyl-terminal peptide. Prohormone fractions, as well as active peptides also, are found in opossum plasma, suggesting that the prohormone forms may be synthesised by cells in the intestinal mucosa and secreted into the circulation. At least some of the prohormone forms may be activated in the target tissue.

Although the precise role uroguanylin plays in the regulation of kidney function is unknown at present, its analogue, *E. coli* heat-stable enterotoxin, has been shown to produce an increase in sodium excretion in isolated, perfused rat kidney (Lima *et al.* 1992). Likewise, heat-stable enterotoxin 5–17 causes a twenty-fold increase in urine cGMP and sustained two-fold increase in sodium excretion when administered to *Didelphis* intravascularly (Forte *et al.* 1989, 1994a) and opossum uroguanylin has been shown to have similar effects on the isolated perfused rat kidney (M C Fonteles personal communication). Epithelial cells of the kidney from the North American alligator (*Alligator mississippiensis*) also have been shown to express receptors for guanylin/uroguanylin and the intestine contains peptides that activate guanylate cyclase (Forte *et al.* 1994b) which suggests an early, fundamental link between the gut and kidney. It is now becoming increasingly apparent that three inter-related systems are involved in fluid-volume homeostasis. The primary system is the renin-angiotensin-aldosterone system which links the kidney to the zona glomerulosa of the adrenal cortex. Two ancillary systems also appear to exist which supplement this classic endocrine loop. The first of these includes members of the natriuretic peptide family which include atrial natriuretic peptide (ANP) and brain natriuretic peptide (BNP) that link the cardiovascular system to the kidney (Garbers 1991; Koller *et al.* 1991; Schulz *et al.* 1989). The second, more recently discovered, system is the guanylin/uroguanylin family of peptides which appears to link the intestinal tract and kidney. Both latter systems consist of peptides that act through various types of guanylate cyclase receptors to effect their influence on fluid-volume homeostasis.

To date *Didelphis virginiana* has played a key role not only in predicting the existence of a newly discovered family of guanylate cyclase-stimulating peptides involved in controlling fluid/electrolyte homeostasis in the intestinal tract, but also in the initial isolation and purification of the newest member of this family, uroguanylin (Hamra *et al.* 1993b). Furthermore, the opossum model continues to play an important role in establishing an endocrine link between the intestinal tract and kidney in the regulation of electrolyte-fluid homeostasis. In this regard uroguanylin appears to be a strong candidate as an intestinal natriuretic factor that exerts a distant action on the renal tubules

In vitro autoradiography utilising *Didelphis* demonstrated receptor guanylate cyclases for these endogenous peptides in a variety of epithelial tissues which was not possible using other species. Since the initial description of their distribution in *Didelphis* (Krause *et al.* 1990) these receptors have been demonstrated in a variety of epithelial tissues of other species utilising different techniques (Schultz *et al.* 1992). Why *Didelphis* exhibits such an abundance of demonstrable GC-C receptors related to these endogenous peptides is unknown. It may be that, in the case of this particular species, all receptor subtypes and perhaps as yet undetermined receptors have the ability to recognise and bind the probe used ($^{125}I_{1-19}$ heat-stable enterotoxin) which does not occur in other species. In the case of kidney receptor density, it may be related to the fact that *Didelphis* has fewer nephrons for its size than most other mammals (Maluf 1987; Rytand 1938). To date the most effective means of identifying the morphological location of the GC-C receptor is with ^{125}I-heat-stable enterotoxin. Iodinated enterotoxin molecules remain the probes of choice because iodination of tyrosine 9 in guanylin molecules appears to interfere with the biological activity of these radioligands resulting in poor binding of ^{125}I-guanylin to receptors. For this reason, radiolabelled forms of guanylin/uroguanylin are not used and labelled enterotoxins currently are the best radioligands for demonstrating guanylin/uroguanylin receptor binding sites. Antibodies to the GC-C receptor have now been developed and future studies will re-evaluate the current ^{125}I-enterotoxin binding data to determine if the observed binding is indeed the result of binding with only the GC-C receptor. As better techniques are developed it is hoped that a more precise distribution of guanylin- and uroguanylin-specific receptors, as well as those of perhaps other closely related unknown receptors, will be determined.

Didelphis and the cell line derived from its kidney (the OK cell line) have proved to be invaluable models for gaining insights into the evolving story of the guanylin family of peptides, their receptors, and mechanisms of action. Use of the opossum model predicted the presence of the endogenous peptide guanylin and shortly thereafter led to the discovery of the second member of this family of peptides, uroguanylin, which was first isolated and purified from the opossum. Future use of the opossum model will be important in elucidating the endocrine link between the intestinal mucosa and the kidney as well as the mechanism of action of uroguanylin in the regulation of nephron function in fluid and electrolyte homeostasis. It is anticipated that other regulatory mechanisms will be discovered for these hormones by future studies using the opossum. Included are their roles in reproduction , liver function and their influence on salt and water transport in airway epithelium. Clearly, *Didelphis* has provided many new insights into the guanylin/uroguanylin field of research.

References

Anderson M P, Sheppard D N, Berger H A and Welsh M J (1992). Chloride channels in the apical membrane of normal and cystic fibrosis airway and intestinal epithelia. *American Journal of Physiology*, 263, L1–L14.

Baxter P S, Goldhill J, Hardcastle J, Hardcastle P T and Taylor C J (1988). Accounting for cystic fibrosis. *Nature*, 335, 211.

Bear C E, Li C, Kartner N, Bridges R J, Jensen T J, Ramjessingh M and Riordan J R (1992). Purification and functional reconstitution of the cystic fibrosis transmembrane conductance regulator (CFTR). *Cell*, 68, 809–818.

Black R E, Brown K H, Becker S and Yunus M (1982a). Longitudinal studies of infectious diseases and physical growth of children in rural Bangladesh: I. Patterns of morbidity. *American Journal of Epidemiology*, 115, 305–314.

Black R E, Brown K H, Becker S, Abdul Alim A R M and Huq I (1982b). Longitudinal studies of infectious diseases and physical growth of children in rural Bangladesh: II. Incidence of diarrhea and association with known pathogens. *American Journal of Epidemiology*, 115, 315–324.

Burgess M N, Bywater R J, Cowley C M, Mullan N A and Newsome P M (1978). Biological evaluation of a methanol-soluble, heat-stable *Escherichia coli* enterotoxin in infant mice, pigs, rabbits, and calves. *Infection and Immunity*, 21, 526–531.

Cassel D and Pfeuffer T (1978). Mechanism of cholera toxin action: covalent modification of the guanyl nucleotide-binding protein of the adenylate cyclase system. *Proceedings of the National Academy of Sciences USA*, 75, 2669–2673.

Cetin Y, Kuhn M, Kulasksiz H, Adermann K, Bargsten G, Grube D and Forssmann W-G (1994). Enterochromaffin cells of the digestive system: cellular source of guanylin, a guanylate cyclase-activating peptide. *Proceedings of the National Academy of Sciences USA*, 91, 2935–2939.

Cohen M B, Mann E A, Lau C, Henning S J and Giannella R A (1992). A gradient in expression of the *Escherichia coli* heat-stable enterotoxin receptor exists along the villus-to-crypt axis of rat small intestine. *Biochemical and Biophysical Research Communications*, 186, 483–490.

Currie M G, Fok K F, Kato J, Moore R J, Hamra F K, Duffin K L and Smith C E (1992). Guanylin: an endogenous activator of intestinal guanylate cyclase. *Proceedings of the National Academy of Sciences USA*, 89, 947–951.

De Sauvage F J, Keshav S, Kuang W-J, Gillett N, Henzel W and Goeddel D V (1992). Precursor structure, expression, and tissue distribution of human guanylin. *Proceedings of the National Academy of Sciences USA*, 89, 9089–9093.

Dharmsathaphorn K, Mandel K G, Masui H and McRoberts J A (1985). Vasoactive intestinal polypeptide-induced chloride secretion by a colonic epithelial cell line. *Journal of Clinical Investigation*, 75, 462–471.

Donta S T, Wallace R B, Whipp S C and Olarte J (1977). Enterotoxigenic *Escherichia coli* and diarrheal disease in Mexican children. *Journal of Infectious Diseases*, 135, 482–485.

Drewett J G and Garbers D L (1994). The family of guanylyl cyclase receptors and their ligands. *Endocrine Reviews*, 15, 135–162.

Eldeib M M R, Parker C D, Veum T L, Zinn G M and White A A (1986). Characterization of intestinal brush border guanylate cyclase activation by *Escherichia coli* heat-stable enterotoxin. *Archives of Biochemistry and Biophysics*, 245, 51–65.

Fan X H, Hamra F K, Pace V M, Krause W J, Freeman R H and Forte L R (1994). Distribution of uroguanylin and guanylin mRNAs in opossum tissues. *Federation of American Societies for Experimental Biology Journal*, 8, A552.

Fawcett D W (1994). Bloom and Fawcett. *A Textbook of Histology.* 12th edn, Chapman and Hall, New York, London.

Field M, Graf L H, Laird W J and Smith P L (1978). Heat-stable enterotoxin of *Escherichia coli: in vitro* effects on guanylate cyclase activity, cyclic GMP concentration, and ion transport in small intestine. *Proceedings of the National Academy of Sciences USA*, 75, 2800–2804.

Field M, Rao M C and Chang E B (1989). Intestinal electrolyte transport and diarrheal disease. *New England Journal of Medicine*, 321, 800–806.

Forte L R, Eber S L, Turner J T, Freeman R H, Fok K F and Currie M G (1993a). Guanylin stimulation of Cl^- secretion in human intestinal T^{84} cells via cyclic guanosine monophosphate. *Journal of Clinical Investigation*, 91, 2423–2428.

Forte L R, Hamra F K, Fan X H, Krause W J, Currie M G and Freeman R H (1994a) Uroguanylin: an intestinal peptide hormone with natriuretic activity. *Gastroenterology*, 106, A808.

Forte L R, Krause W J and Freeman R H (1988). Receptors and cGMP signaling mechanism for *E. coli* enterotoxin in opossum kidney. *American Journal of Physiology*, 255, F1040–F1046.

Forte L R, Krause W J and Freeman R H (1989). *Escherichia coli* enterotoxin receptors: localization in opossum kidney, intestine, and testis. *American Journal of Physiology*, 257, F874–F881.

Forte L R, Krause W J, Freeman R H, Francis S H and Corbin J D (1992a). Regulation of intestinal chloride channels by *E. coli* heat-stable enterotoxin: molecular mechanisms. In *Recent Advances in Toxinology Research*. Eds P Gopalakrishnakone and C K Ten. National University of Singapore, Singapore. Vol. 3. pp 485–500.

Forte L R, Krause W J and Freeman R H (1993b). Guanylin bioactivity in human intestinal and opossum kidney cells. In *Advances in Second Messenger and Phosphoprotein Research*. Eds B L Brown and P R M Dobson. Raven Press, New York. Vol. 28. pp 133–138.

Forte L R, Krause W J, Hamra F K, Currie M G and Freeman R H (1994b). Alligator kidney has receptors for guanylin/uroguanylin and intestine contains peptides that activate guanylate cyclase. *Federation of American Societies for Experimental Biology Journal*, 8, A553.

Forte L R, Thorne P K, Eber S L, Krause W J, Freeman R H, Francis S H and Corbin J D (1992b). Stimulation of intestinal $C1^-$ transport by heat-stable enterotoxin: activation of a cAMP-dependent protein kinase by cGMP. *American Journal of Physiology*, 263, C607–C615.

Frantz J C, Jaso-Friedman L and Robertson D C (1984). Binding of *Escherichia coli* heat-stable enterotoxin to rat intestinal cells and brush border membranes. *Infection and Immunity*, 43, 622–630.

Freeman R H, Forte L R, Hamra F K, Currie M G and Krause W J (1994). Uroguanylin: a potential natriuretic peptide hormone produced by the intestine. *Federation of American Societies for Experimental Biology Journal*, 8, A552.

Gabriel S E, Brigman K N, Koller B H, Boucher R C and Stutts M J (1994). Cystic fibrosis heterozygote resistance to cholera toxin in the cystic fibrosis mouse model. *Science*, 266, 107–109.

Garbers D L (1991). The guanylyl cyclase-receptor family. *Canadian Journal of Physiology and Pharmacology*, 69, 1618–1621.

Garcia K C, De Sauvage F J, Struble M, Henzel W, Reilly D and Goeddel D V (1993). Processing and characterization of human proguanylin expressed in *Escherichia coli*. *Journal of Biological Chemistry*, 268, 22397–22401.

Gari'epy J, Judd A K and Schoolnik G K (1987). Importance of disulfide bridges in the structure and activity of *Escherichia coli* enterotoxin ST1b. *Proceedings of the National Academy of Sciences USA*, 84, 8907–8911.

Garrett K L, Grounds M D and Beilharz M W (1992). Nonspecific binding of nucleic acid probes to Paneth cells in the gastrointestinal tract with *in situ* hybridization. *Journal of Histochemistry and Cytochemistry*, 40, 1613–1618.

Giannella R A (1981). Pathogenesis of acute bacterial diarrheal disorders. *Annual Review of Medicine*, 32, 341–357.

Giannella R A, Luttrell M and Thompson M (1983). Binding of *Escherichia coli* heat-stable enterotoxin to receptors on rat intestinal cells. *American Journal of Physiology*, 245, G492–G498.

Greenburg R N and Guerrant R L (1981). *E. coli* heat-stable enterotoxin. *Pharmacology and Therapeutics*, 13, 507–531.

Goldstein J L, Sahi J, Bhuva M, Layden T J and Rao M C (1994). *Escherichia coli* heat-stable enterotoxin-mediated colonic $C1^-$ secretion is absent in cystic fibrosis. *Gastroenterology*, 107, 950–956.

Guerrant R L, Hughes J M, Chang B, Robertson D C and Murad F (1980). Activation of intestinal guanylate cyclase by heat-stable enterotoxin of *Escherichia coli*: studies of tissue specificity, potential receptors, and intermediates. *Journal of Infectious Diseases*, 142, 220–228.

Guandalini S, Rao M C, Smith P L and Field M

(1982). cGMP modulation of ileal ion transport: *in vitro* effects of *Escherichia coli* heat-stable enterotoxin. *American Journal of Physiology,* 243, G36–G41.

Hakki S, Robertson D C and Waldman S A (1993). A 56kDa binding protein for *Escherichia coli* heat-stable enterotoxin isolated from the cytoskeleton of rat intestinal membranes does not possess guanylate cyclase activity. *Biochimica et Biophysica Acta*, 1152, 1–8.

Hamra F K, Eber S L, Krause W J, Freeman R H, Smith C E, Duffin K L, Siegel N R, Currie M G and Forte L R (1993a). Opossum guanylin: purification from urine, structure and activity. *Federation of American Societies for Experimental Biology Journal,* 7, A215.

Hamra F K, Forte L R, Eber S L, Pidhorodeckyj N V, Krause W J, Freeman R H, Chin D T, Tompkins J A, Fok K F, Smith C E, Duffin K L, Siegel N R and Currie M G (1993b). Uroguanylin: structure and activity of a second endogenous peptide that stimulates intestinal guanylate cyclase. *Proceedings of the National Academy of Sciences USA,* 90, 10464–10468.

Hamra F K, Krause W J, Freeman R H, Currie M G and Forte L R (1994). Purification and properties of prohormone forms of uroguanylin and guanylin. *Federation of American Societies for Experimental Biology Journal,* 8, A552.

Hughes J M, Murad F, Chang B and Guerrant R L (1978). Role of cyclic GMP in the action of heat-stable enterotoxin of *Escherichia coli*. *Nature,* 271, 755–756.

Huott P A, Liu W, McRoberts J A, Giannella R A and Dharmsathaphorn K (1988). Mechanism of action of *Escherichia coli* heat-stable enterotoxin in a human colonic cell line. *Journal of Clinical Investigation,* 82, 514–523.

Inoue T, Kobayashi T, Fujisawa Y, Kawamoto Y and Miyama A (1984). Effect of heat-stable enterotoxins of *Escherichia coli* and *Yersinia enterocolitica* on cyclic guanosine 3',5'-monophosphate levels of cultured cells. *Federation of European Microbiological Societies. Microbiology Letters,* 22, 219–221.

Inoue T, Okamoto K, Moriyama T, Takahashi T, Shimizu K and Miyama A (1983). Effect of *Yersinia enterocolitica* ST on cyclic guanosine 3',5'-monophosphate levels in mouse intestines and cultured cells. *Microbiology and Immunology,* 27, 159–166.

Kato J, Wiegand R C and Currie M G (1993). Characterization of the structure of pre-proguanylin. In *Advances in Second Messenger and Phosphoprotein Research* Eds B L Brown and P R M Dobson. Raven press, New York. Vol. 28. pp 139–142.

Khare S, Wilson D, Wali R K, Tien X–Y, Bissonnette M, Niedziela S M, Bolt M J B, Sitrin M D and Brasitus T A (1994). Guanylin activates rat colonic particulate guanylate cyclase. *Biochemical and Biophysical Research Communications,* 203, 1432–1437.

Kita T, Smith C E, Fok K F, Duffin K L, Moore W M, Karabatsos P J, Kachur J F, Hamra F K, Pidhorodeckyj N V, Forte L R and Currie M G (1994). Characterization of human uroguanylin: a member of the guanylin peptide family. *American Journal of Physiology,* 266, F342–F348.

Köckerling A and Fromm M (1993). Origin of cAMP-dependent Cl^- secretion from both crypts and surface epithelia of rat intestine. *American Journal of Physiology*, 264, C1294–C1301.

Koller K J, Lowe D G, Bennett G L, Minamino N, Kangawa K, Matsuo H and Goeddel D V (1991). Selective activation of the B natriuretic peptide receptor by c-type natriuretic peptide (CNP). *Science*, 252, 120–123.

Koyama H, Goodpasture C, Miller M M, Teplitz R L and Riggs A D (1978). Establishment and characterization of a cell line from the American opossum *(Didelphis virginiana). In vitro,* 14, 239–246.

Krause W J, Cullingford G L, Freeman R H, Eber S L, Richardson K C, Fox K F, Currie M G and Forte L R (1994). Distribution of heat-stable enterotoxin/guanylin receptors in the intestinal tract of man and other mammals. *Journal of Anatomy,* 184, 407–417.

Krause W J and Cutts J H (1992). Development of the digestive system in the North American opossum (*Didelphis virginiana*). *Advances in Anatomy Embryology and Cell Biology*, 125, 1–151.

Krause W J, Cutts J H and Leeson C R (1977). The postnatal development of the alimentary canal in the opossum. III. Small intestine and colon. *Journal of Anatomy,* 123, 21–45.

Krause W J, Freeman R H and Forte L R (1990). Autoradiographic demonstration of specific binding sites for *E. coli* enterotoxin in various epithelia of the North American opossum. *Cell and Tissue Research,* 260, 387–394.

Krause W J, Freeman R H, Eber S L, Hamra F K, Fok K F, Currie M G and Forte L R (1995). Distribution of *Escherichia coli* heat-stable enterotoxin/guanylin/uroguanylin receptors in the avain intestinal tract. *Acta Anatomica*, 153, 210–219.

Krause W J, Forte L R, Eber S L, Freeman R H and Cullingford G L (1993). Localization and distribution of guanylin/heat-stable enterotoxin receptors in the intestinal mucosa of man and other vertebrates. *Gastroenterology*, 104, A835.

Kuhn M, Adermann K, Jähne J, Forssmann W-G and Rechkemmer G (1994). Segmental differences in the effects of guanylin and *Escherichia coli* heat-stable enterotoxin on Cl^- secretion in human gut. *Journal of Physiology*, 479, 433–440.

Kuhn M, Raida M, Adermann K, Schulz-Knappe P, Gerzer R, Heim J-M and Frossmann W-G (1993). The circulating bioactive form of human guanylin is a high molecular weight peptide (10.3kDa). *Federation of European Biochemical Societies Letters*, 318, 205–209.

Kuno T, Kamisaki Y, Waldman S A, Gariépy J, Schoolnik G and Murad F (1986). Characterization of the receptor for heat-stable enterotoxin from *Escherichia coli* in rat intestine. *Journal of Biological Chemistry*, 261, 1470–1476.

Lima A A M, Monteiro H S A and Fonteles M C (1992). The effects of *Escherichia coli* heat-stable enterotoxin in renal sodium tubular transport. *Pharmacology and Toxicology*, 70, 163–167.

Maluf N S R (1987). Kidney of the great Indian rhino *Rhinoceros unicornis*, Linnaeus. *American Journal of Anatomy*, 180, 403–421.

Martin G F and Wang X M (1997). Development and developmental plasticity of descending spinal pathways in the North American opossum, *Didelphis virginiana*. In *Marsupial Biology: Recent Research and New Perspectives*. Eds. N R Saunders and L A Hinds. UNSWP: Sydney. 356–379.

Ozaki H, Sato T, Kubota H, Hata Y, Katsube Y and Shimonishi Y (1991). Molecular structure of the toxic domain of heat-stable enterotoxin produced by a pathogenic strain of *Escherichia coli*. *Journal of Biological Chemistry*, 266, 5934–5941.

Picciotto M R, Cohn J A, Bertuzzi G, Greengard P and Nairn A C (1992). Phosphorylation of the cystic fibrosis transmembrane conductance regulator. *Journal of Biological Chemistry*, 267, 12742–12752.

Rao M C, Orellana S A, Field M, Robertson D C and Giannella R A (1981). Comparison of the biological actions of three purified heat-stable enterotoxins: effects on ion transport and guanylate cyclase activity in rabbit ileum *in vitro*. *Infection and Immunity*, 33, 165–170.

Riordan J R, Rommens J M, Kerem B-S, Alon N, Rozmahel R, Grzelczak Z, Zielenski J, Lok S, Plavsic N, Chou J-L, Drumm M L, Iannuzzi M C, Collins F S and Tsui L-C (1989). Identification of the cystic fibrosis gene: cloning and characterization of complementary DNA. *Science*, 245, 1066–1073.

Ruey-Bing Y, Foster D C, Garbers D L and Fülle H-J (1995). Two membrane forms of guanylyl cyclase found in the eye. *Proceedings of the National Academy of Sciences USA*, 92, 602–606.

Rytand D A (1938). The number and size of mammalian glomeruli as related to kidney and to body weight, with methods for their enumeration and measurement. *American Journal of Anatomy*, 62, 507–520.

Sack D A, Merson M H, Wells J G, Sack R B and Morris G K (1975). Diarrhoea associated with heat-stable enterotoxin-producing strains of *Escherichia coli*. *Lancet*, 2, 239–241.

Schulz S, Chrisman T D and Garbers D L (1992). Cloning and expression of guanylin. Its existence in various mammalian tissues. *Journal of Biological Chemistry*, 267, 16019–16021.

Schulz S, Green C K, Yuen P S T and Garbers D L (1990). Guanylyl cyclase is a heat-stable enterotoxin receptor. *Cell*, 63, 941–948.

Schulz S, Singh S, Bellet R A, Singh G, Tubb D J, Chin H and Garbers D L (1989). The primary structure of a plasma membrane guanylate cyclase demonstrates diversity within this new receptor family. *Cell*, 58,1155–1162.

Taylor C J, Baxter P S, Hardcastle J and Hardcastle P T (1988). Failure to induce secretion in jejunal biopsies from children with cystic fibrosis. *Gut*, 29, 957–962.

Thomas D D and Knoop F C (1983). Effect of heat-stable enterotoxin of *Escherichia coli* on cultured mammalian cells. *Journal of Infectious Diseases*, 147, 450–459.

Tien X-Y, Brasitus T A, Kaetzel M A, Dedman J R

and Nelson D J (1994). Activation of the cystic fibrosis transmembrane conductance regulator by cGMP in the human colonic cancer cell line, Caco-2. *Journal of Biological Chemistry,* 269, 51–54.

Tilly B C, Winter M C, Ostedgaard L S, O'Riordan C, Smith A E and Welsh M J (1992). Cyclic AMP-dependent protein kinase activation of cystic fibrosis transmembrane conductance regulator chloride channels in planar lipid bilayers. *Journal of Biological Chemistry,* 267, 9470–9473.

Waldman S A and Murad F (1987). Cyclic GMP synthesis and function. *Pharmacological Reviews*, 39, 163–196.

Welsh M J (1987). Electrolyte transport by airway epithelia. *Physiological Reviews,* 67, 1143–1184.

Welsh M J, Smith P L, Fromm M and Frizzell R A (1982). Crypts are the site of intestinal fluid and electrolyte secretion. *Science,* 218, 1219–1221.

Weymer A, Huott P, Liu W, McRoberts J A and Dharmsathaphorn K (1985). Chloride secretory mechanism induced by prostaglandin E_1 in a colonic epithelial cell line. *Journal of Clinical Investigation,* 76, 1828–1836.

Wheeler E J and Wheeler J K (1964). Comparative study of Paneth cells in vertebrates. *Anatomical Record,* 148, 350.

White A A, Krause W J, Turner J T and Forte L R (1989). Opossum kidney contains a functional receptor for the *Escherichia coli* heat-stable enterotoxin. *Biochemical and Biophysical Research Communications,* 159, 363–367.

Wiegand R C, Kato J and Currie M G (1992a). Rat guanylin cDNA: characterization of the precursor of an endogenous activator of intestinal guanylate cyclase. *Biochemical and Biophysical Research Communications*, 185, 812–817.

Wiegand R C, Kato J, Huang M D, Fok K F, Kachur J F and Currie M G (1992b). Human guanylin: cDNA isolation, structure and activity. *Federation of European Biochemical Societies Letters*, 311, 150–154

SECTION

5

DEVELOPMENTAL NEUROBIOLOGY

CHAPTER 18

Hugh Tyndale-Biscoe and the developmental neurobiology of vision at the Australian National University

Richard Mark

Introduction

Many neurobiologists have had the idea that study of the developing mammalian nervous system might have much to gain by admitting, on an equal footing, the other great group of mammals, the Metatheria (for example, Rocha-Miranda & Lent 1978; Tyndale-Biscoe & Janssens 1988; Mark & Marotte 1992). Embracing such catholicism has three benefits. Where there are systematic differences between the way things are laid out or function, they offer insights into the common purpose. This is the comparative approach and was the main concern of naturalists who first studied the exotic Australian fauna (Owen 1837). Another aspect of comparative studies which can be of help in understanding development is that the growth and maturation of many marsupial young is slower than that of comparable placental mammals, sometimes very much so (Cockburn & Johnson 1988). This means that events that appear to happen simultaneously in placentals may be weeks apart in marsupials with the result that chronological sequences become obvious (Harman & Beazley 1989). With such insight it is often possible to re-appraise the development in placentals and see the same succession emerge even though events may occur more rapidly and be less clearly staged (LaVail *et al.*1991).

The third advantage depends on the metatherian invention of pouchlife, fuelled by lactation which takes the place of prolonged intrauterine gestation. The pouch young is a free-living fetus and if the essentials of the pouch environment, warmth and fluid intake, are provided, it may be removed, anaesthetised for operation or recording and, be replaced in the pouch, re-attached to the teat and allowed to continue to develop normally. The usefulness of this approach depends not on the difference between metatherian and eutherian development, but on their essential similarity. We have argued (Mark & Marotte 1992) that the similarities are so striking that the study of pouch young can give insights into the early stages of mammalian brain development, which in the common laboratory animals occurs in normally inaccessible intrauterine stages. The whole thrust of this tactic is to take advantage of the accessibility of the pouch young for recording or intervention in development. Either to use techniques in marsupials just as easily applied to the usual laboratory animals, or to document in a new marsupial species things that are already well understood in more conventional laboratory models, is rather to miss the point.

SETTING UP THE COLONY

Setting up the colony is a simple idea, the overall sense of which is easy to see, but it requires a very large investment in the acquisition of basic biological knowledge, especially of the anatomy and normal physiology of the animal, with comparatively little initial gain in the form of original published work. There is also a great deal of practical animal husbandry to be learned in order to bring it to reality. Shortly after I moved to the Research School of Biological Sciences in the late 1970s, Lauren Marotte and I went to see Hugh Tyndale-Biscoe with a view to starting up a program on early brain development, based on a marsupial model. Our first concern was to see whether surgical operations on the brain and eye of the early pouch young were possible, with adequate long-term post-operative survival. Hugh provided animals from his tammar wallaby (*Macropus eugenii*) colony. The first operation, removal and replacement of an eye cup, was a failure, but the patient survived, was re-attached to the teat, thrived and went on to live a normal, if monocular, life.

Having established that early intervention could be made to work, the next problem was finding a method of supplying pouch young of exactly known age, on demand. The obvious way was to use Renfree and Tyndale-Biscoe's method of reactivation of a dormant blastocyst in the tammar by cessation of suckling (Renfree 1979). This is how most pouch young of exactly known age are produced in our colony: the removal of a pouch young for one project activates the dormant blastocyst leading to the timed birth for the next. It also seemed a good idea at that time to try to find out more about the mechanisms of lactational and seasonal reproductive quiescence in the tammar. This led to the recruitment of Steve McConnell as a PhD student, who was jointly supervised by Hugh and me. Since melatonin was pivotal for the control of seasonal quiescence, Steve spent some time in Adelaide with Bob Seamark validating an assay for tammar melatonin. Several important endocrinological papers and, incidentally, the award of the Bolliger Prize of the Australian Mammal Society in 1983 to McConnell came out of this work (McConnell & Hinds 1985; McConnell & Tyndale-Biscoe 1985; McConnell 1986; McConnell *et al.* 1986; Tyndale-Biscoe *et al.* 1986.) These findings raised the possibility of using artificial changes in day length to bring on oestrus and a timed pregnancy. However this would have required too much in the way of special buildings and handling of animals outside of their natural settings. There was also the possibility of using artificial schedules of melatonin administration to simulate daylength changes. Eventually bromocriptine was used (Tyndale-Biscoe & Hinds 1984) to overcome the seasonal reproductive quiescence in winter by activating dopaminergic mechanisms inhibiting prolactin secretion. Even so births are still not programmable in November and December, prior to the beginning of breeding at the summer solstice. The neurological mechanisms of measuring daylength and its links to reproductive cycles still provide a tempting challenge, bringing together visual science and endocrinology.

At about that time I received a strategic plan grant from the Australian National University to build holding facilities for marsupials. Since so much was known about the tammar and how to keep it, especially its proven reliable breeding outdoors in the Canberra climate, there was really no choice but to confine the colony to this species. In near-plague numbers on Kangaroo Island, South Australia, there was no trouble in getting the help of farmers on the island to catch the animals. The tammars were then flown, together with a large sheaf of permits, in a light plane directly to Canberra. The colony has prospered, has been augmented several times with animals from Kangaroo Island, but is now self-sustaining. At the peak of its usage in the summer of 1994 it was producing adults and pouch young of a known age, on a known day, on demand for 6 concurrent research projects.

FIRST STUDIES ON VISUAL DEVELOPMENT

With the first animals installed we had enough back-up to begin the study of the development of the visual system using ^{3}H-proline injections into the eye as a tracer, in order to set the timetable for future studies (Wye-Dvorak 1984). By this technique the first optic axons were not seen to reach the brain until 12 days postnatal. The whole development of central connections along the primary visual pathway occurs postnatally and may be examined in the same detail and using the same techniques right through development. The connections of the optic nerve are many, including a highly laminated lateral geniculate nucleus (9 laminae in the tammar, 11 in the grey kangaroo, against 3 in the cat, Sanderson *et al.* 1984), and an accessory optic system (Hayhow 1966; Wye-Dvorak 1984; Coleman & Beazley 1988). The anatomical layout of the destinations of optic nerve fibres is similar to that seen in placental mammals. The last brain structure to become innervated by optic nerve fibres in development, even though it is closest to the eye, is the suprachiasmatic nucleus. This nucleus is known to be involved in the measurement of daylength and thus is of reproductive importance.

SPERRY'S EYE-ROTATION EXPERIMENT

With a known timetable of the development of central pathways it was possible to take the next step and examine the development of the pathway using experimental intervention. Following on from an old experiment by Sperry (1963), which provided one of the mainstays of his theory of precise developmental specification of neural connections, we decided to try the effects of rotating the eye cup on the day of birth so as to mis-align the microscopic pathways in the optic stalk along which optic nerve fibres grow. Sperry's experiment was done using adult newts, which share with other amphibians and fish the ability to regenerate the optic nerve with the restoration of vision. Since the optic nerve will not regenerate in mammals it had not been possible to do the same experiment on them in order to see whether the far-reaching conclusions Sperry (1963) claimed for the development of the nervous system of the lower vertebrate were applicable to all animals. Unlike total removal, the operation of eye rotation proved to be possible surgically in pouch young on the day of birth, provided the central artery of the retina was not broken, in which case the eye died. The aim was to investigate whether fibres from a rotated eye cup were still able to form their correct connections in the primary visual centres of the lateral geniculate nucleus and the superior colliculus. These centres contain a precise point-to-point projection of the retina and, thus, of the visual world.

Animals grew up until eye opening at which time the rotated eye was filled with horseradish peroxidase (HRP) and a laser lesion was made in a localised region of the peripheral retina. The lesioned axons did not transport the label so their central termination sites in the brain were left unlabelled, whereas terminals of optic nerve fibres arising elsewhere in the eye were densely filled with HRP reaction product (Flett *et al.* 1988). By careful analyses of the geometry of eye rotation, the location of the retinal laser lesion and the location of the filling defect centrally in the superior colliculus or the lateral geniculate nucleus, it was possible to conclude that in all cases the connections made by optic nerve fibres from the rotated eye were exactly those that would have been expected if fibres had found their terminal zones by cues derived from the location of their cell body in the retina, and nothing else (Marotte & Mark 1988). Retinotopic cues, and not the position of the rotated eye in the orbit, supplied the information for making central connections even if the result were a rotated visual field, which would presumably be a great behavioural disadvantage. This is in accord with Sperry's principles.

A chance result of these experiments was several cases in which the optic nerve fibres missed their normal route into the cranium and entered instead by the route of the oculomotor nerve. This brought the optic nerve fibres into the brainstem caudal to their destination rather than rostral. Nevertheless fibres were found running rostrally from the site of entry and they terminated in the normal retino-recipient nuclei of the midbrain. This shows that even if fibre growth is quite abnormal, selectivity of synapse formation can still ensure that erroneus connections are not made.

The question of whether the connections made by optic nerve fibres might be modified by experience of this abnormal visual geometry was examined by delaying the operation of eye rotation until eye opening at about 140 days, not birth. At this time all the connections from the eye to the brain had been made so rotation of the eye simply altered the position of the retina in relation to the body co-ordinates and did not directly affect neural connectivity. Animals were allowed to grow into young adults, giving them several months of visual experience. In order to find out whether there had been any adaptive change in the pattern of synaptic connections between optic nerve fibres and the central neurons we used electrophysiological mapping of the projection of the visual field onto the superior colliculus. Electrophysiological mapping was more suited to the problem than an anatomical method of tracing the connections as it allowed the complete map to be determined in one animal, rather than being pieced together from the results from many animals. This is because the exact amount of eye rotation differed in each animal and small compensatory effects might have been swamped by averaging. The mapping method was greatly improved by adapting a new computer algorithm using the thin plate spline technique to interpolate from measured recording points. We first established the details of the normal projection and measured precisely its orientation and the variation in magnification of different parts of the visual field (Mark *et al.* 1993b, Figure 18.1).

In animals in which the eye had been rotated the map of the visual field on the colliculus appeared, at first sight, to be distorted (Figure 18.1). However, the inverse projection of the surface of the superior colliculus projected into visual space gave a much

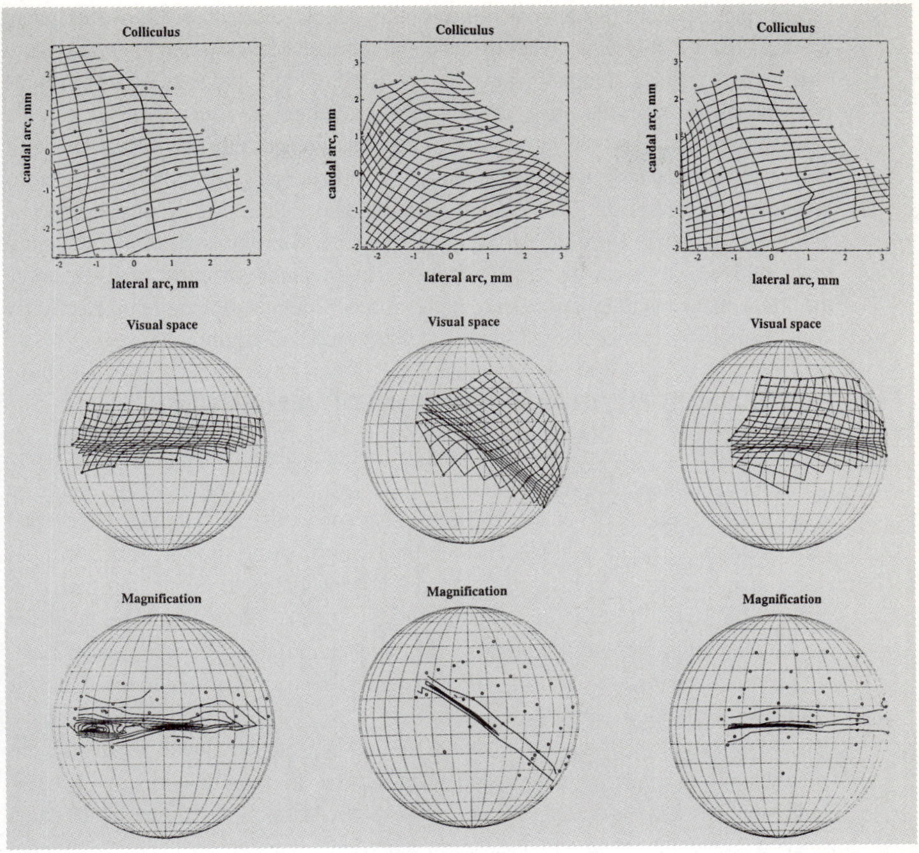

Figure 18.1 Effect of eye rotation and mathematical re-rotation on the map of the visual field on the contralateral superior colliculus.
Left column: Normal animal.
Upper panel: The map of the visual field as conventionally displayed. The horizontal meridian is the darker line running approximately rostro-caudally, nasal field rostrally, superior field medially.
Middle panel: Inverse mapping of the collicular surface on the visual field with the nasal field to the left.
Lower panel: Magnification, that is the density of contour lines in the above panel calculated according to Mark *et al.* (1993b).
Centre column: As above except the eye had been rotated clockwise about the optic axis to a final position of about $30^{o} \pm 5^{o}$ to the horizontal at the age of 151 days, that is, a few days after eye opening. Map made twelve months later. Note that the distortion in the conventional map in the upper panel resolves into a coherent rotation when viewed in visual space. The normally horizontal line of greatest magnification, corresponding to the visual streak of high ganglion cell density in the retina shows a rotation of 35°.
Right column: Mathematical re-rotation of the projection in space and its re-projection onto the collicular surface as illustrated in the **top panel** re-establishes a normal map.
Note: The apparent distortion of the map from the rotated eye is explained by geometrical factors. The fact that the projection from the rotated eye can be restored to normal by the application of one command affecting all points on the map means that its detailed internal structure has not been affected by the abnormal visual experience and that the connections between optic nerve fibres and collicular neurons remain faithful to retinal co-ordinates.
Source: James *et al.* (1993).

simpler picture. With this approach it was clear that the internal structure of the map was conserved, but it was apparently coherently rotated and projected at an angle into visual space. Using a computer algorithm for rotating objects in space it was possible to re-rotate data from the projection from the rotated eye in visual space back to the normal disposition and then to re-project this back to the collicular surface where it produced a normal map, with the details such as a higher magnification along the horizon very well preserved. The amount of re-rotation required to produce a normal retinotopic map was then compared, in each animal, with the amount of anatomical rotation of the eye (James *et al.* 1993). In each case agreement was very good. Therefore the connections made by optic nerve fibres in the midbrain were quite unaffected by visual experience in these circumstances of gross rotational squint.

THE EMERGENCE OF RETINOTOPY IN DEVELOPMENT

The early stages in achieving precise retinotopy in the midbrain projection were studied by Marotte (1990) using the same combined laser lesion and HRP technique as described above (in a series of animals from 26 days onwards). Even though a lesion had been made in the retina, a filling defect in the terminal label in the colliculus was not visible in animals younger than 40 days and it was incomplete for some time thereafter (Figure 18.2). This indicated that, unlike the adult, ganglion cells outside the region of the lesioned retina were also projecting their axons into the region occupied by axons of cells in the lesioned area. That is, the initial ingrowth of optic axons to the colliculus had not led them to precisely the appropriate spot. Was this because the collicular location of axons in the early stages was rather more loosely organised with respect to their neighbours from similar retinal locations than was the case later on? Or was it because some axons made gross errors? This was answered by Marotte (1993) by back-labelling ganglion cells in the retina with localised injections into their terminal zones in the caudal colliculus at progressively earlier ages, with a precision that would be difficult to approach in an intrauterine fetus. The answer was unequivocal. At no stages did ganglion cell axons make gross errors of growth. What might be called inappropriately projecting ganglion cells were always in a halo, diminishing in size with age, around those cells with axons ending in what would become the appropriate site of termination in the adult projection. Those more widespread terminal branches may be removed by axon pruning, cell death, or both. It seemed that, even in the very earliest stages, axons of ganglion cells are distributed in the colliculus and perhaps even grow selectively along pathways, in a manner specified by the retinal location of the cells bodies, albeit less precisely than in the adult.

Such growth can be visualised by labelling retinal ganglion cell axons from a localised spot in the retina by inserting small pledgets of Gelfoam soaked in the fluorescent lipophilic dye, DiI (l,l'-dioctadecyl-3,3,3',3'-tetramethylindocarbocyanine perchlorate) into the retina of pouch young from the scleral surface (Ding & Marotte 1993; 1994; Figure 18.3). Whole mounts of the flattened colliculus examined by fluorescence microscopy show parallel bundles of optic axons coursing in a rostro-caudal direction over the colliculus. From the earliest stages of ingrowth they are always in a rough retinotopic order.

From about 40 days of age, the labelling pattern changed. Axons made localised, highly branched arborisations confined to the appropriate, retinotopic position in the colliculus. The axon arborisations were not found in non-appropriate terminal regions, although they did become relatively smaller by the time of eye opening and the more coarsely distributed axons disappeared.

ELECTROPHYSIOLOGY OF OPTIC INNERVATION OF THE COLLICULUS

At this stage in the project the advantage of working with the marsupial pouch young became even more apparent. It is technically very difficult to make electrophysiological recordings from an intrauterine fetus. With some developmental work on the design of an artificial pouch and experimentation with different anaesthetic regimes, we developed a protocol that permitted stable long-term recording and stimulation in the central nervous system of pouch young (Gummer & Mark 1994). Even before the photoreceptors of the retina have differentiated (Spira & Marotte 1989) and long before eye opening the function of the visual system can be tested by direct electrical stimulation of optic nerve fibres in the eye and recording of the evoked potential along the visual pathway (Freeman & Mark 1993). Evidence for conduction in the optic nerve goes back to 16 days (Reece *et al.* 1995) but potentials referable to the onset of synaptic transmission between optic nerve fibres and neurons of the superior colliculus appear quite suddenly at about 40 days, just when the arbors in the appropriate terminal zones first appear (Mark *et al.* 1993a). These potentials change sign with increasing depth in the colliculus, showing that they are locally produced. Some examples of the potentials recorded at progressively later ages are shown in Figure 18.4.

The analysis of evoked potentials can be carried a stage further by the technique of current source density analysis (Mitzdorf 1985) which yields information about the latency, intensity and depth in the colliculus of excitatory synaptic currents generated in collicular neurons by optic nerve synapses (James & Mark 1994). With this additional computational technique it can be shown that many of the features of the adult projection, such as the most rapidly conducting fibres terminating most deeply in the colliculus, are present from the very beginning (Freeman & Mark 1993). Descending connections between the cerebral cortex and the superior colliculus do not contribute to the collicular evoked potential until just before the time of eye opening (see below).

Taking the anatomical and physiological results together the innervation of the colliculus divides into two successive stages. In the first stage fibres grow into the colliculus in a rough retinotopic order and accumulate in increasing numbers from about 4 days until 40 days. Anatomically many fibres are tipped with growth cones and they show little side-branching. They are capable of carrying impulses, but show no clear evidence that they can transmit to collicular cells.

At about 42 days, two things happen. Anatomically, axons labelled in the colliculus from localised dye deposits in the eye begin to develop dense arborisations. They only appear in the retinotopically correct part of the colliculus. Physiologically, the evoked potential becomes larger and, more importantly, is found to change in sign with depth, which means that the signals are locally generated within the colliculus, by synaptic currents set up in post-synaptic collicular neurons. Terminal arborizations and synaptically evoked potentials are first found in the rostral colliculus. In the next few weeks they are both found further caudally until, by about 65 days, axons have well-developed terminal arborisations, appropriately located in the colliculus according to the site of retinal deposit, and all electrophysiological

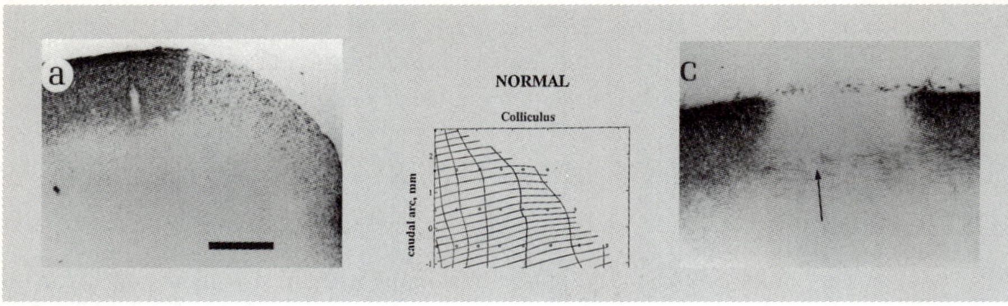

Figure 18.2 Unstained coronal sections through the upper layers of the superior colliculus contralateral to an eye injected with horseradish peroxidase (HRP) showing filling defects produced by lesions in the nasal retina in animals: (a)56, (b)77 and (c)93 days. Note the increase in sharpness of the borders and the decrease in label within the defects with age, indicating an increase in precision of the retinotopic projection.
Scale bar=200 µm for a) to c).

Note: Arrows in (b) and (c) indicate fibres of passage in the stratum opticum.

signals are in the form of reversing waves, increasing in amplitude and complexity with age. (Mark *et al.* 1993).

The two stages then are: directed growth, which is coarsely retinotopic, followed by the formation of terminal arborisations which is more precisely retinotopically disposed and is accompanied by the onset of synaptic transmission.

CORTICAL CONNECTIONS

The growth of nerve fibres into the cerebral cortex and the formation of synapses have been studied with a similar combination of anatomical tracing (Sheng *et al.* 1990, 1991) and electrophysiological recording (Mark & Waite 1995) as were used on the superior colliculus. This pathway has an additional link not present in the retino-collicular pathway. Optic nerve fibres, some of them branches of axons that also terminate in the colliculus, form similarly retinotopically ordered connections with neurons of the lateral geniculate nucleus of the thalamus. These thalamic relay neurons then connect to the cortex, retaining their retinotopic organisation so that the map of the visual world is laid out on the cortex as it is on the colliculus. Localised injections of HRP into the cortex were made at various stages of development. This results in retrograde labelling of the cell bodies of geniculate neurons that send an axon to the cortex and anterograde labelling of fibres that arise in the cortex and terminate in the geniculate (Sheng *et al.* 1990). Injections into the lateral geniculate nucleus label cell bodies of descending axons from the cortex to the geniculate and show the distribution of ascending fibres from the geniculate to the cortex (Sheng *et al.* 1991; Figure 18.5). Fibres from the geniculate reach the cortex by 15 days, very shortly after optic fibres first reach the geniculate. They ramify widely through much of the depth of the developing cortex, well before the cortical layers have differentiated and are retinotopically organized from early stages. Descending connections to the geniculate, that is back-filled neurons in the cortex, are seen later at 40 days.

Both these anatomical findings are at odds with what, until recently, was accepted as the general schema of development of the visual cortex. This has the thalamocortical

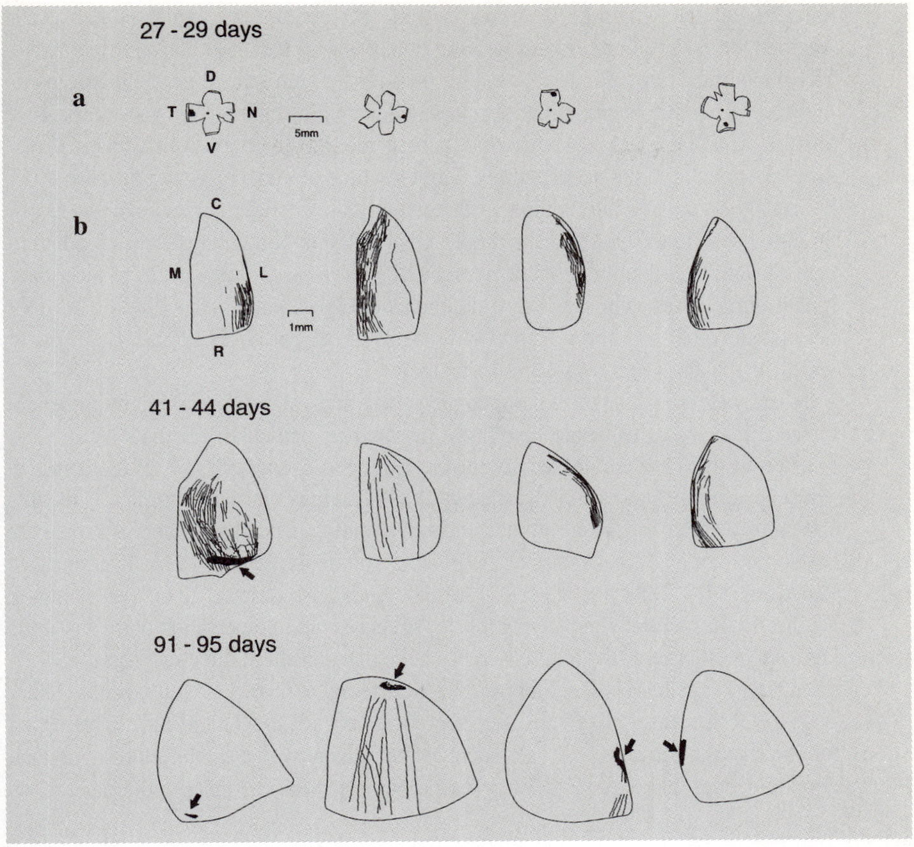

Figure 18.3 The distribution of labelled retinal axons and terminal zones in contralateral superior colliculus after deposits of fluorescent lipophilic dye (DiI) in temporal, nasal, dorsal, and ventral quadrants of the retina is shown at various ages.
(a)*Camera lucida* drawings of retinal whole mounts at 27 to 29 days after birth. DiI deposits are in solid black. Spots mark the optic nerve head. Deposits were in similar positions for other ages.
Key: T: temporal N: *nasal* D: dorsal V: ventral

(b)*Camera lucida* drawings of whole mounts of the superior colliculus. Sizes vary slightly at similar ages because of flattening to produce the wholemounts. Not every labelled axon is drawn, but the rostro-caudal and mediolateral extent of the labelled area and any solitary axons outside it are shown. Terminal zones are solid black and arrowed.

Prior to 41 days axons had few branches and showed a coarse retinotopy, with axons from temporal retina rostrally, nasal axons reaching to the caudal pole, dorsal axons laterally and ventral axons positioned medially.
At 41 to 44 days terminal branching was first seen after labelling temporal, but neither nasal, dorsal, nor ventral axons. This was in retinotopically correct rostral superior colliculus.
By 55 days axons from other retinal quadrants had begun to form terminal zones in their retinotopically correct positions.
By 91 to 95 days discrete terminal zones were present.
(Mark *et al.* 1993)
R: rostral C: caudal M: medial L: lateral
Scale bar (1 mm) in b applies to all ages.

fibres initially growing into the cortex beneath the cortical plate, which contains cells destined for the adult cortex, and forming synapses with a transient layer, the subplate. Thalamocortical fibres are said to wait in this layer for a protracted period before growing into the deeper layers of the developing cortex (Shatz & Luskin 1986; Ghosh & Shatz 1992). There is no evidence that the thalamocortical fibres in the wallaby cortex are ever confined to the deeper cortex. From the time of initial ingrowth they are widely distributed up to a thin compact cell zone at the top of the cortical plate and continue to grow up to this region as the developing layers of the mature cortex differentiate beneath it. Secondly, subplate neurons have been proposed to act as pioneer neurons leading cortical fibres down to the thalamus. (McConnell *et al.* 1989). This does not appear to be so in the wallaby where axons of cells of layers VI and V are first to reach the thalamus and do so simultaneously.

If these were true differences between marsupials and placentals in the development of the cerebral cortex this would be of great importance, particularly as the structure and function of the adult tammar cortex shows nothing to distinguish it from the cortex of similar placental mammals (Vidyasagar et al 1992). However, recent studies in rodents also show that thalamic axons do not undergo a waiting period in the deeper part of the developing cortex (Catalano *et al.* 1991; Erzurumlu & Jhaveri 1992; Kageyama & Robertson 1993; Miller *et al.* 1993; see also Saunders, chapter 22 in this volume). Further, in the hamster a pioneering role for subplate cells in innervation of the thalamus has been questioned as the first cells to reach the thalamus from the visual cortex are primarily layer VI cells. Very few subplate cells were found to project to the thalamus (Miller *et al.* 1993). It seems unlikely that the differences between rodent and cat can be related to differences in the rates of development as the wallaby, with a similar mode of cortical development as that of rodents, has a protracted development similar to the cat.

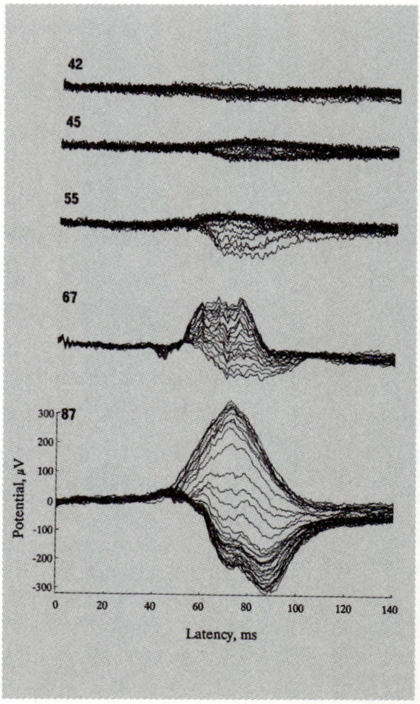

Figure 18.4 The potentials evoked in the superior colliculus of anaesthetised tammar pouch young of increasing ages. Each trace is the average of 10 sweeps and between each trace the recording electrodes was advanced by 40 µm deeper into the superior colliculus.
At 42 days, there was a slight response.
At 45 days and older the response was a negative wave on the surface (downwards deflection) that reversed steadily with depth to positive (upwards deflection) when the electrode recorded form deeper within the colliculus. . The waves increase still further in size and complexity with age until reaching the adult form at about the time of eye opening (140 days).

Note: The responses illustrated are superimposed on a common baseline.

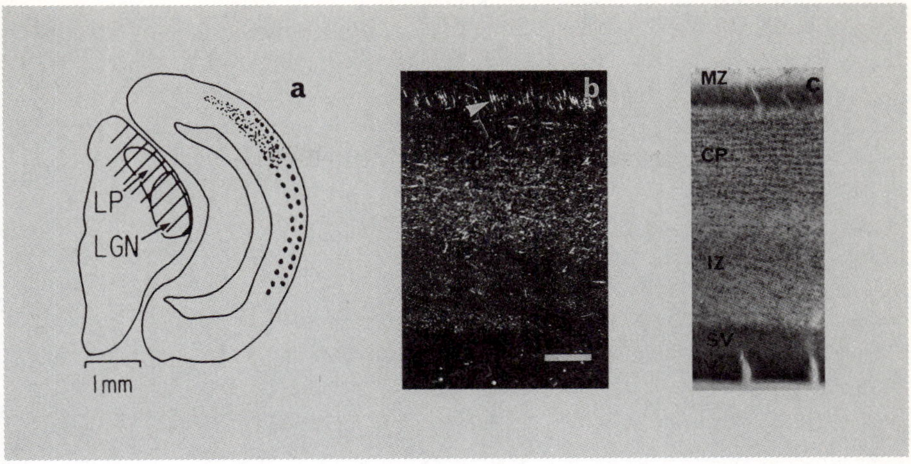

Figure 18. 5 (a) *Camera lucida* drawing of a coronal section through the brain of a 45-day pouch young after an injection of HRP conjugated to wheat germ agglutinin (WGA) (shading) into the lateral geniculate nucleus (LGN) and lateral posterior nucleus (LP). Two bands of retrogradedly labelled cells (large dots) are first seen around this age (40 to 45 days). One of the bands is in the lower part of the cortical plate (CP), while the upper band is within the compact cell zone at the top of the CP, (see (b) and (c)). Anterograde label extends up to this zone (see (b)).

(b) Dark field view of an adjacent section stained with thionin through the dorsal region of visual cortex shown in (a).

(c) Bright field view of an adjacent section stained with thionin through the dorsal region of visual cortex shown in (a). The upper band of labelled cells is present (arrowhead) and anterograde label is present throughout the intermediate zone (IZ) and most of the CP except for the compact cell zone.

Key: MZ: marginal zone SV: subventricular zone V: ventricular zone

Scale bar=100 μm.

Electrophysiological recording in the pouch young can, as in the superior colliculus, show when, and at what level, in the cortex synaptic activity is generated by stimulation of the optic nerve. Recordings from the cortex were begun at day 50 but the first electrical responses to stimulation of the optic nerve did not appear until 62 days. Thus, there is a long delay between the ingrowth of axons from the thalamic relay nuclei to the cortex, which begins at about 15 days, and the first cortical evoked potential at 62 days.

The pathway from the facial whiskers to the cortex follows a similar route to that from the eye, through a thalamic relay station, the nucleus ventralis posteromedialis, the axons from which terminate in the somatosensory cortex adjacent to the visual cortex. The timetable of formation of the thalamocortical pathway is similar to that for the visual system in that ascending nerve fibres appear in the cortex as early as 15 days (Leamey *et al.* 1993 and see Waite & Weller chapter 19 in this volume) but a somatosensory-evoked potential can only be recorded from this cortex at the same time as in the visual system, on day 62. The fact that cortical responsiveness comes on simultaneously in the two pathways suggests that there may be a common trigger (Mark & Waite 1995).

By the use of autoradiographic techniques to find the birthdates of neurons forming the cortical layers (Marotte & Sheng 1995) it can be shown that in the cortex at 62 days, layer IV has just emerged from the compact cell zone at the top of the cortical plate. (The layers of the cortex develop inside-out with the deepest, layer VI, developing first). Layer IV is the principal receiving layer for thalamocortical afferents in both visual and somatosensory systems. Current source density analyses of the evoked potential in the visual cortex from stimulation of the optic nerve show the synaptic activity to be generated in the superficial layers of the cortex (R Mark, unpublished). Taken together these anatomical and physiological findings suggest that the signal for synapse formation between afferent fibres and cortical neurons is the differentiation of the main recipient cortical layer, layer IV. Note again, as in the colliculus, the long period between the first axons reaching the cortex from the thalamus at about 15 days and the first signs of synapse formation between the fibres and cortical neurons at 62 days. The descending axons from cortical cells of layer V to the superior colliculus are first shown to be present anatomically at about 80 days. The potential in the colliculus thought to be generated by cortico-collicular fibres does not appear in current source density analysis until between 130 and 140 days. Once again the innervation seems to have two distinct phases, the growth of fibres clearly separated in time from the formation of functional synapses.

GENERAL PRINCIPLES: THE TWO-STAGE PROCESS

The pathways described, from the optic nerve to the superior colliculus and lateral geniculate nucleus, the connection of the geniculate to the cortex and the descending pathways from the cortex to the lateral geniculate nucleus and to the colliculus, are all retinotopically organised. Adjacent points on the retina are represented as synaptic connections formed by adjacent ganglion cell terminals in the colliculus and geniculate. The same topographic relationships hold for the ascending and descending pathways. Below is a speculative synthesis of how these connections might be made.

Connections in all the links described above seem to be made in the same way. First fibres grow into approximately the correct position and they accumulate for a surprisingly long period. The morphology of the early fibres does not suggest that they form synapses in a specialised waiting zone. They enter the territory in which they are destined to synapse, but do not branch widely. The fibres, in their first growth period, must respond to spatial cues in the matrix in which they grow, to provide the initial rough retinotopy and other specificities of connectivity. The fact that they remain apparently unconnected with their targets while accumulating during the first phase implies that the mechanism requires that all fibres remain in the same state. Perhaps the initial coarse retinotopy is a property of the whole projection and combines fibre-to-fibre interaction prior to the selection of synaptic sites with some faily crude polarity markers. Perhaps newly grown fibres can only sense their position by exchanging information with more mature fibres.

The newly grown fibres then only ramify locally and form functional synapses in response to a postsynaptic signal, which may be due to the differentiation or migration of the population of target cells. In the cortex this signal appears to coincide with the emergence of layer IV, the main termination of visual afferent fibres, from the compact cell zone at the top of the cortical plate.

As soon as there are morphological signs of synapse formation, such as densely branching arbors in the appropriate terminal zones, there is immediate establishment of functional synaptic transmission. That this occurs, even in the absence of meaningful natural activation of receptors and long before function of the synapses could be useful in the behavioural sense, suggests that this process has some role in development. It may act by refining the retinotopic maps by competitive interactions between synapses activated spontaneously (Mark 1974). Blockage of sodium conductance and, therefore, of impulse activity in prenatal kitten optic nerves prevents the segregation of developing terminals into eye-specific bands in the geniculate (Shatz & Stryker 1988); or impulses and synaptic transmission may be incidental. Synapse formation may be necessary to enable connecting neurons to exchange developmental signals of other kinds, including larger molecules such as growth factors.

FUTURE DIRECTIONS

Much in the previous sections remains to be proven. Marsupial research has gone a long way to defining the problems that need answers by identifying a common pattern and sequence in the formation of nerve pathways in the visual system in particular. It has the possibility of providing many of the insights that will come from the next level of analysis. Electrophysiological techniques, as applied to the whole animal, are the most likely source of future advances. It will not take much more in the way of technical refinement of current methods to be able to make long-term records of single unit responses from implanted electrodes in unanaesthetised pouch young. This will give definitive answers to some contentious current questions, particularly about the significance of spontaneous activity of the nervous system during its development.

Acknowledgments

Lauren Marotte has been an equal partner in running this program from the beginning. We are grateful to Bob Jackson, Rainer Geppert, Amanda Devlin and Kevin Williams for setting-up and managing the tammar colony. This chapter only outlines one project in the program. This and others have all stemmed from the work of a series of graduate students:

PhD McConnell, SJ (1984) The pineal gland and reproduction in the tammar wallaby (*Macropus eugenii*); Mayner, L (1985) The anatomy of the sensori-motor neocortex, thalamus and their interconnections in the tammar wallaby (*Macropus eugenii*); Wye-Dvorak, J (1985) Primary visual projections in the tammar wallaby (*Macropus eugenii*); Guppy, A (1986) Comparative acoustical and physiological studies of hearing and directionality in vertebrates; Comans, PE (1987) Development of motoneurons in the lumbar spinal cord of the wallaby; Sheng, X-M (1989) Development of cortical connections in the visual system of the wallaby (*Macropus eugenii*); Ding, Y (1995) The initial innervation of the superior colliculus in relation to retinal development and topography in the wallaby (*Macropus eugenii*); Liu, GB (enrolled 1992) Development of auditory and visual projections in the wallaby (*Macropus eugenii*); Leamey, C (enrolled 1993; Co-operative scholar, UNSW) Development of morphology, connectivity and function in somatosensory thalmus of the wallaby; Lim, C (enrolled 1994) In vitro studies of the development of optic innervation of the wallaby superior colliculus; Hemmi, J (enrolled 1994) Visual acuity of the tammar (*Macropus eugenii*). MSc Flett, DL (1986) Topography of retinal projections to the colliculus and geniculate in the wallaby (*Macropus eugenii*); Faulks IJ (to submit) Somatotopic organization of the ventrobasal thalamus of the wallaby (*Macropus eugenii*). BSc (Honours) Freeman, TCB (1992) Curent source density analysis of the development of optic innervation of the superior colliculus of the wallaby (*Macropus eugenii*).

References

Catalano S M, Robertson R T and Killackey H P (1991). Early ingrowth of thalamocortical afferents to the neocortex of the prenatal rat. *Proceedings of the National Academy of Sciences USA*, 88, 2999–3003.

Cockburn A and Johnson C N (1988). Patterns of Growth. In *The Developing Marsupial. Models for Biomedical Research*. Eds C H Tyndale-Biscoe and P A Janssens. Springer-Verlag, Berlin. pp 28–40.

Coleman L-A and Beazley L D (1988). The accessory optic system of the wallaby, *Setonix brachyurus*: Anatomy in normal animals and after early eye removal. *Journal of Comparative Neurology*, 273, 359–376.

Ding Y and Marotte L R (1993) Retinotopic order of optic axons during the formation of the retinocollicular projection in the wallaby *Macropus eugenii*. *Proceedings of the Australian Neuroscience Society*, 4, 44.

Ding Y and Marotte L R (1994) Retinotopy in the developing optic nerve of the wallaby (*Macropus eugenii*). *Proceedings of the Australian Neuroscience Society*, 5, 122.

Erzurumlu R S and Jhaveri S (1992). Emergence of connectivity in the embryonic rat parietal cortex. *Cerebral Cortex*, 2, 336–352.

Flett D L, Marotte L R and Mark R F (1988). Retinal projections to the superior colliculus and dorsal lateral geniculate nucleus in the tammar wallaby (*Macropus eugenii*): I. Normal topography. *Journal of Comparative Neurology* 271, 257–273.

Freeman T and Mark R F (1993). Onset of electrophysiological responses to stimulation of the optic nerve in the developing superior colliculus of the wallaby (*Macropus eugenii*): Further evidence of a two stage process of formation of topographically ordered connections. *Proceedings of the Australian Neuroscience Society*, 4, 44.

Ghosh A and Shatz C J (1992). Pathfinding and target selection by developing geniculocortical axons. *Journal of Neuroscience*, 12, 39–55.

Gummer A W and Mark R F (1994). Patterned neural activity in brainstem auditory areas of a pre-hearing mammal, the tammar wallaby (*Macropus eugenii*). *NeuroReport*, 5, 685–688.

Harman A M and Beazley L D (1989). Generation of retinal cells in the wallaby Setonix brachyurus. *Neuroscience*, 28, 219–232

Hayhow W R (1966). The accessory optic system in the marsupial phalanger, Trichosurus vulpecula: An experimental degeneration study. *Journal of Comparative Neurology*, 126, 653–672.

James A C and Mark R F (1994). Current source density in the wallaby superior colliculus viewed on space-time has three phases of activity. *Proceedings of the Australian Neuroscience Society*, 5, 213.

James A C, Mark R F and Sheng X-M (1993). Geometry of the projection of the visual field upon the superior colliculus of the wallaby (*Macropus eugenii*): II. Stability of the projection after prolonged rearing with rotational squint. *Journal of Comparative Neurology*, 330, 315–323.

Kageyama G H and Robertson R T (1993). Development of geniculocortical projections to visual cortex in rat: Evidence for early ingrowth and synaptogenesis. *Journal of Comparative Neurology*, 335, 123–148.

LaVail M M, Rapaport D H and Rakic P (1991). Cytogenesis in the monkey retina. *Journal of Comparative Neurology*, 309, 86–114.

Leamey C A, Waite P M E, Marotte L R and Mark R F (1993). The development of the trigeminal pathway from periphery to cortex in the wallaby (*Macropus eugenii*). *Proceedings of the Australian Neuroscience Society*, 4, 103.

Mark R F (1974). *Memory and Nerve Cell Connections: criticisms and contributions from developmental neurophysiology*. Clarendon Press, Oxford.

Mark R F and Marotte L R (1992). Australian marsupials as models for the development of the mammalian visual system. *Trends in Neurosciences*, 15, 51–57.

Mark R F and Waite P M E (1995). Onset of evoked activity in the visual and somatosensory cortices in the wallaby. *Proceedings of the Australian Neuroscience Society*, 6, 190.

Mark R F, Freeman T C B, Ding Y and Marotte L R (1993a). Two stages in the development of a mammalian retinocollicular projection. *NeuroReport*, 5, 117–120.

Mark R F, James A C and Sheng, X-M (1993b). Geometry of the projection of the visual field upon the superior colliculus of the wallaby (*Macropus eugenii*): I. Normal projection. *Journal of Comparative Neurology*, 330, 303–314.

Mark R F, Freeman T C B, Ding Y and Marotte L R (1993). Two stages in the development of a mammalian retinocollicular projection. NeuroReport 5, 117–120.

Marotte L R (1990). The development of retinotopy in projections from the eye to the dorsal lateral geniculate nucleus and superior colliculus of the wallaby (*Macropus eugenii*). *Journal of Comparative Neurology,* 293, 524–539.

Marotte L R (1993). Location of retinal ganglion cells contributing to the early imprecision in the retinotopic order of the developing projection to the superior colliculus in the wallaby (*Macropus eugenii*). *Journal of Comparative Neurology,* 331, 1–13.

Marotte L R and Mark R F (1988). Retinal projections to the superior colliculus and dorsal lateral geniculate nucleus in the tammar wallaby (*Macropus eugenii*): II. Topography after rotation of an eye prior to retinal innervation of the brain. *Journal of Comparative Neurology,* 271, 274–292.

Marotte L R and Sheng X-M (1995). The relationship between developing visual cortical layers and thalamocortical afferents in the wallaby (*Macropus eugenii*). *Proceedings of the Australian Neuroscience Society,* 6, 80.

McConnell S J (1986). Seasonal changes in the circadian plasma melatonin profile of the tammar *Macropus eugenii*. *Journal of Pineal Research,* 3, 119–125.

McConnell S J and Hinds L A (1985) Effect of pinealectomy on plasma melatonin, prolactin and progesterone concentrations during seasonal reproductive quiescence in the tammar *Macropus eugenii*. *Journal of Reproduction and Fertility,* 75, 433–440.

McConnell S J and Tyndale-Biscoe C H (1985) Response in peripheral plasma melatonin to photoperiod change and the effects of exogenous melatonin on seasonal quiescence in the tammar (*Macropus eugenii*). *Journal of Reproduction and Fertility,* 73, 529–538.

McConnell S J, Tyndale-Biscoe C H and Hinds L A (1986) Change in duration of elevated melatonin is the major factor in photoperiod response of the tammar *Macropus eugenii*. *Journal of Reproduction and Fertility,* 77, 623–632.

McConnell S K, Ghosh A and Shatz C J (1989). Subplate neurons pioneer the first axon pathway from the cerebral cortex. *Science,* 245, 978–982.

Miller B, Chow L and Finlay B (1993). The early development of thalamocortical and corticothalamic projections. *Journal of Comparative Neurology,* 335, 16–41.

Mitzdorf U (1985). Current source density method and application in cat cerebral cortex: investigation of evoked potentials and EEG phenomena. *Physiological Reviews,* 65, 37–100.

Owen R (1837) On the structure of the brain in marsupial animals. *Philosophical Transactions of the Royal Society of London,* 127, 87–96.

Reece L J, Lim C H, Ho S M, James A C and Mark R F (1995). *In vitro* analysis of development at the retinocollicular synapse in the tammar wallaby. *Proceedings of the Australian Neuroscience Society,* 6, 137.

Renfree, M B (1979). Initiation of development of diapausing embryo by mammary denervation during lactation in a marsupial. *Nature,* 278, 549–551.

Rocha-Miranda C E and Lent R (Eds) (1978). *Opossum Neurobiology: The opossum as an experimental model for the study of the mammalian nervous system.* Academia Brasileira de Ciencias, Rio de Janeiro

Sanderson K J, Haight J R and Pettigrew J D (1984). The dorsal lateral geniculate nucleus of macropodid marsupials: Cytoarchitecture and retinal projections. *Journal of Comparative Neurology,* 224, 85–106.

Shatz C J and Luskin M B (1986). The relationship between the geniculocotical afferents and their corical target cells during development of the cat's primary visual cortex. *Journal of Neuroscience,* 6, 3655–3668.

Shatz C V and Stryker M P (1988). Prenatal tetrodotoxin infusion blocks segregation of retinogeniculate afferents. *Science,* 242, 87–89.

Sheng X-M, Marotte L R and Mark R F (1990). Development of connections to and from the visual cortex in the wallaby (*Macropus eugenii*). *Journal of Comparative Neurology,* 300, 196–210.

Sheng X-M, Marotte L R and Mark R F (1991). Development of the laminar distribution of thalamocortical axons and corticothalamic cell bodies in the visual cortex of the wallaby (*Macropus eugenii*). *Journal of Comparative Neurology,* 307, 17–28.

Sperry R W (1963). Chemoaffinity in the orderly growth of nerve fiber patterns and connections. *Proceedings of the National Academy of Sciences USA*, 50, 703–710.

Spira A W and Marotte L R (1989). Histological and electron microscopic milestones in the development of the retina of a marsupial wallaby (*Macropus eugenii*). *Anatomy and Embryology,* 179, 571–585.

Tyndale-Biscoe C H and Hinds L A (1984). Seasonal patterns of circulating progesterone and prolactin and response to bromocriptine in the female tammar *Macropus eugenii*. *General and Comparative Endocrinology*, 53, 58–68.

Tyndale-Biscoe C H and Janssens P A (1988). *The Developing Marsupial. Models for Biomedical Research.* Springer-Verlag, Berlin.

Tyndale-Biscoe C H, Hinds L A and McConnell S J (1986). Seasonal breeding in a marsupial. Opportunities of a new species for an old problem. *Recent Progress in Hormone Research,* 42, 471–512.

Vidyasagar T R, Wye-Dvorak J, Henry G H and Mark R F (1992). Cytoarchitecture and visual field representation in area 17 of the tammar wallaby (*Macropus eugenii*).*Journal of Comparative Neurology*, 325, 291–300.

Wye-Dvorak J (1984) Postnatal development of primary visual projections in the tammar wallaby (*Macropus eugenii*). *Journal of Comparative Neurology,* 228, 491–508.

CHAPTER 19

DEVELOPMENT OF SOMATOSENSORY PATHWAYS FROM THE WHISKERS

Phil M E Waite and W Lee Weller

INTRODUCTION

The present chapter describes our current understanding of the organisation and development of the somatosensory system in marsupials. To date most progress in this area has been made on the development of receptors in the face and the pathways associated with these inputs and this chapter therefore concentrates on these advances. There are two probable reasons for the preponderance of facial research. Firstly, curiosity about the afferent inputs important for the establishment of early suckling reflexes has led to investigations of oral receptors. Secondly, fruitful research on the vibrissae and their associated central pathways in rodents (Woolsey & Van der Loos, 1970; for recent reviews see Davies 1988; Killackey *et al.* 1990; Waite & Tracey 1995) has stimulated similar studies in a number of other species, including several marsupials. Investigations have shown that both the peripheral receptors and central somatosensory relays from the face and limbs are similarly organised to those in eutherians (Hazlett *et al.* 1972; Weller 1972; Hamilton & Johnson 1973; Pubols *et al.* 1973; Reynolds *et al.* 1985). However, the protracted extrauterine development of some marsupials is proving useful for answering questions on mechanisms and timing of development of sensory maps which have not been easy to address in rodents.

PERIPHERAL RECEPTORS AND THEIR INNERVATION

In marsupials, as in other mammals, the facial skin and oral and nasal mucosa, as well as specialised structures, such as the teeth and vibrissae, are innervated by the fifth cranial nerve, the trigeminal. This sensory input is analogous to a dorsal root ganglion for spinal inputs, with peripheral processes extending to the receptors and a central root, the trigeminal root, entering the brainstem (Figure 19.1a). At birth the ganglion in the tammar forms a prominent cell mass ventrolateral to the developing brainstem

(Figure 19.1b). Several studies have indicated that the facial skin in marsupials is richly innervated by birth. Thus in the snout of the newborn South American opossum *(Mondelphis domestica)* Jones and Munger (1985) showed that nerve fibres and free nerve endings were present in both the dermis and epidermis, with Merkel cells in the basal epidermis. Similarly, nerve fibres and Merkel cells are seen around the mouth in the newborn native cat *(Dasyurus hallucatus)*, brushtail possum *(Trichosurus vulpecula)* and brown bandicoot *(Isoodon macrourus)* (Gemmell *et al.* 1988). Both silver staining and the carbocyanine dye (DiI) labelling of the trigeminal ganglion at birth in the tammar wallaby *(Macropus eugenii)* show dense innervation of the maxillary skin (Figures 19.1c and d).

The facial skin of most mammals contains some specialised hairs, the sinus hairs or vibrissae, which usually lie in horizontal rows along the muzzle (Figures 19.2a and b). The structure and innervation of these vibrissal follicles (Figure 19.2c) in several marsupials (Hollis & Lyne 1974; Loo & Halata 1991; Marotte *et al.* 1992) are similar to those in eutherians (Rice *et al.* 1986). In both the rat and tammar, the larger follicles receive over 200 myelinated nerve fibres (Klein *et al.* 1988; Marotte *et al.* 1992). The follicles contain a variety of receptors such as Merkel cells, lamellated corpuscles and free nerve ends. Response properties such as receptive fields, adaptation rates and direction sensitivity are similar in the North American opossum (Pubols *et al.* 1973) to those described in rodents (Zucker & Welker 1969; Lichtenstein *et al.* 1990).

Like other hairs, vibrissal follicles develop as a downgrowth of the epidermis (Davidson & Hardy 1952). A recent study has investigated follicle development and innervation in the tammar wallaby (Waite *et al.* 1994). A solid epidermal peg is already present and innervated in the tammar by birth (Figures 19.1c and d). The development of the follicle shows a similar sequence to that in rodents, but lasts 18 to 20 weeks, compared with approximately 3 weeks in rats (Munger & Rice 1986). The sequence of receptor development is also similar, but slower, with Merkel cells being the first receptors to appear, at around 7 weeks, while lanceolate endings are first seen at four months. In all species examined, development and innervation of vibrissal follicles precedes intervibrissal hairs (Davidson & Hardy 1952; Munger & Rice 1986; Mosconi & Rice 1993; Waite *et al.* 1994). Again Merkel cells appear before lanceolate or circumferential innervation.

TRIGEMINAL NUCLEAR COMPLEX

Compared with the facial skin and the somatosensory thalamus and cortex there have been relatively few studies on the trigeminal nuclear complex of marsupials. The adult organisation and somatotopy of the brainstem, including the trigeminal nucleus, has been described for the South American opossum (Oswaldo-Cruz & Rocha-Miranda 1968; Weller 1971). A recent study on the tammar is the first to investigate the adult organisation and development of this complex in a diprotodont (Waite *et al.* 1994). In all studies, the overall organisation of the marsupial trigeminal nuclear complex is similar in extent and arrangement to that described in a variety of mammals (Olszewski 1950; Torvik 1956; Darian-Smith 1973; Ma 1991). The complex lies in the lateral brainstem and extends from the mid pons to the upper cervical cord. It is divided rostro-caudally into 4 subnuclei: principalis (the main sensory nucleus) and oralis, interpolaris and caudalis (the spinal trigeminal nuclei).

In the tammar and brushtail possum the main nucleus is large and distinct and has been shown to project to the ventrobasal (VB) thalamus (Leamey *et al.* 1993; Waite *et al.* 1994; Weller unpublished observation). As in eutherians (Gobel & Dubner 1969; Ma 1991; Noriega & Wall 1991) it contains closely packed, mainly medium-sized cells, often in clusters. The marsupial spinal nuclei are also similar to those in eutherians. Oralis is the least clear, with low cell density and occasional very large cells. Interpolaris is irregular in outline and distinctly heterogeneous. This is partly because it is traversed by vagal fibres, as in the cat and primate, but not in the rat. Caudalis has the familiar laminar structure (Crosby & Yoss 1954; Torvik 1956; Bates & Killackey 1985) closely analogous to that in the dorsal horn. As in the rat, all the spinal nuclei in the tammar contain rostrocaudally running intranuclear fibre bundles (Bates & Killackey 1985; Gobel & Purves 1972; Waite *et al.* 1994). Somatotopy in the opossum is similar to other placental species with a large representation for the mouth, lips, vibrissae, and nose (Weller 1971).

For marsupials the development of the trigeminal nuclear complex has so far only been described in the tammar. The development has been extensively studied in rodents, mainly because of the appearance of a vibrissal-related pattern in principalis, interpolaris and caudalis during early postnatal life (Bates & Killackey 1985; Ma 1993). In newborn tammars the complex can be recognised as a region of increased cell density with both Nissl stains (Figure 19.1b) and mitochondrial enzyme reactions (for example, cytochrome oxidase, CO). By 10 days postnatal, the main nucleus is clearly discernible between the trigeminal tract and the motor nucleus (Figure 19.3a). Throughout the complex, cell density and CO reactivity are higher than surrounding regions. Rostrocaudal subdivisions start to emerge by 30 days and are well discerned by 40 days. In cross-section the main nucleus has the almond shape seen in the adult, (Fig 19.3b) and CO staining shows 5 lines of increased reactivity running obliquely across the nucleus, characteristic of the 5 vibrissal-related lines seen in rodents. This is the earliest stage at which these lines are clear, although hints are present from as early as 10 days. By 60 days obliquely oriented patches of increased reactivity are seen in interpolaris (Figure 19.3c) and caudalis. Thus a vibrissal-related pattern is apparent in the developing subnuclei principalis, interpolaris and caudalis, as in rodents. For the main nucleus, the patches persist into adulthood (Figure 19.3d) although the pattern is less distinct than in young animals. Patches have also been described in adult mice, cats and primates (Nomura & Mizuno 1986 ; Ma 1991; Noriega & Wall 1991), although they are not seen in adult rats (Belford & Killackey 1979). Similar patches are seen in principalis, interpolaris and caudalis in the adult brushtail possum (Weller 1983). These patches of high mitochondrial enzyme activity are considered to reflect regions of high peripheral innervation density and synaptic activity (Wong-Riley 1989). In the possum, rodent, cat and primate, patches are also seen in the dorsal column nuclei (Belford & Killackey 1978; Nomura & Mizuno 1986; Noriega & Wall 1991; Weller 1983); in primates these have been shown to correspond to terminations from the digits (Florence *et al.* 1989).

For the tammar, trigeminal afferent fibres are present throughout the nuclear complex by birth (Waite *et al.* 1994). Labelling of the newborn trigeminal ganglion with DiI, shows extensive labelling of the trigeminal root and tract, extending from the root entry zone (Figure 19.4a) to the upper cervical cord. Individual collaterals can be seen entering the main nucleus (Figure 19.4b) and also projecting into all levels of the spinal nuclei. Gentle touching of the lip or vibrissal pad in the neonatal tammar wallaby (P M E Waite, unpublished observation) produces reflex head turning and mouth opening, suggesting not only

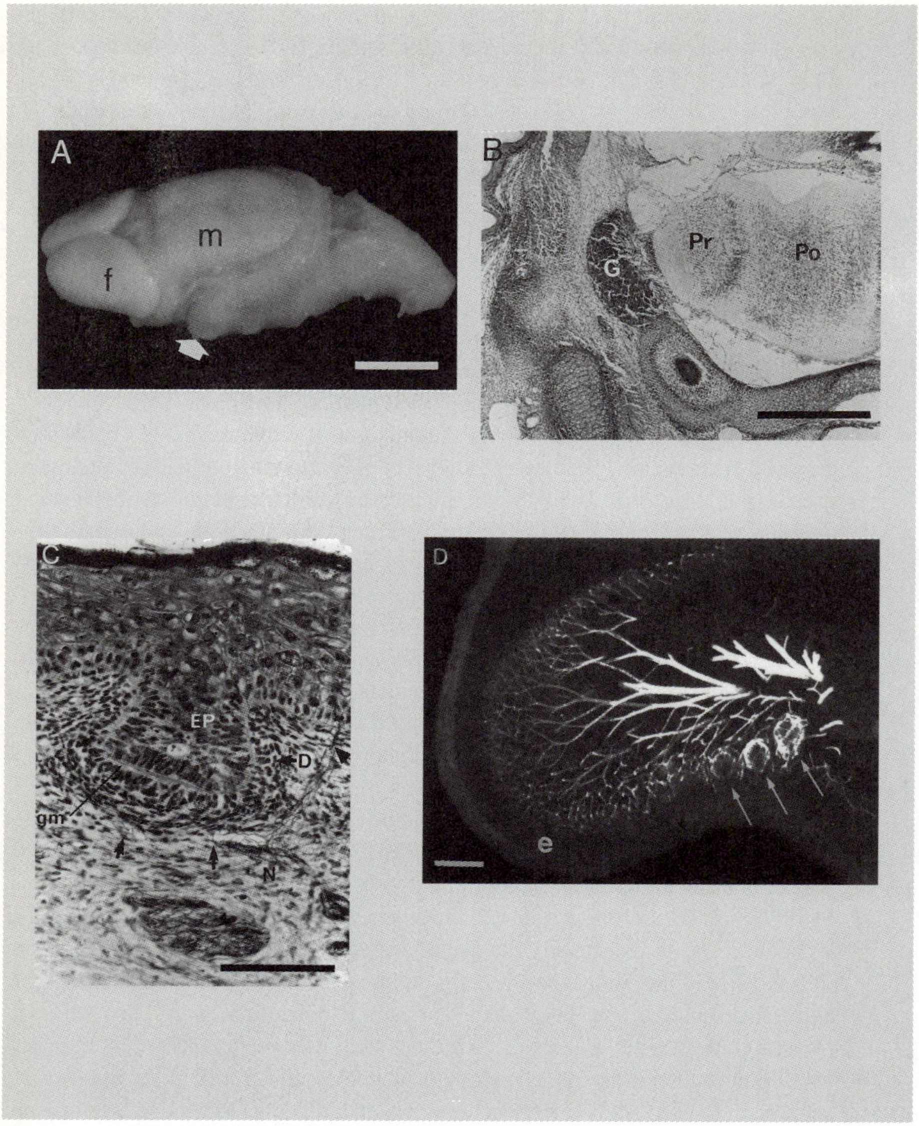

Figure 19.1

The possum (a) and tammar (b to d) at birth.

(a) Dorso-lateral view of the possum brain showing the immature state of the forebrain (f), the midbrain (m) and the site of entry (arrow) of the trigeminal root at the pontine flexure. Scale bar=1 mm.

(b) Coronal section through the tammar head showing the trigeminal ganglion (G) adjacent to the pons (Po). The section shows the site of entry of the trigeminal root into the principal trigeminal nucleus (Pr). Scale bar=0.5 mm.

(c) Silver-stained section of a developing vibrissal follicle showing an epidermal peg (EP) extending into the dermis surrounded by dermal condensations (D). Nerve bundles (N) approach the base of the dermal sheath (arrows) and also branch to innervate the basal layer of the epidermis (arrowhead). Basal lamina or glassy membrane (gm). Scale bar=100 μm.

(d) Confocal image (4 scans at 25 μm intervals) of a 100 μm section through the tammar maxillary process, following DiI application to the trigeminal ganglion. Labelled nerve fibres can be seen branching to supply the epidermis (e) and several developing vibrissal follicles (arrows). Scale bar=200 μm.

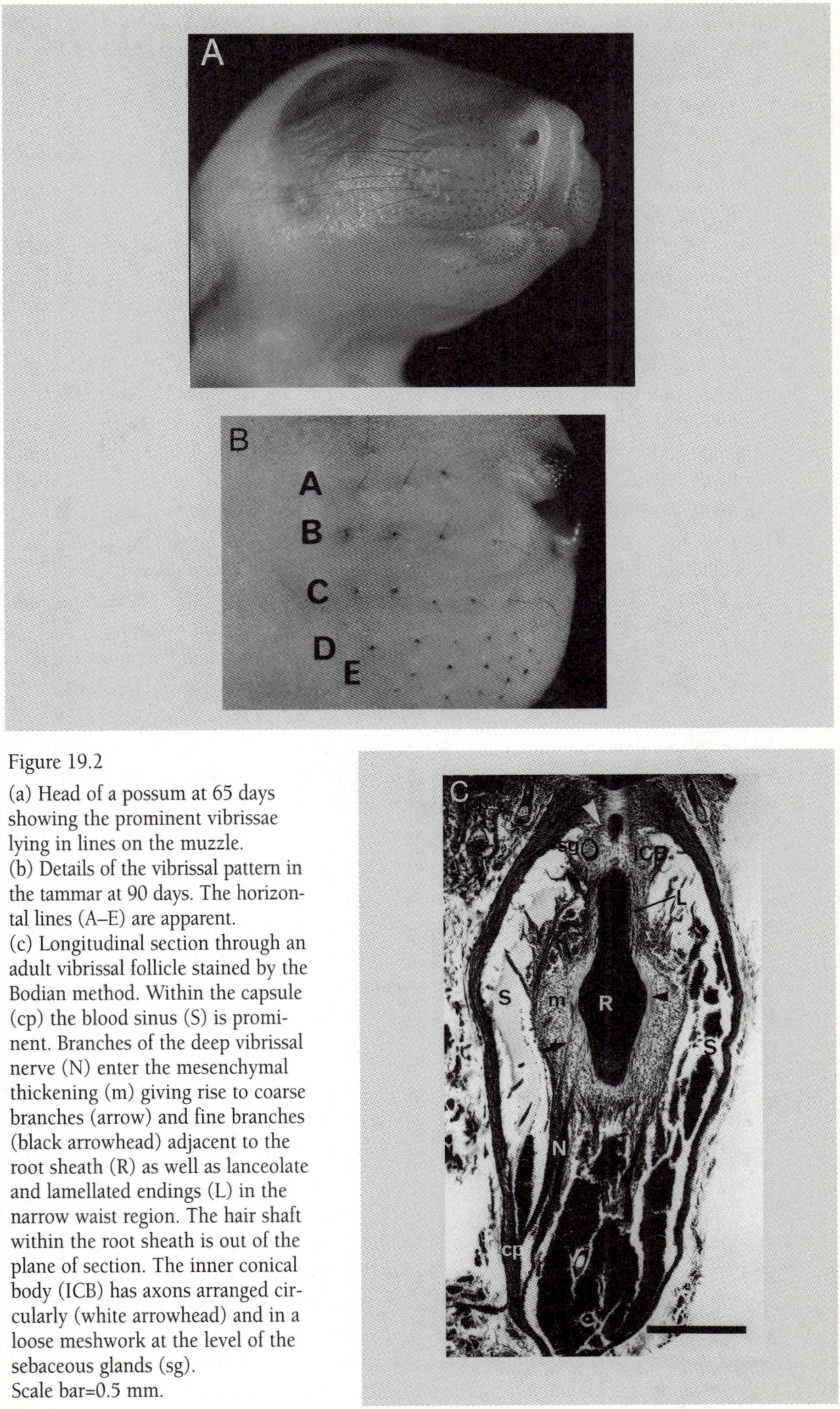

Figure 19.2

(a) Head of a possum at 65 days showing the prominent vibrissae lying in lines on the muzzle.
(b) Details of the vibrissal pattern in the tammar at 90 days. The horizontal lines (A–E) are apparent.
(c) Longitudinal section through an adult vibrissal follicle stained by the Bodian method. Within the capsule (cp) the blood sinus (S) is prominent. Branches of the deep vibrissal nerve (N) enter the mesenchymal thickening (m) giving rise to coarse branches (arrow) and fine branches (black arrowhead) adjacent to the root sheath (R) as well as lanceolate and lamellated endings (L) in the narrow waist region. The hair shaft within the root sheath is out of the plane of section. The inner conical body (ICB) has axons arranged circularly (white arrowhead) and in a loose meshwork at the level of the sebaceous glands (sg).
Scale bar=0.5 mm.

that sensory inputs are already innervating the skin, but also that reflex connections to jaw and neck muscles are already functional. However, this has still to be investigated fully.

VENTROBASAL THALAMUS

The ventrobasal thalamus (VB) is part of the thalamus receiving inputs from the body surface. It is commonly divided into the ventro-posterolateral nucleus (VPL), receiving spinal inputs, and the ventro-posteromedial nucleus (VPM) receiving inputs from the face.

Although studies on the development of the marsupial ventrobasal thalamus are sparse, several workers have described the organisation, responses and projections in adults. The North American opossum *Didelphis virginiana* has been reported to have poorly differentiated thalamic nuclei (Oswaldo-Cruz & Rocha-Miranda 1967). However, differences in marsupial subgroups clearly exist, with an intermediate degree of differentiation in the native cat (Haight & Neylon 1981) and relatively good differentiation in the possum (Haight & Neylon 1978b). In both the North and South American opossum (*Didelphis virginiana* and *D. aurita*) and tammar, electrophysiological recordings have shown that the overall somatotopy of the ventrobasal nucleus is similar to that in eutherians (Pubols & Pubols 1966; Pubols 1968; Sousa *et al.* 1971; Bombardieri *et al.* 1975; Faulks & Mark 1982). Results from lesion experiments and horseradish peroxidase (HRP) labelling also support the idea of a generally comparable organisation in marsupials and eutherians (Hazlett *et al.* 1972; Haight & Neylon 1978a; Gray *et al.* 1981). Projections from the dorsal column nuclei in the possum (Rockel *et al.* 1972) are also similar to those in eutherian mammals.

For Australian marsupials, cytoarchitectural studies show VPM and VPL are well differentiated (Rockel *et al.* 1972; Mayner 1989a). VPL forms a crescent-shaped nucleus, lying medial and dorsal to the external medullary lamina. VPM is clearly visible in Nissl-stained sections as a region of more densely packed cells lying medial to VPL (Rockel *et al.* 1972; Leamey *et al.* 1996). In possum VPM, the cells have been described as lying in discrete clusters in 5 curved bands (Weller 1980) analogous to the vibrissal-related barrels described in rodents (Van der Loos 1976).

In one of the few studies on the development of the somatosensory thalamus Sanderson and Weller (1990a) showed neurogenesis begins around birth in the possum and is complete by 12 days. The nuclei can first be identified about 33 days and are well defined by 41 days. Results in the tammar suggest a similar timecourse for differentiation (Leamey *et al* 1996). Afferent fibres from the main sensory trigeminal nucleus reach the thalamus between 10 days and 15 days (Leamey *et al.* 1993; 1996), and are functionally active from their initial arrival (Leamey and Ho 1996).

SOMATOSENSORY CORTEX

For all marsupials studied, cytoarchitechtonic features of the somatosensory cortex are similar to eutherians (Lende 1963a; Adey & Kerr 1954; Mayner 1989b). Six horizontal layers are recognised with layer IV being particularly thick and prominent with closely packed cells. Early descriptions of the functional organisation of the sensory and motor cortices in the North and South American species of opossum and in the tammar reported a complete overlap of sensory and motor areas (Lende 1963a; 1963b; 1963c; Magalhaes-Castro & Saraiva 1971). While tracing studies have confirmed this overlap for the North American opossum (Donoghue & Ebner 1981), sensory and motor areas are partially divergent in most Australian marsupials examined so far. Thus

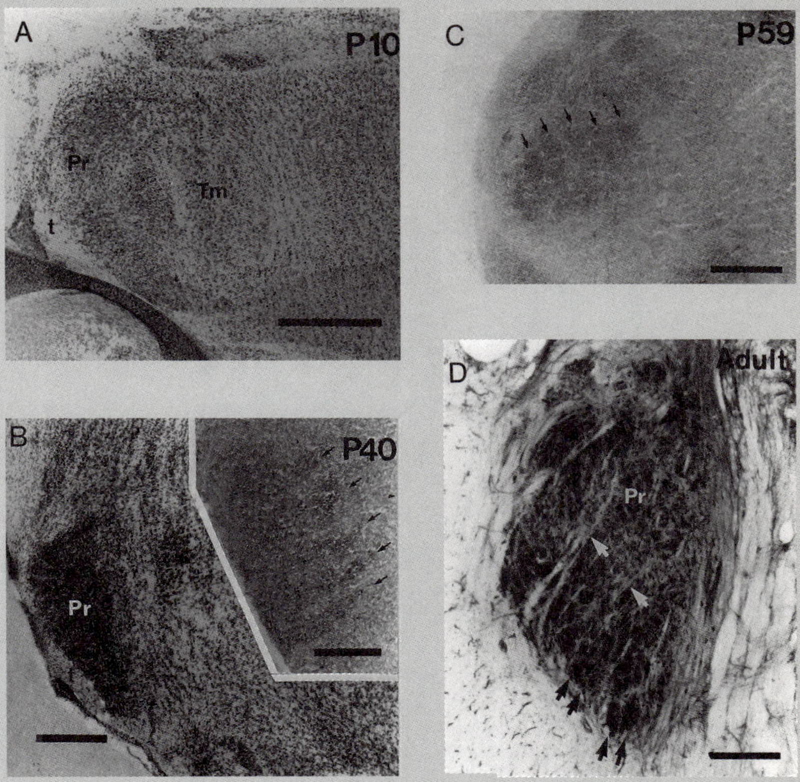

Figure 19.3

Transverse sections through subnuclei of the developing sensory trigeminal nucleus of the tammar.
(a) Nissl-stained section through nucleus principalis (Pr) at 10 days. The nucleus consists of densely packed cells between the trigeminal tract (t) and the trigeminal motor nucleus (Tm).
Scale bar=0.5mm.
(b) The principal trigeminal nucleus (Pr) at 40 days. The almond shape, characteristic of the adult d) is now apparent medial to the tract (t).
Scale bar=0.5 mm.
Inset: A higher power view of principalis reacted for cytochrome oxidase; 5 oblique lines of increased reactivity (arrows) can be seen.
Scale bar=250 μm.
(c) Succinic dehydrogenase (SDH)-stained section through subnucleus interpolaris at 59 days. Five patches (arrows) extend obliquely across the nucleus.
Scale bar=0.5 mm.
(d) Transverse section stained with SDH through the principal trigeminal nucleus (Pr) in the adult. Dense patches of reactivity are seen aligned obliquely across the nucleus (black arrows). Bundles of fibres (white arrows) also traverse the nucleus obliquely.
Scale bar=0.5 mm.

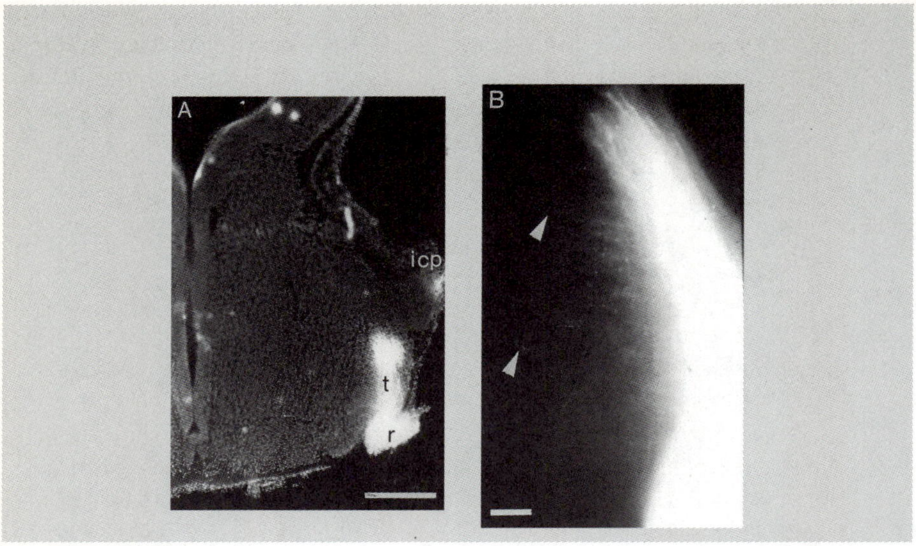

Figure 19.4 (a) Coronal section through the pons in a tammar at birth. The trigeminal ganglion was labelled with DiI; bisbenzimide was used as a counterstain to show the cytoarchitecture. The trigeminal root (r) and tract (t) are densely labelled, (icp) inferior cerebellar peduncle. Scale bar=500 μm.
(b) Higher power of the section shown in a) (without exposure for the counterstain) to show fine collaterals (arrowheads) extending medially from the tract into the principal trigeminal nucleus. Scale bar=50 μm.

more precise mapping techniques have shown less than complete overlap for the native cat (*Dasyurus viverrinus*), a polyprotodont, and the Tasmanian pademelon (*Thylogale billardieri*), a diprotodont (Weller *et al.* 1977). Partial overlap was also demonstrated in the brushtail possum by HRP labelling studies (Haight & Neylon 1978a, 1979). The degree of overlap described is similar to that in rats (Hall & Lindholm 1974) with coextensive motor and sensory regions for the hindlimb, intermediate overlap for the forelimb and very little convergence for the head.

Studies in the possum and tammar wallaby show a particularly large area devoted to inputs from the face, compared with other body regions (Lende 1963a, Adey & Kerr 1954). This facial representation was studied in detail by Weller (1972; 1993) in the brushtail possum. 'Barrels' consisting of clumps of cells with increased density (Figure 19.5c) are seen in layer IV in a number of phalangeridae (gliders and possums) and are analogous to the whisker-related barrels in rodents. Indeed in both rats (Dawson & Killackey 1987) and possums (Weller 1993) barrels are also seen in cortical areas associated with other somatic inputs (for example, fore paws and hind paws) and probably correlate with regions of high peripheral innervation density.

Details of the 'barrel' structure varies in different species. In possums, barrels consist of a uniformly dense cluster of cells (Figures 19.6a and b), similar to the rabbit and porcupine (Woolsey *et al.* 1975; Rice *et al.* 1985) while in rodents they are generally described as hollow (for example, mouse) or partially filled (for example, rat, Woolsey & Van der Loos 1970). The significance of these differences is unknown; it does not vary with brain size or whisking behaviour. Barrel centres are sites of synaptic interaction between thalamic terminals and dendrites of cortical cells, as reflected by the high levels of mitochondrial enzymes (for example, succinic dehydrogenase, SDH; Figure 19.6c) and electron microscopy. Thus the degree of 'hollowness' may simply reflect the relative density of synaptic contacts and the extent to which cells are pushed laterally into the barrel wall. In the adult tammar, barrels cannot be distinguished with routine Nissl stains, although they are apparent with mitochrondrial enzyme histochemistry (Waite *et al.* 1991). This suggests that the pattern of thalamic terminations and synapses in this species is not associated with changes in density of neuronal somata. In the possum and rodent, barrels are surrounded by a cell sparse septa, rich in myelinated fibres (Figure 19.6d) and radial dendrites.

Development of the somatosensory cortex has been studied in several marsupials and follows the same basic 'inside-out' pattern described in eutherians (Reynolds *et al.* 1985; Sanderson and Weller 1990b). In the possum and tammar practically all cortical development occurs postnatally with neurogenesis extending for about two months.

The development of a vibrissal-related pattern in the somatosensory cortex has been investigated in the tammar using mitochrondrial enzyme histochemistry (SDH; Waite *et al.* 1991). Patches associated with the vibrissae are first seen at 85 days (Figure 19.5a) and a well developed barrel field can be recognised by 90 days (Figure 19.5b). As in eutherian animals (Rice *et al.* 1985), the pattern is clearer in young animals than in adults, probably because of the development of other intracortical and cortico-cortical connections. However, no Nissl-stained pattern, corresponding to the SDH pattern, could be recognised at any age in the tammar. By comparison, Nissl staining shows distinct groups of neurons (barrels) by 65 days in possums.

The timing of patch formation may provide clues about the mechanisms controlling their development. The first hint of patches follows closely after the first appearance of layer IV (70 to 80 days), the main target cells of thalamic afferents. The onset of functional activity evoked in the cortex in response to stimulation of the whiskers also occurs at about this time (Waite *et al.* 1991).

The time of arrival of thalamic terminations from ventrobasal thalamus is currently under investigation, but preliminary results indicate that fibres are in the developing cortex as early as 15 days (Leamey *et al.* 1993), some 70 days before patches develop. This suggests that onset of patch formation correlates more closely with events in the target tissue than with the arrival of thalamic afferents and this has led to the hypothesis that local maturation of target cells may regulate onset of aggregations at each level (Waite *et al.* 1994).

LESIONING STUDIES

The importance of the periphery for the development of cortical and subcortical barrels has been well studied in rodents (Van der Loos & Woolsey 1973; Belford & Killackey 1979). Lesions to the whiskers or their afferent nerve supply in newborn rodents prevents the development of barrels at all levels. There is a critical period for this effect; lesions after day 5, when barrels have developed, cause no disruption (Weller & Johnson 1975). The critical period differs for each level and correlates with the time of patch formation for that level (that is until 1 to 2 days postnatal for brainstem, 3 to 4 days postnatal for thalamus, 5 days postnatal for cortex).

The effects of whisker lesioning have been studied in one marsupial, the possum (Weller 1979). As in rodents, lesioning during a critical postnatal period (birth to 49 days) prevented normal cortical barrel formation; after this time, no disruption to the

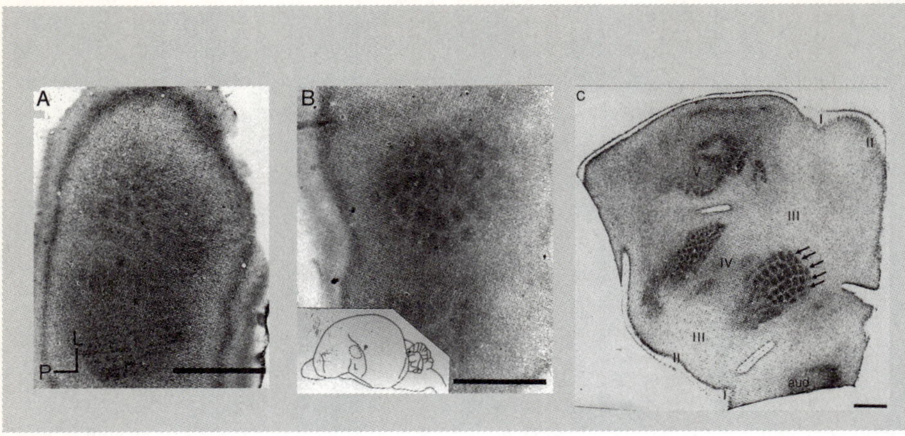

Figure 19.5

(a) Tangential section stained with SDH through the whisker area of the somatosensory cortex in a tammar at 85 days. Aggregations are beginning to form posterolaterally in the whisker area. P, posterior; L, lateral. Scale bar=1 mm.
(b) Tangential section through the somatosensory cortex of a 91-day tammar, stained with SDH. Aggregations are seen in a pattern closely corresponding to the arrangement of the facial whiskers. Scale bar=1 mm. Inset: Lateral view of the brain at 91 days; the dotted line shows the region within which evoked potentials were recorded following stimulation of the whiskers and where SDH aggregations were present.
(c) Nissl-stained tangential section through the flattened cortex of an adult possum. All cortical layers except layer VI are present (I to V). The darkest parts, belonging to granular layer IV, are segmented into cell-dense barrels by cell-sparse septa. In one region the barrels are larger and arranged in rows (arrows). The dark region in the lower left field (aud) corresponds to auditory cortex. Nissl-stain (thionin), 150 μm thick. Scale bar=2 mm.

barrels occurred. Lesions of 1 or 2 rows at 49 days showed loss of corresponding rows of barrels and expansion of adjacent rows (Figure 19.7). Lesions at 53 days caused no disruption. Disruption in subcortical centres was only analysed in a single possum, in which all the mystacial whiskers were lesioned at birth. In that animal the thalamic pattern was altered in the same way as the cortical pattern. Thalamic and cortical barrels corresponding to the removed whisker follicles were absent and the region devoted to the mandibular representation was increased (Figures 19.7c–j). Later whisker lesions (at 7 days to 43 days) resulted in altered patterns that were difficult to interpret without functional mapping studies.

These lesioning results in the possum indicate that, as in rodents, the periphery provides an essential signal before and during the time when barrels are developing but, once formed, barrel structures do not regress. Taken together with the results above on the control of timing, it would appear that both the input fibres and the target tissue provide important signals for formation of aggregations at each level.

DISCUSSION

While discussion of each level of the pathway has been included in the appropriate sections above, consideration of the pathway as a whole is worthwhile. For all essential features, the organisation of the facial skin and the sensory pathway through the

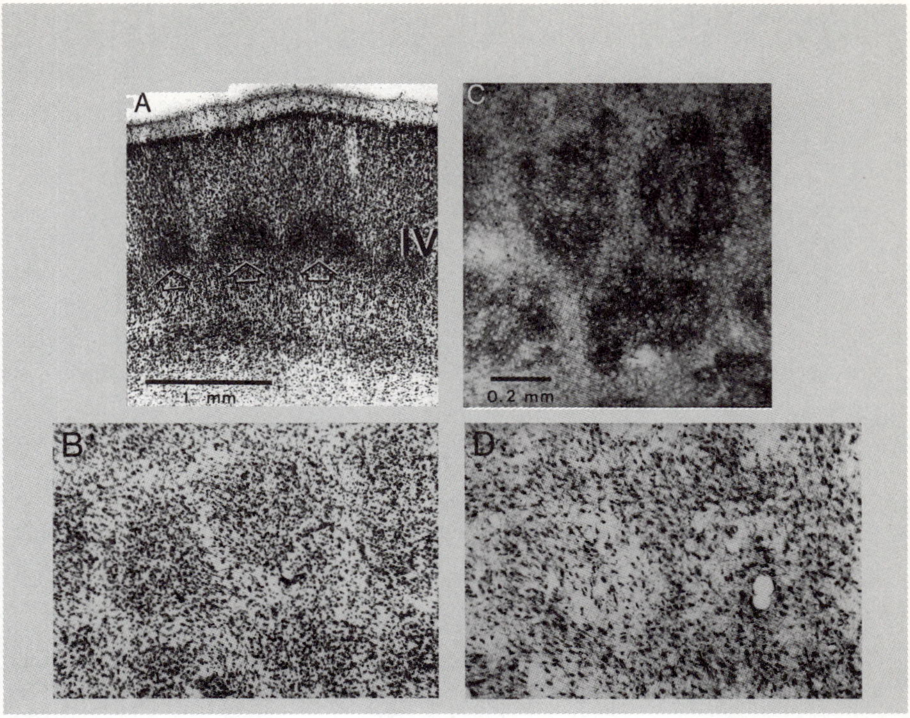

Figure 19.6

Structure of cortical barrels in brushtailed possums.
(a) In radial sections through somatosensory cortex, layer IV is non-uniformly cell-dense — here three barrels are obvious (arrows); Nissl stain (thionin), 150 μm thick.
(b) The non-uniform cell density of layer IV is most striking in sections cut parallel to the surface of the flattened cortex. Cell-dense barrels are separated from each other by cell-sparse septa; Nissl stain (thionin), 150 μm thick.
(c) The cell-dense barrels show higher levels of activity of mitochondrial enzymes than the surrounding septa; succinate dehydrogenase histochemistry, 50 μm thick.
(d) Septa have many more radially oriented myelinated axons than barrels; modified Mahon's myelin stain, 50 μm thick.

trigeminal ganglion to the brainstem is strikingly similar for marsupials and eutherians. Organisation of spinal inputs to the gracile and cuneate nuclei also correlate closely with those in eutherians (Hamilton & Johnson 1973). For the forebrain, the extent of differentiation of thalamic nuclei and sensory and motor cortices differ across marsupial groups, showing least separation in American polyprotodonts and being most clearly differentiated in the diprotodonts. However, these differences are no more than exists across eutherian species, and the separation of somatosensory and motor cortex in the tammar wallaby is similar to that described in the rat (Hall & Lindholm 1974).

The striking feature of the marsupial somatosensory pathway relates to its prolonged postnatal development. In the mouse approximately 12 days separates the first appearance of vibrissae as dermal condensations on the muzzle at E14 (Yamakado & Yohro 1979) and the development of cortical barrels at 3 to 5 days (Rice & Van der Loos 1977). Two-thirds of the pathway development occurs *in utero*. For marsupials, limited information on neurogenesis and somatosensory pathway development is available only for the possum and tammar (Sanderson & Weller, 1990a,b; Leamey *et al.* 1993; Waite *et al.* 1994). In the tammar, the start of the vibrissal follicle development and most neurogenesis in the trigeminal ganglion and nucleus occurs prenatally (gestation 28 days). Reflexes involving trigeminal inputs and jaw musculature are functional at birth (Waite, personal observation). However the follicles continue to mature until 120 days, and the trigeminal nucleus until around 200 days. For the thalamus and cortex all neurogenesis is postnatal, occurring over the first 10 days in the thalamus and for at least 50 days in the cortex (L R Marotte, personal communication). In the possum, cortical barrel development is a little earlier (by 65 days) than in the wallaby, but neurogenesis has a similar timecourse (brainstem <5 days; thalamus <5 to 11 days; somatosensory cortex <5 to 60 days). Thus both species show an early development of the brainstem, presumably necessary for suckling, and a delayed maturation of forebrain neurons and pathways.

This protracted development is being used to advantage to address questions about pattern formation in the trigeminal pathway. What controls the time at which the vibrissial-related pattern is seen at each level? Is it set by a single controlling signal or is it independent at each level? If independent, does it relate to the time of arrival of input fibres or is it controlled by maturation of target tissue? The rapid tempo of development in rodents make these various possibilities difficult to unravel.

Results for the tammar trigeminal nucleus and cortex suggest very different timecourses for these 2 levels, with patch formation starting in the brainstem at 40 days and the cortex 85 days. This suggests that rather than a single controlling signal, development may be occurring independently at each site. Moreover, both levels show a long delay between arrival of afferents (brainstem prenatal, Waite *et al.* 1994; somatosensory cortex 10 to 15 days, Leamey *et al.* 1993) and pattern development (brainstem 40 days, Waite *et al.* 1994; cortex 85 days, Waite *et al.* 1991). While some afferents may make important early connections, for instance, for essential brainstem reflexes, the development of patterns probably correlates with a burst of synaptogenesis and terminal arborisation as discussed. This long delay between afferent arrival and pattern formation indicates that it is unlikely to be the afferent fibres *per se* which set the time for pattern development. The synchrony between cortical pattern formation at 85 days and the appearance of layer IV (70 to 80 days), containing the main target neurons of the thalamocortical fibres, suggests that the control 'clock' may reside with the target cells.

FUTURE DIRECTIONS

To date, studies on marsupial somatosensory systems have been dominated by anatomy. Functional studies, when undertaken, have been mainly used to address questions of somatotopy. Few studies have looked at the functioning of the sensory pathways in development and this is an obvious direction for future work. When do the brainstem, thalamus and cortex become active and how do early responses compare with those in mature animals? Some progress has been made in the visual system using current source density (CSD) analysis in the superior colliculus to study the onset of activity (see Mark chapter 18 in this volume). Similar CSD techniques are currently being used in the somatosensory and visual cortices (R P Mark & P M E Waite, in preparation). Slice preparations are also being developed and are likely to prove useful for looking at the initial development of functional connections. These slices may help to

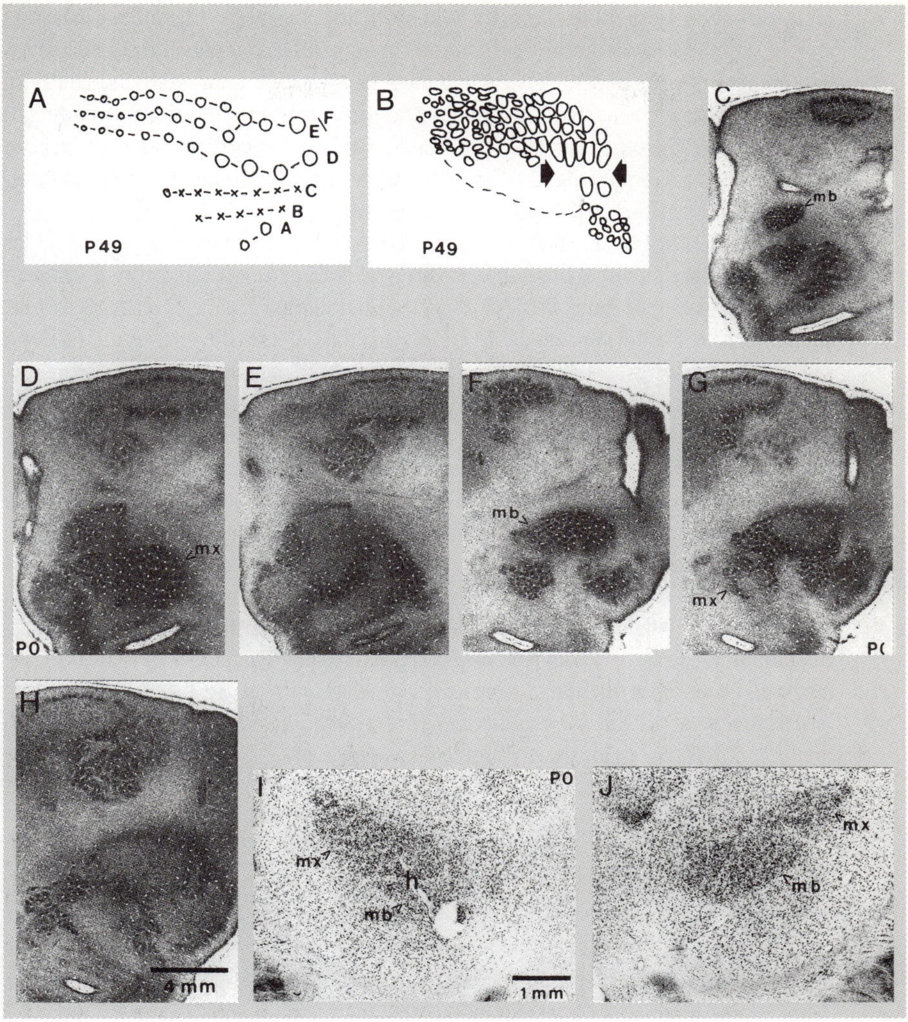

Figure 19.7

Results of removing whiskers follicles during pouch development of brushtailed possums: (a) and (b) 49 days; (c) to (j) newborn). (a) Two rows of large follicles were removed from one side of the snout of a 49-day possum (rows b and c — only follicle C8 was spared) that was allowed to survive to an age of 15 months. Each circle represents a single whisker follicle found in the dissected skin. The Xs mark the positions of follicles removed. (b) Drawing of part of the corresponding barrel pattern. A defect in the pattern (arrows) is present where Row B and Row C barrels should have been. (c) to (e) Successive sections of the control cortex. The pattern was normal, notably the parallel rows of large barrels (mx) and the group of smaller barrels (mb) that represent the whiskers on the chin; Nissl stain (thionin), 150 µm thick. (f) to h) Successive sections of the experimental cortex. All large follicles were removed from one side of the snout of a newborn possum (but 2 follicles out of about 50 were spared). The usual parallel rows of large barrels are missing (mx) and the barrels representing the chin whiskers (mb) are obviously enlarged; Nissl stain (thionin), 150 µm. thick (i) A section through the somatosensory thalamus (control side) of the same possum. The cell-dense part (h) represents the possum's head. Some of the cells are arranged in clusters and rows (mx). They receive sensory information from the large mystacial whisker follicles. The others (mb) receive information from the chin; Nissl stain (thionin), 100 µm thick. (j) A section through the experimental side of the somatosensory thalamus. The rows and groups of cells are missing (mx) and the region receiving information from the chin is enlarged. These thalamic differences parallel those seen in the cortex (c) to (i); Nissl stain (thionin), 100 µm thick.

address the role of spontaneous activity in early synaptogenesis. In addition, labelling studies have shown descending paths from the somatosensory cortex to the thalamus and brainstem in the opossum (Martin & West 1967) and possum (Martin *et al.* 1971). These paths are presumably related to descending controls on the activity ascending in the pathway, as in other systems, but their function and development has not been studied. It is conceivable that the development of these descending paths could influence pattern formation and maturation of the system.

The extrauterine marsupial development allows early manipulation of the system far more easily than in eutherians. This has been used to advantage in the visual system (James *et al.* 1993; Marotte & Mark 1988) but limited use has been made of this for somatosensory paths (Johnson *et al.* 1972; Weller 1979). Future studies are likely to see more extensive use of lesioning to address questions of neural plasticity and regeneration after early injury. The factors which allow sprouting and regeneration in early development, or prevent its occurrence at later stages, are important for a better understanding of recovery of function after injury; the unique features of marsupial development provide a useful model.

Acknowledgments

Phil Waite would like to thank Professor Richard Mark and Dr Lauren Marotte for developing the tammar colony and for sharing a longstanding and successful collaboration on studies on tammar somatosensory development. Ms Cathy Leamey has also made a valuable contribution to this research. Lee Weller acknowledges the help of Noel Swan for the possum lesioning studies.

References

Adey W R and Kerr D I B (1954). The cerebral representation of deep somatic sensibility in the marsupial phalanger and the rabbit: an evoked potential and histological study. *Journal of Comparative Neurology*, 100, 597–626.

Bates C A and Killackey H P (1985). The organization of the neonatal rat's brainstem trigeminal complex and its role in the formation of central trigeminal patterns. *Journal of Comparative Neurology*, 240, 265–287.

Belford G R and Killackey H P (1978). Anatomical correlates of the forelimb in the ventrobasal complex and the cuneate nucleus of the neonatal rat. *Brain Research*, 158, 450–455.

Belford G R and Killackey H P (1979). The development of vibrissae representation in subcortical trigeminal centers of the neonatal rat. *Journal of Comparative Neurology*, 188, 63–74.

Bombardieri R A, Johnson J I and Campos G B (1975). Species differences in mechanosensory projections from the mouth to the ventrobasal thalamus. *Journal of Comparative Neurology*, 163, 41–64.

Crosby E C and Yoss R E (1954). The phylogenetic continuity of neural mechanisms as illustrated by the spinal tract of V and its nucleus. *Research Publication of Association for Nervous and Mental Diseases* 33, 174–208.

Darian-Smith I (1973). The trigeminal system. In *Handbook of Sensory Physiology*, Vol. 2: *Somatosensory System*. Ed. A Iggo. Springer-Verlag, New York pp. 271–314

Davidson P and Hardy M H (1952). The development of mouse vibrissae *in vivo* and *in vitro*. *Journal of Anatomy*, 86, 342–356.

Davies A M (1988). The trigeminal system: an advantageous model for studying neuronal development. *Development, Supplement*, 103, 175–183.

Dawson D R and Killackey H P (1987). The organization and mutability of the forepaw and hindpaw representations in the somatosensory cortex of the neonatal rat. *Journal of Comparative Neurology*, 256, 246–256.

Donoghue J P and Ebner F F (1981). The organization of thalamic projections to the parietal cortex of the virginia opossum. *Journal of Comparative Neurology*, 198, 365–388.

Faulks I J and Mark R F (1982). The somatotopic organization of the thalamic ventrobasal complex in a wallaby, the tammar *(Macropus eugenii)*. *Australian Physiological and Pharmacological Society*, 13, 179P.

Florence S L, Wall J T and Kaas J H (1989). Somatotopic organization of inputs from the hand to the spinal gray and cuneate nucleus of monkeys with observations on the cuneate nucleus of humans. *Journal of Comparative Neurology*, 286, 48–70.

Gemmell R T, Peters B and Nelson J (1988). Ultrastructural identification of Merkel cells around the mouth of the newborn marsupial. *Anatomy and Embryology*, 177, 403–408.

Gobel S and Dubner R (1969). Fine structural studies of the main sensory trigeminal nucleus in the cat and rat. *Journal of Comparative Neurology*, 137, 459–494.

Gobel S and Purves M B (1972). Anatomical studies of the organization of the spinal V nucleus: the deep bundles and the spinal V tract. *Brain Research*, 48, 27–44.

Gray T S, Hazlett J C and Martin G F (1981). Organization of projections from the gracile, medial cuneate and lateral cuneate nuclei in the North Americiam opossum. *Brain, Behaviour and Evolution*, 18, 140–156.

Haight J R and Neylon L (1978a). The organization of neocortical projections from the ventroposterior thalamic complex in the marsupial brush-tailed possum, *Trichosurus vulpecula*: a horseradish peroxidase study. *Journal of Anatomy*, 126, 459–485.

Haight J R and Neylon L (1978b). An atlas of the dorsal thalamus of the marsupial brush-tailed possum, *Trichosurus vulpecula*. *Journal of Anatomy*, 126, 225–245.

Haight J R and Neylon L (1979). The organization of neocortical projections from the ventrolateral thalamic nucleus in the brush-tailed possum, *Trichosurus vulpecula*, and the problem of motor and somatic sensory convergence within the mammalian brain. *Journal of Anatomy*, 129, 673–694.

Haight J R and Neylon L (1981). A description of the dorsal thalamus of the marsupial native cat, *Dasyurus viverrinus* (Dasyuridae). *Brain, Behaviour and Evolution*, 19, 155–179.

Hall R D and Lindholm E P (1974). Organization of motor and somatosensory neocortex in the albino rat. *Brain Research*, 66, 23–38.

Hamilton T C and Johnson J I (1973).

Somatotopic organization related to nuclear morphology in the cuneate-gracile complex of opossums *Didelphis marsupialis virginiana*. *Brain Research*, 51, 125–140.

Hazlett J C, Dom R and Martin G F (1972). Spino-bulbar, spino-thalamic and medial lemniscal connections in the American opossum, *Didelphis marsupialis virginiana*. *Journal of Comparative Neurology*, 146, 95–118.

Hollis D E and Lyne A G (1974). Innervation of vibrissa follicles in the marsupial *Trichosurus vulpecula*,. *Australian Journal of Zoology*, 22, 263–276.

James A C, Mark R F and Sheng X-M (1993). Geometry of the projection of the visual field onto the superior colliculus of the wallaby (*Macropus eugenii*). II. Stability of the projection after prolonged rearing with rotational squint. *Journal of Comparative Neurology*, 330, 315–323.

Johnson J I, Hamilton T C, Hsung J -C and Ulinski P S (1972). Gracile nucleus absent in adult opossums after leg removal in infancy. *Brain Research*, 38, 421–424.

Jones T E and Munger B L (1985). Early differentiation of the afferent nervous system in glabrous snout skin of the opossum, (*Monodelphis domesticus*). *Somatosensory Research*, 3, 169–184.

Killackey H P, Jacquin M J and Rhoades R W (1990). Development of somatosensory system structures. In *Development of Sensory Systems in Mammals*. Ed JR Coleman. John Wiley, New York. pp 403–429.

Klein B G, Renehan W E, Jacquin M F and Rhoades R W (1988) Anatomical consequences of neonatal infraorbital nerve transection upon the trigeminal ganglion and vibrissa follicle nerves in the adult rat. *Journal of Comparative Neurology*, 268, 469–488.

Leamey C A and Ho S M (1996). Anatomy and receptor pharmacology of the onset of function in the somatosensory thalamus of the wallaby during development. *Proceedings of the Australian Neuroscience Society*, 7, 38.

Leamey C A, Waite P M E, Marotte L R and Mark R F (1993). The development of the trigeminal pathway from periphery to cortex in the wallaby (*Macropus eugenii*). *Proceedings of Australian Neuroscience Society*, 4, 103.

Leamey C A, Marotte L R and Waite P M E (1996). Time course of development of the wallaby trigeminal pathway. II Brainstem to thalamus and the emergence of cellulor aggregations. *Journal of Comparative Neurology*, 364, 494–514.

Lende R A (1963a). Sensory representation in the cerebral cortex of the opossum (*Didelphis virginiana*). *Journal of Comparative Neurology*, 121, 395–403.

Lende R A (1963b). Motor representation in the cerebral cortex of the opossum (*Didelphis virginiana*). *Journal of Comparative Neurology*, 121, 405–415.

Lende R A (1963c). Cerebral cortex: a sensorimotor amalgam in the Marsupialia. *Science*, 141, 730–732.

Lichtenstein S H, Carvell G E and Simons D J (1990). Responses of rat trigeminal ganglion neurons to movements of vibrissae in different directions. *Somatosensory and Motor Research*, 7, 47–65.

Loo S-K and Halata Z (1991). Innervation of hairs in the facial skin of marsupial mammals. *Journal of Anatomy*, 174, 207–219.

Ma P M (1991). The barrelettes—architectonic vibrissal representations in the brainstem trigeminal complex of the mouse. I. Normal structural organization. *Journal of Comparative Neurology*, 309, 161–199.

Ma P M (1993). The barrelettes—architectonic vibrissal representations in the brainstem trigeminal complex of the mouse. II. Normal post-natal development. *Journal of Comparative Neurology*, 327, 376–397.

Magalhães-Castro B and Saraiva P E S (1971). Sensory and motor representation in the cerebral cortex of the marsupial *Didelphis azarae azarae*. *Brain Research*, 34, 291–299.

Marotte L R and Mark R F (1988). Retinal projections to the superior colliculus and dorsal lateral geniculate nucleus in the tammar wallaby (*Macropus eugenii*): II. Topography after rotation of an eye prior to retinal innervation of the brain. *Journal of Comparative Neurology*, 271, 274–292.

Marotte L R, Rice F L and Waite P M E (1992). Morphology and innervation of facial vibrissae in the tammar wallaby. (*Macropus eugenii*). *Journal of Anatomy*, 180, 401–417.

Martin G F and West H J (1967). Efferent neocortical projections to sensory nuclei in the brainstem of the opossum (*Didelphis virginiana*). *Journal of The Neurological Sciences*, 5, 287–302.

Martin G F, Megirian D and Roebuck A (1971). Corticobulbar projections of the marsupial phalanger *Trichosurus vulpecula*. I. Projections to the pons and medulla oblongata. *Journal of Comparative Neurology*, 142, 275–296.

Mayner L (1989a). A cyto-architectonic description of the thalamus of the tammar wallaby, *Macropus eugenii*. *Brain, Behaviour and Evolution*, 33, 342–355.

Mayner L (1989b). A cyto-architectonic study of the cortex of the tammar wallaby, *Macropus eugenii*. *Brain, Behaviour Evolution*, 33, 303–316.

Mosconi T M and Rice F L (1993). Sequential differentiation of sensory innervation in the mystacial pad of the ferret. *Journal of Comparative Neurology*, 333, 309–325.

Munger B L and Rice F L (1986). Successive waves of differentiation of cutaneous afferents in rat mystacial skin. *Journal of Comparative Neurology*, 252, 404–414.

Nomura S and Mizuno N (1986). Histochemical demonstration of vibrissae-representing patchy patterns of cytochrome oxidase activity within the trigeminal sensory nuclei in the cat. *Brain Research*, 380, 167–171.

Noriega A L and Wall J T (1991). Parcellated organization in the trigeminal and dorsal column nuclei of primates. *Brain Research*, 565, 188–194.

Olszewski J (1950). On the anatomical and functional organization of the spinal trigeminal nucleus. *Journal of Comparative Neurology*, 92, 401–413.

Oswaldo-Cruz E and Rocha-Miranda C E (1967). The diencephalon of the opossum in stereotaxic coordinates. I. The epithalamus and dorsal thalamus. *Journal of Comparative Neurology*, 129, 1–38.

Oswaldo-Cruz E and Rocha-Miranda C E (1968). *The brain of the opossum (Didelphis marsupialis)*. Universidade Federal do Rio de Janeiro.

Pubols B H (1968). Retrograde degeneration study of somatic sensory thalamocortical connections in brain of Virginia opossum. *Brain Research*, 7, 232–251.

Pubols B H and Pubols L M (1966). Somatic sensory representation in the thalamic ventrobasal complex of the Virginia opossum. *Journal of Comparative Neurology*, 127, 19–34.

Pubols B H, Donovick P J and Pubols L M (1973). Opossum trigeminal afferents associated with vibrissa and rhinarial mechanoreceptors. *Brain, Behaviour Evolution*, 7, 360–381.

Pubols B H, Pubols L M, DiPette D J and Sheely J C (1976). Opossum somatic sensory cortex: a microelectrode mapping study. *Journal of Comparative Neurology*, 165, 229–246.

Reynolds M L, Cavanagh M E, Dziegielewska K M, Hinds L A, Saunders N R and Tyndale-Biscoe C H (1985). Postnatal development of the telencephalon of the tammar wallaby (*Macropus eugenii*). *Anatomy and Embryology*, 173, 81–94.

Rice F L and Van der Loos H (1977). Development of the barrels and barrel field in the somatosensory cortex of the mouse. *Journal of Comparative Neurology*, 171, 545–560.

Rice F L, Gomez C, Barstow C, Burnet A and Sands P (1985). A comparative light microscopic analysis of the development of the primary somatosensory cortex: interspecies similarities during barrel and laminar development. *Journal of Comparative Neurology*, 236, 477–495.

Rice F L, Mance A and Munger B L (1986). A comparative light microscopic analysis of the sensory innervation of the mystacial pad. I. Innervation of vibrissal follicle-sinus complexes. *Journal of Comparative Neurology*, 252, 154–174.

Rockel A J, Heath C J and Jones E G (1972). Afferent connections to the diencephalon in the marsupial phalanger and the question of sensory convergence in the "posterior group" of the thalamus. *Journal of Comparative Neurology*, 145, 105–130.

Sanderson K J and Weller W L (1990a). Neurogenesis in a marsupial: the brush-tailed possum (*Trichosurus vulpecula*). II. Sensorimotor pathways. *Brain, Behaviour and Evolution*, 35, 339–349.

Sanderson K J and Weller W L (1990b). Gradients of neurogenesis in possum neocortex. *Developmental Brain Research*, 55, 269–274.

Sousa A P B, Oswaldo-Cruz E and Gattass R (1971). Somatotopic organization and response properties of neurons of the ventrobasal complex in the opossum. *Journal of Comparative Neurology*, 142, 231–248.

Torvik A (1956). Afferent connections to the sensory trigeminal nuclei, the nucleus of the solitary tract and adjacent structures. *Journal of Comparative Neurology*, 106, 51–141.

Van der Loos H and Woolsey T A (1973). Somatosensory cortex: structural alterations following early injury to sense organs. *Science*, 179, 395–398.

Van der Loos H (1976). Barreloids in mouse somatosensory thalamus. *Neuroscience Letters* 2, 1–6.

Waite P M E, Marotte L R and Mark R F (1991). Development of whisker representation in the cortex of the tammar wallaby, *Macropus eugenii. Developmental Brain Research*, 58, 35–41.

Waite P M E, Marotte L R and Leamey C A (1994). Timecourse of development of the wallaby trigeminal pathway. I. Periphery to brainstem. *Journal of Comparative Neurology*, 350, 75–95.

Waite P M E and Tracey D T (1995) Trigeminal sensory systems. In *The Rat Nervous System*. 2nd Edn. Ed G Paxinos. Academic Press, New York. pp 705–724.

Weller W L (1971). *Patterns of Mechanoreceptor Projections to the Trigeminal Nuclei of the Opossum*. MS thesis, Michigan State University, Michigan

Weller W L (1972). Barrels in somatic sensory neocortex of the marsupial *Trichosurus vulpecula* (brush-tailed possum). *Brain Research*, 43, 11–24.

Weller W L (1979). Experimental alteration of neocortex by removal of vibrissa follicles from pouch young brush-tailed possums, *Trichosurus vulpecula. Journal of Anatomy*, 128, 659.

Weller W L (1980). 'Barreloids'in a subdivision of the ventral posterior thalamic nucleus in the brush-tailed possum, *Trichosurus vulpecula. Journal of Anatomy*, 130, 214.

Weller W L (1983). Vibrissa-related spatial patterns of succinic dehydrogenase activity in the somatosensory system of brush-tailed possums, *Trichosurus vulpecula. Journal of Anatomy*, 136, 663.

Weller W L (1993). SmI cortical barrels in an Australian marsupial, *Trichosurus vulpecula* (brush-tailed possum): structural organization, patterned distribution, and somatotopic relationships. *Journal of Comparative Neurology*, 337, 471–492.

Weller W L and Johnson, J I (1975). Barrels in cerebral cortex altered by receptor disruption in newborn, but not in five-day-old mice (Cricetidae and Muridae). *Brain Research*, 83, 504–508.

Weller W L, Haight J R, Neylon L and Johnson J I (1977). A re-assessment of the mechanoreceptor projections to cerebral neocortex in marsupial wallabys (*Thylogale*). *Journal of Anatomy*, 124, 531–532.

Wong-Riley M T T (1989). Cytochrome oxidase: an endogenous metabolic marker for neuronal activity. *Trends in Neurosciences*, 12, 94–101.

Woolsey T A and Van der Loos H (1970). The structural organization of layer IV in the somatosensory region (SI) of mouse cerebral cortex. *Brain Research*, 17, 205–242.

Woolsey T A, Welker C and Schwartz R H (1975). Comparative anatomical studies of the SmI face cortex with special reference to the occurrence of 'barrels' in layer IV. *Journal of Comparative Neurology*, 164, 79–94.

Yamakado M and Yohro T (1979). Subdivision of mouse vibrissae on an embryological basis, with descriptions of variations in the number and arrangement of sinus hairs and cortical barrels in BALB/c (nu/+; nude, nu/nu) and hairless (hr/hr) strains. *American Journal of Anatomy*, 155, 153–174.

Zucker E and Welker W I (1969). Coding of somatic sensory input by vibrissae neurons in the rat's trigeminal ganglion. *Brain Research*, 12, 138–156.

CHAPTER 20

PROTEINS IN PLASMA, CEREBROSPINAL FLUID AND BRAIN DURING POSTNATAL DEVELOPMENT IN MARSUPIALS

Katarzyna M Dziegielewska, Graham W Knott and Norman R Saunders

INTRODUCTION

The mammalian brain develops in a well-ordered fashion, the mechanism of which has been subject of many studies over several years. After neural tube closure, the anterior end develops into three primary vesicles, the most anterior of which then gives rise to the lateral ventricles. More caudally, but continuous with the forebrain, are the mid-brain and hind-brain, each of which has its own fluid-filled spaces, which later in development become the IIIrd and IVth ventricles. The fluid filling the ventricles is called the cerebrospinal fluid (CSF) (see Davson *et al.* 1987 and Saunders 1992 for reviews). The brain develops in its own environment, separated from blood by interfaces in which cerebral endothelial cells and the choroid plexus epithelial cells are joined by tight junctions. These interfaces are an important component of mechanisms referred to as the 'blood-brain barrier' (Saunders 1992).

Many studies have shown that the blood-brain and blood-CSF barriers to proteins are well developed in the adult, and also in eutherian fetuses, at least as far back in gestation as it has been possible to test experimentally in these species (see Saunders 1992 for references).

In the adult there is a fairly free access of macromolecules from the CSF to the brain (Davson *et al.* 1987) but the concentration of protein in the adult CSF is characteristically low (20 to 40 mg/100ml). In fetal CSF, on the other hand, the protein concentration is always higher and can be as high as 50 times that in the adult (Dziegielewska & Saunders 1988).

The protein concentrations within the CSF have been measured in many different species and in all those studied so far a peak of protein concentration occurs early in brain development (see Table 20.1). In the eutherian species studied the peak of CSF protein concentration appears in the fetal stages of development. This is compared with the marsupial species (tammar wallaby, *Macropus eugenii* and South American grey short-tailed opossum *Monodelphis domestica*), born at an extremely early stage of development, in which the peak occurs postnatally (Table 20.1).

The origin and possible functions of the high protein concentration in fetal CSF have been a subject of many studies (for review see Dziegielewska & Saunders 1988; Saunders 1992). The individual proteins found in CSF are immunologically identical to those found in the plasma (for example, Dziegielewska *et al.* 1980; Dziegielewska & Saunders 1988). Their origin in the CSF has been proposed to come from two main sources:
(a) transfer (passive and/or active) from plasma; and
(b) synthesis *in situ* either by the choroid plexus cells or the brain itself.
This review first discusses the nature of proteins in marsupial CSF in development and in the adult; this is followed by a review of recent results investigating the mechanism of transfer of protein from the blood into the CSF. Finally, in a short paragraph, the distribution of some plasma proteins within the brain is discussed.

PROTEINS IN THE CSF AND PLASMA IN *MONODELPHIS DOMESTICA* AND *MACROPUS EUGENII*

The only two marsupial species in which proteins in CSF have been studied during postnatal development are the tammar and *Monodelphis*. Both animals are born at an extremely immature stage of brain development, the tammar after 27 days gestation (see Dziegielewska *et al.* 1986) and *Monodelphis* after 14 days (Saunders *et al.* 1989), as discussed by Saunders, Chapter 22, in this volume. In both species total protein and albumin concentrations in plasma and CSF have been estimated (Dziegielewska *et al.* 1986, 1989). Results for total protein concentration, albumin concentration and their ratios between CSF and plasma in the developing *Monodelphis* are illustrated in Figures 20.1 and 20.2.

The results from the tammar for total protein, albumin and fetuin (the first fetal protein to be described, Pedersen 1944) concentrations in plasma and CSF, and the ratios for all three protein measurements are illustrated in Figures 20.3, 20.4 and 20.5.

In both the tammar and *Monodelphis* plasma, the total protein concentration increased steadily from low levels at birth (about 0.4g/100ml in *Monodelphis* and about 1.0g/100ml in the tammar) to elevated values in the adult (6 to 8g/100ml; see Figures 20.1 and 20.3). Albumin concentration in plasma of the tammar also increased progressively, from about 300 mg/100ml in the newborn to about 4g/100ml in the adult (Dziegielewska *et al.* 1986; Dziegielewska *et al.* 1989). The pattern of change of plasma albumin concentration with age in *Monodelphis* was strikingly different from that in the tammar (cf Figures 20.2 and 20.4). In *Monodelphis* there was a marked rise in plasma albumin concentration between 5 days and 6 to 7 days, followed by a sharp decline by 11 days, after which there was a fairly steady increase, as in the tammar. This transient rise in plasma albumin in *Monodelphis* appears to be due to transfer of maternal albumin in the milk across the wall of the gut (Knott 1995; Knott *et al.* 1997).

In the CSF, on the other hand, the total CSF protein concentration increased after birth to reach a peak at about 15 days in the tammar and 6 to 8 days in the *Monodelphis*. After the peak concentration (over 400mg/100ml in both species), the total protein concentration declined with increasing age to reach about 65 mg/100ml in the later stage 210 days pouch young tammar (Dziegielewska, *et al.* 1986) and 23mg/100ml in

able 20.1. Total protein concentrations in CSF of 7 species

SF PROTEIN ONCENTRATION ng/100ML)	SHEEP	PIG	RAT	RABBIT	CHICKEN	TAMMAR	MONODELPHIS
eak during evelopment	1143 ± 83 (31 days)	961 ± 95 (31 days)	317 ± 15 (22 days)	529 ± 18 (20 days	523 ± 16 (11 days)	405 ± 21 (42 days)	499 ± 25 (23 days)
dult CSF	26 ± 2	31 ± 3	24 ± 8	29 ± 3	141 ± 15	66 ± 3	23 ± 3.8
ge (in days) ost-conception birth	150	115	22	32	21	28	14

otes:
.) The comparison is between the levels of protein at the peak of CSF concentration and the levels in the dult.
) Timing of the peak (in days) after conception is shown in parentheses.
) Mean ± sem.
ources:
alues for sheep, pig, rat, rabbit, chicken and tammar are from review, Dziegielewska & Saunders (1988).
alues for *Monodelphis* are from Dziegielewska *et al.* (1989).

adult *Monodelphis* (Dziegielewska *et al.* 1989). CSF/plasma ratios for both total protein and albumin in both species declined with age.

Another plasma protein, fetuin, has also been measured in the tammar plasma (Jones *et al.* 1991). Its levels, both in plasma and in the CSF, remained stable up to 45 days, but then the CSF concentration dropped dramatically from nearly 20mg/100ml to less than 1mg/100ml by 75 days. The plasma concentration of fetuin increased with age after 75 days, to reach about 200 mg/100ml in the late pouch young (see Figure 20.5).

Apart from albumin and fetuin, other proteins that were identified in the tammar pouch young by their cross-reactivity with monospecific antibodies to human proteins were: α-fetoprotein (AFP), transferrin, α_2 macroglobulin, α_1-anti-trypsin, and fibrinogen. These proteins were identified in both plasma and CSF. Low levels of immunoglobulins were identifiable from the newborn stage in both *Monodelphis* and the tammar.

TRANSFER OF PROTEINS FROM BLOOD INTO THE CSF

Developing brain is surrounded by a highly specific and specialised fluid, the cerebrospinal fluid (CSF). The main sites of CSF production are the epithelial cells forming the choroid plexuses, found in all four ventricles (Davson *et al.* 1987). These epithelial cells are joined by tight junctions which form the basis of the blood-CSF barrier (for review see Saunders 1992). Additionally, endothelial cells of cerebral blood vessels are also joined by tight junctions, forming the blood-brain barrier proper.

Many studies have shown that the blood-brain and blood-CSF barriers to proteins

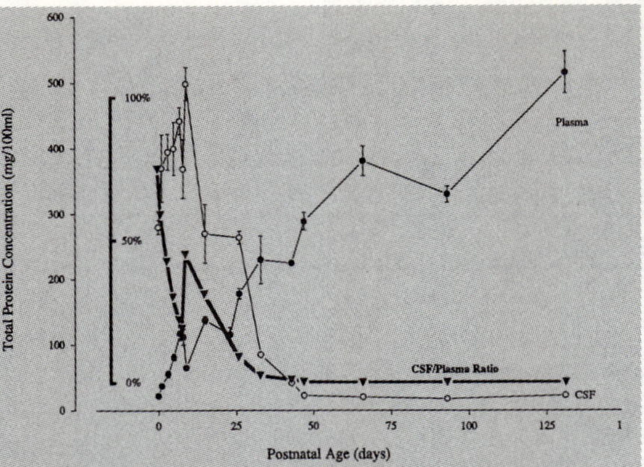

20.1

Figure 20.1
Total protein concentration (mg/100ml) in plasma (• right ordinate) and CSF (o left ordinate) of postnatal *Monodelphis domestica* at vartious postnatal ages (abscissa, days).
Note: Mean ± sem. CSF/plasma ratio (%, thick line) — concentration in CSF over concentration in plasma x100.
Source: Knott (1995).

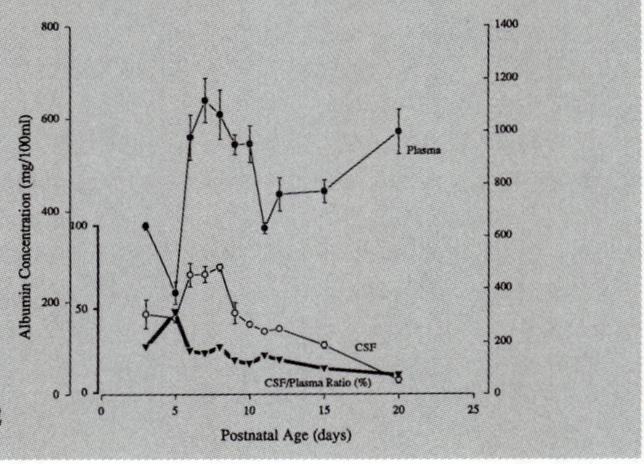

20.2

Figure 20.2
Albumin (endogenous) concentration (mg/100ml) in plasma (• right ordinate) and CSF (o left ordinate) of *Monodelphis domestica* at various postnatal ages (abscissa, days).
Note: Mean values ± sem. Also shown is the CSF/plasma ratio (%, thick line).
Source: Knott (1995).

are very effective in the adult (Davson *et al.* 1987). Experimental evidence from the last 10 to 15 years has demonstrated that the fetal brain is also protected from the rest of the circulation by highly developed and regulated blood-brain and blood-CSF barrier mechanisms (see Saunders 1992 for review).

In addition to blood-brain and blood-CSF barriers, the developing brain seems to have an extra barrier between the CSF and the brain itself, the CSF-brain barrier (Fossan *et al.* 1985). This barrier, formed by strap junctions between the cells of the ventricular zone, is developmentally regulated and is present only during earlier stages of brain development, when the protein concentrations in the CSF are high (Fossan *et al.* 1985; Saunders 1992).

Individual proteins within the CSF are immunologically identical to those found in plasma (Dziegielewska & Saunders 1988). Their transfer from plasma to CSF has been shown to be by passive diffusion in the adult (Davson *et al.* 1987) and by a mixture of specific transfer and underlying passive transfer in younger brains (Dziegielewska *et al.* 1991; Habgood *et al.* 1992; Saunders 1992). The quantitative contribution from these two sources seems to be different at different stages of brain development. Thus, in the youngest stages so far investigated (E60 in the sheep, Dziegielewska *et al.* 1979, 1991), the specific component of the transfer mechanism seems to account for about half of the albumin present in the CSF. In older brains, the contribution of the specific transfer declines (Dziegielewska *et al.* 1979) and in the adult all of the albumin present in the CSF can be accounted for by passive diffusion alone.

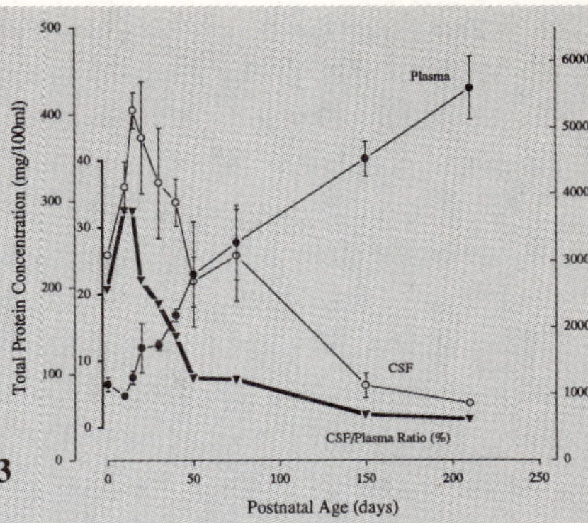

20.3

Figure 20.3

Total protein concentration (mg/100 ml) in plasma (• right ordinate) and CSF (o left ordinate) of postnatal *Macropus eugenii*, (tammar wallaby).
Note: Abscissa, postnatal age in days from birth. CSF/plasma ratio (%, thick line).
Source: Dziegielewska et al. 1986; Jones *et al.* (1991).

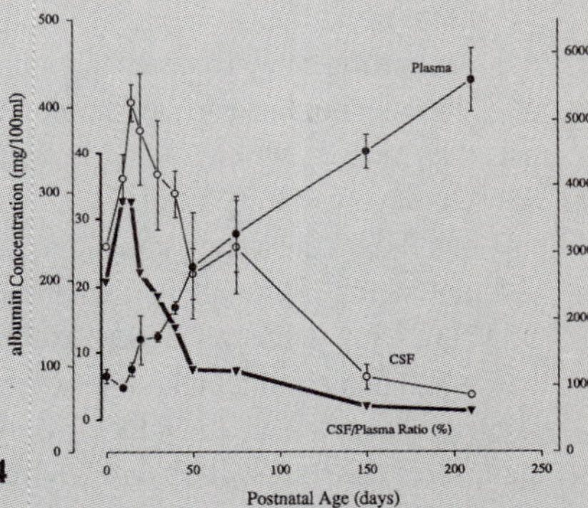

20.4

Figure 20.4

Albumin concentration (mg/100 ml) in plasma (• right ordinate) and CSF (o left ordinate) of postnatal *Macropus eugenii*.
Note: Abscissa, postnatal age in days from birth. CSF/plasma ratio (%, thick line).
Source: Dziegielewska et al. 1986; Jones *et al.* (1991).

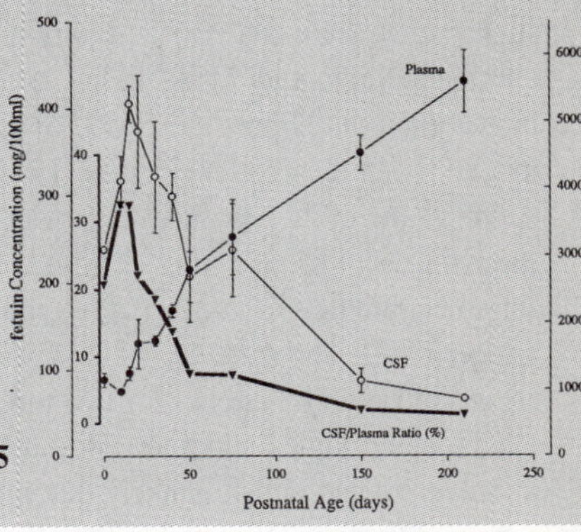

20.5

Figure 20.5

Fetuin concentration (mg/100 ml) in plasma (• right ordinate) and CSF (o left ordinate) postnatal *Macropus eugenii*.
Note: Abscissa, postnatal age in days from birth. CSF/plasma ratio (%, thick line).
Source: Dziegielewska et al. 1986; Jones *et al.* (1991).

E60 in the sheep (gestation is 150 days) is, in terms of CSF protein concentration, already after the peak which, in the sheep, occurs between E30 and E40. Physiological experiments on such young fetuses are technically very difficult. We have used marsupial young as our experimental animals as the accessibility of the pouch young, together with the extreme immaturity of its brain at birth, provides an opportunity to test the effectiveness of the barriers to protein and transfer mechanisms at much earlier stages of brain development than was possible previously.

The peaks of total protein and of albumin concentrations in CSF occur at about 5 to 8 days in *Monodelphis* (Figures 20.1 and 20.2 on page 348). We have investigated the transfer of albumin from plasma into CSF under steady-state conditions (for discussion of importance of steady-state condition for such experiments see

Dziegielewska *et al*, 1991; Habgood *et al.* 1992) at the peak (7 to 8 days) and later (32 to 36 days). The young were injected intraperitoneally with a standard volume of a 20% solution of bovine serum albumin (BSA), or human serum albumin (HSA), or succinylated BSA (succ-BSA) as previously described in detail for young rats (Habgood, 1990; Habgood *et al.* 1992). Animals were left for 16 hours to 20 hours, after which they were carefully removed from the mother; individual animals were killed with an overdose of halothane anaesthetic prior to the sampling of CSF and blood (see Habgood *et al.* 1992 for details). Total protein concentrations were measured using the method of Bradford (1976) and individual albumin concentration were estimated using the radial immunodiffusion technique of Mancini as described in detail previously (for example, Dziegielewska *et al.* 1991; Habgood *et al.* 1992).

The CSF/plasma ratios for *Monodelphis* endogenous albumin (MSA) and injected foreign albumins at the two ages are illustrated in Figure 20.6. The concentration of naturally occurring MSA is assumed to be at steady-state. The CSF/plasma steady-state ratios for all albumins early in development were considerably higher in younger animals than later on. For MSA at 7 to 8 days, steady-state ratio was 25.9% ± 1.0 and for 32 to 36 days it was 1.65% ± 0.1.

At 7 to 8 days the steady state CSF/plasma ratios for MSA and injected BSA and HSA were not significantly different whereas that for succ-BSA was 30% ($P<0.001$) below the MSA steady state level (Figure 20.6). By 32 to 36 days the ratios for MSA, BSA and HSA had fallen considerably to between 1.5 and 3.0% (Figure 20.6). We have not so far tested succ-BSA in this older group, but we would predict from results in other species that the ratio will be similarly very low and not obviously different from that of the unmodified albumin. Preliminary results (Knott *et al.* 1997) at intermediate ages suggest that there is a transient period of development when BSA is transported less effectively than either MSA or HSA and that the ratio for succ-BSA may be much less, indicating a developmental period of specific transfer of albumin as previously found in fetal sheep (Dziegielewska *et al.* 1991) and neonatal rats (Habgood *et al.* 1992).

The young of the *Monodelphis* provided, for the first time, an animal model in which mechanisms of protein transfer from blood into CSF could be studied at such an early stage of brain development, that is, at around the time of the peak of protein concentration in CSF. The reduction of the CSF/plasma ratio due to chemical modification was argued to provide evidence for two mechanisms of entry into the CSF (Habgood *et al.* 1992). The chemical modification was suggested to cause disruption of species specific active transfer and it is likely that most of succ-BSA is entering the CSF purely by passive diffusion. The passive component of the protein transfer depends on the age; the highest contribution being made by the passive route was at the youngest age examined (7 to 8 days, Figure 20.6). This high level could be the reason why BSA seems to reach similar ratio to MSA at the youngest stage examined, but becomes lower than MSA at the older age groups.

Thus the ability to distinguish between, and transfer, different species of albumin, appears to occur only for a particular period in development (Dziegielewska *et al.* 1991; Habgood *et al.* 1992). The only other study that examined protein transfer across the blood - CSF barrier in a marsupial species used the tammar (Dziegielewska *et al.* 1988). This species has a protracted period of development in the pouch. The peak in CSF protein concentration occurs at about 15 days, but the peak albumin concentration is much later at about 40 days (Dziegielewska *et al.* 1986). The CSF/plasma ratio

Figure 20.6
CSF/plasma ratios (%) for endogenous albumin and for exogenous albumin sampled 15 hours after IP injection in *Monodelphis domestica* in two different age groups.
Note: Error bars show sem:
* significant difference from MSA ($p < 0.05$).
** significant difference from MSA ($p < 0.001$).

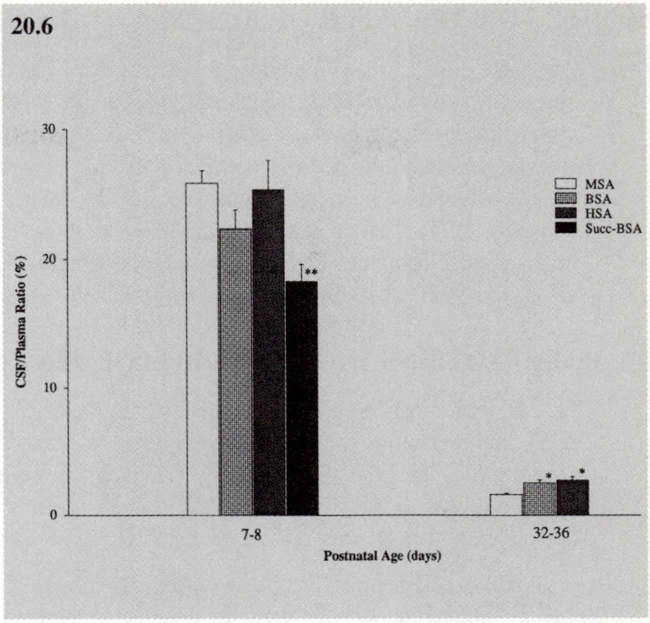

for endogenous tammar albumin at 20 days (16.4 ± 3) and 60 days (12.3 ± 1.0) are not significantly different. In pouch young injected intraperitoneally with HSA, by 24 hours (assumed steady state) the CSF/plasma ratios were not significantly different from those given above for tammar albumin. We have not tested chemically modified albumin (for example, succ-BSA or HSA) in the tammar. However, it seems likely that the prolonged developmental period (at least to 60 days) during which the CSF/plasma ratios for both tammar albumin and HSA remain very high compared with the adult (0.9%, Dziegielewska *et al.* 1986) indicates that there is a specific developmentally regulated transfer mechanism for albumin from plasma to CSF as in other species.

This study of protein penetration in *Monodelphis* is the first, and so far the only one investigating the mechanism of protein transfer from blood into CSF around the time of the peak of protein concentration in the CSF and subsequently. In this species, during the time when the peak protein concentration is developing within the CSF, only a relatively small proportion of albumin is derived from the plasma by the species-specific active component of the transfer system, more than 60% is derived by the non-discriminatory route (Knott *et al.* 1997). It is not clear whether this non-discriminatory route corresponds to passive diffusion as suggested for other species later in development or whether it represents a less sensitive/specific stage of active transfer. During this early period the mechanism does not appear to distinguish between albumin from different species. Discrimination of different albumins occurs immediately after the peak of total protein concentration in the CSF and then disappears again in older *Monodelphis*, as has been shown for eutherian species (Dziegielewska *et al.* 1991; Habgood *et al.* 1992). However, both total protein concentration and albumin concentration in CSF remained elevated for a considerably longer period in the tammar than in *Monodelphis*, which is perhaps a reflection of the much more protracted period of pouch development in the tammar (Dziegielewska *et al.* 1986).

ROUTE FOR TRANSFER OF ALBUMIN FROM PLASMA TO CSF

The route of transfer of albumin from plasma to CSF has been studied mainly in fetal sheep. At both the light microscopical (Dziegielewska *et al.* 1991) and the electronmicroscopical level (Balslev *et al.* 1997) it has been shown, using immunocytochemistry that transfer of albumin occurs across the choroid plexuses; at an ultrastructural level the route is within the tubuloendoplasmic reticular (TER) system described by Møllgård and Saunders (1977). That the same process also occurs in developing marsupials is indicated by similar light microscopical immunocytochemical observations in the tammar (Dziegielewska *et al.* 1988) and in *Monodelphis* (Knott *et al.* 1997).

PLASMA PROTEINS IN BRAIN DEVELOPMENT

Several studies have shown that proteins, immunologically identical to plasma proteins, are present in the developing brain in a number of different species (see Dziegielewska & Saunders, 1988 for references). Immunocytochemical techniques have been used to study the distribution of these proteins in the developing brain of several eutherian species, including man (Dziegielewska *et al.* 1987). One fetal

Table 20.2. Distribution of fetuin in postnatal tammar dorsolateral neocortex and choroid plexus

STRUCTURE		AGE					
		nb	3d	5d	9d	15d	20d
ventricular zone	cells	+	+ +	+++	(+)	(+)	—
subventricular zone	cells	0	0	—	—	+	—
intermediate zone	cells	0	0	0	0	(+)	—
	fibres	0	0	0	0	(+)	—
subplate zone	cells	0	0	0	0	++++	—
	fibres	0	0	0	0	(+)	—
cortical plate	cells	0	0	++	++	++	—
primordial	cells	—	+	++	splits into marginal zone and		
plexiformlayer	fibres	+	++	(+)	subplate zone		
marginal zone	cells	0	0	++	++	++	—
	fibres	0	0	(+)	+	+	—
choroid plexus	cells	+++	+++	+++	+++	+++	+++
plasma		++++	++++	++++	++++	++++	+++
CSF		++++	++++	++++	++++	++++	++

Key:
nb newborn; other ages indicated are days postnatal
\+ positive staining (range: + (weak staining) to + + + + (very strong staining);— staining absent
0 region not yet developed
Note: The cortical plate is first apparent at 5 to 6 days postnatal, particularly in the more lateral part of the dorsolateral neocortex. The migration of cells into the primordial plexiform layer (ppl) at 5 to 6 days postnatal appears to divide the ppl into marginal zone and subplate zone.
Source: Jones *et al.* (1991).

glycoprotein, fetuin, has been established as a marker for cortical plate cells in all animal brains studied so far, including marsupials (Jones *et al.* 1991; Adam *et al.* 1993). The data for fetuin expression in the developing neocortex of the tammar are summarised in Table 20.2 (from Jones *et al.* 1991) and illustrated in Figure 20.7 together with the only information on *Monodelphis* brain that is currently available.

The presence of fetuin in early postnatal fibre systems, in some cells of the primordial plexiform layer and in early cortical plate cells, is similar to that of fetuin in fetal eutherian animals (see Brown *et al.* 1992 for references). Additionally, the sequence of appearance of fetuin positive cells in the tammar (Jones *et al.* 1991) and, to a lesser degree, in the *Monodelphis* (Adam *et al.* 1993) coincides with the transient population of early cortical plate cells described previously in fetal cat (Luskin & Shatz, 1985) and sheep (Saunders *et al.* 1992). From detailed studies of the development of the tammar telencephalon (Reynolds *et al.* 1985), using ^{3}H-thymidine to establish cells' birth dates, it seems most likely that fetuin-positive cells are born sometime between the day of birth and 2 days (Jones *et al.* 1991) in this species.

The role of fetuin in brain development is not known and the proportion of fetuin-positive cells in the early neocortex varies in different species: in some eutherian mammals (sheep, cow) the majority of early cortical plate cells are fetuin-positive, whereas in both marsupial species (tammar and *Monodelphis*) the positive cells are more sparsely distributed. Additionally, the staining in marsupial brains lasts for much shorter periods of time than that of fetuin in eutherian brains (Jones *et al.* 1991). The earlier disappearance of fetuin from cortical plate cells of the tammar and *Monodelphis*, as opposed to sheep and cow, could reflect the differences in cortical plate development in marsupial

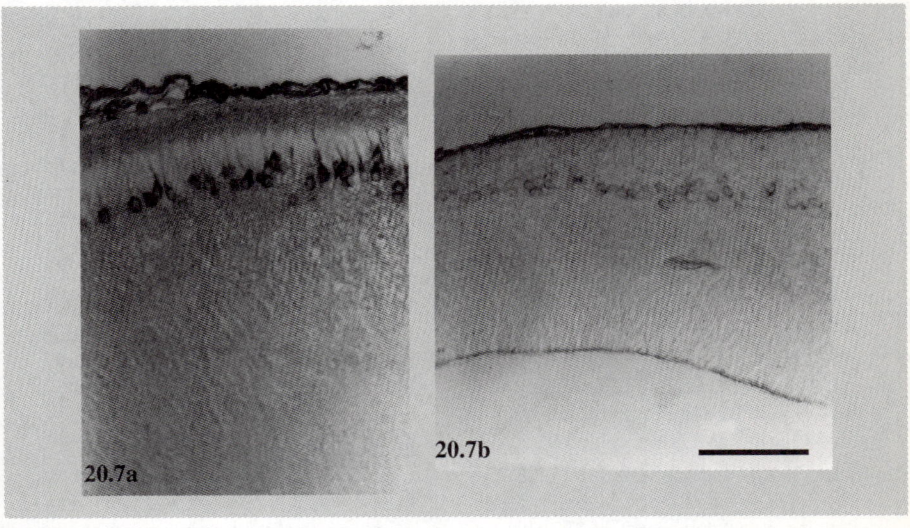

Figure 20.7

Distribution of fetuin in (a) the developing dorsolateral neocortex of the 15-day tammar and (b) 8-day *Monodelphis* .

Scale bar= 100 μm.

Notes: (1) Sections were stained for fetuin using the PAP method (see Jones *et al.* (1991) and Adam *et al.* (1993) for details). (2) The pial surface is uppermost. (3) Note darkly stained (fetuin-positive) cells in the early cortical/subplate region in both species.

and eutherian species (Reynolds *et al.* 1985; Reynolds & Saunders 1988). This, in turn, may be associated with the generally more mature appearance of the subplate and developing layer VI in the marsupial compared with eutherians (see Reynolds *et al.* 1985; Reynolds & Saunders 1988; Saunders, Chapter 22 in this volume). The demonstration of the presence of fetuin in early cortical plate cells in marsupials opens up the possibility of direct experimental manipulation of the system to study the importance of this fetal glycoprotein in brain development. Such an approach would not have been possible in eutherian fetuses at the time of fetuin appearance in the cortical plate cells due to the difficulties in accessing such young fetuses *in utero*.

Acknowledgments

This work was supported by the Wellcome Trust, Medical Research Council, UK and Action Research for the Crippled Child, UK (1983 to 1992) and by the Australian Research Council and National Health and Medical Research Council, Australia (1992 to 1996).

References

Adam E, Dziegielewska K M, Saunders N R and Schumacher U (1993). Neuraminic acid specific lectins as markers of early cortical plate neurons. *International Journal of Developmental Neuroscience*, 11, 451–460.

Balslev Y, Dziegielewska K M, Møllgård K and Saunders N R (1997). Intercellular barriers to and transcellular transfer of albumin in the fetal sheep brain. *Anatomy and Embryology* 95, 229–236.

Bradford M M (1976). A rapid and sensitive method for the quantitation of microgram quantities of protein utilising the principle of protein-dye binding. *Analytical Biochemistry*, 72, 248–254.

Brown W M, Saunders N R, Møllgård K and Dziegielewska K M (1992). Fetuin-an old friend revisited. *BioEssays*, 14, 749–755.

Davson H, Welch K and Segal M B (1987). *The Physiology and Pathophysiology of the Cerebrospinal Fluid.* Churchill Livingstone, Edinburgh.

Dziegielewska K M and Saunders N R (1988). The development of the blood-brain barrier: Proteins in fetal and neonatal CSF, their nature and origins. In *Handbook of Human Growth and Developmental Biology.* Eds E Meisami and P S Timiras, CRC Press, Boca Raton. Vol. 1, Pt A, pp 169–191.

Dziegielewska K M, Evans C A N, Malinowska D H, Møllgård K, Reynolds J M, Reynolds M L and Saunders N R (1979). Studies of the development of brain barrier systems to lipid insoluble molecules in fetal sheep. *Journal of Physiology*, 292, 207–231.

Dziegielewska K M, Evans C A N, Fossan G, Lorscheider F L, Malinowska D H, Møllgård K, Reynolds, M L, Saunders N R and Wilkinson S (1980). Proteins in cerebrospinal fluid and plasma of fetal sheep during development. *Journal of Physiology*, 300, 441–455.

Dziegielewska K M, Hinds L A, Sarantis M E P, Saunders N R and Tyndale-Biscoe C H (1986). Proteins in cerebrospinal fluid and plasma of the pouch young tammar wallaby (*Macropus eugenii*) during development. *Comparative Biochemistry and Physiology*, 83B, 561–567.

Dziegielewska K M, Møllgård K, Reynolds M L and Saunders N R (1987). A fetuin-related glycoprotein (α_2-HS) in human embryonic and fetal development. *Cell and Tissue Research*, 248, 33–41.

Dziegielewska K M, Hinds L A, Møllgård K, Reynolds M L and Saunders N R (1988). Blood-brain, blood-cerebrospinal fluid and cerebrospinal fluid-brain barriers in a marsupial (*Macropus eugenii*) during development. *Journal of Physiology*, 403, 367–388.

Dziegielewska K M, Habgood M, Jones S E, Reader M. and Saunders N R (1989). Proteins in cerebrospinal fluid and plasma of postnatal *Monodelphis domestica* (grey short-tailed opossum). *Comparative Biochemistry and Physiology,* 92B, 569–576.

Dziegielewska K M, Habgood M D, Møllgård K, Stagaard M and Saunders N R (1991). Species-specific transfer of plasma albumin from blood into different cerebrospinal fluid compartments in the fetal sheep. *Journal of Physiology,* 439, 215–237.

Fossan G, Cavanagh M E, Evans C A N, Malinowska D H, Mollgard K, Reynolds M L and Saunders N R (1985). CSF-brain permeability in the immature sheep fetus: a CSF-brain barrier. *Developmental Brain Research*, 18, 113–124.

Habgood M D (1990). *Barriers of the Developing Brain*. PhD thesis, University of Southampton, Southampton.

Habgood M D, Sedgwick J E C, Dziegielewska K M and Saunders N R (1992). A developmentally regulated blood-cerebrospinal fluid transfer mechanism for albumin in immature rats. *Journal of Physiology,* 456, 181–192.

Jones S E, Christie D L, Dziegielewska K M, Hinds L A and Saunders N R (1991). Developmental profile of a fetuin-like glycoprotein in neocortex, cerebrospinal fluid and plasma of post-natal tammar wallaby *(Macropus eugenii). Anatomy and Embryology*, 183, 313–320.

Knott G W (1995). *Barriers of the Developing Brain: In Vivo and In Vitro Studies*. PhD thesis. University of Tasmania, Tasmania.

Knott G W, Dziegielewska K M, Habgood M D, Li Z and Saunders N R (1997). Albumin transfer across the choroid plexus of postnatal S. American opossum (*Monodelphis domestica*). *Journal of Physiology* 499, 179–194.

Luskin M B and Shatz C J (1985). Studies of the earliest generated cells of the cat's visual cortex: cogeneration of subplate and marginal zones. *Journal of Neuroscience,* 5, 1062–1075.

Møllgård K and Saunders N R (1977). A possible transepithelial pathway via endoplasmic reticulum in foetal sheep choroid plexus. *Proceedings of the Royal Society of London B.* 199, 321–326.

Pedersen K O (1944). Fetuin, a new globulin isolated from serum. *Nature*, 154, 575.

Reynolds M L and Saunders N R (1988). Differentiation of the neocortex. In *The Developing Marsupial. Models for Biomedical Research.* Eds C H Tyndale-Biscoe and P A Janssens. Springer-Verlag, Berlin. pp 101–116.

Reynolds M L, Cavanagh M E, Dziegielewska K M, Hinds L A, Saunders N R and Tyndale-Biscoe C H (1985). Postnatal development of the telencephalon of the tammar wallaby (*Macropus eugenii*). An accessible model of neocortical differentiation. *Anatomy and Embryology*, 173, 81–94.

Saunders N R (1992). Ontogenetic Development of Brain Barrier Mechanisms. In *Handbook of Experimental Pharmacology. Physiology and Pharmacology of the Blood Brain Barrier.* Ed M W B Bradbury. Springer-Verlag, Berlin. Vol. 103, pp 327–369.

Saunders N R, Adam E, Reader M and Møllgård K (1989). *Monodelphis domestica* (grey short-tailed opossum): an accessible model for studies of early neocortical development. *Anatomical Embryology*, 180, 227–236.

Saunders N R, Habgood M D, Ward R A and Reynolds M L (1992). Origin and fate of fetuin-containing neurons in the developing neocortex of the fetal sheep. *Anatomy and Embryology,* 186,

CHAPTER 21

DEVELOPMENT AND DEVELOPMENTAL PLASTICITY OF DESCENDING SPINAL PATHWAYS IN THE NORTH AMERICAN OPOSSUM, *DIDELPHIS VIRGINIANA*

George F Martin and Xian Ming Wang

INTRODUCTION

It is generally accepted that descending spinal axons do not grow around a lesion in adult mammals, including man. However, it is reasonable to predict that such growth might occur during development when neurons are programmed for axonal elongation and the environment is permissive for it. That prediction is supported by the finding that cortical axons grow around a lesion of their spinal pathway in neonatal rats (Bernstein & Stelzner 1983; Schreyer & Jones 1982; Bates & Stelzner 1993) and cats (Bregman & Goldberger 1982, 1983a,b) and around lesions of the medullary pyramids in newborn hamsters (Kalil & Reh 1979, 1982) and cats (Tolbert & Der 1987). It is interesting, however, that brainstem axons show no evidence for comparable plasticity at the same age (Prendergast & Stelzner 1976; Bregman & Goldberger 1982, 1983a,b). It might be argued that brainstem axons never have the ability to circumvent a lesion, but based on the results of developmental studies we have entertained another possibility. In eutherian mammals, brainstem axons grow into the spinal cord prenatally (Bregman & Goldberger 1982; Lakke & Marani 1991; Kudo *et al.* 1993) well before axons from the cerebral cortex (Donatelle 1977; Schreyer & Jones 1982; Theriault & Tatton 1989) and by birth they may have simply lost their potential for plasticity. We predicted, therefore, that brainstem axons will grow around a lesion of their spinal pathway if it is made at earlier stages of development than that present at birth in rats, cats, and hamsters. In order to test that hypothesis we have employed the North American opossum, *Didelphis virginiana,* which is born in a very immature state (Figure 21.1), 12 to 13 days after conception (Hartman 1952). In the opossum, many brainstem axons grow into the spinal cord postnatally (see, for example, Cabana &

Figure 21.1 Photograph of a newborn opossum. Crown-rump length 10mm. Source: Reproduced from Martin *et al.* (1993), courtesy of Pergamon Press Ltd.

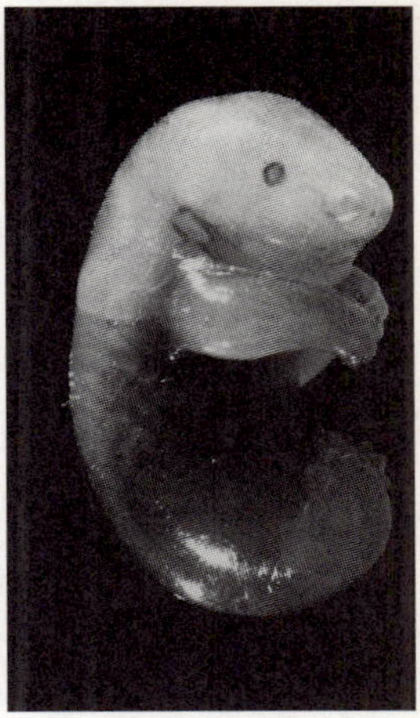

Martin 1981; 1984; Martin *et al.* 1993) making it possible to lesion them at fetal stages of development without intrauterine surgery. Using opossums, we have shown that brainstem axons are indeed able to grow around a lesion of their spinal pathway and that the critical period for plasticity ends earlier than that for corticospinal axons (Martin & Xu 1988; Xu & Martin 1989; Wang *et al.* 1994).

Growth of descending spinal axons through a lesion also fails to occur in adult mammals. There is increasing evidence, however, that it takes place during development. Such growth has been reported *in vitro* after transection of the neonatal spinal cord in the grey short-tailed South American opossum, *Monodelphis domestica* (Treherne 1992; Woodward *et al.* 1993) or the fetal rat (Saunders *et al.* 1992) and *in vivo* after a comparable lesion in the North American opossum (Martin *et al.* 1994) and also in *Monodelphis domestica* (Saunders *et al.* 1995 and Chapter 22 in this volume). In the following account, we review the results of *in vivo* experiments using the North American opossum which document that descending spinal axons grow around and through a lesion during development and that loss of developmental plasticity results from diminished growth potential with age, as well as the maturation of the environment through which growth must occur. Since our experimental design and the interpretation of our results are based on developmental studies, we review those results first.

THE DEVELOPMENT OF DESCENDING SPINAL PATHWAYS

In order to determine whether supraspinal and propriospinal axons have reached lumbar levels of the cord by birth, we have used the retrograde transport of the fluorescent marker Fast Blue (FB) (Martin *et al.* 1993). Pouch-young opossums were obtained from females captured in the wild and from animals bred at The Ohio State University. The day of birth was considered to be postnatal day 0*. The age of the pups from captured animals was estimated by comparing their snout-rump length and external features with those of animals at known ages (Cutts *et al.* 1978). In order to operate on the pouch young, the mothers were anaesthetised with intramuscular injections of Ketamine (1.2 ml, 100 mg/ml) and Metofane inhalation. After relaxation of the pouch sphincter, the pups were anaesthetised individually by hypothermia or Metofane inhalation and the lumbar cord was exposed for injections of the marker (0.05/0.30 µl in a 5% aqueous solution). In some of the cases operated at 19 days, the spinal cord was hemisected rostral and contralateral to the injection. All injections were made by pressure using a micropipette (50 μm tip) attached to a 1 μl Hamilton syringe. After the injections were completed and the surgical exposure closed, the pups were returned to the vivarium with their mother for 1 to 5 days before being anaesthetised again and perfused through the heart with saline followed by either a 0.2 M citrate buffer-10% formaldehyde solution or 4% ice-cold buffered paraformaldehyde. The brain and spinal cord were subsequently removed and stored in the fixative. Sections from the injection site, the spinal cord rostral to the injection, and the brain were cut on a freezing microtome (30 μm) and mounted onto glass slides using a non-fluorescing medium so that they could be examined using a Leitz Orthoplan photomicroscope equipped for epifluorescence. The size and geometry of the injection and lesion (when present) and the locations of labelled neurons were recorded on drawings of the appropriate sections using an X–Y plotter interfaced with the microscope stage by position transducers. FB labelled neurons were identified using filter cube A (excitation wavelength=340–380 nm) of the Ploem epi-illumination system and their locations were described using the terminology of Oswaldo-Cruz and Rocha-Miranda (1968) for the opossum brain.

The brain of newborn opossums is very immature (Figure 21.2 and insert, Fig, 21.3a). For example, red nuclei cannot yet be identified in the mesencephalon (Figure 21.2a) and inferior olivary nuclei are not present in the caudal brainstem (Figure 21.2b). When FB was injected into the lumbar cord the day of birth and the pups were maintained for 2 to 5 days for retrograde transport, only a few propriospinal neurons were labelled rostral to the injection and they were limited to the ventral part of the developing grey matter. Supraspinal labelling was also sparse, but it was reliably present within the presumptive coeruleus complex and the medial pontine reticular nucleus (Figures 21.2a and 21.3). In some cases, labelled neurons were also found within the ventral part of the rostral hypothalamus, the interstitial nucleus of the medial longitudinal fasciculus, the lateral vestibular nucleus, the medullary reticular formation, and the medullary raphe. Differences in the number and location of labelled neurons may have been due to individual variation and/or differences in the size and location of the injection. When injections were made at 2 days postnatal, propriospinal and supraspinal labelling was more extensive. In the spinal cord rostral to the injections, labelled neurons were still most numerous in the ventral grey matter, but some were present in the intermediate zone between the developing dorsal and ventral horns. At brainstem levels, labelled neurons were found within the presumptive paraventricular and lateral nuclei of the hypothalamus, the interstitial nucleus of the medial longitudinal fasciculus, the reticular formation of the pons and medulla, the lateral and inferior vestibular nuclei, the presumptive coeruleus complex, and the medullary raphe. By 4 days, propriospinal labelling was similar to that present at 2 days, but additional labelling was found at brainstem levels within the medial vestibular nucleus, the solitary complex, and the reticular formation deep to the dorsal column nuclei. Propriospinal labelling from a case injected at 7 days is shown in Figure 21.4 and it can be seen that labelled neurons are present in both the dorsal and ventral horns. In cases injected at 9 to 12 days, labelled neurons were present, for the first time, in the red nucleus.

The results of the above studies suggested that rubral axons reach the spinal cord later than axons from other areas of the brainstem. In order to verify that observation we also studied rubrospinal development using the orthograde transport of horseradish peroxidase conjugated to wheat germ agglutinin (WGA-HRP) (Cabana & Martin

**Dr Martin designates the day of birth as P1 (or PD1) in his published papers. Here the dates are altered to the convention that the day of birth is zero (P O).*

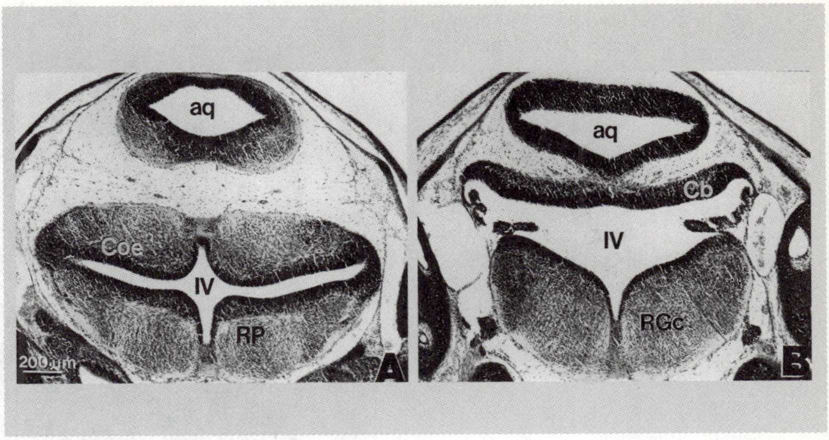

Figure 21.2 Nissl-stained sections, (a) rostral,
(b) caudal, through the brainstem of a newborn opossum. The cerebral aqueduct (aq), the cerebellum (Cb), the
presumptive coeruleus complex (Coe), the fourth ventricle (IV), the gigantocellular reticular nucleus (RGc), and the pontine reticular nucleus (RP) are indicated.
Scale bar =200 µm.

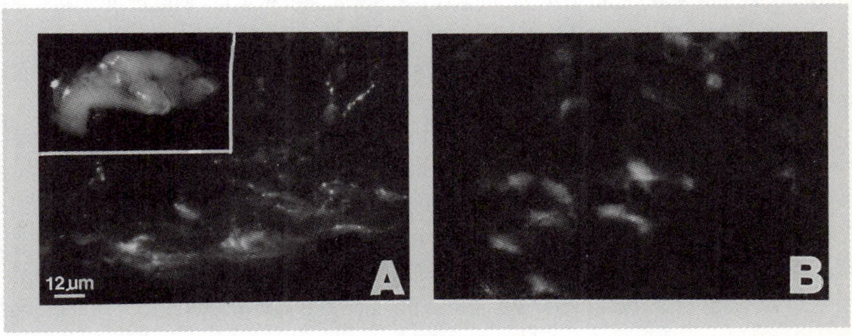

Figure 21.3 Neurons labelled in (a) the presumptive coeruleus complex and
(b) medial pontine reticular nucleus by an injection of the fluorescent marker Fast Blue (FB) into the lumbar spinal cord of a newborn opossum.
Inset: Lateral view of the gross brain in a newborn.
Scale bar=12 µm.

1986; Martin *et al.* 1986, 1988). In those experiments, the animals were anaesthetised and operated as described above, but the surgical exposure was made over the mesencephalon so the marker could be injected into the red nucleus. After an appropriate survival time for transport, the pups were re-anaesthetised and killed by perfusion so that the brain and spinal cord could be removed and sectioned with a freezing microtome. The sections were subsequently processed histochemically for HRP according to Mesulam (1978). When rubral injections were made at 6 days, labelled axons were present in the dorsal part of the contralateral lateral funiculus (the location of rubrospinal axons in adult opossums, Martin & Dom 1970; Cabana & Martin

Figure 21.4 Fluorescence photomicrograph of the neuronal labelling produced in the spinal cord rostral to bilateral lumbar injections of FB at 7 days. The lateral (LF) and ventral (VF) funiculi are indicated and the arrow points to labelled axons in thefasciculus gracilis. Scale bar=100 μm.

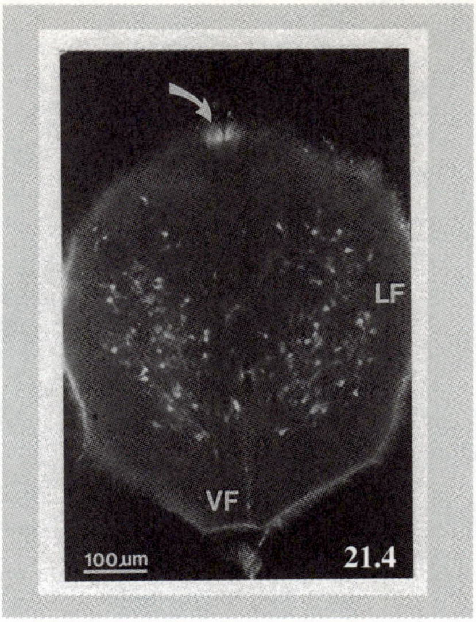

1986) at cervical and thoracic levels; but, as predicted by the retrograde tracing results, they were not found in the lumbosacral cord. Rubral injections labelled axons within the marginal zone of the lumbar cord at 12 days, but it was not until about 23 days that labelling was found in those areas of the grey matter which are innervated by rubral axons in adult opossums (Martin & Dom 1970; Cabana & Martin 1986). The results of orthograde tracing experiments, therefore, verified that rubral axons do not reach lumbar levels of the spinal cord in the opossum until well after birth.

The 19 day pups subjected to lumbar injections of FB and hemisection of the cord on the contralateral side rostral to the injections were used to establish laterality of the brainstem-spinal projections described above (Martin *et al.* 1993). Almost all of the labelling in the dorsal part of the lateral vestibular nucleus and the medial part of the pontine reticular formation was ipsilateral to the injection, whereas that within the red nucleus and the inferior vestibular nucleus was contralateral to it. Labelling in most other areas of the brainstem and within the hypothalamus was bilateral with either an ipsilateral or contralateral predominance. A comparable pattern of labelling was seen when the same experiment was performed in adult opossums. Such results suggest that the laterality of brainstem-spinal projections is established early in development.

Immunohistochemical studies for serotonin and tyrosine hydroxylase have also provided evidence for early growth of brainstem axons into the spinal cord (DiTirro *et al.* 1983; Pindzola *et al.* 1990; Martin *et al.* 1991). Serotoninergic neurons are present within the medullary raphe and adjacent reticular formation on the day of birth (Figure 21.5a) and it is reasonable to assume that some of them give rise to the spinal axons which express the same phenotype at that age (Figure 21.5b). Although serotoninergic axons are relatively numerous in the cervical cord the day of birth, they are sparse at lumbosacral levels. At both cervical (Figure 21.5b) and lumbar levels, they are limited primarily to the marginal zone. Subsequent growth into the grey matter follows rough rostral to caudal gradients and it is qualitatively complete by 49 days. Neurons that immunostain for tyrosine hydroxylase, the rate limiting enzyme in catecholamine biosynthesis, are also present in the opossum's brainstem at birth (Pindzola *et al.* 1990). They are found within the hypothalamus, the presumptive coeruleus complex, and the ventrolateral medulla and it is likely that some of them give rise to the axons which immunostain for the same enzyme in the spinal cord. Tyrosine hydroxylase immunoreactive axons are sparse and limited to the marginal zone of the spinal cord at day of birth. By 2 days, a few such axons are present in the grey matter and by 43 days, they are found in all of the areas innervated in the adult animal.

Cortical axons do not reach lumbosacral levels in either the adult or developing animal (Martin & Fisher 1968; Cabana & Martin 1985; Martin & Cabana 1985). In order to establish when cortical axons first reach spinal levels we made injections of fluorescent markers into the rostral cervical cord. Cortical neurons were not labelled by such injections until approximately 29 days (Cabana & Martin 1985). The first neurons labelled were found caudal to the orbital sulcus in presumptive primary somatosensory/motor cortex (Lende 1963a, b). At later stages of development, comparable injections labelled neurons over a larger area of cortex than could be labelled in adult animals suggesting that transient projections exist.

We have also studied corticospinal development using the orthograde transport of WGA-HRP (Cabana & Martin 1985). In those experiments, the pups were anaesthetised and operated as described above, but the injections were placed in those areas of cortex which were labelled in the retrograde tracing experiments. After an appropriate survival for axonal transport, the pups were re-anaesthetised and killed prior to perfusion so that the brain and spinal cord could be removed and sectioned for processing by HRP histochemistry (Mesulam 1978). Even when multiple injections were made, axons were not labelled in the spinal cord until approximately 29 days postnatal (Cabana & Martin 1985). The first axons labelled at spinal levels were limited to those regions of the dorsal and lateral funiculi which contain corticospinal axons in the adult animal. Labelled axons were not present in the grey matter until several days later and they did not reach their caudal extent until about 37 days. At some stages of development, labelled axons were present in areas of the white and grey matter which did not contain them in adult animals, suggesting that overgrowth and subsequent pruning occurs.

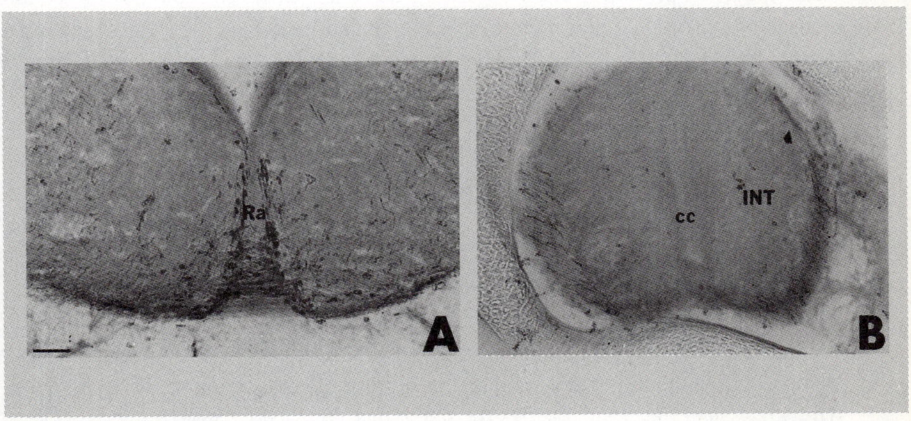

Figure 21.5 Photomicrographs of serotonin immunoreactivity in (a) the caudal brainstem and (b) cervical cord of a newborn opossum. The central canal (cc), the intermediate zone (INT) and the raphe (Ra) are indicated. Scale bar in (a)=80 μm.

The results of the studies described above provide an overview of descending spinal development in the opossum making it possible to conclude that:
(a) some propriospinal and supraspinal axons are already present in the lumbar cord by birth;
(b) axons from the medial pontine reticular formation are among the first from the brainstem to innervate the lumbar cord although they are followed quickly by axons from other areas including the lateral vestibular nucleus;
(c) rubral axons do not grow into the lumbar cord until 9 to 12 days, and d) axons from the cerebral cortex never innervate the lumbar cord and do not even grow into cervical levels until about 29 days postnatal.
The early growth of reticular and vestibular axons into the spinal cord has also been described in zebrafish (Kimmel 1982; Metcalf *et al.* 1986), certain amphibians (Forehand & Farel 1982; Nordlander *et al.* 1985; van Mier & ten Donkelaar 1984), the chick (Glover & Petursdottir 1991), the rat (Kudo *et al.* 1993), and the South American opossum, *Monodelphis domestica* (Wang *et al.* 1992) and the relatively late development of the rubrospinal tract has been documented in the clawed toad (van Mier & ten Donkelaar 1984), the chick (Okado & Oppenheim 1985), the rat (Lakke & Marani 1991; Kudo *et al.* 1993), and the South American opossum (Wang *et al.* 1992). In those species which possess a corticospinal tract, cortical innervation of the spinal cord occurs well after the development of brainstem-spinal connections (for example, the rat, Donatelle 1977; Schreyer & Jones 1982; Oudega *et al.* 1994; the cat, Theriault & Tatton 1989; Bregman & Goldberger 1982; 1983a,b; and the South American opossum, Wang *et al.* 1992).

EVIDENCE FOR GROWTH OF RUBROSPINAL AXONS AROUND A LESION OF THEIR SPINAL PATHWAY

We began studying brainstem-spinal plasticity by asking whether rubral axons in the North American opossum grow around a lesion during development. Rubral axons were chosen for study because they do not reach caudal levels of the cord until well after birth. Pups of different ages were anaesthetised and operated in the pouch (as described in the previous section) so that the thoracic spinal cord could be exposed for unilateral transection of the rubrospinal tract. After approximately 30 days, they were re-anaesthetised and re-operated so that injections of FB or WGA-HRP could be made at least two segments caudal to the lesion and ipsilateral to it. The intent of the injections was to label any neurons in the contralateral red nucleus (Figure 21.9b) that supported axons caudal to the lesion. In the opossum, as in other species, the rubrospinal tract is almost entirely crossed (Xu & Martin 1991). The pups were maintained for 2 to 3 days after injection before being terminally anaesthetised and perfused with an appropriate fixative so that the spinal cord and brain could be removed. Frozen sections through the lesion, the injection and the red nucleus were either mounted onto glass slides using a non-fluorescent mounting medium (FB cases) or processed for HRP according to Mesulam (1978) prior to mounting. Two adult opossums were also anaesthetised and subjected to lesioning of the rubrospinal tract using sterile techniques. Thirty days later they were re-anaesthetised so that WGA-HRP could be injected 2 segments caudal and ipsilateral to the lesion. After 2 to 3 days, they were terminally anaesthetised and perfused prior to removal of their spinal cord and brain,which were then sectioned, and processed for HRP. When the lesion was made in adult opossums and older pups, neurons were never labelled in the contralateral red nucleus, but when it was made before 25 days, such labelling was always present (Martin & Xu 1988; Xu & Martin 1989). The results from animals between 25 and 53 days were variable. It was our tentative interpretation that the labelled neurons supported axons which had grown around the lesion during the 30 day survival.

As a further test for the existence of rubrospinal plasticity, we employed the orthograde transport of WGA-HRP (Martin & Xu 1988; Xu & Martin 1989). Pouch-young and adult opossums were subjected to lesions of the rubrospinal tract at thoracic levels, as described above, and 38 to 60 days later, they were re-anaesthetised and re-operated so that WGA-HRP could be injected into the red nucleus contralateral to it. The pups were maintained for about 48 h for orthograde transport of the injected marker before being anaesthetised again and perfused so that the spinal cord and brain could be removed, sectioned, and processed for HRP. When the lesion was made at 53 days or in the adult animal, labelled axons could be traced to the lesion site, but not caudal to it. When the lesions were made between 17 and 32 days, however, rubrospinal axons were labelled caudal to the lesion (Figure 21.6). The position of labelled axons at the lesion site and caudal to it varied. In one case, the lesion extended across the midline and labelled axons were located at the edge of the spinal cord at the lesion site and within the dorsal funiculus caudal to it (17 days, Figures 21.6 and 21.7). When the lesion was less extensive, labelled axons were present within grey matter left at the lesion site and within the dorsal part of the lateral funiculus, their normal position, caudal to it (19 and 32 days, Figure 21.6). In the youngest pups, labelled axons extended to at least lumbar levels and in all cases, they terminated in areas of the grey matter which were normal for them.

Based on the results of the retrograde and orthograde tracing experiments we concluded that rubral axons grow around a lesion of their spinal pathway during early development and that a critical period exists for that plasticity. We then asked whether growth around the lesion results from regeneration of cut axons, the arrival of relatively late growing axons, or both. In order to address that question, anaesthetised pups were subjected to unilateral or bilateral injections of FB into the rostral lumbar or caudal thoracic cord at 17, 25, 35, and 49 days as described above. Approximately 4 days later, they were re-anaesthetised and re-operated so the rubrospinal tract could be cut unilaterally, 4 or 5 segments rostral to the injections. In some cases the previously injected FB was removed by gentle suction. When the injection was unilateral, the lesion was made ipsilateral to it. The animals were maintained for about 30 days before being anaesthetised a third time for injection of Diamidino-yellow (DY), usually bilaterally, between the site of the first injection and lesion. Four to five days later the pups were killed by an overdose of anaesthetic and perfused so that the spinal cord and brain could be removed for sectioning with a freezing microtome. The sections were mounted onto glass slides and coverslipped for viewing with a fluorescence microscope. Using the A cube of the Ploem illumination system (excitation wavelength=340 to 380 nm), FB could be identified in the cytoplasm of labelled neurons by its bright blue fluorescence and DY could be identified in the nucleus by its yellow colour. When this experiment was performed during the critical period for rubrospinal plasticity, only a few neurons were labelled by FB in the contralateral red nucleus. We interpreted these results to suggest that most rubrospinal neurons died after axotomy and in a subsequent study, we showed that approximately 75% of them failed to survive a comparable lesion (Xu & Martin 1992). In contrast, many neurons

Figure 21.6

Plot of rubrospinal labelling (arrows) present well above the lesion (T1, C1 and C5), 5 mm rostral to the lesion, at the deepest part of the lesion, 5 mm caudal to the lesion, and at lumbar levels in 3 cases subjected to lesions at 17, 32 and 19 days postnatal, and injections of horseradish peroxidase conjugated to wheat germ agglutinin (WGA-HRP) into the contralateral red nucleus approximately 30 days later. The cerebral aqueduct (aq), the oculomotor nucleus (OcM), the cerebral peduncle (ped), and the red nucleus (RN) are indicated in the drawings of midbrain sections.
Source: Reproduced by Xu and Martin (1989), courtesy of Wiley-Liss, a division of Wiley and Son, Inc, New York.

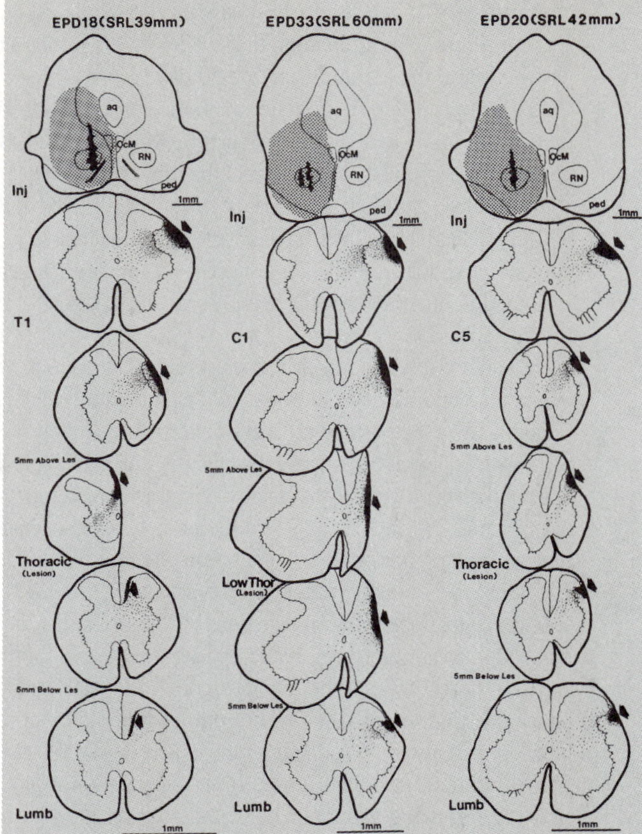

21.6

Figure 21.7

Darkfield photomicrographs of rubrospinal labelling at (a) the lesions site and (b) caudal to it in the case plotted on the left in Figure 21.8. The dorsal funiculus (DF) is indicated for reference.
Source: Reproduced in modified form from Martin and Xu (1988), courtesy of Elsevier Science Publishers BV, Amsterdam.

21.7

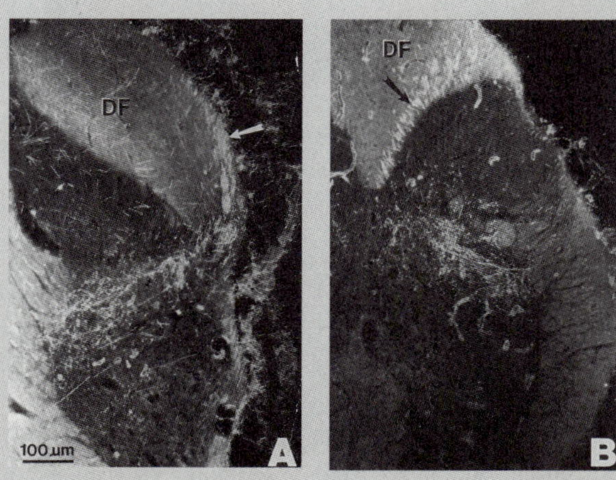

21.8

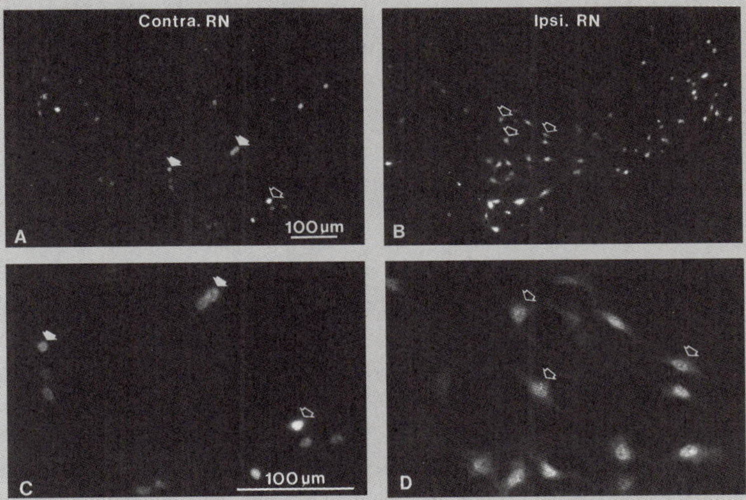

Figure 21.8 Fluorescence photomicrographs of neurons labelled by Fast Blue or Diamidino-Yellow (DY) alone or by both markers in the red nucleus contralateral and ipsilateral to the lesion from a case subjected to the double-labelling/lesion paradigm described in the text.
(c) and d) are higher magnifications of (a) and (b) respectively. On the contralateral side, most of the neurons were labelled by DY alone—two examples are indicated by solid arrows in (a) and (c). One double-labelled neuron is present in the field (open arrow in (a) and (c)). On the ipsilateral side, all types of labelled neurons can be seen, although most are double labelled (open arrows in (b) and (d)).
Scale bar =100 µm.
Source: Reproduced from Xu and Martin (1991), courtesy of Wiley-Liss, a division of Wiley and Sons, Inc. New York.

were labelled by DY alone (Figures 21.8a and c). It seems reasonable to suggest that the axons of such neurons were not present in the caudal cord at the time of the FB injection, but that they grew around the lesion to reach that level during the 30 day survival. A few double-labelled neurons were present, however (Figure 21.8c), even in the cases subjected to FB removal at the time of lesioning. We assumed that such neurons survived axotomy, as evidenced by the presence of FB and that they supported axons which grew caudal to the lesion to incorporate DY. Based on these results, we have concluded that the rubral axons which grew around the lesion at the ages studied were primarily late growing. Regeneration of cut axons does occur, but it is probably limited by the massive cell death which occurs after axotomy.

Evidence for growth of reticulospinal and vestibulospinal axons around a lesion of their pathway

We also asked whether axons from the medial pontine reticular nucleus and the dorsal part of the lateral vestibular nucleus (Figure 21.9a) grow around a lesion and, if so, whether the critical period(s) for such growth ends earlier than that for rubrospinal axons (Wang *et al.* 1994). We chose these particular axons for study because, as described above, they innervate the spinal cord before rubral axons (Cabana & Martin 1981, 1984; Martin *et al.* 1993) and their spinal projections are almost exclusively ipsilateral (Martin *et al.* 1975; Martin *et al.* 1993). It was our working hypotheses that all brainstem axons grow around a lesion of their pathway at some stage of development and that axons which reach the spinal cord first lose that ability before axons which arrive at a later date.

In the experiments designed to study reticulospinal and vestibulospinal plasticity, the spinal cord of anaesthetised pups was hemisected during the critical period for rubrospinal plasticity and approximately 30 days later bilateral injections of FB were made about 4 segments caudal to the lesion. The pups were maintained for 4 to 5 days after surgery before being killed by an overdose of anaesthetic and perfused so the brain and spinal cord could be removed and sectioned, as described above. Sections through the medial pontine reticular nucleus, the lateral vestibular nucleus, and the red nucleus were examined for labelled neurons using a fluorescence microscope and the A cube of the Ploem illumination system. When the hemisection was made at 19 days, labelled neurons were not present in either the medial pontine reticular (Figure 21.10b) or lateral vestibular nucleus (Figure 21.10d) ipsilateral to the lesion (the experimental side) although they were present on the contralateral (control) side (Figures 21.10a and c). As expected from previous studies, neurons were labelled in the red nucleus contralateral to the lesion (the experimental side; Figure 21.10e) although they were not as numerous as on the ipsilateral (control) side (Figure 21.10f). We concluded that reticular and vestibular axons had not grown around the lesion, at least not to the level of the injection, although rubral axons had done so. When the hemisection was made at 4 or 11 days, however, pontine reticular and lateral vestibular neurons were labelled ipsilateral to the lesion. The results from a 4 day case are documented in Figure 21.11 where it can be seen that labelled neurons were present on the experimental side in both the pontine reticular (Figure 21.11b) and lateral vestibular (Figure 21.11d) nuclei, although they were not as numerous as on the control side (Figures 21.11a and c). As expected, rubral neurons were labelled bilaterally (Figures 21.11e and f). We concluded that reticular and vestibular axons grow around a lesion of their pathway during development, but that the critical period for such plasticity ends earlier than that for rubrospinal axons. In the above experiments, cases were removed from the study if the hemisection was not complete at the time of sacrifice.

Evidence for growth of selected supraspinal axons through a lesion of their spinal pathway

In most of the above experiments it appeared that supraspinal axons grew around the lesion, not through it. The first definitive evidence for growth through a lesion was obtained in orthograde tracing experiments designed to help document reticulospinal and vestibulospinal plasticity (Martin *et al.* 1994). Two of the pups subjected to hemisection of the thoracic cord (T-7 to T-8) at 4 days were maintained until adulthood

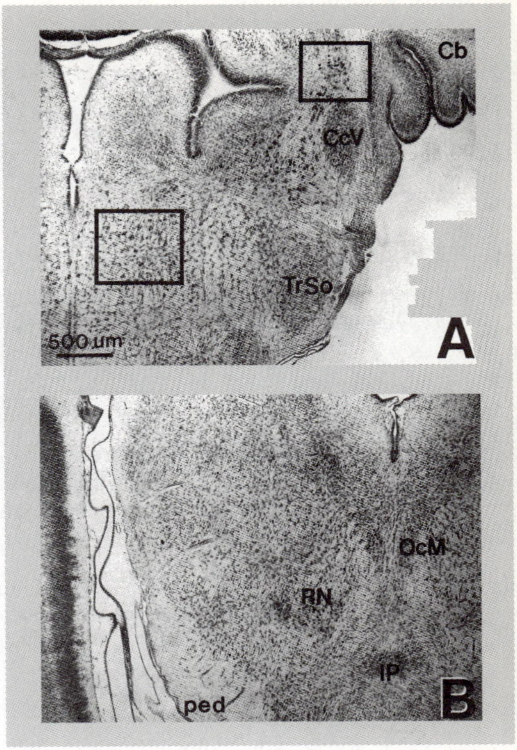

Figure 21.9
Light field photomicrographs of Nissl-stained sections through (a) the pons and (b) midbrain of an animal killed at 40 days. The rectangles in (a) outline those portions of the pontine reticular (left side) and lateral vestibular (right side) nuclei analyzed in the reticulospinal, vestibulospinal plasticity studies. The cerebellum (cb), the ventral cochlear nucleus (CcV), the interpeduncular nucleus (IP), the oculomotor nucleus (OcM), the cerebral peduncle (ped), the red nucleus (RN) and the oral spinal trigeminal nucleus (TrSo) are indicated. Source: Reproduced from Wang *et al.* (1994), courtesy of Wiley-Liss, a division of Wiley and Sons, Inc. New York.

when they were anaesthetised and re-operated (using sterile techniques) so that stereotaxic injections of the fluorescent marker Fluoro-Ruby could be made into either the pontine reticular or lateral vestibular nucleus ipsilateral to the lesion. The animals were maintained for approximately 30 days before being killed and perfused so that the brain and spinal cord could be removed for sectioning. Sections through the injection site, the brainstem caudal to the injection, and the spinal cord were mounted onto glass slides and examined using a Leitz Orthoplan fluorescence microscope equipped with filter cube N2 of the Ploem illumination system (excitation wavelength=530 to 560 nm). In the spinal cord rostral to the lesion, labelled axons were present in the ventral funiculus ipsilateral to the injection (Figures 21.12a and b) and within laminae VII and VIII of the grey matter, primarily on the same side. In contrast to what was expected, however, labelled axons were present in partially reconstructed cord on the side of the lesion (Figures. 21.12c and e); none were found on the contralateral side. As predicted from the retrograde tracing results, labelled axons were also present caudal to the lesion (Figure 21.12d). In both cases, labelled axons could be traced to T-11. We tentatively concluded that the spinal cord is capable of regeneration at the lesion site when hemisections are made at 4 days and that supraspinal axons grow through the lesion, not around it. A Nissl-stained section through the thoracic cord of a 4 day pup is shown in Figure 21.12f. It is not yet clear why the spinal cord regenerates after hemisection at 4 days in some cases, but not others.

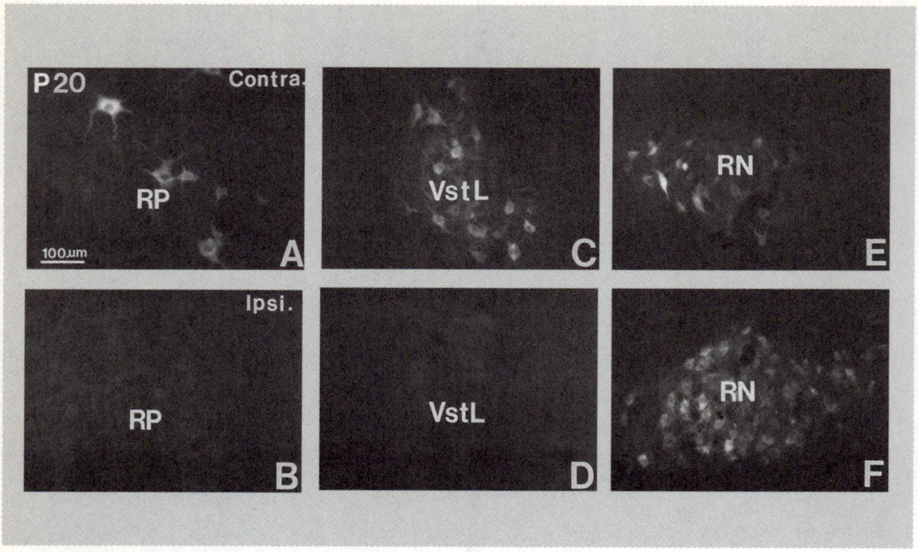

Figure 21.10 Fluorescence photomicrographs from (a) (b) the medial part of the pontine reticular nucleus, (c) (d) the dorsal part of the lateral vestibular nucleus and (e) (f) the red nucleus contralateral (Contra.) and ipsilateral (Ipsi.) to the lesion in a pup subjected to hemisection of the cord at 19 days and bilateral injections of FB caudal to the lesion 30 days later.
Scale bar =100 µm.
Source: Reproduced from Wang *et al.* (1994), courtesy of Wiley-Liss, a division of Wiley and Sons, Inc. New York.

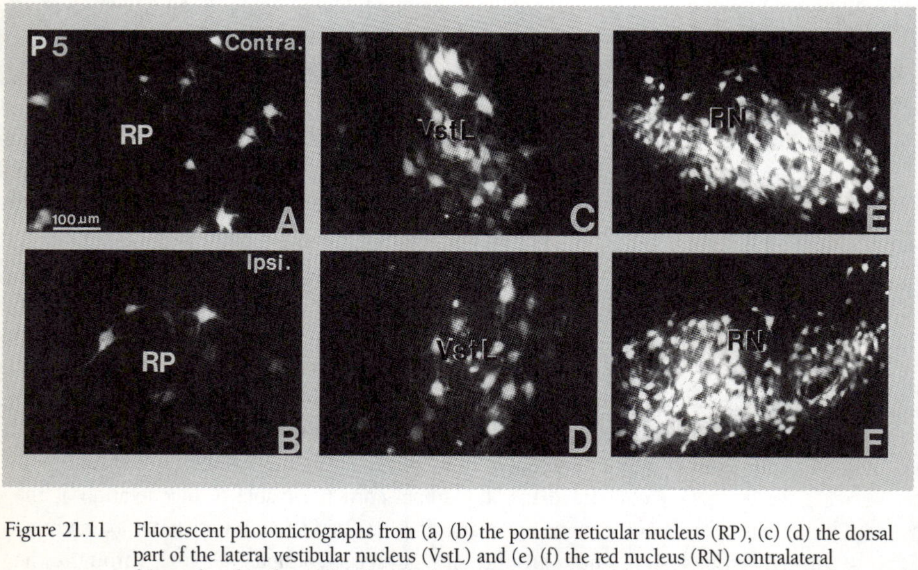

Figure 21.11 Fluorescent photomicrographs from (a) (b) the pontine reticular nucleus (RP), (c) (d) the dorsal part of the lateral vestibular nucleus (VstL) and (e) (f) the red nucleus (RN) contralateral (Contra.) and ipsilateral (Ipsi.) to the lesion in a pup subjected to hemisection of the cord at 4 days and bilateral injections of FB caudal to the lesion 30 days later.
Scale bar =100 µm.
Source: Reproduced from Wang *et al.* (1994), courtesy of Wiley-Liss, a division of Wiley and Sons, Inc. New York.

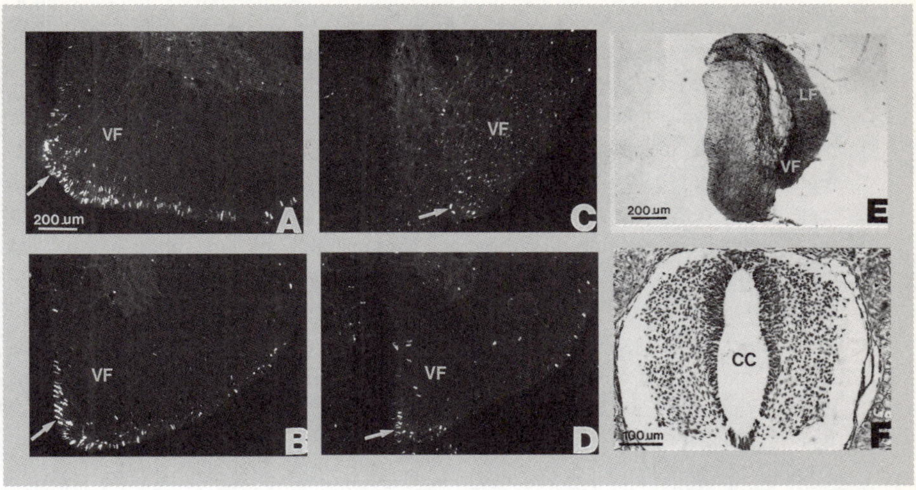

Figure 21.12 (a) to (d) Fluorescence photomicrographs of axons labelled by an injection of Fluoro-Ruby into the lateral vestibular nucleus of an adult opossum subjected to a hemisection of the mid-thoracic cord at 4 days. Labelled axons, a few of which are indicated by arrows, are present in the ventral funiculus (VF) ipsilateral to the injection at (a) cervical and (b) rostral thoracic levels, and in the same area at (c) the lesion site and (d) caudal to it.
(e) Rephotographed section (c) from which the coverslip was removed so that it could be stained for Nissl substance and rephotographed at lower power using light field optics. The lateral (LF) and ventral (VF) funiculi are indicated.
(f) A Nissl-stained section through the thoracic cord of a 4-day-old pup, the age at the time of lesioning in the case documented in (a) to (e). The central canal (cc) is indicated.
Scale bar for (a) to (e)=200 µm; for (f)=100 µm.

EVIDENCE FOR GROWTH OF SUPRASPINAL AND PROPRIOSPINAL AXONS THROUGH A COMPLETE TRANSECTION OF THE SPINAL CORD

Supraspinal axons grow across a transection of the spinal cord *in vivo* in developing birds (Hasan *et al.* 1991; 1993; Shimizu *et al.* 1990) and amphibians (Beattie *et al.* 1990; Forehand & Farel 1982), and to some extent in adult lizards (for example Duffey *et al.* 1990; 1992), salamanders (for example Davis *et al.* 1989; 1990), teleost fish (for example Bentley & Zottoli 1993; Coggeshall *et al.* 1982; Sharma *et al.* 1993) and lampreys (for example Cohen 1988; Davis and McClellan 1993; 1994; Selzer 1977; Wood & Cohen 1979). *In vitro* evidence for comparable growth has been reported in neonatal opossums (the South American opossum, *Monodelphis domestica*) and fetal rats (Saunders *et al.* 1992; Treherne *et al.* 1992; Woodward *et al.* 1993) so it seemed reasonable to predict that it would also occurs *in vivo.* In order to test that prediction, we transected the mid-thoracic cord of anaesthetised pups at 4 days (Martin *et al.* 1994). Some of the pups were killed with an overdose of anaesthetic and perfused shortly after lesioning so that sagittal sections of the spinal cord could be cut and stained for Nissl substance to verify the presence of a complete transection (Figure 21.13a). The remaining animals were returned to the vivarium with their mother for approximately 30 days to provide time for growth of axons across the lesion. After the appropriate survival, the pups were re-anaesthetised and re-operated so that bilateral injections of FB could be made several segments caudal to the lesion. They were maintained 5 to 7 days for retrograde transport before being killed by an overdose of anaesthetic and perfused so that their spinal cord and brain could be removed for sectioning. Frozen sections through the lesion site, the injection sites, the spinal cord rostral to the lesion and the brainstem were mounted onto glass slides and examined using a fluorescence microscope as described above. It could be seen grossly that regeneration had occurred at the lesion site in all cases. In fact, the site of transection could only be identified because it was slightly thinner than normal (Figure 21.13b). After assuring ourself that the sections through the lesion did not contain FB, the cover slips were removed and they were stained for Nissl substance. Recognizable spinal cord was present at the lesion site although it was abnormal in appearance (Figure 21.13c). In the spinal cord rostral to the lesion (Figure 21.13d) and in the brainstem (Figure 21.14), neurons were labelled in all of the areas expected from the results of comparable injections in age-matched, unlesioned controls. Counts of labelled neurons in the medial pontine reticular nuclei, and the lateral vestibular nuclei (Figures 21.14e and f) were comparable in experimental and control cases. The number of labelled profiles in the red nuclei of experimental cases was less than in controls.

In subsequent experiments, we transected the thoracic cord at 11 days and made injections of FB caudal to the lesion 30 days later, as described above. When the animals were killed 5 to 7 days after the injections, the spinal cord was very thin at the lesion site and histologically (Nissl preparations), it contained relatively few neurons and little recognisable architecture. Most of the cells at the lesion site appeared to be connective tissue in origin and/or glia. Labelled neurons were present in the spinal cord rostral to the lesion and within the brainstem, however, although they were much fewer in number than in the 4 day cases and restricted in location. Most noticeable was the lack of labelling within the medial pontine reticular nuclei and the dorsal part of the lateral vestibular nuclei. Neurons were labelled bilaterally in the red nuclei, although they were not as numerous as in the controls. When the transection was performed at 19 days, labelled neurons were sparse rostral to the lesion and in the brain- stem they were limited mainly to the raphe and adjacent reticular formation, the coeruleus complex and the red nucleus where they were most numerous. In pups subjected to transection at about 25 days, a few neurons were labelled in the raphe and red nucleus. A more detailed account of this work has been published (Wang *et al.* 1996).

We have concluded from the above experiments that regeneration of relatively normal structure occurs after complete transection of the spinal cord at very early stages of development and that propriospinal and supraspinal axons grow across the lesion. The results of recent double-labelling experiments indicate that at least some of the growth across the lesion resulted from regeneration of cut axons. As noted earlier, axons from the pontine reticular formation and lateral vestibular nucleus reach the spinal cord well before axons from the red nucleus. It is interesting, therefore, that the critical (permissive) period for growth of pontine reticular and lateral vestibular axons through the lesion site ends earlier than for comparable growth of rubrospinal axons.

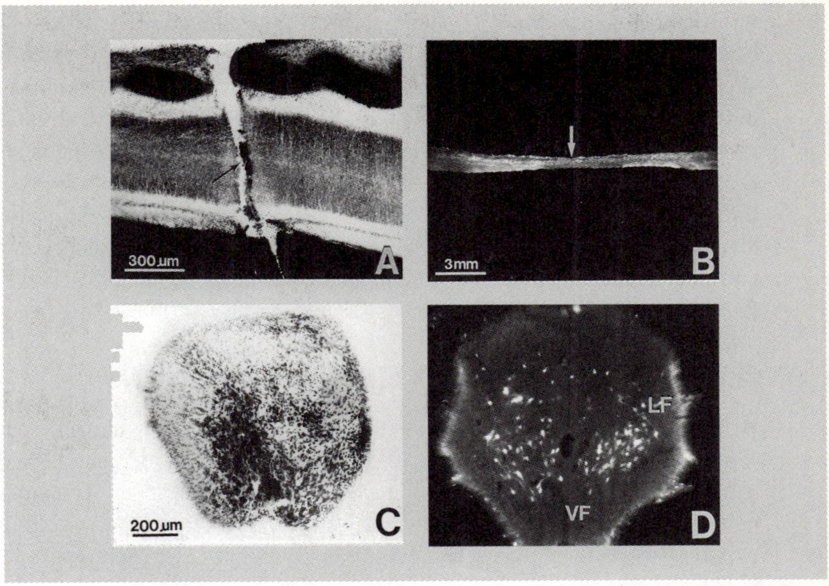

Figure 21.13 (a) The lesion site (arrow) in a pup subjected to transection of the mid-thoracic cord at 4 days and killed shortly thereafter, shown microscopically (Nissl-stained, sagittally cut section).
(b) The lesion site from an animal which survived 30 days after transection, shown grossly (Nissl-stained section).
(c) The lesion site from an animal which survived 30 days after transection, shown microscopically (Nissl-stained section). Note that recognisable cytoarchitecture is present although it is not normal.
(d) A fluorescence photomicrograph of the propriospinal labelling rostral to the transection produced by bilateral injections of FB caudal to the lesion. The injections were made 30 days after lesioning. The lateral (LF) and ventral (VF) funiculi are indicated on one.
Scale bar for (a)=300 µm; for (b)=3mm; for (c) and (d)=200 µm.

POSSIBLE EXPLANATIONS FOR FAILURE OF SUPRASPINAL AXONS TO GROW AROUND OR THROUGH A LESION AFTER THE END OF THE CRITICAL PERIOD

It is tempting to conclude from the results described that failure of descending spinal axons to grow around or through a lesion results from decreased growth potential with age (the role of intrinsic growth determinants of regeneration has been reviewed by Fawcett 1992). It is possible, however, that maturation of the environment through which growth must occur also plays a role. Since central nervous system myelin contains proteins which inhibit axonal growth (Caroni & Schwab 1988a; b; Schnell & Schwab 1990; Kapfhammer *et al.* 1992; Mukhopadhyay *et al.* 1994), we asked whether the end of the critical period for growth of supraspinal axons around or through a lesion correlates temporally with the development of myelin (Ghooray & Martin 1993a). When the spinal cords of developing opossums were immunostained for myelin basic protein (MBP), immunoreactivity was found in the ventral funiculus before the lateral funiculus (Figure 21.15), but in each funiculus its appearance correlated roughly with the end of the critical period for growth around or through a lesion of the brainstem axons contained within it. These results suggested that the development of myelin may be a factor in loss of developmental plasticity. A cause and effect relationship between myelin formation and failure of supraspinal axons to grow through a lesion has been claimed for the chick *in vivo* (Keirstead *et al.* 1992) and a temporal correlation between developmental expression of myelin-associated neurite growth inhibitors and failure of axons to grow across regenerated cord *in vitro* has been documented for the South American opossum (Schwab *et al.* 1993; Varga *et al.* 1995).

CONCLUDING REMARKS AND FUTURE DIRECTIONS

We have shown that propriospinal and supraspinal axons grow into the spinal cord very early in development and that they do so sequentially. Although detailed birth date studies have not been reported in *Didelphis*, it is possible that the neurons which innervate the spinal cord first are the ones which are generated the earliest. Based on unpublished ^{3}H-thymidine material from the collection of Dr J Johnson of Michigan State University, it appears that most neurons of the medial pontine reticular formation are generated prior to birth and it will be recalled that axons which originate within that area are among the first to innervate the spinal cord. At the other end of the spectrum, the cerebral cortex, which matures much later than the brainstem, is the last region to innervate the spinal cord. It is possible, of course, that the delay between birth date and generation of a spinal axon is greater for some neurons than others and that birth date alone does not explain the developmental sequence revealed by our studies. This issue should be addressed in future experiments.

It now seems clear that brainstem axons grow around a lesion of their spinal pathway during development in opossums and that such plasticity results primarily from growth of late arriving axons. It also appears that the opossum's spinal cord is able to regenerate after injury during early development and that propriospinal and supraspinal axons grow through the lesion. Comparison of the results obtained from developmental studies and those derived from studies of developmental plasticity suggest that a correlation exists between developmental history and the end of the critical period for plasticity. Axons which grow into the spinal cord first appear to lose the ability to grow around or through a lesion before axons which arrive at a later date.

Several issues should be addressed in future studies of developmental plasticity. It will certainly be important to identify the relative roles played by diminished growth potential with age and the changing environment through which growth must occur in loss of plasticity. Studies designed to address the environment should consider the possible significance of extracellular matrix molecules (for example, laminin, fibronectin, heparin sulfate proteoglycans) as well as myelin-associated inhibitory proteins. Supraspinal neurons die in large numbers after axotomy in immature mammals and it is possible that they are particularly susceptible to loss of trophic factors. If trophic factors rescue axotomised neurons during development, as suggested for rubrospinal neurons in neonatal rats (Diener-Ostfield *et al.* 1993), it may be possible to increase the number of neurons available to support regenerating axons. It will also be of interest to determine whether axons which grow around or through a lesion of their spinal pathway support normal function at later stages of development and in the

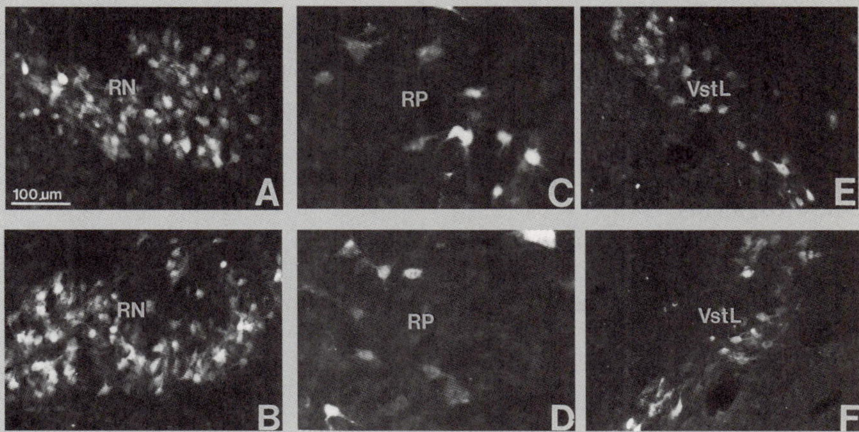

Figure 21.14 Fluorescence photomicrographs of labelled neurons in (a) (b) the red nuclei (RN), (c) (d) the medial pontine reticular nuclei (RP) and (e) (f) the dorsal part of the lateral vestibular nuclei (VstL) from a case subjected to transection of the thoracic cord at 4 days and bilateral injections of FB several segments caudal to the lesion 30 days later.
Note: Labelling is bilaterally symmetrical in all three areas.
Scale bar =100 µm.

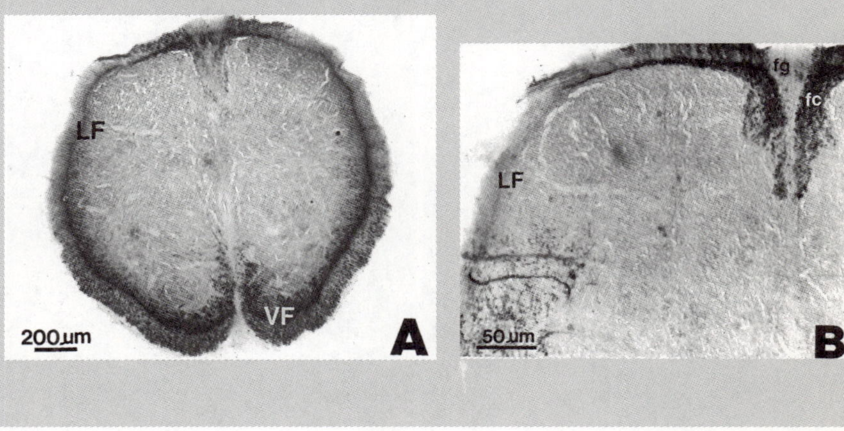

Figure 21.15 Photomicrographs of myelin basic protein (MBP) immunoreactivity in sections through (a) the thoracic and (b) the cervical cord of a 25-day-old opossum.
Note: MBP is present in the ventral funiculus (VF), but not the lateral funiculus (LF) at thoracic levels (a). Even at cervical levels, evidence for MBP is not present in the dorsal part of the lateral funiculus (b). fg = fasciculus gracilis fc = fasciculus cuneatus.

adult animal. Results from our laboratory using *Didelphis* (Basso *et al.* 1996) indicate that axons which grow through lesions made at 4 days postnatal support remarkably normal use of the hindlimbs at maturity. Comparable results have also been presented for the South American opossum, *Monodelphis domestica*. (Saunders *et al.* 1994, 1995).

Studies on developmental plasticity may be useful clinically. If the rules which govern growth of spinal axons around or through a lesion during development can be identified, it may be possible to recapitulate comparable plasticity in the injured spinal cord of adult mammals, including humans.

Acknowledgments

The authors would like to thank Ms Mary Ann Jarrell for surgical assistance, tissue processing, plotting, and typing the manuscript, as well as Mr. Karl Rubin for photographic help. The studies described herein were supported by United States Public Health Service grants NS-25095 and NS-10165 to Dr Martin.

References

Basso D M, Terman J R, Wang X M, Martin G F and Bresnahan (1996). Opossums develop some stepping after spinal cord transection (TX) on postnatal day (PD) 8 but develop normal locomotion after TX on PD 5. *Society for Neuroscience Abstracts*, 22, 1489.

Bates C A and Stelzner D J (1989). Regeneration of corticospinal axons in the neonatal rat. *Anatomical Record*, 223, 12A.

Bates C A and Stelzner D J (1993). Extension and regeneration of corticospinal axons after early spinal injury and the maintenance of corticospinal topography. *Experimental Neurobiology*, 123, 106–117.

Beattie M S, Bresnahan J C and Lopate G (1990). Metamorphosis alters the response to spinal cord transection in *Xenopus laevis* frogs. *Journal of Neurobiology*, 21, 1108–1122.

Bentley A P and Zottoli S J (1993). Central nervous system lesion triggers inappropriate pathway choice in adult vertebrate system. *Brain Research*, 630, 333–336.

Bernstein D R and Stelzner D J (1983). Plasticity of the corticospinal tract following midthoracic spinal injury in the neonatal rat. *Journal of Comparative Neurology*, 221, 382–400.

Bregman B S and Goldberger M E (1982). Anatomical plasticity and sparing of function after spinal cord damage in neonatal cats. *Science*, 217, 553–555.

Bregman B S and Goldberger M E (1983a). Infant lesion effect: I. Development of motor behavior following neonatal spinal cord injury in cats. *Developmental Brain Research*, 9, 103–117.

Bregman B S and Goldberger M E (1983b). Infant lesion effect. III. Anatomical correlates of sparing and recovery of function after spinal cord damage in newborn and adult cats. *Developmental Brain Research*, 9, 137–154.

Cabana T and Martin G F (1981). The origin of brainstem-spinal projections at different stages of development in the North American opossum. *Developmental Brain Research*, 2, 163–168.

Cabana T and Martin G F (1984). Developmental sequences in the origin of descending spinal pathways. Studies using retrograde transport techniques in the North American opossum (*Didelphis virginiana*). *Developmental Brain Research*, 15, 247–263.

Cabana T and Martin G F (1985). Corticospinal development in the North American opossum: Evidence for a sequence in the growth of cortical axons in the spinal cord and for transient projections. *Developmental Brain Research*, 23, 69–80.

Cabana T and Martin G F (1986). The development of the rubrospinal tract. An experimental study using the orthograde transport of WGA-HRP in the North American opossum. *Developmental Brain Research*, 30, 1–11.

Caroni P and Schwab M E (1988). Two membrane protein fractions from rat central myelin with inhibitory properties for neurite growth and fibroblast spreading. *Journal of Cell Biology*, 106, 1281–1288.

Caroni P and Schwab M E (1988). Antibody against myelin associated inhibitor of neurite growth neutralizes non-permissive substrate properties of CNS white matter. *Neuron*, 1, 85–96.

Coggeshall R E, Birse S G and Youngblood C S (1982). Recovery from spinal transection in fish. *Neuroscience Letters*, 32, 259–264.

Cohen A H (1988). Regenerated fibers of the lamprey spinal cord can coordinate fictive swimming in the presence of curare. *Journal of Neurobiology*, 19, 193–198.

Cutts J H, Krause W J and Leeson C R (1978). General observations on the growth and development of the pouch-young opossum, *Didelphis virginiana*. *Biology of the Neonate*, 33, 264–272.

Davis B M, Duffy M T and Simpson Jr S B (1989). Bulbospinal and intraspinal connections in normal and regenerated salamander spinal cord. *Experimental Neurology*, 103, 41–51.

Davis B M, Ayers J L, Koran L, Carlson J, Anderson M C and Simpson S B Jr. (1990). Time course of salamander spinal cord regeneration and recovery of swimming: HRP retrograde pathway tracing and kinematic analysis. *Experimental Neurology*, 108, 198–213.

Davis G R and McClellan A D (1993). Time course of anatomical regeneration of descending brainstem neurons and behavioral recovery in spinal-transected lamprey. *Brain Research*, 602, 131–137.

Davis G R and McClellan A D (1994). Extent and time course of restoration of descending brainstem projections in spinal cord-transected lamprey. *Journal of Comparative Neurology*, 344, 65–82.

Diener-Ostfield P, DiStefano P S, McAtee M M and Bregman B S (1993). Trophic and trophic influence of exogenous NT-3 and BDNF after neonatal spinal cord lesions. *Society for Neuroscience Abstracts*, 19, 1105.

DiTirro F J, Martin G F and Ho R H (1983). A developmental study of substance-P, somatostatin, enkephalin and serotonin immunoreactive elements in the spinal cord of the North American opossum. *Journal of Comparative Neurology*, 213, 241–261.

Donatelle J M (1977). Growth of the corticospinal tract and the development of placing reactions in the postnatal rat. *Journal of Comparative Neurology*, 175, 207–232.

Duffy M T, Simpson S B Jr, Liebich D R and Davis B M (1990). Origin of spinal cord axons in the lizard regenerated tail: supernormal projections from local spinal neurons. *Journal of Comparative Neurology*, 293, 208–222.

Duffy M T, Liebich D R, Garner L K, Hawrych A, Simpson Jr S B and Davis B M (1992). Axonal sprouting and frank regeneration in the lizard tail spinal cord: correlation between changes in synaptic circuitry and axonal growth. *Journal of Comparative Neurology*, 316, 363–374.

Fawcett J W (1992). Intrinsic neuronal determinants of regeneration. *Trends in Neurosciences*, 15, 5–8.

Forehand C J and Farel P B (1982). Spinal cord development in anuran larvae. II. Ascending and descending pathways. *Journal of Comparative Neurology*, 209, 495–508.

Forehand C J and Farel P B (1982). Anatomical and behavioral recovery from the effects of spinal cord transection: dependence on metamorphosis in anuran larvae. *Journal of Neuroscience*, 2, 654–662.

Ghooray G T and Martin G F (1993a). The development of myelin in the spinal cord of the North American opossum and its possible role in loss of rubrospinal plasticity. A study using myelin basic protein and galactocerebroside immunohistochemistry. *Developmental Brain Research*, 72, 67–74.

Ghooray G T and Martin G F (1993b). The development of radial glia and astrocytes in the spinal cord of the North American opossum, *Didelphis virginiana*. An immunohistochemical study using anti-vimentin and anti-glial fibrillary acidic protein. *Glia*, 9, 1–9.

Glover J C and Petursdottir G (1991). Regional specificity of developing reticulospinal, vestibulospinal, and vestibulo-ocular projections in the chick embryo. *Journal of Neurobiology*, 22, 353–376.

Hartman C G (1952). *Possums*. University of Texas Press, Austin.

Hasan S J, Nelson B H, Valenzuela J I, Keirstead H S, Shull S E, Ethell D W and Steeves J D (1991). Functional repair of transected spinal cord in embryonic chick. *Restorative Neurology and Neuroscience*, 2, 137–154.

Hasan S J, Keirstead H S, Muri G D and Steeves J D (1993). Axonal regeneration contributes to repair of injured brainstem-spinal neurons in embryonic chick. *Journal of Neuroscience*, 13, 492–507.

Kalil K and Reh T (1979). Regrowth of severed axons in the neonatal central nervous system: establishment of normal connections. *Science*, 205, 1158–1161.

Kalil K and Reh T (1982). A light and electron microscopic study of regrowing pyramidal tract fibers. *Journal of Comparative Neurology*, 211, 265–275.

Kapfhammer J P, Schwab M E and Schneider G E (1992). Antibody neutralization of neurite growth inhibitors from oligodendrocytes results in expanded pattern of postnatally sprouting retinocollicular axons. *Journal of Neuroscience*, 12, 2112–2119.

Keirstead H S, Hasan S J, Muir G D and Steeves J D (1992). Suppression of the onset of myelination extends the permissive period for the functional repair of embryonic spinal cord. *Proceedings of the National Academy of Sciences USA*, 89, 11664–11668.

Kimmel C B (1982). Reticulospinal and vestibulospinal neurons in the larvae of a teleost fish, *Brachidanio rerio*. In *Descending pathways to the spinal cord, Progress in Brain Research*. Eds H G J M Kuypers and G F Martin, Elsevier, Amsterdam. Vol. 57, pp 1–23.

Kudo N, Furukawa F and Okado N (1993). Development of descending fibers to the rat embryonic spinal cord. *Neuroscience Research*, 16, 131–141.

Lakke E A J F and Marani E (1991). Prenatal descent of rubrospinal fibers through the spinal cord of the rat. *Journal of Comparative Neurology*, 314, 67–78.

Lende R A (1963a). Sensory representation in the cerebral cortex of the opossum (*Didelphis virginiana*). *Journal of Comparative Neurology*, 121, 395–403.

Lende R A (1963b). Motor representation in the cerebral cortex of the opossum (*Didelphis virginiana*). *Journal of Comparative Neurology*, 121, 405–415.

Martin G F and Fisher A M (1968). A further evaluation of the origin, the course and the termination of the opossum corticospinal tract. *Journal of Neuroscience*, 7, 177–187.

Martin G F and Dom R (1970). The rubrospinal tract of the opossum, *Didelphis virginiana*. *Journal of Comparative Neurology*, 138, 19–30.

Martin G F, Beattie M S, Bresnahan J C, Henkel C K and Hughes H C (1975). Cortical and brainstem projections to the spinal cord of the American opossum, *Didelphis marsupialis virginiana*. *Brain, Behavior and Evolution*, 12, 270–310.

Martin G F and Cabana T (1985). Cortical projections to superficial laminae of the dorsal horn and to the ventral horn of the spinal cord in the North American opossum. Studies using the orthograde transport of WGA-HRP. *Brain Research*, 337, 188–192.

Martin G F, Cabana T and Hazlett J C (1986). The development of rubrospinal, cerebellorubral and corticorubral connections in the North American opossum. Evidence for asynchronism. *Neurochemical Pathology*, 5, 221–236.

Martin G F, Cabana T and Hazlett J C (1988). The development of selected rubral connections in the North American opossum. *Behavioral Brain Research*, 28, 21–28.

Martin G F and Xu X M (1988). Evidence for developmental plasticity of the rubrospinal tract. Studies using the North American opossum. *Developmental Brain Research*, 39, 303–308.

Martin G F, Ho R H, Xu X M and Pindzola R R (1991). The development of descending spinal pathways in the North American opossum. In *Developmental Neurobiology of Breathing*. Eds GG Haddad and JP Farber, Marcel Dekker, New York. pp 177–198.

Martin G F, Ghooray G T, Ho R R, Pindzola R R and Xu X M (1991). The origin of serotoninergic projections to the lumbosacral spinal cord at different stages of development in the North American opossum. *Developmental Brain Research*, 58, 203–213.

Martin G F, Pindzola R R and Xu X M (1993). The origins of descending projections to the lumbar cord at different stages of development in the North American opossum. *Brain Research Bulletin*, 30, 303–317.

Martin G F, Terman J R and Wang X M (1994). Evidence for growth of supraspinal axons through hemisected and transected spinal cord in the developing opossum. *Society for Neuroscience Abstracts*, 20, 1270.

Mesulam M-M (1978). Tetramethyl benzidine for horseradish peroxidase neurohistochemistry: A non-carcinogenic blue reaction product with superior sensitivity for visualizing neural afferents and efferents. *Journal of Histochemistry and Cytochemistry*, 26, 106–117.

Metcalfe W K, Mendelson B and Kimmel C B (1986). Segmental homologies among reticulospinal neurons in the hindbrain of the zebrafish larva. *Journal of Comparative Neurology*, 251, 147–159.

Mukhopadhyay G, Doherty P, Walsh F S, Crocker P R and Filbin M T (1994). A novel role for myelin-associated glycoprotein as an inhibitor of axonal regeneration. *Neuron*, 13, 757–767.

Norlander R H, Baden S T and Ryba T M (1985). Development of early brainstem projections to the tail spinal cord of *Xenopus*. *Journal of Comparative Neurology.* 231, 519–529.

Okado N and Oppenheim R (1985). The onset and development of descending pathways to the spinal cord in the chick embryo. *Journal of Comparative Neurology*, 232, 143–161.

Oswaldo-Cruz E and Rocha-Miranda C E (1968). *The Brain of the Opossum* (*Didelphis virginiana*): *A Cytoarchitectonic Atlas in Stereotaxic Coordinates.* Universidade Federal do Rio de Janeiro, Rio de Janeiro.

Oudega M, Varon S and Hagg T (1994). Distribution of corticospinal motor neurons in the postnatal rat: Quantitative evidence for massive collateral elimination and modest cell death. *Journal of Comparative Neurology*, 347, 115–126.

Pindzola R R, Ho R H and Martin G F (1990). The development of catecholaminergic projections to the spinal cord in the North American opossum, *Didelphis virginiana*. *Journal of Comparative Neurology*, 24, 399–417.

Prendergast L J and Stelzner D (1976). Changes

in the magnocellular portion of the red nucleus following thoracic hemisection in the neonatal and adult rat. *Journal of Comparative Neurology*, 166, 163–172.

Saunders N R, Balkwill P, Knott G, Habgood M D, Mollgard K, Treherne J M and Nicholls J G (1992). Growth of axons through a lesion in the intact CNS of fetal rat maintained in long-term culture. *Proceedings of the Royal Society of London B*, 150, 171–180.

Saunders N R, Weller L, Deal A, Knott G W (1994). Recovery from spinal cord injury in neonatal South American opossums. *Society for Neuroscience Abstracts*, 20, 1269.

Saunders N R, Deal A, Knott G W, Varga Z M and Nicholls J G (1995). Repair and recovery, following spinal cord injury in a neonatal marsupial (*Monodelphis domestica*). *Clinical and Experimental Pharmacology and Physiology*, 22, 518–526.

Schnell L and Schwab M E (1990). Axonal regeneration in the rat spinal cord produced by an antibody against myelin-associated neurite growth inhibitors. *Nature*, 343, 269–272.

Schreyer D J and Jones E G (1982). Growth and target finding by axons of the corticospinal tract in prenatal and postnatal rats. *Neuroscience*, 7, 1837–1853.

Schreyer D J and Jones E G (1983). Growing corticospinal axons by-pass lesions of neonatal rat spinal cord. *Neuroscience*, 9, 31–40.

Schwab M E, Bandtlow C E, Varga Z and Nicholls J (1993). Developmental expression of myelin-associated neurite growth inhibitors correlates with loss of regeneration after spinal cord lesions in the opossum. *Society for Neuroscience Abstracts*, 19, 682.

Selzer M (1977). Mechanisms of functional recovery and regeneration after spinal cord transection in larval sea lampreys. *Journal of Physiology*, 277, 395–408.

Sharma S C, Jadhao A G and Prasada Rao P D (1993). Regeneration of supraspinal projection neurons in the adult goldfish. *Brain Research*, 620, 221–228.

Shimizu I, Oppenheim R W, O'Brien M and Shneiderman A (1990). Anatomical and functional recovery following spinal cord transection in the chick embryo. *Journal of Neurobiology*, 21, 918–937.

Theriault E and Tatton W G (1989). Postnatal redistribution of pericruciate motor cortical projections within the kitten spinal cord. *Developmental Brain Research*, 45, 219–237.

Tolbert D L and Der T (1987). Redirected growth of pyramidal tract axons following neonatal pyramidotomy in cats. *Journal of Comparative Neurology*, 260, 299–311.

Treherne J M, Woodward S K A, Varga Z M, Ritchie J M and Nicholls J G (1992). Restoration of conduction and growth of axons through injured spinal cord of neonatal opossum in culture. *Proceedings of the National Academy of Science USA*, 89, 431–434.

van Mier P and ten Donkelaar H J (1984). Early development of descending pathways from the brain to the spinal cord in *Xenopus laevis*. *Anatomy and Embryology*, 170, 295–306.

Varga Z M, Schwab M E and Nichols J G (1996). Suppression of regeneration after lesion of mammalian spinal cord in culture by myelin-associated neurite growth inhibitory proteins. *Proceedings of the National Academy of Science USA*, 92, 10959–10963.

Wang X M, Xu X M, Qin Y Q and Martin G F (1992). The origins of supraspinal projections to the cervical and lumbar spinal cord at different stages of development in the gray short-tailed Brazilian opossum, *Monodelphis domestica*. *Developmental Brain Research*, 68, 203–216.

Wang X M, Qin Y Q, Xu X M and Martin G F (1994). Developmental plasticity of reticulospinal and vestibulospinal axons in the North American opossum, *Didelphis virginiana*. *Journal of Comparative Neurology*, 349, 288–302.

Wang X M, Terman J R and Martin G F (1996). Evidence for growth of supraspinal axons through the lesion after transection of the thoracic spinal cord in the developing opossum, *Didelphis virginiana*. *Journal of Comparative Neurology*, 371, 104–115.

Wood M R and Cohen M J (1979). Synaptic regeneration in identified neurons of the lamprey spinal cord. *Science*, 206, 344–347.

Woodward S K A, Treherne J M, Knott G W, Fernandez J, Varga Z M and Nicholls J G (1993). Development of connections by axons growing through injured spinal cord of neonatal opossum in culture. *Journal of Experimental Biology*, 176, 77–88.

Xu X M and Martin G F (1989). Developmental plasticity of the rubrospinal tract. Studies using the North American opossum. *Journal of Comparative Neurology*, 279, 368–387.

Xu X M and Martin G F (1991). Ipsilaterally projecting rubrospinal neurons in adult and developing opossums. *Anatomical Record*, 231, 538–547.

Xu X M and Martin G F (1992). The response of rubrospinal neurons to axotomy at different stages of development in the North American opossum. *Journal of Neurotrauma*, 9, 93–105.

CHAPTER 22

MARSUPIALS AS MODELS FOR STUDIES OF DEVELOPMENT AND REGENERATION OF THE CENTRAL NERVOUS SYSTEM

Norman R Saunders

INTRODUCTION

Much of the work described in this chapter is due to an association with Hugh Tyndale-Biscoe that began in 1983. At that time my research group was mainly working on barriers in the developing brain, particularly in the sheep fetus. We used that experimental preparation because it was possible to study exchange of materials between blood and brain or cerebrospinal fluid (CSF) at earlier stages of brain development and under better controlled physiological conditions than was possible in commoner laboratory animals such as postnatal rats. However we had reached the limits of what was technically possible at around 40 days gestation in the sheep. It was clear that there were interesting questions to be asked about the relation between the proteins in CSF, their origins and their contribution to earlier stages of brain development than could be studied in fetal sheep, because of their fragility and small size. Kate Dziegielewska and I used the occasion of the International Union of Physiological Sciences Congress in Sydney in 1983 to spend time in Canberra with Hugh Tyndale-Biscoe and Lyn Hinds. Our research interests in the developing nervous system were very different from their work on reproduction in tammars. Nevertheless that visit and several subsequent ones were extremely productive and convinced us of the value of neonatal marsupials for studying development of the nervous system. On moving to Southampton in 1986, Kate Dziegielewska and I wanted to have a more proximate source of marsupials. Tammars were impractical and Hugh advised us to try the South American grey short-tailed opossum (*Monodelphis domestica*, referred to here as *Monodelphis*) an animal that I think neither of us had heard of. We were able to obtain a few breeding pairs from Mark Ferguson in Manchester and are delighted that he agreed to contribute (see Chapter 15) to this book in honour of Hugh. The present chapter and that by Dziegielewska *et al.* (Chapter 20) summarise our work with

Monodelphis domestica so far; these chapters also include relevant material from our studies with Hugh and Lyn in the tammar, set in the more general context of use of marsupials as models for studies of development and recovery from injury in the central nervous system (CNS).

Marsupials are born at an extremely early stage of development of their nervous system. They are able only to breathe and suck and they have just sufficient locomotor ability to climb from the site of birth to attach themselves to the teat within the pouch or on the abdominal wall in the case of pouchless marsupials. The immaturity of the nervous system of neonates of marsupial species means that a much greater proportion of CNS development occurs *ex utero* and hence is more accessible to observation and experimental manipulation, than is the case for the young of eutherian species, which would have to be studied within the uterus. The young of all marsupial species studied so far are extremely immature at birth, although some are even less mature than others (see Reynolds & Saunders 1988). In contrast, the stage of development of the CNS at birth in different eutherian species varies widely depending upon the species (cf. mouse and sheep).

Even if we compare those eutherian species that have the most immature neonates with marsupial newborns, it is clear that marsupials are born at a stage of CNS development that is equivalent to an early fetal stage in eutherians, for example, fetal mouse neocortex at E12 to E13 equates approximately to the stage of development of that brain region in the newborn of the marsupial species so far studied. Neonatal marsupials are also considerably more robust than the equivalent eutherian fetuses. This, combined with their immaturity, makes the newborn of marsupial species ideal for many types of study of the developing CNS in mammals. A further advantage of marsupial species is that a few systems in their CNS are functionally effective well before other parts of the CNS, that in the adult are present in the same region. For example, the brainstem neural substrate for rhythmic ventilation of the lungs (Nicholls *et al.* 1990; Zou 1994) the circuits required for suckling and those in the spinal cord required for locomotion to the pouch or teat following birth, can be studied more or less in isolation from the more complex circuits that develop in these regions of the more mature CNS.

The chapters in this section of the book deal with several major neurobiological problems that have been studied in marsupials in the last 10 years or so. They will only be mentioned in the present article in outline, in the context of broader issues of development and recovery from injury in the CNS of marsupials. The topics covered in this chapter are development of the neocortex, especially in opossums and tammar wallabies (also dealt with partly in Mark's article, Chapter 18) and development and recovery from injury in spinal cord of the opossum *Monodelphis* (which is complementary to Martin and Wang's article on the same problem studied in the North American opossum, *Didelphis virginiana*, see Chapter 21 in this volume)

AVAILABILITY OF MARSUPIAL SPECIES FOR NEURODEVELOPMENTAL AND CNS REPAIR STUDIES

Developmental studies that depend upon animals caught in the wild suffer from a number of serious disadvantages. The supply of animals, both in number and age range, is likely to be unpredictably variable. Ages of such pouch young can generally only be estimated by comparison with growth tables of, for example, body weight or crown-rump length); this is clearly likely to be less accurate than dates obtained from the birth date recorded in a laboratory-bred colony. The genetic origin and diversity of caught animals will be much greater and impossible to control than is the case in a captive colony. However, colonies of marsupials specifically bred for experimental purposes have only been successful for a few species. Amongst Australian species the most successful has been the tammar wallaby (*Macropus eugenii*), largely due to the work of Hugh Tyndale-Biscoe (see Prologue in this volume and Tyndale-Biscoe & Janssens 1988). A feature of the tammar, which may be an advantage in some experimental circumstances, but a disadvantage in others, is that this species normally only produces one young at a time. A definite disadvantage of the tammar, which appears to be shared by all other Australian marsupials is that they are seasonal breeders (see Chapter 18 in this volume; Tyndale-Biscoe & Janssens 1988). As described by Mark (see Chapter 18) the tammar has been used extensively by his group for studies of the development of the visual system. The other Australian marsupial that has been used extensively for CNS developmental studies, also of the visual system, is the related species, the quokka, *Setonix brachyurus*. This work has been carried out by Lyn Beazley and her colleagues in Western Australia (for example, Harman & Beazley 1989; Beazley *et al.* 1995; Dunlop 1990) using animals bred in a colony in the Department of Zoology, University of Western Australia, Nedlands.

A study of neocortical development in the possum (*Trichosurus vulpecula*) has been published by Sanderson & Weller (1990) , based on animals caught in the wild. A substantial colony of fat-tailed dunnarts (*Sminthopsis crassicaudata*) has been in existence in Adelaide for many years (Bennett *et al.* 1990) but only preliminary studies of its CNS development have been carried out (Figures 22.1, 22.2 and 22.3; N R Saunders, unpublished studies). Smaller colonies of a few other species have also been established: Northern brown bandicoot (*Isoodon macrourus*, established by Gemmell 1982, 1990); striped-faced dunnart (*Sminthopsis macroura*, Selwood & Wooley 1991). No studies of regeneration or repair of the nervous system using an Australian species seem to have been published.

In the Americas, especially North America, two species of opossum have been used increasingly for studies of development of various physiological systems including the nervous system. These are *Didelphis virginiana*, the North American opossum, referred to here as *Didelphis* and *Monodelphis*, already mentioned above. An atlas of the adult brain of *Didelphis virginiana* and the related South American opossum, *Didelphis aurita,* has been published (Oswaldo-Cruz & Rocha-Miranda 1968). Early hopes that *Didelphis* would become established as a laboratory animal that could be bred as required (McCrady 1938) do not seem to have been fulfilled; although Jurgelski and colleagues (for example, Jurgelski & Porter 1974) published detailed descriptions of the conditions required for successful breeding of *Didelphis* in the early 1970s, there were still only 10 laboratories using the opossum as a major animal model in the late 1970s (Jurgelski, 1979). As indicated by articles in this volume (Krause *et al.* Chapter 17; Martin & Wang, Chapter 21 in this volume) most current studies involving *Didelphis* have had to be carried out with a combination of specifically bred animals and ones that have been caught in the wild. In contrast, the much smaller and more tractable *Monodelphis* has been successfully established as a laboratory species, very much thanks due to the efforts of John VandeBerg (see VandeBerg 1983 and Chapter 14 in this volume). In addition to his large colony in San Antonio, there are now many colonies in different parts of North America, several of which are used for neurodevelopmental studies. There are three colonies in

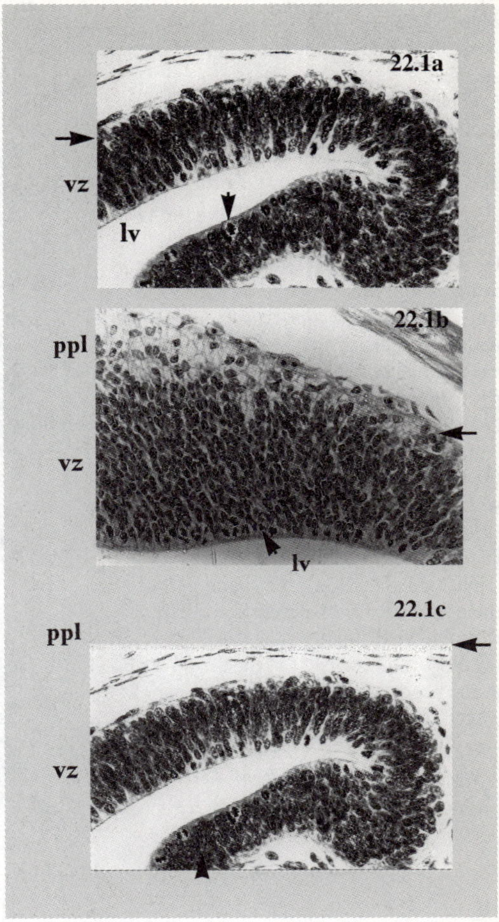

Figure 22.1
Neocortex on day of birth in 3 marsupial species
(a) *Sminthopsis crassicaudata* (fat-tailed dunnart).
(b) *Monodelphis domestica* (South American grey short-tailed opossum).
(c) *Macropus eugenii* (tammar wallaby).
Key:
lv=lateral ventricle
vz=ventricular zone
ppl=primordial plexiform layer
Note: Arrows indicate outer surface of neural tissue that is the site at which pia-arachnoid will develop. Arrowheads indicate examples of mitotic cells. Note that the brain of *Sminthopsis* is much smaller and is the least developed of the 3 species at birth. There is almost no ppl in *Sminthopsis*. The ppl is the first differentiating layer of neocortex into which the earliest migrating neurons are inserted. Subsequent migration of presumptive neurons into the ppl forms the cortical plate (cf. Figures 22.2 and 22.3) Coronal sections of dorsal neocortex from brains fixed in Bouin's fixative and embedded in paraffin wax. Sections (5–7 μm), stained with haematoxylin and eosin.
Scale bar=50 μm.

Europe (Southampton, Basel and Oxford) that originated from Mark Ferguson's colony in Manchester (see Chapter 15). There is now one colony of this species in Australia, at the University of Tasmania in Hobart, that came from animals derived from the Southampton colony. All of these animals in turn, together with those now in experimental colonies in North America, came from a small number of animals caught in Catanga, Brazil, in the late 1970s and taken to the USA, with subsequent additions of small numbers of animals caught in the wild (see VandeBerg, 1983). As well as breeding relatively easily in captivity, this species has the further advantages of multiple young and in having no pouch, the young thus being even more accessible to experimental manipulation than in species with pouches. Also, under laboratory conditions they will breed all year-round, although recent studies in the wild suggest that they are normally seasonal breeders (Bergallo & Cequeira 1994). *Monodelphis* also appears to be notably disease-free. For example in the colonies in Southampton, Basel and Hobart, no animals have died that were not culled or used for experiments. The only surprise is that this species, which has been available in North America for experimental purposes since the late 1970s (see VandeBerg 1983) has not caught on as an experimental species until comparatively recently.

Figure 22.2
Neocortex 5 days after birth in 3 marsupial species.
(a) *Sminthopsis crassicaudata* (fat-tailed dunnart).
(b) *Monodelphis domestica* (South American grey short-tailed opossum). Cortical plate (cp) has just begun to appear at 5 days in *Monodelphis* and the tammar.
(c) *Macropus eugenii* (tammar wallaby).
Scale bar=50 μm.
Key:
lv=lateral ventricle
vz=ventricular zone
ppl=primordial plexiform layer
cp=cortical plate
Notes:
(1) Coronal sections of dorsal neocortex were prepared as in Figure 22.1.
(2) cp is particularly evident in the lateral part of the *Monodelphis* section (right side of section); the lateral part of the dorsal neocortex develops earlier than the more medial parts.
(3) Note the expansion of the dorsal wall of the neocortex that has occurred mainly by cell division within the ventricular zone (vz). The ppl is now clearly evident in the *Sminthopsis* section (cf. Figure 22.1a)

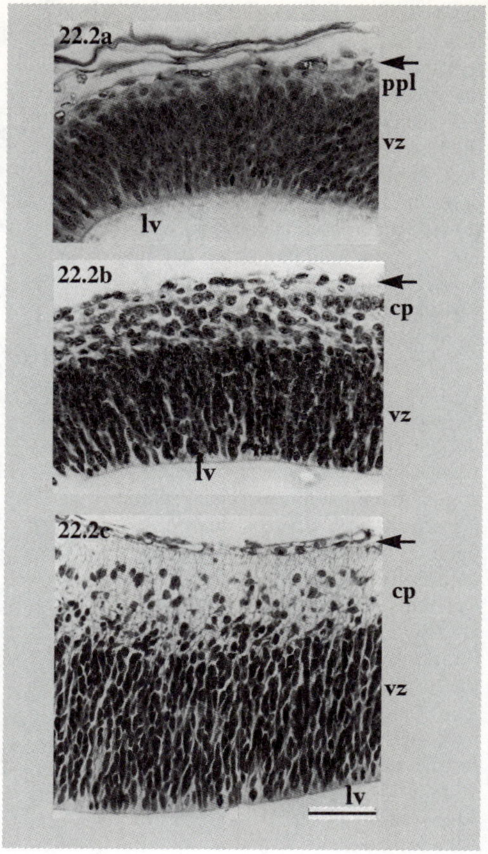

DEVELOPMENT OF THE NEOCORTEX

Study of the development of the neocortex has become a major research enterprise in eutherian mammals and is too large a topic to do justice to in a short chapter (see Marin-Padilla 1988; Kostovic & Rakic 1990; Allandoerfer & Shatz 1994 for reviews). However, work on marsupial species is providing interesting insights into general problems of early neocortical development that are proving difficult to tackle in eutherians because of the inaccessibility of the fetus. Although the general pattern of development of the marsupial neocortex has been shown to be the same as in eutherians (for example, Reynolds *et al.* 1985; Sanderson & Weller 1990), there are some significant differences, particularly with respect to timetable, that may provide answers to some currently controversial points. The neocortex of both eutherians and marsupials develops from the neuroepithelium, which lines the primitive ventricles when they first develop from the recently closed neural tube. In the embryonic forebrain vesicle the only other layer is the primordial plexiform layer (Figure 22.1); this is a cell-sparse narrow zone, which lies outside the neuroepithelium (also known as the ventricular zone). The layers that are characteristic of the fetal (eutherian) or neonatal (marsupial) neocortex are formed by migration of cells to form initially the cortical plate (Figures 22.2, 22.3 and 22.4); these presumptive neurons are generally held to be postmitotic. They

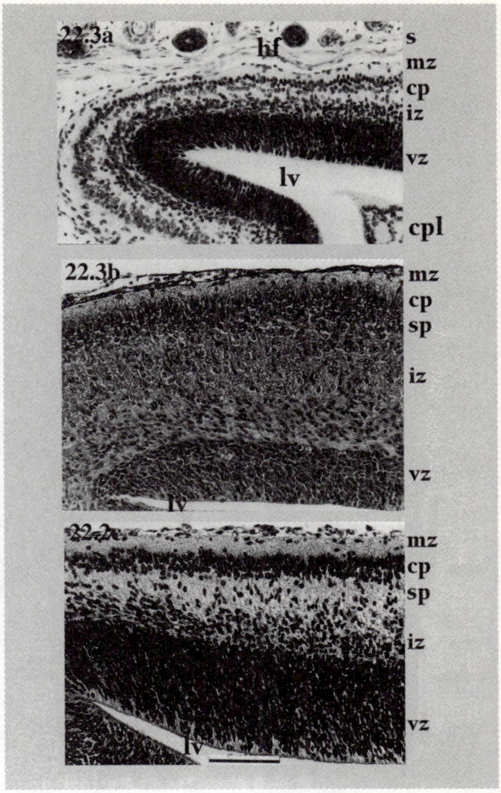

Figure 22.3
Neocortex 14 to 15 days after birth in 3 marsupial species.
(a) *Sminthopsis crassicaudata* (fat-tailed dunnart).
(b) *Monodelphis domestica* (South American grey short-tailed opossum).
(c) *Macropus eugenii* (tammar wallaby).
Scale bar=100 µm.
Key:
lv=lateral ventricle
vz=ventricular zone
ppl=primordial plexiform layer
cp=cortical plate
s=skin
hf=hair follicle
mz=marginal zone
iz=intermediate zone
cpl=choroid plexus
Notes:
(1) Coronal sections of dorsal neocortex from brains prepared as in Figure 22.1.
(2) Note change of scale from Figures 22.1 and 22.2.

migrate in an inside-out sequence that results in the youngest neurons being outside (nearer the brain surface) the older neurons. Subsequently the characteristic layers (II to VI) differentiate, the innermost layer VI first and the outermost layer, which is derived from the cortical plate, layer II, differentiates last.

A major unresolved problem in the development of the neocortex concerns the relation between incoming axons from, for example, thalamic nuclei, which will eventually form functionally effective synapses with neurons of some layers of the developed neocortex, particularly layer IV. When the first thalamic axons grow into the cortex this occurs at or before the first cells, destined to become neurons, migrate out from the ventricular zone. This is well before layer IV cells have migrated to the cortical plate, let alone differentiated into their characteristic layer. However, the timetable on which these processes occur appears to be very different in different species, although part of this may be due to the rate at which development occurs. Thus the rat brain develops relatively rapidly, so that processes that are temporally dispersed in slowly growing brains, such as those of the cat or monkey, may be compressed into the same timespan in the rat brain without necessarily being related. Also some processes that occur over a prolonged period in a slowly growing brain, may occur so fast in a rapidly growing brain as to be not clearly discernible. Marsupial brains may provide some answers to questions about how the early development of the neocortex is organised. This is partly because there are some interesting differences in the precise patterns of development of their brains, but also because studies in developing marsupial brains

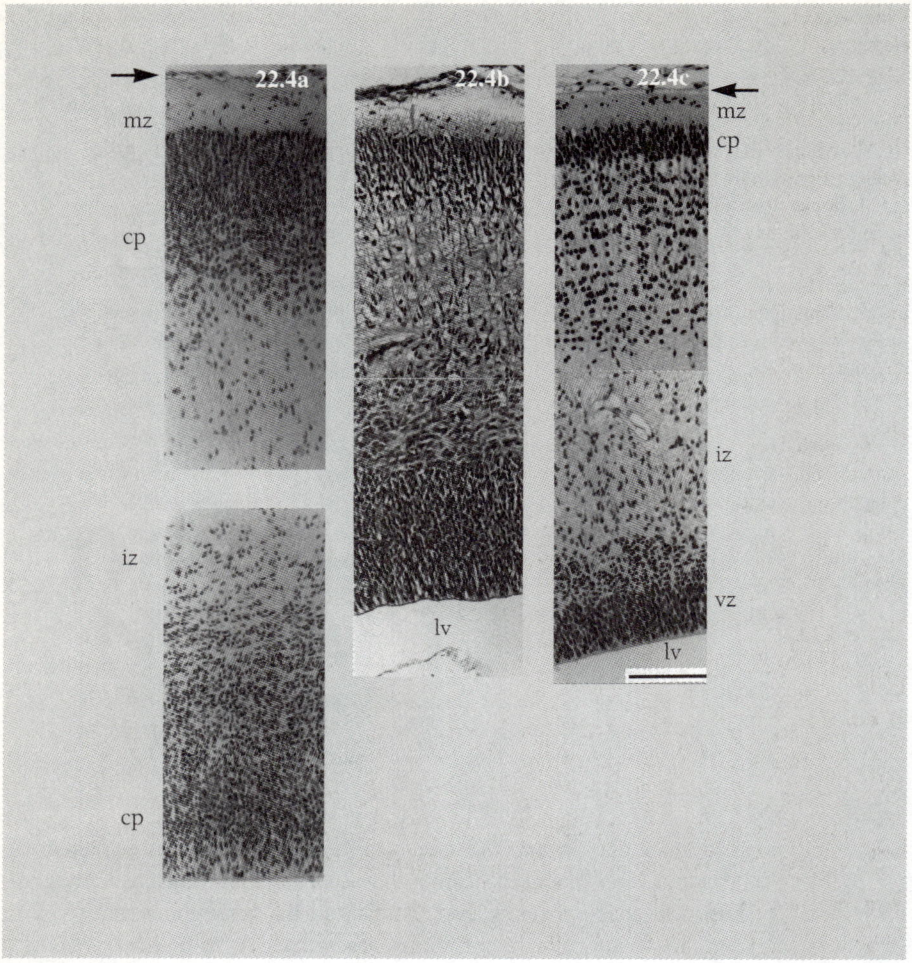

Figure 22.4 Comparison of developing neocortex in eutherian and marsupial species.
(a) Fetal sheep, 50 days gestation (E50).
(b) Fetal cat (E38).
(c) Tammar wallaby (38 days postnatal).
Scale bar=50 μm.
Key:
mz=marginal zone
cp=cortical plate
iz=intermediate zone
vz=ventricular zone
lv=lateral ventricle
Notes:
(1) Arrows indicate forming pia-arachnoid.
(2) Coronal sections of dorsal neocortex are from brains fixed in Bouin's fixative and embedded in paraffin wax.
(3) Sections (5 to 7 μm) were stained with haematoxylin and eosin.
(4) Fetal sheep brain is much larger than the other two. The gap is approximately 520 μm of the section which could not be fitted on the plate.

have produced some unexpected findings. For example, it is widely believed that synapses only appear in the developing neocortex after the cortical plate has begun to form and then only on either side of the cortical plate rather than within it. However, Balslev *et al.* (1992) have shown in *Monodelphis* that a few synapses are present in the primordial plexiform layer, before the first cells of the cortical plate have migrated (Figure 22.5) , as is described in more detail below.

Development of the neocortex has been extensively studied in the tammar (Reynolds *et al.* 1985; Reynolds & Saunders, 1988; Jones *et al.* 1991; Sheng *et al.*1990, 1991; Mark & Marotte, 1992; Mark Chapter 18 in this volume) and the related quokka (see Beazley *et al.* 1995 for references). The work by Mark's group (reviewed in Mark & Marotte, 1992; Chapter 18 this volume) and by Beazley's group (Beazley, *et al.* 1995) has mainly been on the visual system.

All marsupial species appear to share the common characteristic that the whole of the development of the neocortex takes place after birth, in contrast to eutherian mammals in which a significant proportion of neocortical development takes place *in utero*, although there are considerable differences in the stage of cortical development at birth in different eutherian species, which depend on the length of gestation. But even species such as the rat or mouse, that are born in an immature state compared with many other eutherian species, have all of the layers of their neocortex established at birth. This is in contrast to marsupials (Figure 22.1, page 385), all of which have only an 'embryonic' neocortex at birth, which consists of a narrow outer primordial plexiform layer and an inner, much deeper ventricular zone, the cells of which divide and migrate to form 5 of the 6 layers of the mature neocortex, as described above. There is some variation between marsupial species in the degree of development of the primordial plexiform layer at birth (Figure 22.1). For example, *Sminthopsis* has only an extremely narrow outer primordial plexiform layer with very few postmitotic cells that have migrated into it; whereas in the tammar this region is deeper in proportion to the thickness of the wall of the cerebral vesicle and contains more postmitotic cells. The equivalent stage of neocortical development in rats is around embryonic day E13 to E14. In a larger, more slowly growing, eutherian species with a well-developed neocortex at birth, such as the sheep, the stage of brain development that is equivalent to a marsupial species occurs in the sheep fetus at around E30 (term in the sheep is 150 days from conception). The immaturity of the neocortex in marsupials at birth makes them ideal for studies of the early stages of its development such as cell division, migration, differentiation and connectivity. Formation of the cortical plate, which is the developmental forerunner of 5 of the 6 layers of the adult neocortex, occurs entirely after birth (Figures 22.2 and 22.3, pages 386 and 387), for example, it is first apparent in the tammar at 5 to 6 days postnatal (Figure 22.2; Reynolds *et al.* 1985) and in *Monodelphis* at 3 to 4 days postnatal (Figure 22.2; Saunders *et al.* 1989) and *Didelphis* at 7 days (Krause & Saunders 1994). Comprehensive reviews of studies of the development of the neocortex in eutherian species have been published (Marin-Padilla 1988; Allandoerfer & Shatz 1994). Although less information is available about marsupial species, it is quite clear, as indicated above, that the general pattern of development of the neocortex is similar in eutherians and in marsupials. Early stages of neocortex formation are illustrated for two marsupial species in Figure 22.1 (page 385). After the initial two layered 'embryonic' stage of the first few postnatal days, cells migrating from the ventricular zone are inserted into the primordial plexiform layer to

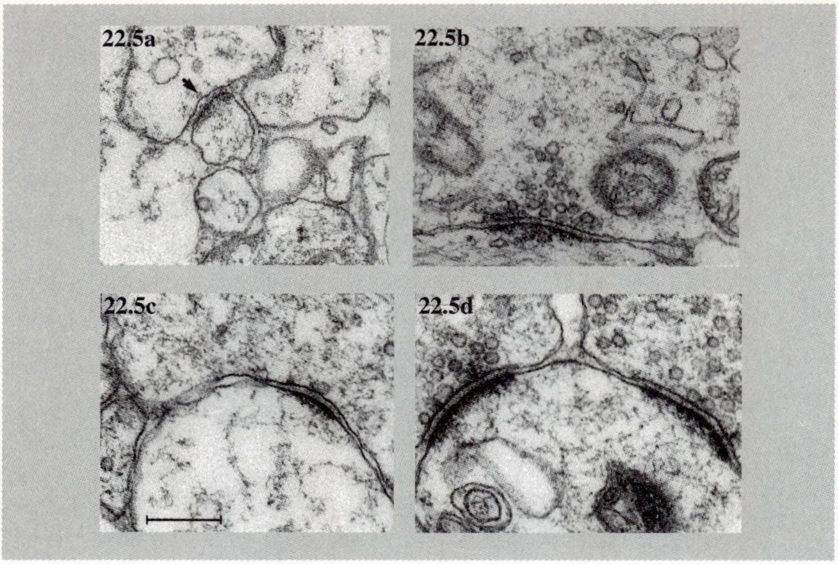

Figure 22.5 Early synapse formation in primordial plexiform layer of neonatal *Monodelphis*. a) and b) are from a newborn animal. Note the presence of the symmetrical membrane thickenings and lack of synaptic vesicles in a presumed presynaptic element ((a), arrow). Synapses in the upper part of the marginal zone of the developing neocortex of 5-day-old and 10-day-old *Monodelphis* are shown in (c) and (d).
Scale bar=0.2 µm.
Source: Balslev *et al.* (1992).

form the cortical plate which then has the cell sparse marginal zone (which becomes layer I) outside it and the subplate zone deep to it (Figures 22.2 and 22.3). The equivalent stage of development in the cat (Luskin & Shatz 1985) occurs around 31 to 33 days after conception (E31 to E33, term is at E65) E35 in the sheep (term is 150 days), approximately E45 in the monkey (term is 165 days) and E16 in the rat (term is 22 days). In all species the cortical plate is expanded by further addition of migrating cells its outer border (inside-out sequence), the inner cells of the cortical plate gradually maturing and acquiring the characteristics of layers II to VI of the adult neocortex. Particular attention has been paid to the developmental history of the first cells to be inserted into the primordial plexiform layer. By the addition of less mature cells outside this first layer of cells the first cells of the cortical plate come to lie in a zone referred to as the subplate (see Allandoerfer & Shatz 1994). As the neocortex develops these cells finish up in the lower part of layer VI (the deepest of the cortical layers) and in the white matter that is deep to layer VI). At least in the cat, Shatz and her colleagues (see Allandoerfer & Shatz 1994) have convincingly shown that this population of neurons is transient and have largely, if not entirely, disappeared after birth. Similar data have been obtained for the fetal sheep, which has the advantage that the fetal protein, fetuin, has been shown to be a marker for these subplate cells (Saunders *et al.* 1992a) and is synthesised by these cells (Dziegielewska *et al.* 1993). A comparison of the fetal neocortex in cat and sheep with that of the postnatal tammar is shown in Figure 22.4. Comparison of the tammar at 38 days postnatal in this figure with the tammar brain sections in Figures 22.1, 22.2 and 22.3, illustrates how the whole of neocortical development occurs after birth in marsupials. The comparison of the two eutherian species with the tammar in Figure 22.4 (page 388) also stresses how much of neocortical development occurs *in utero* in eutherians, paticularly given that the fetal cat is not born until E65 and the fetal sheep until E150.

It has been proposed by Molliver *et al.* (1973) and subsequently elaborated by Shatz (see Allandoerfer & Shatz 1994) that these transient subplate cells provide a scaffold for the incoming thalamocortical axons. These appear to form transient synapses with the subplate cells, while 'waiting' to enter the cortical plate at a later stage when the layer IV, with which these afferent make definitive contacts, is developing. Quite why there should be such a delay is not known. In other species, for example, the rat, the situation is less clear. Some authors claim that there is no waiting period (for example, Catalano *et al.* 1991). Studies in marsupials may be able to contribute to this problem, although there are some difficulties of terminology in defining the cortical plate and subplate in these species. Although the general inside-out sequence of cell migration in the formation of the layers of the neocortex is similar in eutherians and marsupials, the developing marsupial neocortex is strikingly different in its microscopical appearance in a number of ways. Firstly, the cells of the subplate and those destined to become layer VI appear to mature much more rapidly than at equivalent stages in eutherians (see Figure 22.4 and cf. tammar, Reynolds *et al.* 1985 and sheep, Saunders *et al.* 1992a). Also the cells in layer VI and later forming layers have a degree of order not seen in eutherian species so far studied. This is particularly striking in the tammar in which rows of neurons are lined up in register parallel to the surface of the brain (Figure 22.4 and Reynolds *et al.* 1985). However these differences from the developing eutherian neocortex have led to difficulties in defining the subplate. In our original studies (Reynolds *et al.* 1985) we thought that the whole depth of the more mature layers of neurons in the innermost zone of the neocortex corresponded to the subplate. This is almost certainly incorrect. On the other hand Mark and his colleagues (Sheng *et al.* 1991; Mark & Marotte 1992) consider that all of these cell layers constitute the cortical plate, which they divide into an outer compact cell zone (cortical plate in Reynolds *et al.* 1985) and an inner less compact zone containing the more mature cells (subplate zone, in Reynolds *et al.* 1985), with a narrow subplate zone deep to this layer. The less compact zone contains numerous incoming fibres from thalamic nuclei, which are present from very early stages of cortical plate development; this leads Mark and his colleagues to conclude that the waiting time in marsupials does not exist or is very short and that, unlike eutherians, the cortical plate is invaded very early in development by thalamocortical afferents. These are therefore assumed to make their definitive synaptic contacts much earlier than in eutherians (although the relevant electronmicroscopical studies that would be needed to show the presence of synapses do not appear to have been carried out). This is not a trivial point as it has important implications for interpretation and understanding of results from eutherian studies. In fact it seems likely that these difficulties are resolved by the study of Jones *et al.* (1991) who used an antibody to the tammar equivalent of the fetal protein, fetuin, to identify the subplate cells in the tammar and examine their developmental history. As mentioned above fetuin has been shown to be a marker for subplate neurons in fetal sheep and also in many other species (see Dziegielewska & Brown 1995). A comparison of fetuin staining in the fetal sheep and postnatal tammar is shown in Figure 22.6. Jones *et al.* (1991) found that fetuin staining of subplate cells was only present between 5 days and 20 days after birth, even in the least mature parts of the developing neocortex. This is in marked contrast to the fetal sheep, which has a similar-sized brain. In the sheep, fetuin is present in the first cells to be inserted into the primordial plexiform plate at around E35 and subsequently are found in the subplate at E40 to E60 and in the presumptive layers VI and white matter at E80. Even at E125, shortly before birth (at E150) a few fetuin-positive neurons could be found in layer VI and the presumptive white matter deep to layer VI. Thymidine labelling studies showed that the fetuin-positive subplate cells die (Saunders *et al.* 1992a) probably mainly after E80, as had previously been shown for the fetal cat neocortex by Luskin & Shatz (1985). The latter study was confined to thymidine labelling as an antiserum to cat fetuin is not available, although there is indirect evidence from lectin histochemistry using a lectin extracted from *Sambucus nigra*. This binds specifically to the terminal carbohydrate (neuraminic acid) on fetuin and stains subplate cells in the fetal cat (Adam *et al.* 1993). Systematic thymidine and fetuin staining studies have not been carried out in the tammar, so it is not certain that the disappearance of fetuin-staining by 20 postnatal in the neocortex of this species is synonymous with cell death; but given the results from the sheep fetus, this seems likely. Thus it seems that in the tammar the subplate cells are only present for a few days (5 to 15 days in the first parts of the neocortex to develop a cortical plate and up to 20 days postnatal in the later developing regions) and that the period in which the thalamocortical afferents make transient synaptic contact with the subplate cells is extremely brief when compared with eutherian species that have comparable brain sizes and rates of brain growth (for example, sheep, cat, Figure 22.4, page 389). The neurons of presumptive layer VI and those in more superficial layers in the tammar, mature much more rapidly than in eutherians such as the sheep (Reynolds *et al.* 1985). They are therefore presumably able to accept synaptic contacts from the incoming thalamocortical fibres at a much earlier time than in eutherians and the subplate cells having performed their transient function, die. Thus the more mature neurons deep to the cortical plate proper (Mark's compact cell zone) are just that, that is, more mature and no longer part of the cortical plate: they are already forming connections with incoming nerve fibres. Why this process should be so precocious in the tammar (and presumably other marsupials) is not clear. It may be, as suggested by Reynolds *et al.* (1985) that it is a reflection of the much earlier birth of the tammar. It may be an evolutionary development associated with the much greater sensory input and possibilities for locomotor activity that one can suppose is part of pouch life compared with the cushioned watery environment of the eutherian fetus.

Thus studies of the developing neocortex in marsupials may lead to a more general understanding of such development in mammals in general. The fact that the timescale appears to be so different in marsupials and that they are so much more accessible to experimental manipulation means that future studies are likely to contribute a lot more useful information and insights into the mechanisms involved in early neocortical development.

Another aspect of neocortical development to which studies in marsupial are contributing is on the problem of when and where synapses first form. The dogma from eutherian studies is that synapses do not appear in the neocortex until after the first appearance of the cortical plate and that once synapses appear they are confined to the marginal zone outside the cortical plate and to the subplate zone deep to the cortical plate (see Vaughn 1989). Studies of synapse formation in neocortex of neonatal

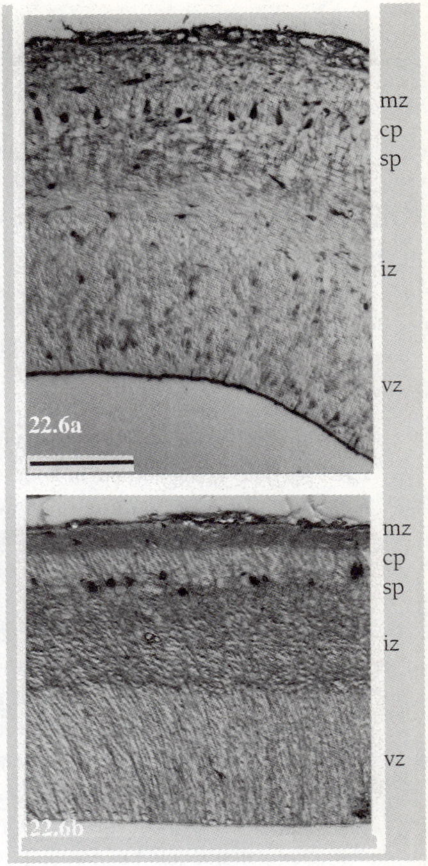

Figure 22.6 Fetuin staining in fetal sheep (a) and postnatal tammar (b) neocortex
Coronal sections of dorsal neocortical wall of E40 fetal (above) sheep and 15-day tammar (below).
Note:
(1) Bouin's fixed material, paraffin sections were stained using the peroxidase-antiperoxidase method and species specific anti-fetuin antisera (see Jones *et al.* (1991) for methods).
(2) Differential interference contrast microscopy, Olympus BMX 50.
(3) Note strong fetuin staining in neurons in subplate of lower cortical plate; also fibre staining in marginal zone and intermediate zone.
Scale bar=100 µm for both figures.

Monodelphis domestica have shown that both these points are in error (Balslev *et al.* 1992). Studies in *Monodelphis* showed that synapses appear in the primordial plexiform layer before the appearance of the cortical plate and that once the latter has appeared a few synapses can be found on cells within the cortical plate itself (Figure 22.5, page 390). Synapses were shown to be present on dendrites and some soma of cells immediately above and below the cortical plate once it had formed. These observations prompted us to re-examine the situation in the developing rat brain (Balslev *et al.* 1996) where we have now found synapses in the primordial plexiform layer as early as E14, something which seems not to have been observed previously.

Once the cortical plate had appeared we not only found synapses above and below it, but also there were a few synapses within the cortical plate at E16; this has not been reported before. The cells of origin of the nerve processes that form these synapses are not known and it remains the case that the bulk of synapses do not form on cortical plate cells until they have matured into discernible layers, especially layer IV. The finding of synapses in the early cortical plate of the fetal rat (Balslev *et al.* 1996) needs to be followed up in other species. If exclusive to the rat (and perhaps other species with small, rapidly growing brains) then it may indeed be the case that in the rat and similar animals there is no waiting time or zone. Even if synapses are found within the cortical plate early in its development in species such as the cat and the sheep, this does not preclude the occurrence of a waiting zone. The number of synapses found in the early fetal rat cortical plate was very small. These might represent the connections of 'pioneer' fibres which mark out territories of the cortical plate that will be subsequently innervated by thalamocortical afferents staking out neuronal territories for systematic innervation once the majority of layer IV cells have reached an appropriate stage of maturity.

REPAIR AND REGENERATION OF THE SPINAL CORD FOLLOWING INJURY

Unlike the peripheral nervous system of vertebrates and the central nervous system (CNS) of lower vertebrates the mammalian CNS has no ability to repair itself by regeneration of functionally effective neurites from the injured axons of damaged neurons. It has been known at least since the time of Ramon y Cajal (1928) that damaged CNS neurons are capable of sprouting in response to injury, but the neurites so produced do not progress across the site of injury and subsequently regress. As discussed by Martin and Wang in Chapter 21 in this volume, individual tracts in the immature mammalian spinal cord appear to have some capacity for regenerative outgrowth, although many neurons, when their axons are injured, may die, depending upon the stage of their development at which they are injured. Examples of spinal cord tracts, which have been shown to have some capacity for regenerative outgrowth are the rubrospinal tract of North American opossums around 20 days postnatal (Xu & Martin 1991) and the corticospinal tract in the neonatal rat (Bates & Stelzner 1993). However, the fibre growth which occurred included a substantial growth of new fibres that were not present at the site of the lesion when it was made. This is only to be expected since, at the age at which the animals were operated upon, their spinal cords still had a considerable amount of growth and development to undergo. Also it is important to note that in these experiments on neonatal animals and in many other such studies (see, for example, Bregman *et al.* 1993) fibre growth, whether new or regenerative, only occurred via adjacent undamaged spinal cord and not across the actual lesion. Thus complete spinal transection, even in an animal as immature as a newborn rat, is not followed by any significant axon growth through the lesion unless the local environment is made more favourable for neurite growth, by, for example, implanting fetal CNS tissue into the site of the lesion.

Experiments involving implantation of fetal CNS into injured spinal cord have been carried out both in adult and neonatal animals; neurite growth occurs into and through the fetal tissue, rather than around the lesion, as occurs in the absence of implanted fetal CNS (Bernstein & Stelzner 1983; Bregman *et al.* 1993). Any functional recovery that occurs in such preparations is much more likely to occur in younger animals. This appears to be because the fibres that grow across the transplanted tissue make effective connections beyond the site of the lesion; for many years this had only been achieved by using partial lesions of the spinal cord (for example, 'overhemisection', see review by Bernstein-Goral & Bregman 1993). However, recently Iwashita *et al.* (1994) published results of experiments in which a segment of lower thoracic cord was resected in neonatal rats and replaced with an equivalent segment from E14 to E17 fetal rats. These animals subsequently showed a significant degree of normal morphological and locomotor development.

At least part of the explanation for the age-related greater recovery following spinal cord lesion appears to be that, after an early stage of spinal development, the cord begins to produce inhibitory peptides which prevent the growth of neurites from the damaged ends of axons. Thus Schnell & Schwab (1990) and Schwab (1991, 1993) have shown that two peptides probably synthesised by oligodendroglia in the more mature spinal cord, prevent neurite outgrowth; but the inhibitory activity can be countered by antibodies against the inhibitory peptides (see also Johnson 1993). It is also likely however, that the ability of immature neurons to respond to injury with neurite outgrowth is much greater than in adult neurons (Fawcett 1992). This is particularly apparent in the studies of the response of chick embryo spinal cord to injury. The groups of Oppenheim (for example, Shimizu *et al.* 1990) and Steeves (for example, Hasan *et al.* 1993) have shown that in the period of E2 to E13 the chick embryo is able to respond to a spinal cord injury by repair and growth, which results in the significant degree of normal function when the chicks hatch. Double-labelling studies appear to show that following spinal cord injury in the chick embryo and also in the neonatal rat with fetal CNS implants (see Bregman *et al.* 1993) or where fibres are able to grow around the site of injury (Bates & Stelzner 1993) a functionally significant proportion of the fiber growth that occurs is due to regeneration of neurites from the injured axons. However, growth of new axons occurring as part of normal development also contributes to the level of functional ability that can be demonstrated in these animals. The article by Martin and Wang (Chapter 21 in this volume) stresses the distinction between the possibility that spinal cord tracts, if injured early in their development, may produce new neurites which grow around the lesion and a still earlier stage when the neurites will grow across the lesion as has been shown using *in vitro* mammalian CNS preparations (Saunders *et al.* 1992b; Treherne *et al.* 1992 and see section on *in vitro* preparations below) and more recently in neonatal *Monodelphis in vivo* (Saunders *et al.* 1995, and below) and in *Didelphis* (see Martin & Wang, Chapter 21 in this volume).

These results were obtained by exploiting the more favourable aspects of the greater ability of the immature newborn marsupial central nervous system to repair itself following injury. Such results show that providing we are able to go early enough in development, even in a mammalian CNS there will be evidence of structural and functional repair following a spinal cord injury, without the need to implant fetal CNS tissue. Similar experiments in the rat would have required intrauterine operations on the fetus, which would have been technically much less feasible. An outline of these *in vivo* experiments is given below; more details can be found in Saunders *et al.* (1995) and Nicholls & Saunders (1996).

EXPERIMENTS *IN VIVO*

Methods

Sterile operations were carried out under anaesthesia and sterile conditions as described in detail elsewhere (Saunders *et al.*. 1995) on neonatal *Monodelphis* aged 4 to 8 days. A few operations were carried out on animals of 12 to 16 days. The operation consisted of a transverse incision through the skin at the level of the first thoracic vertebra (T1) followed by a 10 second forceps crush through the skin incision, which aimed to crush the whole cross-section of the spinal cord via the dorsal surface of the developing vertebral column. The effectiveness of the crush was checked by testing for local reflexes, by electrophysiological testing and by morphological examination, as described below (Figure 22.7; also see section on *in vitro* methods in this chapter, and in more detail in Saunders *et al.* 1995). At different times after operation the young were examined and compared with control (unoperated) animals. Video recordings of locomotor ability were made.

Results and discussion of in vivo experiments

At early stages after operation the animals were clearly paralysed caudal to the level of the lesion, but did show local segmental withdrawal reflexes in response to a mild nociceptive stimulus to the hindlimbs. As recovery and development proceeded the operated animals began to gain use of their hind limbs and show limited locomotor activity involving coordinated rhythmic movements of both forelimbs and hindlimbs. This activity was grossly similar in operated and control animals by 3 to 4 weeks post-operatively, but was limited in scope because the animals were still at an early stage of their development. Preliminary results of morphological studies showed growth of fibres across the crush by 7 to 10 days after operation (Saunders *et al.* 1995 and NR Saunders, unpublished observations). Martin and his colleagues (Chapter 21 in this volume and Wang *et al.* 1996) have obtained similar results in *Didelphis virginiana* pouch young. In their experiments the thoracic spinal cord was transected at 4 days postnatal (counting the day of birth as 0 days). The lesioned cords were studied morphologically at 30 to 40 days after the lesion had been made. It was further shown that fibre growth across the lesion was less for transections made at 11 days postnatal or later, which complements *in vivo* the finding of Varga *et al.* (1995a) that in the *Monodelphis in vitro* spinal cord preparation fibre growth does not occur across a crush made at the upper cervical level if the crush is made at 12 days postnatal or later. The neuronal cells which gave rise to the fibres that grew across the lesion site in the youngest *Didelphis* were identified by using Fast Blue injections made into the cord caudal to the lesion, 30 to 40 days after the operation (Wang *et al.* 1996). Numerous labelled cells were found in the pontine reticular nucleus, the lateral vestibular nucleus and the red nucleus. Behavioural studies in pouch young that have undergone similar spinal transections and survived for longer periods are currently being carried out to determine what degree of functional recovery and development occurs in these animals (Dr G F Martin, personal communication).

In our *Monodelphis* experiments, by the age of weaning (55 to 60 days) the operated animals could walk, run and climb in a manner that was difficult to distinguish from the same activities in controls. Once the animals reached adulthood we carried out more systematic tests involving climbing a 2 cm diameter pole placed at 45° and walking across a metal grid. Ordinary walking movements were well coordinated but involved a raised pelvis and rather rolling movement of the hind part of the body, compared with controls. The operated animals were able to climb the angled pole, but tended to use their prehensile tail to grip the pole than did the controls. Although the operated animals appeared to have more difficulty with this task than the controls, they nevertheless managed (see Saunders *et al.* 1995 for details). Similarly with the grid test, both operated and unoperated animals made mistakes in placing their feet on the bars, the operated animals did so more frequently. In preliminary tests we have also shown that *Monodelphis* adults are excellent swimmers and that animals that had been operated as neonates and survived to adulthood were also able to swim effectively (Figure 22.8). This is an important observation since it discounts the possibility that the walking movements described above were due to triggering of local spinal pattern generators by

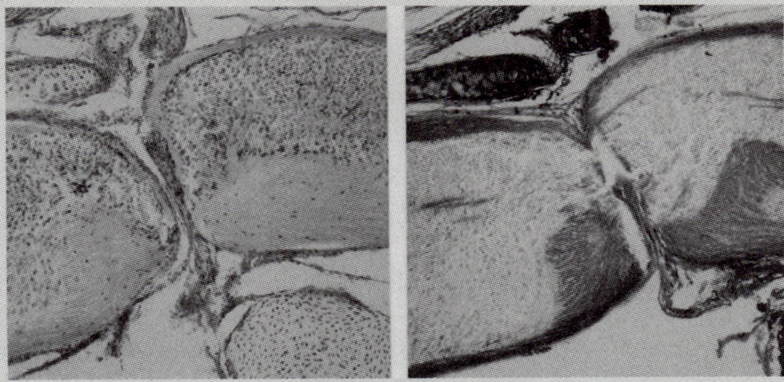

Figure 22.7 Two examples of spinal cord lesions in 6-day-old *Monodelphis*.
Left: 1 hour after crushing, section stained with silver.
Right: 6 hours after crushing, section stained with haematoxylin and eosin.
(Note that damaged ends of the spinal cord have already 'fused', although it will be several days before fibres grow across the site of the crush.)

tactile and proprioceptor information from the limbs. This conclusion is supported by the observation that the operated animals, when adults, were able to climb and support themselves on their hindlimbs when exploring an unfamiliar environment. Overall the operated animals showed a remarkable degree of near-normal locomotor activity. The degree of sensory feedback for both balance and proprioception involved in some of these tasks strongly indicates that a significant degree of effective neural connectivity had occurred across the site of the spinal crush. This has not been investigated in detail yet, but preliminary results of morphological studies of adult animals whose spinal cord had been completely crushed in the first week of life showed an apparently largely normal structure at the site of the lesion (Figure 22.9), as well as above and below it. So far spinal cords from four operated animals that survived to adulthood have been serially sectioned in a transverse plane and the sections silver stained. Such sections were compared with similarly stained sections at equivalent levels from a control animal. (An outline of the morphological methods used is given in the next section on *in vitro* experiments). An example is shown in Figure 22.9 from which it is apparent that the structure of the crushed cord has largely been restored and developed normally. This is a remarkable finding given that the original crush resulted in a temporary gap across the width of the cord of up to 0.5 to 1.0 mm (cf. Figure 22.7, page 397). There are two important outstanding points still to be investigated. The first is to determine the proportion of nerve fibres that grew across the crush that originated by regeneration from axons damaged by the crush and the proportion that grew through the crush for the first time as part of normal spinal cord development. In similar spinal cord lesion experiments in chick embryos (Hasan *et al.* 1993) and in cords of newborn rats with partial lesions or lesions implanted with fetal CNS (Bates & Stelzner 1993; Bregman *et al.* 1993), a substantial proportion (for example, 40%) of the fibres growing across the crush appeared to originate by regeneration. The other point will be to show to what extent the pattern of connections formed by the fibres crossing the crush in either direction corresponds to normal. This is an extremely important point and would be so even if it turned out that all of the fibres growing across the lesion were new rather than regenerating. This is because even if regenerating neurites can be induced to grow within an injured spinal cord, a quite separate problem is whether or not they will be able to innervate functionally useful targets, let alone their normal ones. The finding that, in our experiments, axons are able to grow directly across the cord lesion and make functionally effective connections provides a model for the study of innervation in a system in which the normal pathway has been disrupted.

ISOLATED CNS PREPARATION

In the early stages of our collaborative experiments with Nicholls and his colleagues in Basel, we discovered more or less by accident that the whole CNS of neonatal *Monodelphis* could be dissected out and maintained in tissue culture for many days (Nicholls *et al.* 1990) and showed subsequently that under appropriate tissue culture conditions that the preparations retained good preservation of ultrastructure and continued conduction of evoked action potentials along the length of the spinal cord of the preparation for up to at least 10 days *in vitro* (Stewart *et al.* 1991; Zou *et al.* 1991). It was not initially clear whether this remarkable and unexpected finding was due to a specific property of a marsupial nervous system or whether it was a more general property of an immature mammalian CNS. In order to test this we carried out similar experiments with isolated CNS of immature rats (Saunders *et al.* 1992b) which were chosen to be fetal so as to have a similar stage of nervous system development as neonatal *Monodelphis*. The initial experiments were not very successful, especially when we tried to reproduce the cord crush experiments of Treherne *et al.* (1992). This turned out to be due to something rather simple, namely temperature. The element of good fortune in our original *Monodelphis* experiments turned out to be even greater than we had realised. We arbitrarily set up the opossum experiments at room temperature, which in a Swiss laboratory is rather stable at 24 to 25°C. At this temperature the fetal rat preparations would remain conducting for several days, but if crushed they showed no signs of recovery of conduction of impulses across the crush. Also their ultrastructure was poorly preserved compared with the opossum preparations. However we discovered that the body temperature of adult *Monodelphis* is appreciably lower than that of the rat: 32 to 33°C compared with 37°C. When the culture temperature was raised to 29°C we then obtained similar results for E15 to E16 fetal rats as we had obtained for neonatal *Monodelphis* (Saunders *et al.* 1992b). Since we had shown that the same experiments could be carried out in the more easily available fetal rat, it might have seemed appropriate to switch to that species for later experiments. However, the main advantage of the marsupial was that it was possible to do the *in vivo* spinal cord crush experiments in neonatal *Monodelphis* that would have been exceptionally difficult, if not impossible, in fetal rats *in utero*; therefore for short term regeneration experiments that can be carried out *in vitro* it was important for the experiments to be carried out in the same species so as to be able to make a direct comparison with the species in which the *in vivo* experiments had been performed. There is an additional practical advantage of the marsupial, which is that, since the young are already born at the time of the experiment, it is not necessary to kill the mother, as is the case when obtaining fetal rats for such experiments.

Methods

(a) Operations

Fetal rats and neonatal opossums were used for these experiments. Details of operative procedures are to be found in Stewart *et al.* 1991; Saunders *et al.* 1992b). The young rats (E15, E16, E18 and E21) were removed by caesarean section from anaesthetised pregnant mothers and exsanguinated. Neonatal opossums came from our colonies of *Monodelphis domestica* in the Biocentre, University of Basel and The Central Animal House in the University of Tasmania. The young were given an overdose of anaesthetic and exsanguinated. The CNS was removed by dissection in cold Eagle's Basal Medium (BME GIBCO) bubbled with 5% by volume CO_2 in O_2 and transferred to BME containing 0.2% to 1% fetal calf serum (GIBCO) and 0.2 mg/ml gentamicin (GIBCO) continuously bubbled with 5% CO_2 in O_2 under sterile conditions in 50 ml or 250 ml tissue culture flasks. Some preparations were maintained for periods of up to 10 days *in vitro* as controls. The spinal cord of other preparations was crushed immediately after dissection either with forceps or a 6, 0 thread using a modification of a method first described by Lubinska (1959) for peripheral nerve.

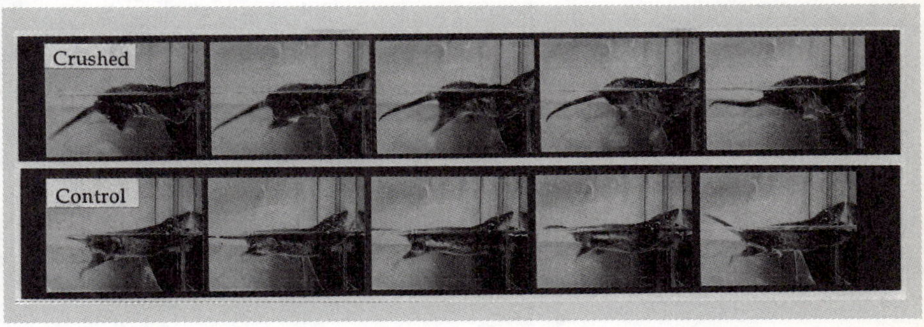

Figure 22.8 Swimming test.
Upper frames: An adult *Monodelphis* whose whose spinal cord was crushed at 6 days postnatal.
Lower frames: An unoperated adult.

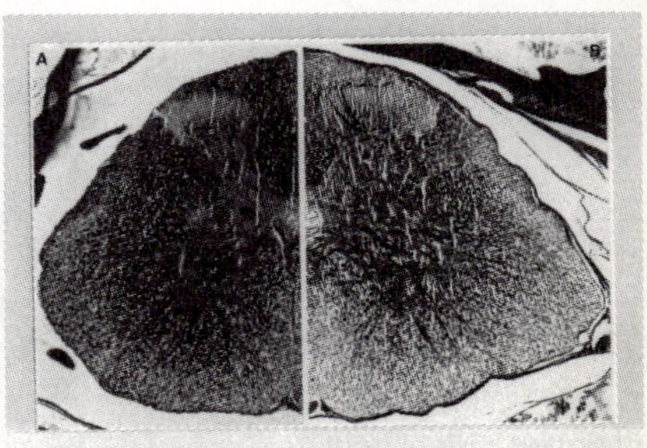

Figure 22.9 Transverse sections of adult *Monodelphis* spinal cord the the level of the first thoracic vertebra (T1). Silver stained, Sevier and Munger (1965), 15 μm sections, Edge Scientific multiple oblique illumination microscope.
Left: control adult.
Right: adult whose spinal cord was crushed at 6 days postnatal. (Note apparently normal structure.)

(b) Electrical recording

The brain was removed from the preparations at the end of the culture period at the level of the upper brainstem and suction electrodes were applied to each end of the spinal cord as described previously (Saunders *et al.* 1992b). When recordings were made from injured preparations, the electrodes were positioned with one above and one below the crush site at distances of several millimetres relative to the crush. For testing the survival and viability of crushed preparations, recordings were made at different positions along the cord with both electrodes either above or below the site of the crush.

(c) Morphological studies

Spinal cords from *in vitro* or *in vivo* studies were fixed for light microscopy in Bouin's solution (overnight at room temperature) or in 4% paraformaldehyde in 0.1 M phosphate buffered saline (PBS, pH 7.4) either overnight at 4°C prior to cryosectioning or for 15 min at room temperature for DiI studies. Neurites were stained with a fluorescent, lipophilic, carbocyanine dye, DiI (l,l'-dioctadecyl-3,3,3',3'-tetramethylindocarbocyanine perchlorate) as described previously (Saunders *et al.* 1992b). Labelling was viewed in whole mount with a standard Rhodamine Filter Set and a long working distance (25X or 10X) objective. Axons labelled with DiI were observed within and extending well beyond the crush. To test the ability of the crushing procedure to produce an effective and complete lesion, six preparations were crushed and labelled as described above, but were then fixed after only 2 h to prevent any subsequent repair.

Bouin's fixed material was washed briefly in tap water, dehydrated in graded alcohols, cleared in xylene and embedded in paraffin wax. Serial sections 2 to 20 µm were cut in either transverse or longitudinal planes. Sections were stained with haematoxylin and eosin or by the silver staining method of Munger (Sevier & Munger 1963). Some sections were stained for specific antigens using immunocytochemical methods (see below). Paraformaldehyde fixed specimens that were used for immunocytochemistry were rinsed in PBS, cut into blocks of 0.5 to 1 mm, immersed in graded sucrose solutions (15 to 30% in PBS) and frozen on dry ice for cryosectioning. Sections of 20 µm thickness were mounted on 1% gelatine-coated slides and air dried for several hours. Preparations were post fixed for 10 min on the slides in 4% PFA/PBS and rinsed in 3 changes of PBS. Nonspecific binding was blocked using 0.1% bovine serum albumin (BSA)/1% normal goat serum in PBS or 5% fetal calf serum (FCS) for 30 min at room temperature. These sections were then used for immunocytochemistry (see below).

(d) Immunocytochemistry

Paraffin-embedded sections were dewaxed and used for peroxidase-antiperoxidase (PAP) immunocytochemistry, as described in detail previously (Møllgård *et al.* 1994). The following antibodies (dilutions indicated in brackets) were used. Antibodies against (a) radial glia and/or astrocytes: mouse anti-swine vimentin (1:10, DAKOPATTS, DK), rabbit anti-cow S-100 (1:500, DAKOPATTS, DK), rabbit anti-cow glial fibrillary acidic protein (GFA) (1:250, DAKOPATTS, DK); (b) oligodendrocytes: rabbit anti-human myelin basic protein (MBP) (1:500, DAKOPATTS, DK); (c) neurons: rabbit anti-human neuron specific enolase (NSE) (1:2000, DAKOPATTS, DK).

Results of in vitro experiments

Previous studies using both fetal rat and postnatal opossum preparations have shown that the structure of these preparations is well maintained under *in vitro* conditions for up to 10 days in culture, even when examined at the electron microscopical level (Stewart *et al.* 1991; Saunders *et al.* 1992b). Crushing consistently produced a complete block of impulse conduction across the crush immediately after making the lesion and for several days subsequently. In two initial series of experiments (Treherne *et al.* 1992; Saunders *et al.* 1992b), no crushed preparations showed evidence of conducted impulses across the crush at 1 to 2 days after crushing (opossum, *n*=4; fetal rat, *n*=7). Over the next few days an increasing proportion of crushed preparations showed recovery of conducted action potentials across the crush, so that by 5 days almost all were conducting (opossum, 6/7; fetal rat, 5/5). In later studies about 80% of rat preparations that survived were conducting impulses across the crush by 4 to 6 days later. Growing fibres within and beyond the crush were detected using DiI labelling or silver staining. In a more extensive series of experiments the effect of age on the response to crushing was investigated in neonatal opossum. This showed that spinal cords from progressively animals a decreasing proportion of preparations showed neurite growth and recovery of conduction through the crush (made at a lower cervical level); by 13 to 14 days postnatal no cords showed recovery of impulse conduction across the crush. Because there is a steep rostro-caudal gradient of development in the spinal cord of *Monodelphis*, crushes placed more caudally (lumbar region) at 12 days were followed by recovery of conduction in preparations in which cervical lesions showed nc recovery (Varga *et al.* 1995a)

MORPHOLOGY: LIGHT MICROSCOPY

Maturation of neuronal and glial cell markers during the period when successful fibre outgrowth was obtained in both fetal rat and neonatal opossum preparations, was assessed using appropriate cell markers and immunocytochemistry (see Methods, above). The lower cervical spinal cord is composed of an inner ventricular zone (neuroepithelial cells), an intermediate zone containing large motor neurons, single glial cells and small blood vessels, and an outer marginal zone with dense fibre bundles and a few small glial cells. Significant continued growth and development of the spinal cord occurs *in vitro* over several days in the opossum (see, for example, Figs 2 and 3 in Møllgård *et al.* 1994). Over the 3 day period from 3 to 6 days postnatal, there was an increase in overall size of the cord and a marked increase in staining of neurons by NSE, especially in the ventral horns of the cord. There was strong NSE staining of axons in the large ventral and lateral funiculi as well as some staining of the much smaller dorsal funiculus. There was also a considerable increase is staining of radial glial cell processes by vimentin in all three funiculi (dorsal, lateral and ventral). The general level and distribution of the staining for the two markers was similar at 6 days postnatal *in vivo* and for 3 day old preparations after 3 days in culture. Several astrocytic somata were stained with anti-vimentin, especially at the outer margin of the ventral horn. The staining pattern with anti-GFA was generally similar to that for vimentin, but the number of S-100 positive radial glial processes and endfeet was less than for anti-vimentin or anti-GFA staining. In culture over the same 3 day period the size and gross morphology of the cultured cords was surprisingly similar to that of the *in vivo* cords at 6 days postnatal. A similar degree of neuronal and glial cell maturation was apparent *in vitro* as occurred *in vivo*. Further details can be found in Møllgård *et al.* (1994). At later ages astrocytes (staining for GFA) became more numerous especially in the presumptive grey matter, whereas the radial glial processes had declined in prominence by 14 days. Myelination begins in the ventral medulla around 7 days and progresses caudally to reach the lower thoracic and lumbar segments after 14 days (Varga *et al.* 1995a).

MORPHOLOGY: ELECTRON MICROSCOPY

Crushed and control spinal cords from preparations cultured *in vitro,* for different periods were examined in the electron microscope, as described in the Methods (above). Providing that preparations were maintained at an appropriate temperature *in vitro* (24 to 25°C for *Monodelphis*, 29 to 33°C for fetal rat, cultured CNS preparations showed a remarkable degree of preservation of ultrastructure, with continued growth and development (see Saunders *et al.* 1992b; Møllgård *et al.* 1994). Preparations that had been crushed exhibited growth cones and large-calibre axonal varicosities which contained many dense core vesicles (see Saunders *et al.* 1992b and Nicholls *et al.* 1994, for examples).

DISCUSSION

The main finding of these experiments is that both *in vitro* and *in vivo*, complete spinal cord crush in the neonatal *Monodelphis* is followed by growth of nerve fibres through the crush with restoration of action potential conduction. This recovery does not occur after about 12 days postnatal for a lesion of the cord at C7 to 8. So far this has only been shown unequivocally for preparations *in vitro*, because none of the animals operated upon at 12 to 18 days survived. This successful fibre growth and recovery of function, both *in vivo* and *in vitro,* contrasts with the results of earlier studies on immature non-marsupial mammals such as neonatal rodents and adult cats (for example, Bates & Stelzner 1993; Bregman *et al.* 1993; Reier *et al.* 1992; Iwashita *et al.* 1994). In all earlier experiments on immature mammalian spinal cord, injury was followed by fibre growth around the lesion, unless the lesion was implanted with fetal CNS. The present finding that *in vitro* preparations of both *Monodelphis* and rat CNS can respond to a crush injury by growth of fibres and conduction of action potentials through the crush, shows that these responses to injury are a function of immaturity of the CNS, rather than to some intrinsic feature of the marsupial nervous system. In this respect the mammalian nervous system appears to be similar to that of the chick embryo. As shown by Oppenheim and his colleagues (for example, Shimizu *et al.* 1990) and Steeves and his colleagues (for example, Hasan *et al.* 1993) the spinal cord of the chick embryo can respond to injury by fibre growth and a considerable degree of functional development and recovery, providing that the injury is made very early in development. Also, like the chick embryo, providing the spinal cord lesion is made early in development, a considerable degree of recovery and development of normal locomotor function occurs in *Monodelphis* following complete spinal cord crush in the first few days of life. In neonatal rats, significant functional recovery following complete spinal cord transection only occurred when fetal spinal cord was inserted at the site of the lesion (Iwashita *et al.* 1994). Thus it can be predicted from our *in vitro* results that, in order to obtain functional recovery in immature rats without implanting fetal CNS, it would be necessary to operate on the spinal cord *in utero* at about E15, which would of course be technically difficult and also difficult to obtain survival of the young.

As discussed above an important point is to what extent the fibre growth and recovery following spinal cord injury, whether *in vivo* or *in vitro*, is due to regeneration from injured fibres and to what extent it is explained by growth of new axons that were not present at the site of injury, but which grow subsequently as part of normal growth of the spinal cord. Nicholls and his colleagues have recently shown in cultured *Monodelphis* spinal cord preparations of the type described in the Results section above, that crushed dorsal and ventral column fibres regenerate and that this regeneration is developmentally regulated (Varga *et al.* 1996). Double-labelling studies using the carbocyanine dyes, DiI and DiO (3,3'-dioctadecyl-oxacarbocynine perchlorate) showed that a substantial proportion of dorsal column axons growing across a crush injury originated from pre-existing axons, that is were regenerating fibres rather than newly developing ones (Varga *et al*, 1995b). Additional evidence for the ability of injured axons to regenerate following a crush injury was provided by *in vitro* experiments in which dorsal root ganglion neurons and their central processes were first labelled with DiI and then crushed. The subsequent growth of the injured fibres was then documented in time-lapse video (Varga *et al.* 1995b). These experiments together with those cited in the introduction to this section indicate that injured immature spinal cord axons are capable of a regenerative response. However, in the studies of recovery of function following spinal cord injury *in vivo* it remains to be determined how much the regenerating fibres contribute to the functional recovery that has been shown to occur.

Even if none of the growth had turned out to be due to regeneration, the growth of new fibres through the crush would have been a new and interesting finding compared with earlier work. The present work makes it clear that providing the spinal cord is injured early enough in its development, then it is capable of growing nerve fibres through a lesion, as occurs in chick embryos (Shimizu *et al.* 1990) and in lower vertebrates (for example, Lurie & Stelzner 1991).

The changes in immature spinal cord that permit neurite growth and effective innervation of target neuron populations and result in functional recovery and development early but not later on, are likely to be multiple and complex, including down regulation of the growth potential of neurons as they develop (perhaps correlated with successful target innervation), down regulation of growth factors and guidance cues and appearance of inhibitory factors of the type described by Schwab (1993). The importance of developmentally regulated inhibitory factors in postnatal *Monodelphis* spinal cord has recently been established (Varga *et al.* 1995a, b). Once the cellular and molecular factors involved in successful regeneration by injured immature nerve fibres have been unravelled, it may then be possible to devise ways of transforming injured adult spinal cord back to a more immature state that would promote neurite growth and functionally effective innervation, which in turn would lead to functional recovery in patients with such injuries.

CONCLUSION

The work discussed in this chapter illustrates the intrinsic interest of studies of CNS development in marsupials. It is also clear that marsupial neonates are valuable models for studying certain aspects of nervous system development that would otherwise be inaccessible within the uterus of eutherians. The early stages of the research described here were begun in collaboration with Hugh Tyndale-Biscoe. Without his initial involvement and subsequent advice and encouragement little of it would have been undertaken.

FUTURE DIRECTIONS

The main fields outlined in this chapter are poised for major advances over the next few years, particularly by the application of molecular biological techniques. In the developing neocortex the mechanism by which incoming afferent fibres from thalamic nuclei make temporary and subsequently definitive connections with neurons of the neocortex is very likely to be worked out. Studies in marsupials are likely to contribute significantly, partly because of the accessibility of the young at the stage of development when these important events are happening. But also because as outlined here, marsupials are, on the one hand, sufficiently similar to eutherians to be a useful model, but on the other hand are different in ways that may give clues to the underlying biological processes. The fundamental problems of what determines which neuroepithelial (ventricular zone) cells become which neurons and find their way to the appropriate neocortical layer will also be elucidated over the next 10 years.

The problem of why CNS neurons are so poorly able to respond to injury is likely to be with us for a long time. It seems probable that it will be immensely complicated to reach a sufficient understanding of the processes involved, in order to be able to propose therapeutic strategies that can be applied in patients with spinal cord injuries. Not only do we have to have a better understanding of why injured neurons fail to grow new axons, especially in the adult, we also have to understand the processes that are involved in axons finding their appropriate targets. It would be less than useful if means were found to obtain significant axon growth from the site of injury towards the target sites, if inappropriate targets were innervated. It is not clear why non mammalian vertebrates are able to show significant regenerative repair of their CNS following injury, but mammalian vertebrates, except when very immature, are not. It may be an evolutionary consequence of the increased complexity of the mammalian CNS. After all, if every time we bumped our heads, our neurons started sprouting and making new connections, the world would be a very confusing place! Nevertheless, the work described here using *Monodelphis domestica* and that described by Martin and Wang in Chapter 21 of this volume, using *Didelphis*, holds out great promise for future understanding, in which marsupials will play a leading role as experimental models. However, my guess is that the processes that have to be understood and manipulated are so complex that it may be 10 to 20 years before sufficient progress has been made for the currently gloomy outlook for patients with spinal cord injuries to be improved.

Acknowledgments

Part of this work was made possible by grants from The Levehulme Trust, UK, The Wessex Medical Trust, UK, The Ramaciotti Foundations, Australia and the Australian Research Council. I should like to thank my colleagues in Hobart, Kate Dziegielewska, Graham Knott, Tracey Walls and Karen Lomasney and colleagues in Basel, John Nicholls and Zoltan Varga for their contributions to much of the work described in this chapter.

References

Adam E, Dziegielewska K M, Saunders N R and Schumacher U (1993). Neuraminic acid specific lectins as markers of early cortical plate neurons. *International Journal of Developmental Neuroscience*, 11, 451–460.

Allandoerfer K L and Shatz C J (1994). The subplate, a transient neocortical structure: its role in the development of connections between thalamus and cortex. *Annual Reviews of Neuroscience*, 17, 185–218.

Balslev Y, Saunders N R and Møllgård K (1992). Onset of neocortical synaptogenesis in neonatal *Monodelphis domestica* (South American Grey Short-tailed Opossum). *Synapse*, 10, 267–270.

Balslev Y, Saunders N R and Møllgård K (1996). Synaptogenesis in the developing brain: An electron microscopical study *Acta Anatomica*, 156, 2–10.

Bates, C A and Stelzner, D J (1993). Extension and regeneration of corticospinal axons after early spinal injury and the maintenance of corticospinal topography. *Experimental Neurology*, 123, 106–117.

Beazley L D, Dunlop S A, Chelvanyagam D K and Ross W M (1995). Wiring up the visual system. *Clinical and Experimental Pharmacology and Physiology*, 22, 550–558.

Bennett J H, Breed W G, Hayman D L and Hope R M (1990). Reproductive and genetical studies with a laboratory colony of the Dasyurid marsupial *Sminthopsis crassicaudata*. *Australian Journal of Zoology*, 37, 207–222.

Bergallo H G and Cerqueira R (1994). Reproduction and growth of the opossum *Monodelphis domestica* (Mammalia: Didelphidae) in northeastern Brazil. *Journal of Zoology*, 232, 551–563.

Bernstein D R and Stelzner D J (1983). Plasticity of the corticospinal tract following midthoracic spinal injury in the postnatal rat. *Journal of Comparative Neurology*, 221, 382–400.

Bernstein-Goral H and Bregman B S (1993). Spinal cord transplants support the regeneration of axotomized neurons after spinal cord lesions at birth: A quantitative double-labelling study. *Experimental Neurology*, 123, 118–132.

Bregman B S, Kunkel-Bagden E, Reier P J, Dai H N, McAtee M and Gao D (1993). Recovery of function after spinal cord injury: Mechanisms underlying transplant-mediated recovery of function differ after spinal cord injury in newborn and adult rats. *Experimental Neurology*, 123, 3–16.

Catalano S M, Robertson R T and Killackey H P (1991). Early ingrowth of thalamocotical afferents to the neocortex of the prenatal rat. *Proceedings of the National Academy of Sciences USA*, 88, 2999–3003.

Dunlop S A (1990). Early development of retinal ganglion cell dendrites in the marsupial *Setonyx brachyurus*, quokka. *Journal of Comparative Neurology*, 293, 425–447.

Dziegielewska K M, Reader M, Matthews N, Brown W M, Møllgård K and Saunders N R (1993). Synthesis of the foetal protein fetuin by early developing neurons in the immature neocortex. *Journal of Neurocytology*, 22, 266–272.

Dziegielewska K M and Brown W (1995). *Fetuin*. Springer-Verlag, New York.

Fawcett J W (1992). Intrinsic neuronal determinants of regeneration. *Trends in Neurosciences*, 15, 5–8.

Gemmell R T (1990). Longevity and reproductive capability of *Isoodon macrourus* in captivity. In *Bandicoots and Bilbies*. Eds J H Seebeck, P R Brown, R I Wallis and C M Kemper. Australian Mammal Society and Surrey Beatty and Sons Pty Ltd, Sydney. pp 213–217.

Gemmell R T (1982). Breeding bandicoots in Brisbane (*Isoodon macrourus*; Marsupialia, Peramelidae). *Australian Mammals*, 5, 187–193.

Harman A M and Beazley L D (1989). Generation of retinal cells in the wallaby, *Setonix brachyurus*. *Neuroscience*, 28, 219–232.

Hasan S J, Keirstead H S, Muir G D and Steeves J D (1993). Axonal regeneration contributes to repair of injured brainstem-spinal neurons in embryonic chick. *Journal of Neuroscience*, 13, 492–507.

Iwashita Y, Kawaguchi S and Murata M (1994). Restoration of function by replacement of spinal cord segments in the rat. *Nature*, 367, 167–170.

Johnson,A R (1993). Contact inhibition in the failure of mammalian CNS axonal regeneration. *BioEssays*, 15, 807–813.

Jones S E, Christie D L, Dziegielewska K M, Hinds L A and Saunders N R (1991). Developmental profile of a fetuin-like glycoprotein in neocortex, cerebrospinal fluid and plasma of post-natal tammar wallaby (*Macropus eugenii*). *Anatomy and Embryology*, 183, 313–320.

Jurgelski W Jr and Porter M E (1974). The opossum (*Didelphis virginiana kerr*) as a biomedical model. III Breeding the opossum in captivity: methods. *Laboratory Animal Science*, 24, 412–425.

Jurgelski W (1979). The marsupial as a laboratory animal. *Institute of Laboratory Animal Resources News*, 22, 18–21.

Kostovic I and Rakic P (1990). Developmental history of the transient subplate zone in the visual and somatosensory cortex of the macaque monkey and human brain. *Journal of Comparative Neurology*, 297, 441–470.

Krause W J and Saunders N R (1994). Brain growth and neocortical development in the opossum. *Annals of Anatomy*, 176, 395–407.

Lubinska L (1959). Region of transitions between preserved and regenerating parts of myelinated nerve fibres. *Journal of Comparative Neurology*, 113, 315–335.

Lurie D I and Selzer M E (1991). Axonal regeneration in the adult lamprey spinal cord. *Journal of Comparative Neurology*, 306, 409–416.

Luskin M B and Shatz C J (1985). Studies of the earliest generated cells of the cat's visual cortex: cogeneration of subplate and marginal zones. *Journal of Neuroscience*, 5, 1062–1075.

McCrady E Jr (1938). The embryology of the opossum. *American Anatomical Memoirs*, 16, 1–233.

Marin-Padilla M (1988). Early ontogenesis of the human cerebral cortex. In *Cerebral Cortex*. Eds A Peters and E G Jones. Plenum Press, New York. Vol. 7, pp 1–34.

Mark R F and Marotte L R (1992). Australian marsupials as models for the developing mammalian visual system. *Trends in Neurosciences*, 15, 51–57.

Møllgård K, Balslev Y, Stagaard Janas M, Treherne J M, Saunders N R and Nicholls J G (1994). Development of spinal cord in the isolated CNS of a neonatal mammal (the opossum *Monodelphis domestica*) maintained in longterm culture. *Journal of Neurocytology*, 23, 151–165.

Molliver M E, Kostovic I and van der Loos H (1973). The development of synapses in cerebral cortex of the human fetus. *Brain Research*, 50, 403–407.

Nicholls J G, Stewart R R, Erulkar S D and Saunders N R (1990). Reflexes, fictive respiration and cell division in the brain and spinal cord of the newborn opossum, *Monodelphis domestica*, isolated and maintained *in vitro*. *Journal of Experimental Biology*, 152, 1–15.

Nicholls J G, Vischer H, Varga Z, Erulkar S and Saunders N R (1994). Repair of connections in injured neonatal and embryonic spinal cord in vitro. *Progress in Brain Research*, 103, 263–269.

Nicholls J G and Saunders N R (1996). Regeneration of immature mammalian spinal cord after injury. *Trends in Neurosciences*, 19, 229–234.

Oswaldo-Cruz E and Rocha-Miranda C E (1968). *The Brain of the Opossum (Didelphis marsupialis). A Cytoarchitectonic Atlas in Stereotaxis Coordinates.* Instituto de Biofisica, Universidade Federal do Rio de Janeiro, Rio de Janiero.

Ramon y Cajal S (1928). *Degeneration and Regeneration of the Nervous System. History of Neuroscience* No 5, (transl. R M May). Eds J De Felipe and E G Jones (1991). Oxford University Press, Oxford.

Reier P J, Stokes B T, Thompson F J and Anderson D K (1992). Fetal cell grafts into resection and contusion/compression injuries of the rat and cat spinal cord. *Experimental Neurology*, 115, 177–188.

Reynolds M L, Cavanagh M E, Dziegielewska K M, Hinds L A, Saunders N R and Tyndale-Biscoe C H (1985). Postnatal development of the telencephalon of the tammar wallaby (*Macropus eugenii*). An accessible model of neocortical differentiation. *Anatomy and Embryology*, 173, 81–94.

Reynolds M L and Saunders N R (1988). Differentiation of the neocortex. In *The Developing Marsupial. Models for Biomedical Research.* Eds C H Tyndale-Biscoe and P Janssens. Springer-Verlag, Berlin. pp 101–116.

Sanderson K J and Weller W L (1990). Gradients of neurogenesis in possum neocortex. *Developmental Brain Research*, 55, 269–274.

Saunders N R, Adams E, Reader M and Møllgård K (1989). *Monodelphis domestica* (grey short-tailed opossum): an accessible model for studies of early neocortical development. *Anatomy and Embryology*, 180, 227–236.

Saunders N R, Habgood M D, Ward R A and Reynolds M L (1992a). Origin and fate of fetuin-containing neurons in the developing neocortex of the fetal sheep. *Anatomy and Embryology*, 186, 477–486.

Saunders N R, Balkwill P, Knott G, Habgood M D, Møllgård K, Treherne J M and Nicholls J G

(1992b). Growth of axons through a lesion in the intact CNS of fetal rat maintained in long-term culture. *Proceedings of the Royal Society of London B*, 250, 171–180.

Saunders N R, Deal A, Knott G W, Varga Z M and Nicholls J G (1995). Repair and recovery, following spinal cord injury in a neonatal marsupial (*Monodelphis domestica*). *Clinical and Experimental Pharmacology and Physiology*, 22, 518–526.

Schnell L and Schwab M E (1990). Axonal regeneration in the rat spinal cord produced by an antibody against myelin-associated neurite growth inhibitors. *Nature* , 343, 269–272.

Schwab M E (1991). Nerve fibre regeneration after traumatic lesions of the CNS; progress and problems. *Philosophical Transactions of the Royal Society of London B*, 331, 303–306.

Schwab M E (1993). Experimental aspects of spinal cord regeneration. *Current Opinion in Neurology and Neurosurgery*, 6, 549–553.

Selwood L and Woolley P A (1991). A timetable of embryonic development, and ovarian and uterine changes during pregnancy, in the stripe-faced dunnart, *Sminthopsis macroura* (Marsupialia: Dasyuridae). *Journal of Reproduction and Fertility*, 91, 213–227.

Sevier A C and Munger B L (1965). A silver stain for paraffin sections of neural tissue. *Journal of Neuropathology and Experimental Neurology*, 24, 130–135.

Sheng, X-M, Marotte, L R and Mark, R F (1990). Development of connections to and from the visual cortex in the wallaby (*Macropus eugenii*). *Journal of Comparative Neurology*, 300, 196–210.

Sheng X-M, Marotte L R and Mark R F (1991). Development of the laminar distribution of thalamocortical axons and corticothalmic cell bodies in the visual cortex of the wallaby. *Journal of Comparative Neurology*, 307, 17–38.

Shimizu I, Oppenheim R W, O'Brien M and Shneiderman A (1990). Anatomical and functional recovery following spinal cord transection in the chick embryo. *Journal of Neurobiology*, 21, 918–937.

Stewart R R, Zou D J, Treherne J M, Møllgård K, Saunders N R and Nicholls J G (1991). The intact central nervous system of the newborn opossum in long-term culture: Fine structure and GABA-mediated inhibition of electrical activity. *Journal of Experimental Biology*, 161, 25–41.

Treherne J M, Woodward S K A, Varga Z M, Ritchie J M and Nicholls J G (1992). Restoration of conduction and growth of axons through injured spinal cord of neonatal opossum in culture. *Proceedings of the National Academy of Sciences, USA* 89, 431–434.

Tyndale-Biscoe C H and Janssens P A (eds) (1988). *The Developing Marsupial. Models for Biomedical Research*, Springer-Verlag.

VandeBerg J (1983). The gray short-tailed opossum: a new laboratory animal. *Institute of Laboratory Animal Resources News*, 26, 9–12.

Varga Z M, Brandtlow C E, the late Erulkar S D, Schwab M E and Nicholls J G (1995a). The critical period for repair of CNS of neonatal opossum (*Monodelphis domestica*) in culture: correlation with development of glial cells, myelin and growth inhibitory molecules. *European Journal of Neuroscience*, 7, 2119–2129.

Varga Z M, Schwab and Nicholls, J G (1995b). Myelin-associated neurite growth-inhibitory proteins and suppression of regeneration of immature mammalian spinal cord in culture. *Proceedings of the National Academy of Sciences USA*, 92, 10959–10963.

Varga, Z M, Fernandez, J, Blackshaw, S, Martin, A R, Muller, K J, Adams, W B and Nicholls, J G (1996). Neurite outgrowth through lesions of neonatal opossum spinal cord in culture. *Journal of Comparative Neurology*, 366, 1–13.

Vaughn J E (1989). Review: Fine structure of synaptogenesis in the vertebrate central nervous system. *Synapse*, 3, 255–285.

Wang X M, Terman J R and Martin, G F (1996). Evidence for supraspinal growth of axons through the lesion after transection of the thoracic spinal cord in the developing opossum, *Didelphis virginiana*. *Journal of Comparative Neurology*, 371, 104–115.

Xu X M and Martin G F (1991). Evidence for new growth and regeneration of cut axons in developmental plasticity of the rubrospinal tract in the North American opossum. *Journal of Comparative Neurology*, 313, 103–112.

Zou D-J, Treherne J M, Stewart R R, Saunders N R and Nicholls J G (1991). Regulation of $GABA_B$ receptors by histamine and neuronal activity in the isolated spinal cord of neonatal opossum in culture. *Proceedings of the Royal Society of London B*, 246, 77–82.

Zou D-J (1994). Respiratory rhythm in the isolated central nervous system of newborn opossum. *Journal of Experimental Biology*, 197, 201–213.

SPECIES INDEX

GENERAL INDEX